Apex Predators in the Anthropocene

Apex Predators in the Anthropocene

How Humans Shape Large Carnivore Populations

EDITED BY

Robert A. Montgomery

Professor of Conservation Science, Director of the Centre for National Parks and Protected Areas, and Director of Research and Knowledge Exchange

University of Cumbria

Cumbria

United Kingdom of Great Britain and Northern Ireland (the)

Remington J. Moll

Associate Professor

Department of Natural Resources and the Environment, University of New Hampshire

New Hampshire

United States of America (the)

OXFORD
UNIVERSITY PRESS

OXFORD
UNIVERSITY PRESS

Great Clarendon Street, Oxford, OX2 6DP,
United Kingdom

Oxford University Press is a department of the University of Oxford.
It furthers the University's objective of excellence in research, scholarship,
and education by publishing worldwide. Oxford is a registered trade mark of
Oxford University Press in the UK and in certain other countries.

Published in the United States of America by Oxford University Press
198 Madison Avenue, New York, NY 10016, United States of America

British Library Cataloguing in Publication Data
Data available

Library of Congress Control Number: 2026938301

ISBN 9780198892694
ISBN 9780198892700 (pbk.)

DOI: 10.1093/9780191996665.001.0001

Printed and bound by
CPI Group (UK) Ltd., Croydon, CR0 4YY

The manufacturer's authorized representative in the EU for product safety is
Oxford University Press España S.A. of Parque Empresarial San Fernando de Henares,
Avenida de Castilla, 2 – 28830 Madrid (www.oup.es/en or product.safety@oup.com).
OUP España S.A. also acts as importer into Spain of products made by the manufacturer.

FSC
www.fsc.org
MIX
Paper | Supporting
responsible forestry
FSC® C013604

Contents

List of Figures vii
List of Tables ix
Preface x
List of contributors xiii

Section One The Nature of Coupled Human and Natural Systems Globally **1**

1 **The Role of Humans in Ecological Systems** **3**
Robert A. Montgomery, Anna Rouviere, and Remington J. Moll

2 **Human–Carnivore Interactions in the Pleistocene** **18**
Matt W. Hayward, Blaire Van Valkenburgh, Cassandra K. Bugir, and Chris N. Johnson

3 **Human Disturbances Define Carnivore Ecology in the Anthropocene** **38**
Christopher J. Schell, Justin Brashares, and Lauren Stanton

Section Two The Lethal Effects of Humans on Carnivore Populations **63**

4 **The Impacts of Hunting on Large Carnivore Populations** **65**
Andrés Ordiz, Jon E. Swenson, Cyril Milleret, and Richard Bischof

5 **How Conflict with Humans Changes Carnivores** **82**
Anna Rouviere and Robert A. Montgomery

6 **Lethal Effects of Roads on Large Carnivores** **103**
Mairi K. P. Poisson, Andrew R. Butler, Clara R. Dawson, Adam T. Ford, Taylor S. Readyhough, Jenna K. Scherger, and Remington J. Moll

Section Three The Risk Effects of Humans on Carnivore Populations **121**

7 **Cumulative Costs of Carnivore Behavioural Responses to Humans** **123**
Nikita Frizzelle, Justin P. Suraci, Michael Clinchy, and Liana Y. Zanette

8 Quantifying Anthropogenic Risk Effects on Carnivores **138**
Justine A. Smith and Justin P. Suraci

9 Risk Effects Impact Carnivore Population Persistence **160**
Jonathan N. Pauli, Shotaro Shiratsuru, Matthew M. Smith, and Mauriel Rodriguez Curras

Section Four Implications of Human-Carnivore Interactions on Ecosystem Functioning **179**

10 Cascading Effects of Humans on Trophic Dynamics **181**
Laura R. Prugh, Samantha E.S. Kreling, Taylor R. Ganz, Calum X. Cunningham, Yasmine Hentati, Benjamin K. Sullender, Catherine M. Breen, and Michael Procko

11 The Trait-Mediated Indirect Effects of Human Interaction **197**
Kaggie D. Orrick, Mary K. Burak, and Oswald J. Schmitz

12 The Consequences of Trophic Downgrading, and the Promise of Trophic Upgrading **209**
Kaitlyn M. Gaynor and Nicholas S. Wright

Section Five The Effects of Conservation on Carnivore Populations **229**

13 Contemporary Land-Sparing Initiatives and Intergenerational Equity in Carnivore Conservation **231**
Fikirte Gebresenbet, Siewe Siewe, and Fidelis A. Atuo

14 Measures to Promote Human–Carnivore Coexistence **246**
Christine E. Wilkinson, Harshad Karandikar, and Mitchell W. Serota

15 The Future of Human–Carnivore Interactions in the Anthropocene **269**
Nyeema C. Harris, Evie Jones, Pallavi Ghaskadbi, and Mariela G. Gantchoff

Glossary of Terms 289
Index 291

List of Figures

1.1 Humans outside natural systems (a) and within them (b–e). 5
2.1 (a) Replication of a Cro-Magnon hunter. (b) Recreated scene of butchery and meat drying activities of Cro-Magnon people. (c) Replica of a 25,000-year-old artwork. (d) Matt Hayward attempting to recreate Cro-Magnon art alongside other children. 20
2.2 Image of the replica of the 30–33,000-year-old Panel of the Lions from Chauvet Cave, France, held at the Brno Museum Anthropos. 22
3.1 Conceptual framework illustrating how direct drivers of biodiversity loss impact apex predator ecology at various ecological scales. 40
3.2 Key examples of direct drivers influencing apex predator ecology in the literature. 43
4.1 Summary of direct (lethal) and indirect, risk effects of hunting large carnivores for these species and potentially for their ecological function in nature. 74
5.1 The possible effects of human-induced mortality on carnivore survival rates. 90
5.2 Summary of the possible effects of human lethal interventions on carnivore population size. 90
6.1 Camera trap images of wolves using linear structures in various locations throughout Canada, USA. 104
6.2 Among studies utilizing radio telemetry or GPS tracking methods, human take such as hunting, trapping, and poaching were a much higher contributor to large carnivore mortality than roads. 109
6.3 Studies using radio telemetry or GPS tracking methods (n = 31) were conducted in three continents. Carcass studies relying on roadkill data were conducted in five continents. 113
6.4 The number of publications by country for the 87 papers in our review of the lethal impact of roads on large carnivore species. 114
7.1 Conceptual illustrations of a human-induced landscape of fear on carnivores and their prey. 126
8.1 Example methods for studying risk effects in apex carnivores. 142
8.2 Case Study: risk effects in pumas of the Santa Cruz Mountains, C. A. 150
9.1 The mechanisms by which human risk effects have negative fitness and demographic consequences for carnivore populations. 161
9.2 Human intensity (quantified as the maximum level of HFI that a large carnivore experiences throughout its range) compared across (1) two variables we used as a proxy for persistence and (2) two variables we used as a proxy for behavioural response of carnivores. 169

9.3 Frequency distributions of the two possible perceptions of risk (no risk perceived [blue] or risk perceived [grey]) are plotted, showing an arbitrary value of instantaneous predation risk (black diagonal line) along a gradient. 171
9.4 Predictions from the 'threshold of risk' hypothesis showing the tolerance of a species (FN/FP; x-axis) and the predictability of risk (y-axis) in a given landscape provide a persistence cline for individuals of a population. 171
10.1 Pervasive influence of humans on trophic networks. 182
10.2 Trophic triangles can reduce the strength of linear cascades. 184
10.3 Human alterations of the forage landscape influence trophic relationships and can alter the implications of reduced predator densities. 186
10.4 Humans influence scavenging dynamics through several main pathways (blue = positive, red = negative). 189
11.1 Humans can indirectly affect predators, which have direct and cascading direct and indirect effects on the meso-predator's prey and vegetation in its environment. 199
12.1 Trophic downgrading can alter a myriad of biotic, abiotic, and socio-ecological processes, due to the loss of large carnivores and subsequent changes to the density and behaviour of other species. 211
12.2 The loss of large carnivores can alter ecosystems via trophic cascades. 219
13.1 A diagram illustrating the interconnectedness of people, land, and large carnivores. 241
14.1 A framework depicting the social-ecological factors contributing to human-carnivore interactions (HCI). 247
14.2 A framework for identifying and operationalizing wildlife corridors, comprising ecological and socioeconomic assessments and Free, Prior, and Informed Consent (FPIC) to evaluate the feasibility of establishing wildlife corridors. 251
14.3 Examples of interventions and tools designed to mitigate carnivore–livestock conflict. 253
14.4 Some of the key benefits and challenges of living alongside apex predators. 255
14.5 A brown bear (*Ursus arctos*) in Katmai National Park, Alaska. 258
15.1 Diverse human–carnivore interactions. 270
15.2 A conceptual diagram or 'threat web' of interactions among anthropogenic pressures affecting human–carnivore interactions. 276
15.3 Vulnerability of range contractions for African carnivores modified from Harris et al. (2023a). 279
15.4 Examples of technological advancements aiding in the monitoring and management of conservation. 281

List of Tables

6.1 The 27 large carnivore species included in the literature review 106
6.2 The number of papers, and the number of papers by method type, with study sites in each continent 107
6.3 Reported demographic data of large carnivore mortalities by the studies using radio telemetry or GPS tracking methods 110
11.1 Trait-mediated indirect effects transmitted from humans to predators that influence ecological communities 201
14.1 Mechanisms for improving upon global human–carnivore coexistence strategies 260

Preface

The inception of this book began several years ago when we were engaged in collaborative discussions relating to the diversity of impacts that humans can exert on large mammal populations. At that time, we were conducting research to describe the Eltonian niche of humans. We found the human niche to be substantially broader and more diverse than had previously been described. Our research demonstrated that human impacts on large carnivore populations have often been depicted as being top-down. Humans have a long history of persecuting large carnivore populations to preserve human security or to decrease competition over a shared food supply. Many cultures globally have, for instance, intentionally hunted large carnivores to preserve populations for human consumption or to reduce the probability of carnivore depredation of livestock. However, we found that only a portion of human impacts are top-down and that they also influence trophic systems via bottom-up processes and exert effects that permeate all aspects of food web dynamics. For example, though humans can be voracious hunters and predators of large carnivores, they often display a strong reverence for carnivores with symbols of the species used extensively in marketing and promotional materials to represent strength, valour, athleticism, and courage. Carnivores are a guild of predators that hold very high social value and are regularly identified as among the most charismatic species on Earth. Consequently, carnivores are often the subjects of extensive conservation efforts with campaigns and initiatives to protect and restore their populations around the world. On the other hand, humans also respond to the risks inherent to interactions with large carnivores. Thus, in addition to acting as predators, humans are also vulnerable to large carnivores, with injury or death that can follow. Finally, humans have changed the abiotic and biotic conditions in the environments that large carnivores inhabit. And certain carnivores have demonstrated an ability to benefit from the built environment or from subsidies that are available via the human food supply. Via the process of quantifying the different roles inherent to the Eltonian niche of humans, we were able to more clearly describe the human role in the trophic systems in which they reside.

That was the seed of the idea that has now taken full form in the development of this book. Our interest with this book is to portray the unique and multi-varied contributions of humans on ecological systems, with an emphasis on apex predators (particularly carnivores). Negative interactions with humans can not only result in direct mortality for individual carnivores (i.e. lethal effects), but the process of avoiding human interaction can carry fitness costs (i.e. risk or 'non-lethal' effects—the latter is sometimes considered a misnomer because such effects contribute to mortality). Additionally, as human–carnivore interactions typically occur at the very top of food chains, additional effects can cascade down trophic systems via trait- and density-mediated indirect pathways with implications for sympatric species. Thus, we must articulate the many ways humans shape large carnivore populations and describe the adaptations that carnivores make to live alongside humans. To achieve this vision, we had to first identify the teams of interdisciplinary scientists that we would invite to provide contributions to this book. We developed a couple of symposia to be held at international academic conferences and invited colleagues doing cutting-edge research in this area. Concurrently, we read the work of others doing emergent research in this area. And from the moment that we sent out the invitations, the process of confirming contributors for each chapter could not have been more straightforward. We have successfully recruited colleagues from around the world detailing human–carnivore relationships in Europe, North America, South America, Africa, Asia, and Australia. Furthermore, these colleagues have eloquently written about the natural history, ecology, management, and conservation of these large carnivore populations. We are lucky to have such an amazing group of contributors to this volume, and we thank them sincerely for their efforts.

We are delighted to present Apex Predators in the Anthropocene complete with 15 chapters that capably describe the magnitude and strength of human–large carnivore interactions across five main sections of contribution including; (*i*) the nature of coupled human and natural systems globally, (*ii*) the lethal effects of humans on large carnivore populations, (*iii*) the risk effects of humans on large carnivore populations, (*iv*) the implications of human–large carnivore interactions on ecosystem functioning, and (*v*) the effects of conservation on large carnivore populations. We kick off this section with a chapter (Chapter 1: The Role of Humans in Ecological Systems) that introduces the concept of coupled human and natural systems, beginning with a historical overview that takes the reader from past human impacts on natural systems to those present today. Matt Hayward from the University of Newcastle then (Chapter 2: Human–Carnivore Interactions in the Pleistocene) describes the fact that humans and carnivores have coexisted for millennia, yet debate regarding the dynamics of this coexistence and evidence is scattered throughout the geological record. This chapter critically reviews the predominant hypotheses and evidence for how humans interacted with such species. The final chapter (Chapter 3: Human Disturbances Define Carnivore Ecology in the Anthropocene) in section one is written by Chris Schell of the University of California, Berkeley and builds off of the two former chapter to demonstrate how humans and carnivores interact in an ever-changing world. This chapter focuses on quantifying the ecology of these interactions and offers a framework with case study examples to make the point for the reader.

Andrés Ordiz, from the Universidad de León, provides the first chapter (Chapter 4: The Impacts of Hunting on Large Carnivore Populations) of the second section of the book and analyses the ways in which human hunting pressure not only has affected large carnivore population abundance, but also the very structure of these populations. In the next chapter (Chapter 5: How Conflict with Humans Changes Carnivores) Anna Rouviere from the University of Oxford describes the multiple dimensions of human–carnivore conflict and the mechanisms that underlie this conflict. This chapter attributes much of human–carnivore conflict to instances of livestock depredation and describes the behavioural modifications that carnivores make to reduce conflict with humans. Mairi Poisson from the University of New Hampshire provides the concluding chapter in this section (Chapter 6: Lethal Effects of Roads on Large Carnivores) via a global literature review of road effects on large carnivores. This chapter rounds out the discussion of the lethal effects of humans on apex predators.

A series of three tremendous chapters then capably describe the risk effects that can flow from carnivore interactions with humans. Nikita Frizzelle from the University of Western Ontario kicks off this section (Chapter 7: Cumulative Costs of Carnivore Behavioural Responses to Humans) by describing how carnivores modify their behaviour in response to humans, including changes to their vigilance, feeding behaviour, and movement. Justine Smith from the University of California Davis and Justin Suraci from Conservation Science Partners then builds on this theme in the development of a chapter (Chapter 8: Quantifying Anthropogenic Risk Effects on Carnivores) that describes and analyses field and statistical methods that have proved effective in quantifying the behavioural modifications of carnivores to human activity. Jonathan Pauli from the University of Wisconsin then provides the concluding chapter (Chapter 9: Risk Effects Impact Carnivore Population Persistence) in this section, which highlights the pathways by which human-induced changes in carnivore behaviour can alter both carnivore fitness and, thereby, carnivore populations.

In section four of this book, the implications of human–carnivore interactions on broader ecosystem dynamics via cascading mechanisms are explored. Laura Prugh from the University of Washington provides the first chapter in this section (Chapter 10: Cascading Effects of Humans on Trophic Dynamics), describing how human lethality can trigger effects that can cascade down trophic systems, influencing the magnitude and strength of sympatric species. Kaggie Orrick from Yale University takes the next step (Chapter 11: The Trait-Mediated Indirect Effects of Human Interaction) and articulates that while carnivores often exert strong top-down influence on other species, their behavioural adaptations to humans represent a potentially common trait-mediated pathway by which humans indirectly influence communities. This chapter not only details how humans can initiate these indirect effects but also discusses their implications for community dynamics and ecosystem structure and functioning. The final chapter of this section (Chapter 12: The Consequences of Trophic Downgrading, and the Promise of Trophic Upgrading) is provided by Kaitlyn Gaynor from the University of British Columbia and builds on the seminal trophic downgrading literature from the early 2000s. This chapter quantifies the degree to which trophic systems have been downgraded by human actions across the globe and the implication thereof for large carnivores.

Given these complex and interacting lethal, risk, and indirect anthropogenic effects, nearly every species of large carnivore across the globe is of conservation concern, with the majority of having declining population trends. Thus, the final section of this book focuses on the importance of carnivore conservation. Fikirte Gebresenbet from the University of New Hampshire provides the first chapter of this section (Chapter 13: Contemporary Land-Sparing Initiatives and Intergenerational Equity in Carnivore Conservation) and overviews the human dimensions of carnivore conservation. This chapter emphasizes the current diversity of stakeholder views and how multigenerational cultural and societal identities shape perceptions of carnivore conservation and policy. Christine Wilkinson from the Natural History Museum of Los Angeles County, authors the next chapter (Chapter 14: Measures to Promote Human–Carnivore Coexistence), describing the extensive efforts that have been implemented to conserve and restore large carnivore populations. This chapter emphasizes that anthropogenic impacts on carnivore populations are not uniformly negative. The diversity of measures to promote human coexistence with carnivores are reviewed with specific evidence provided from a number of case study landscapes. The final chapter of this book (Chapter 15: The Future of Human–Carnivore Interactions in the Anthropocene), provided by Nyeema Harris at Yale University, provides a vision for the future of human–carnivore interactions in a dynamic twenty-first century. This chapter details how progressive conservation practices can be implemented to better protect and restore carnivore populations while remaining cognizant of the importance of human livelihood.

In summary, we have greatly enjoyed the process of working with this amazing array of colleagues to develop *Apex Predators in the Anthropocene*. We believe that this book will help to unify ecological thinking on the topic of people and predators, which will in turn set the stage for cutting-edge research and effective conservation action depicting how humans shape animal populations. Once validated on carnivores, the principles featured in this book can be applied to other species of conservation and management concern. Finally, this book will help dispel the notion that humans and carnivores exist in predominantly non-overlapping domains that occasionally intersect and augment a more holistic and interconnected view of people and predators. This point emphasizes the central importance of envisioning the future of ecological research and conservation action to occur across coupled human and natural systems. The ways in which humans seek to control and restore large carnivore populations is a metaphor for the complexity of the ways in which humans shape animal populations. Thus, more broadly, the nature of human–carnivore interactions is a contentious issue globally in which humans simultaneously influence, threaten, and support this group of species and the prey they depend upon. This situation creates complex coupled human–ecological dynamics that will ultimately determine the fate of both the persistence of carnivore populations and the use of natural resources by people. We hope that this book offers a set of tools to contribute to these dialogues so that diverse voices can engage in the decisions that will be made about the best ways for humans to shape large carnivore populations.

Robert A Montgomery and Remington J. Moll

List of contributors

Fidelis Akunke Atuo PhD, Associate Professor, Department of Biological and Environmental Sciences, Southeast Missouri State University, Cape Girardeau, MO, US.

Richard Bischof PhD, Faculty of Environmental Sciences and Natural Resource Management, Norwegian University of Life Sciences, Ås, Norway.

Justin Brashares PhD, Professor, University of California, Berkeley, CA, US.

Catherine Breen PhD, Research Scientist, University of Washington, Seattle, WA, US, Research Scientist, University of Washington, Seattle, WA, US, Research Scientist, University of Washington, Seattle, WA, US and Research Scientist, University of Washington, Seattle, WA, US.

Cassandra K. Bugir PhD, Centre of Conservation Science, University of Newcastle, Newcastle, Australia

Mary Burak PhD, Postdoctoral Research Associate, Institute at Brown for Environment and Society, Brown University, Providence, RI, US and Postdoctoral Research Associate, Department of Ecology, Evolution, and Organismal Biology, Brown University, Providence, RI, US.

Andrew Butler, Natural Resources and the Environment, University of New Hampshire, Durham, NH, US.

Michael Clinchy, Research Associate, Biology Department, Western University, London, Canada.

Calum X. Cunningham PhD, Research Fellow, School of Natural Sciences, University of Tasmania, Hobart, Australia.

Clara R. Dawson, M.S. Student, University of New Hampshire, Durham, NH, US.

Adam Ford PhD, Professor, University of British Columbia, Kelowna, Canada and Professor, University of British Columbia, Kelowna, Canada.

Nikita Renata Frizzelle, Lecturer, Department of Biology, Western University, London, Canada.

Mariela Gantchoff PhD, Assistant Professor, Environmental Biology Program, Department of Biology, University of Dayton, Dayton, OH, US.

Taylor Ganz PhD, Science Lead, The Nature Conservancy, Hailey, ID, US and Research Scientist, School of Environmental and Forest Sciences, The University of Washington, Seattle, ID, US.

Kaitlyn M. Gaynor, Assistant Professor, Zoology and Botany, University of British Columbia, Vancouver, Canada.

Fikirte Gebresenbet PhD, Assistant Professor, Natural Resources and the Environment, University of New Hampshire, Durham, NH, US.

Pallavi Ghaskadbi PhD, Postdoctoral Scholar, Department of Wildlife, Fish and Environmental Sciences, Swedish University of Agricultural Sciences (SLU), Umeå, Sweden.

Nyeema C. Harris PhD, Knobloch Family Associate Professor of Wildlife and Land Conservation, Director of Applied Wildlife Ecology (AWE) Lab, School of the Environment, Yale University, USA.

Matt W. Hayward BSc. (Hons I); PhD, Professor of Conservation Science, and Director of the Centre for Conservation Science, School of Environmental and Life Science, University of Newcastle, Newcastle, Australia.

Yasmine Hentati, PhD Candidate, University of Washington, Seattle, WA, US.

Christopher Johnson PhD, Professor of Wildlife Conservation, School of Natural Sciences, University of Tasmania, Hobart, Australia.

Evie Jones PhD, Nisga'a Fisheries and Wildlife Department, Nisga'a Lisims Government, New Aiyansh, BC, Canada.

Harshad Karandikar PhD, Postdoctoral Scientist, Conservation Science Partners, Truckee, CA, US.

Samantha E. S. Kreling, Postdoctoral Researcher, Purdue University, IN, US.

Cyril Milleret, Postdoctural Researcher, 1CEFE, Univ Montpellier, CNRS, EPHE, IRD, Montpellier, France and Researcher, Faculty of Environmental Sciences and Natural Resource Management, Norwegian University of Life Sciences, Ås, Norway.

Remington Moll, Associate Professor, Department of Natural Resources and the Environment, University of New Hampshire, Durham, NH, US.

Robert Alistair Montgomery PhD, Professor of Conservation Science, Director of the Centre for National Parks and Protected Areas, and Director of Research and Knowledge Exchange, Institute of Science and Environment, University of Cumbria, Ambleside, UK.

Andrés Ordiz PhD, Faculty of Environmental Sciences and Natural Resource Management, Norwegian University of Life Sciences, Norway and Departamento de Biodiversidad y Gestión Ambiental, Universidad de León, Spain.

Jonathan N. Pauli PhD, Professor of Forest and Wildlife Ecology, University of Wisconsin-Madison, Madison, WI, US.

Mairi K. P. Poisson, Graduate Research Assistant, Department of Natural Resources and the Environment, University of New Hampshire, Durham, NH, US.

Michael Procko, Graduate Research Fellow, Utah State University, Logan, UT, US.

Laura R. Prugh PhD, Professor of Quantitative Wildlife Sciences, School of Environmental and Forest Sciences, University of Washington, Seattle, WA, US.

Taylor S. Readyhough, Natural Resources and the Environment, University of New Hampshire, Durham, NH, US.

Mauriel Rodriguez Curras, Postdoctoral Researcher, University of California Berkeley, Berkeley, CA, US.

Anna Rouviere, DPhil Student, Department of Biology, University of Oxford, UK.

Christopher J. Schell PhD, Assistant Professor, Department of Environmental Science, Policy, and Management, University of California Berkeley, Berkeley, CA, US.

Jenna K. Scherger MSc, Graduate Research Assistant, Department of Biology, University of British Columbia Okanagan, Kelowna, Canada.

Oswald Schmitz PhD, Professor, School of the Environment, Yale University, New Haven, CT, US.

Mitchell Serota PhD, Scientist and Manager of Ecology and Nature Tech Solutions, Doris Duke Foundation, New York, NY, US.

Shotaro Shiratsuru PhD, Postdoctoral Researcher, Trinity College Dublin, Dublin, Ireland.

Siewe Siewe Siewe PhD, MSE, Business Development Associate, Cowboy Enterprises, Oklahoma State University, Stillwater, OK, US.

Justine Smith PhD, Associate Professor, Department of Wildlife, Fish, and Conservation Biology, University of California, Davis, Davis, CA, US.

Matthew M. Smith PhD, Research Assistant Professor, Caesar Kleberg Wildlife Research Institute, University of Texas A&M-Kingsville, Kingsville, TX, US.

Benjamin K. Sullender PhD, Director of Geospatial Science, Audubon Alaska, Anchorage, AK, USA, and Research Analyst Lead, School of

Environmental and Forest Sciences, University of Washington, Seattle, WA, USA.

Justin Suraci, Senior Scientist and Director of Wildlife Conservation Science, Conservation Science Partners, Truckee, CA, US.

Jon E. Swenson PhD, Dr.Habil, Professor emeritus of Ecology and Natural Resources Management, Faculty of Environmental Sciences and Natural Resource Management, Norwegian University of Life Sciences, Ås, Norway.

Blaire Van Valkenburgh PhD, Distinguished Research Professor, University of California, Los Angeles, CA, US.

Christine E. Wilkinson PhD, Assistant Curator of Community Science, Research & Collections Department, Natural History Museum of Los Angeles County, CA, US.

Nicholas Wright, PhD Candidate, University of British Columbia, Vancouver, Canada.

Liana Y. Zanette, Professor of Biology, Western University, London, Canada.

SECTION ONE

The Nature of Coupled Human and Natural Systems Globally

CHAPTER 1

The Role of Humans in Ecological Systems

Robert A. Montgomery, Anna Rouviere, and Remington J. Moll

Introduction

Evolutionary evidence identifies that humans have been an independent species for only a few hundred thousand years (Stringer 2016, Galway-Witham and Stringer 2018). And yet, across that time, humans have completely transformed the biotic and abiotic complexion of the Planet. Human populations have exceeded carrying capacity estimates and moved into every biome globally, inhabiting innumerable niches with supreme behavioural plasticity (Cohen 1995, Groucutt et al. 2015). Not only have humans pursued an exceptionally broad prey base, but they have also manufactured their own food supply via agriculture and livestock domestication. Humans have fortified structures, created the built environment, and mechanized industrial production of goods and services. In support of the human population, complex networks have been established in supply chains that connect premier financial institutions in the Global North with tier-four suppliers of raw materials in the Global South. And in the process, humans have altered every ecological system on the planet and forced co-occurring species to contend with their anthropocentric perspectives (Ellis 2011, Ellis et al. 2023). It is clear that there is no other species on Earth like humans and yet, at the core, people are still just members of the Animal Kingdom.

Diverging from an ape-like ancestor approximately seven million years ago, a collection of pre-human ancestors, known collectively as the hominins, began to behave differently to other great apes (Harrison 2010, Langergraber et al. 2012). For starters, they tended to depend more upon standing and walking in upright fashions (Andrews 2020). Via a series of musculoskeletal adaptations, the hominins developed bipedalism which subsequently carried several distinct advantages. Firstly, hominins came down from the trees and began to enter new habitats on foot (Senut et al. 2018). This process enabled the exploration of novel and productive niches to be explored. Secondly, walking in a bipedal posture reduced dorsal exposure to bright sunlight and offered certain energetic advantages (Raichlen and Pontzer 2021). Finally, walking on two feet freed up the hands to be used in carrying various goods (Hewes 1964, Carvalho et al. 2012). While hominin artefacts indicative of tool use don't become evident until stone implements around 3.3 million years ago (Harmand et al. 2015), it has been postulated that tool use may pre-date this. Earlier tools may have been fashioned from wooden materials, that have simply not survived, or taken the form of unmodified stones (Panger et al. 2002). Nevertheless, in these ways, hominins were able to expand their range and carry foodstuffs and other objects along those journeys. And thus, it was these and other adaptations that enabled hominins to radiate in evolutionary complexity and global extent.

As recently as ~200,000 years ago, there were at least three distinct species of hominin. These were *Homo heidelbergensis*, *Homo neanderthalensis*, and *Homo sapiens*. At this time, these hominin

Robert A. Montgomery, Anna Rouviere, and Remington J. Moll, *The Role of Humans in Ecological Systems*. In: *Apex Predators in the Anthropocene*.
Edited by: Robert A. Montgomery and Remington J. Moll, Oxford University Press. © Oxford University Press (2026).
DOI: 10.1093/9780191996665.003.0001

species were not only integral components of the trophic systems that they inhabited but were also fearsome predators. To put it simply, they had to be. These hominin species were in direct conflict with sympatric large carnivores over a shared food supply (Antón, Galobart, and Turner 2005, Espigares et al. 2013). Both hominins and carnivores preferred to hunt large mammalian ungulates. Given that the hominins possessed comparatively modest physical defences, in relation to their large carnivore competitors (i.e. no claws, long canines, or incredible strength), it has long been believed that they survived via the implementation of interference competition strategies. For instance, hominins may have tracked carnivores on the hunt, waited for them to make a kill, and then forced them off the kill site with the use of weaponry. This form of kleptoparasitism required ingenuity and teamwork that compensated for a lack of brute strength. Recent findings, however, have demonstrated that these hominins were likely more formidable than previously thought. Evidence from caves in Germany suggests that Neanderthals were using Eurasian cave lion (*Panthera spelaea*) skins some 190,000 years ago (Russo et al. 2023). Further evidence from 48,000 years ago shows signs of Neanderthals killing with spears and of butchery of cave lions (Cueto et al. 2016, Russo et al. 2023). Thus, to make up for a lack of ferocity, hominins were ingeniously using weapons and, in the process, turning even apex predators into prey.

In reality, the tendency to wield, make, and use tools has been one of the defining characteristics of humans. There are certainly other species that could be classified as tool-users, but none have done so in such a diverse and ingenious array of forms and functions (Seed and Byrne 2010). That ability, facilitated by an inordinately high brain-to-body weight ratio (encephalization rate), has enabled humans to conquer every habitat on Earth (Grove 2017). There is no climate that humans have not overcome, and restrictions against range expansion have been human-induced (e.g. the Antarctic Treaty preventing colonization of Antarctica) rather than resource-limited. Humans occupy deserts, tundra, taiga, icescapes, savannahs, equatorial jungles, and temperate forests. The transformation and depletion of Earth's natural resources at the hand of humans has motivated some to conceptualize how tool use could grow new resources and new societies on other planets in the solar system (Maity and Saxena 2024). But for the time being, human society is bound to Earth where people exert a diversity of impacts on the species and ecosystems therein.

Coupled human and natural systems

The evolutionary context described earlier clearly demonstrates that humans and their hominin ancestors have been integral components of ecological systems for millions of years. Curiously, however, humans have often conceptualized themselves as being separate from nature. Nature has typically been depicted as places where humans go to view wildlife and wild places, rather than places where humans are also resident. It is interesting because, on one level, humans are just another line in the great ape lineage. However, chimpanzees (*Pan troglodytes*) and gorillas (*Gorilla sp.*) are clearly *of* nature, rather than apart from it. So, what is the difference? The difference seems to lie in the massive intellect of humans, which has enabled people to design cities, devise political systems of governance, and invent new forms of technological transportation. In the process, people have had time to think and, in that contemplation, have worshipped different Gods and subscribed to notions of creation and exceptionalism. In contrast, non-human animals do not seem to expend much thought conceptualizing how they came to be on Earth. Although a few primates (chimpanzees) may demonstrate a level of self-recognition (Gallup Jr. 1982), such consciousness does not approach that of humans, whose very taxonomic moniker (*sapiens*) hinges on the idea of a 'thinking' or 'wise' being. So, humans are clearly different, and the places in which they tend to live as well as the consequences of the decisions that they make within those places have been referred to as the human system. In contrast, the places where humans tend not to live (i.e. where species interact in the absence of direct human activities en masse) have been referred to as natural systems, at least by some Western cultures (Figure 1.1). Once again, the nomenclature of 'natural system', at least in the Western viewpoint, reinforces the notion that natural states are those that are devoid

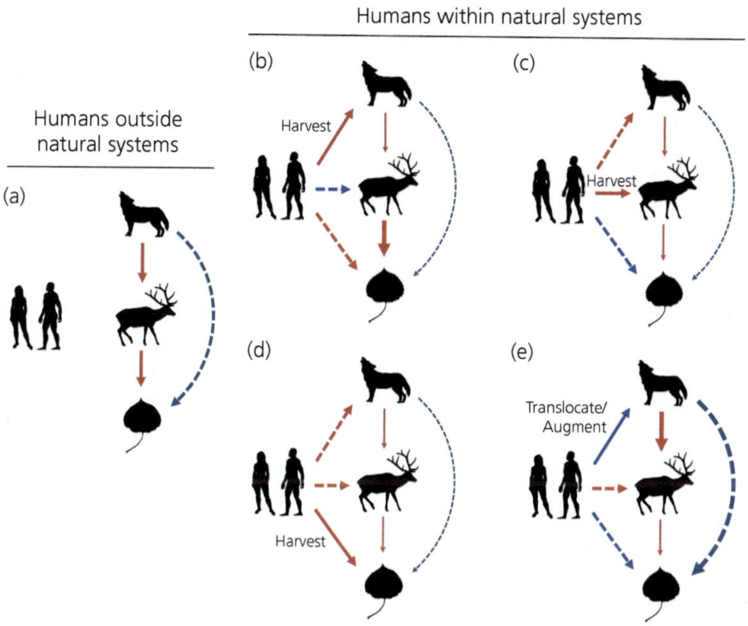

Figure 1.1 Humans outside natural systems (a) and within them (b–e). The red lines depict the negative numerical effects, with arrows indicating the affected population. The blue lines represent the positive numerical effects, whereas the dashed lines are the indirect effects. The width of the lines indicates the magnitude of effect. Panel (a) shows a trophic candace between wolves (*Canis lupus*), elk (*Cervus canadensis*), and quaking aspen (*Populus tremuloides*); panels (b), (c), and (d) show how human harvest of wolves, elk, and aspen, respectively, change trophic dynamics; panels (c) and (d) show similar effects, but (d) would occur with a time lag through bottom-up limitation; panel (e) shows how translocating or augmenting wolf populations could change trophic dynamics.

of people. In contrast, some non-Western cultures (e.g. indigenous peoples of North America prior to European contact) have long held a more intimately intertwined view of human and non-human entities (Martinez 2003, Kimmerer 2013). Thus, with the goal of integrating humans within natural processes, or at least providing a setting in which it is clear that humans can substantively impact nature, a new set of terminology has been developed.

The 'coupled human and natural systems' framework has been developed to facilitate quantification and modelling of the complex effects that humans can exert across the world and the ways nature can provide positive feedback to the social and economic systems upon which people depend (Wang et al. 2018, Liu et al. 2021). This framework has been adopted in science and society in an effort to drive interdisciplinarity and shape fundamental research (Liu et al. 2007). Large international funding bodies, such as the National Science Foundation in the United States, for instance, have developed specific calls for research in the dynamics of coupled human and natural systems (Baerwald, Firth, and Ruth 2016). Outside of frameworks and funding bodies, however, the term coupled human and natural systems can capably describe the environments over which humans and carnivores interact and the consequences of those interactions on people and predators.

While these concepts are more thoroughly described in Chapter 5 (see 'How conflict with humans changes carnivores'), briefly, the interactions of humans and carnivores within these coupled human and natural systems are most often benign. Humans may come into contact with carnivores, or vice versa, and nothing of note occurs even though there has clearly been an interaction. When these interactions yield benefits for one or both parties, however, human-carnivore interactions can be positive (Figure 1.1). These can be instances in which humans gain enjoyment from viewing carnivores or carnivores access various subsidies (e.g. foodstuffs) while inhabiting the built environment (Bateman and Fleming 2012, Rode et al. 2021). On the flip side, when human–carnivore interactions result in negative outcomes for one or both parties, conflict occurs (Zimmermann, McQuinn, and Macdonald 2020). Human-carnivore conflict is a highly pertinent topic globally as it threatens the security and private property of people as well as, via the mechanisms of retaliation, the conservation of carnivores (Kissui 2008, Inskip et al. 2013, Felix et al. 2022, Braczkowski et al. 2023). Given that conflict is one of the primary mechanisms restricting the population growth of large carnivores globally, and that over 75% of the remaining 31 species of large carnivores have declining population trends (Ripple et al. 2014), developing sustainable solutions for conflict is a massive conservation priority.

In creating and applying effective interventions, at least two important considerations should be taken into account. The first is the fact that human–carnivore conflict is a multifaceted problem requiring tailored solutions. For example, interventions seeking to reduce the likelihood of carnivores depredating people's livestock would be quite different from the interventions implemented to decrease the severity of human retaliation against carnivores (Treves, Wallace, and White 2009, Blackwell et al. 2016, Johansson et al. 2016, Beck et al. 2019). In this example, changing the way that humans respond or altering the decision-making process of carnivores are both important components of reducing human–carnivore conflict and yet, the interventions associated with each are completely different. Via this point we emphasize that there needs to be a clear articulation of the aims and objectives of the interventions designed to solve human-carnivore conflict. The second substantive issue is that of scale. It is critical that the interventions correspond to the scale over which human-carnivore conflict can occur (Montgomery et al. 2018, Chaka et al. 2021). Coarsely, there are four relevant scales. The first, which is also the finest, is the household scale. This scale defines the interactions between humans and carnivores that can occur at the scale of an individual household. This may be a home where livestock are threatened by carnivores. Next, there is the community scale where patterns of human–carnivore conflict can be assessed at the level of a collection of households. The third scale is positioned at the landscape level. This scale may correspond to a specific protected area or the extent of the adjacent carnivore population. Finally, the regional scale depicts patterns of human-carnivore conflict at the level of countries, regions, or species range extents. Discordance between the scale at which human-carnivore conflict data is collected and evaluated in the creation of conservation applications is common (Montgomery et al. 2018). Reducing that mismatch is needed to develop more effective interventions designed to solve specific problems within human-carnivore conflict.

Nevertheless, we make the point here that human-carnivore interactions take place within coupled human and natural systems. And within these systems, there is a diversity of roles that humans have and ecological impacts that they can exert. These impacts are generated by human actions, which can involve behaviours such as lethal exploitation, non-lethal activities, landscape modification, or the re-wilding of ecosystem structure (Vitousek et al. 1997, Creel and Rotella 2010, Bakker and Svenning 2018). These actions can trigger direct effects on biodiversity, such as the reduction or increase of plant or animal populations, or changes in the traits of those species. Herein, humans also have indirect effects such as trophic cascades, different forms of competition (e.g. exploitative versus apparent), or even commensalism or mutualism (Hebblewhite et al. 2005, Madin et al. 2016), although in some cases the classical ecological meanings of these terms change when applied to humans, as discussed next. The decisions and actions of humans, of course, can also have systems-level effects that may alter community dynamics, ecosystem services and functions, and indeed, the persistence of species of conservation concern (Western 2001, Milner-Gulland 2012). Thus, humans not only affect trophic systems via top-down processes, but also via bottom-up process as well as triggering changes along all sides of these ecological communities (Muhly et al. 2013). When considering the Eltonian niche of humans in ecological systems, there are five major roles that humans play (Moll et al. 2021). Humans can act as predators, as the constructors of niches, as a hyperkeystone species, as risk-responders, and as pseudomutualists. Here, we describe each of the roles and in the process demonstrate that humans are a very different kind of species in trophic system ecology.

Humans as predators

It has now been well-documented that humans are capable of harvesting biodiversity at rates that are considerably higher than any other predator that the world has ever known (Darimont et al. 2015). Herein, we use the term *harvest* intentionally. Harvest refers to the removal of plants or animals from ecological systems (Figure 1.1). Thus, this term is just as relevant for the collection of corn from an agricultural field as it is for the hunting of a white-tailed deer (*Odocoileus virginianus*). Importantly, humans can also induce lethality in animal species even when harvest does not occur. Thus, the direct lethal effects of humans on animals can include hunting (in both the legal and illegal forms), population culling for management purposes, conflict killings, vehicle strikes, plant or animal trafficking, and built environment strikes (e.g. birds dying when flying into human structures), among others. While the trafficking and trade of exotic plants and animals is a large component of the direct effects of humans on ecological systems, harvest does not always imply lethal exploitation. A central question, however, is whether the pet trade, for instance, represents an act of predation. On the one hand, there is evidence to suggest that it certainly should be. Afterall, the exotic pet trade, at least prior to captive breeding becoming possible, has a numerical impact on animal populations (Ceballos-Mago, González, and Chivers 2010, Harris et al. 2017). As these animals are typically incapable of ever being effectively restored to the wild, these human actions reduce the number of animals in the populations. On the other hand, however, predation defines instances in which two species interact and all or a significant portion of the biomass of one species is consumed by the other (sensu Abrams 2000). Thus, if consumption is inherent, then there are many examples of the direct effects of humans on plant and animal populations that would not be considered acts of predation (Packer et al. 2011, Burgin and Hardiman 2015, Voigt et al. 2022). When taken in this light then, perhaps the predatory abilities of humans have been inadvertently exaggerated.

A reasonable question to ask here is, why should it matter whether humans consume the species that they harvest? Firstly, it matters for the presentation of humans as predators and the types of effects that they might trigger. In predator-prey ecology, for example, the direct killing of prey provides an example of a consumptive effect (CE) (Pessarrodona et al. 2019). Notice that, once again, consumption is inherent to the theory that underpins the interactions of predators and prey. Here, these CEs involve prey death and the numerical reduction of the population, as well as consumption of all or a significant portion of the biomass of the prey species by the predator. The implications of these lethal effects, and the implications of energy flow between predator and prey, have been rendered in elegant population dynamic models such as the Lotka–Volterra models for lynx (*Lynx canadensis*) and snowshoe hare (*Lepus americanus*) (Elton and Nicholson 1942, Gilpin 1973). In the absence of consumption, there would be little functional relationship needed to couple the population dynamics of lynx with those of the hare. Furthermore, because of the behavioural decisions that prey make to avoid predators, predation can also exert non-consumptive effects (NCEs) (Lima and Dill 1990, Lima 1998). These NCEs occur when the accumulation of the trade-offs flowing from the behavioural plasticity of prey under the threat of predation risk carries fitness costs. In many cases, NCEs can be just as influential on animal populations as CEs (Mckee et al. 1997, Sheriff et al. 2020). Thus, consumption is integral to fundamental predator–prey theory as well as accurately quantifying the impact of predation on animal populations.

The context described here makes clear that consumption matters in predator–prey ecology but does not answer the question of whether it changes the way that animals view humans. To evaluate that particular question, we must explore whether animals have evolved defence mechanisms that are induced by humans. Firstly, as hominins have been integral members of ecological systems for millions of years and humans for hundreds of thousands of years, there is adequate evolutionary time for these inducible defences to have evolved. Thereby, the integrative research of Tollrian and Harvell

(1999) defines the four conditions that are necessary for inducible defences:

1. The cues indicating predator presence exhibit variation in spatial or temporal dimensions;
2. The predator cues are accurate indicators of the spatio-temporal risk of predation;
3. The defences that are deployed can effectively reduce that predation risk;
4. The application of these defences carries costs.

Via an extensive review of 187 studies, Montgomery, Macdonald, and Hayward (2020) found that the first three conditions are readily satisfied among research on human-large mammal interactions. However, whether defences have evolved in large mammals that are induced by human predation remains an area of active inquiry because this research has yet to determine whether the behavioural decisions made by animals in relation to humans carry fitness costs. Or to put it another way, it has proved very challenging for researchers to quantify whether humans have NCEs on animal populations. Thus, once again, the issue of consumption is essential to the ways in which humans are conceptualized in ecological systems.

So, we now know that consumption is critical to predator-prey population dynamic modelling as well as the evolution of inducible defences in animals vulnerable to predation. But does consumption matter in the eyes of an animal that, once dead, has no concept of whether its biomass is consumed by the predator? Interestingly, Frid and Dill (2002) explored this very topic in their development of the risk-disturbance hypothesis. Among other things, the risk-disturbance hypothesis postulates that, at least in systems in which animals have been hunted by humans, 'disturbance stimuli and true predatory stimuli may be indistinguishable from the animal's perspective' (Frid and Dill 2002). This is a very interesting concept because while humans are highly capable predators, they kill animals far more often than they consume them, and disturb animals at rates considerably higher than they predate or kill. What Frid and Dill are suggesting is that the magnitude of the behavioural response of animals to humans will be equivalent whether a human is lethal or benign. Thus, the risk-disturbance hypothesis implies that animals view all humans as lethal and respond accordingly. So, does this mean that animals will respond identically to hunters and hikers? Resolving that question would seem to depend on the cues of human predation and their population-level costs over time (see Chapter 9, this volume).

Recall that the first two conditions required for inducible defences to evolve entail cues which must be spatio-temporally variable and provide a reliable indication of predation risk. Interestingly, research has only recently begun to assess the cues of human predation (Carter et al. 2008, Smith et al. 2017, Liu et al. 2023). Firstly, these cues can be visual, auditory, or olfactory in origin. Visual cues include people moving about a landscape, the types of clothing that they wear, and the appearance of various items (e.g. cars, food, possessions) from human systems. Auditory cues include the sounds of people talking, cars running, or other acoustic characteristics of the built environment. Finally, olfactory cues consistent of the scent of humans themselves and may also include fragrances that are clearly indicative of human behaviour (e.g. food cooking, campfires burning, etc.). Importantly, the strength of these cues should be expected to vary as a function of human hunting mode (Thaker et al. 2011, Montgomery et al. 2022).

There are three hunting modes that exist among predators in nature. These include the active, sit-and-wait, and sit-and-pursue hunting modes (Huey and Pianka 1981, Schmitz 2007). An active or coursing predator is, more or less, constantly on the move about their landscape in search of prey. Thus, the cues of active predators are broadly dispersed and do not tend to provide an accurate reflection of imminent predation risk (Schmitz 2008). In contrast, sit-and-wait or ambush predators remain in a location until prey are within striking distance before they attack. Consequently, their cues are highly localized and, when detected by prey, warrant an immediate response (Wikenros et al. 2015). Finally, there is the sit-and-pursue predator which largely remains in place until prey are detected and then begins the hunt. Here, the cues would still be quite localized but exist somewhere in between the other two modes (Preisser, Orrock, and Schmitz 2007, Miller, Ament, and Schmitz 2014). Most predators on Earth have evolved to use just one of these hunting modes. Humans, on the other hand, flexibly use all three and are capable of moving between them with comparative ease (Montgomery et al. 2022, Gaynor et al. 2024). We can only speculate that humans create a very confusing landscape of cues for animals to detect.

Returning to our fundamental question, however, we can explore whether animals should be expected to be able to distinguish between the cues of lethal humans (e.g. hunters) and benign humans (e.g. hikers). For an animal to be capable of doing so, the cues of hunters would need to be different to those of hikers, for example. Let's consider an example of a state forest in Colorado, United States, that has a large hunting season each autumn (e.g. for ungulates) and is also heavily used by recreational hikers. At least from a visual perspective, the cues of these different phenotypes of people would be distinct. The hunter, for instance, would likely be wearing camouflage and carrying a weapon (e.g. bow and arrow or rifle), whereas the hiker might be clad in colourful clothing and would be travelling without a weapon. The auditory and olfactory cues would likely be quite different between these two types of people as well. Hikers might be talking conversationally, whereas hunters are often silent in an effort to promote stealth. Various olfactory elements are often used by hunters in disguise, while hikers may very well use perfume or cologne before hitting the trail. The point here is not that the cues of these two different types of people would absolutely be distinguishable in the eyes of an animal, but rather that the effect of these cues could be tested.

Two studies along these lines showed that African elephants (*Loxodonta africana*) are able to use any of the three types of sensory cues to distinguish between lethal and non-lethal groups of humans (Bates et al. 2007, McComb et al. 2014). In the Amboseli ecosystem of Kenya, two tribes of people regularly interact with elephants. The Kamba people are pastoralists that maintain a practice of peacefully avoiding elephants. The other tribe is the Maasai which have historically hunted elephants with spears. The Maasai people wear traditional dress with the predominant colour being red, while the Kamba wear paler colours (Bates et al. 2007). These studies tested elephants' reactions to acoustic cues by playing recordings of Maasai voices and Kamba voices (McComb et al. 2014), to olfactory cues by exposing elephants to garments worn by Maasai and Kamba people (Bates et al. 2007), and to visual cues by presenting elephants with red cloths and white cloths (Bates et al. 2007). The studies found that elephants reacted much more strongly to all cues associated with the Maasai, which carried potential lethal risks, when compared to the Kamba, which tended only to be capable of disturbing elephants. These studies demonstrate the utility of field-based experiments that can test the risk-disturbance hypothesis.

We also recommend new field-based studies and experiments to quantify the prospective costs that may emerge from animal–human interactions. This recommendation was also discussed in Montgomery, Macdonald, and Hayward (2020) where they wrote: 'We recommend renewed ingenuity in devising novel field experiments that can effectively quantify the fitness costs of large mammal behavioural plasticity in relation to human actions'. This has not yet occurred for two reasons. Firstly, these types of experiments are very challenging to implement in large mammal systems where the animal research subjects are comparatively mobile, elusive, and long-lived. Given that experiments to quantify NCEs have largely been developed in carefully controlled aquaria or terraria (Miller, Ament, and Schmitz 2014, Bourdeau, Bach, and Peacor 2016, Eaton et al. 2016, McMahon et al. 2018), scaling these principles to field systems with large carnivores and their prey is difficult. Secondly, this field of study is still emerging because many researchers continue to conflate NCEs with animal behavioural plasticity (Montgomery, Macdonald, and Hayward 2020). An important point herein is that the behavioural decisions of animals to avoid humans are often inconsequential. An animal detects a human, quickly moves away from them, and resumes their life history. The animal moves from point A to point B with a ~20-second deviation to their routine emerging from the human interaction. That one instance is unlikely to generate an NCE. However, if these types of interactions happen frequently enough where the effects can accumulate, or if the severity of the interaction is extreme enough, then the animal may reduce its foraging (e.g. Smith et al. 2017) or experience higher stress to the point where their survival and reproduction may be compromised. Thus, we recommend field-based experiments and quasi-experimental designs that can elucidate the NCEs that might emerge from human predation (see Chapter 8, this volume). This type of research will be critical to achieve a full appreciation of the ways in which humans affect animals, even large carnivores, when they behave as predators.

Niche constructors

Many animal species have been referred to as ecosystem engineers when they physically modify landscapes to the point of substantially impacting other species. African elephants, for example, regularly topple acacia trees with subsequent impacts that can scale to convert woodlands into savannahs (Laws 1970, Haynes 2012). Beavers (*Castor spp.*) steadily fell trees with their extensive incisors in the building of dams, which alter the trajectory, extent, and impact of river systems (Brazier et al. 2021). Termites (*Isoptera*) not only burrow, but also construct vertical mounds, that can completely change the structure of the ecosystems which they inhabit (Dangerfield, McCarthy, and Ellery 1998). These are very impressive examples of ecosystem engineering that positively pale in comparison to the construction capacities of humans. It is now estimated that almost 3% of the world's terrestrial land is dominated by the human-built environment (CIESIN, IFPRI, and CIAT 2011). Cities, towns, commercial strips, roads, railways, and dams necessary to maintain human society have exponentially grown since the Industrial Revolution. Not only that, but in an effort to promote the viability of human society, people have converted vast swaths of land globally for agriculture and have domesticated numerous species of wild animal, enabling livestock husbandry for food (Wirsenius, Azar, and Berndes 2010, Havlík et al. 2013, van Zanten et al. 2016). In the process of advancing these largely self-serving actions, however, humans have opened up new niches that can be filled by wild animals (Boivin et al. 2016). Furthermore, human actions, including the things that they build, are driving trait-based changes in animals and evolutionary rates that are far swifter than via non-human evolutionary forces (Hendry, Gotanda, and Svensson 2017).

Many of the trait-based evolutionary changes have occurred via anthropogenically-induced population bottlenecks. Take for example habitat degradation, fragmentation, and destruction that have occurred via the expansion of the human-built environment. In these ways, many animals have had their ranges restricted to ecological islands within a broader anthropogenic context (Fahrig 2003, Montgomery et al. 2014). Here, it is quite straightforward to imagine how animals retained in the *natural* portions of these now-coupled human and natural systems would compete over access to the modified niche space available. However, with global nature-positive initiatives that seek to expand protected areas, to connect existing protected areas via the implementation of ecological corridors, and to re-wild ecosystems both florally and faunally, it remains unclear whether animals will adapt to the modified niche space made available via these processes. For example, large carnivores might be able to traverse between protected areas via human-preserved corridors of habitat, but such movements would likely require behavioural adjustments to evolutionarily novel conditions (e.g. tolerating highly fragmented habitat or vehicle strike risks during dispersal). Similarly, protected areas might be physically reachable but situated in novel or sub-optimal climates from an evolutionary perspective. Thus, an important line of inquiry in future will feature the study of the niches that are inhabited by the translocated species and the modification to the niches of the existing species within those systems. Humans are the most dominant species of ecosystem engineer on the planet and there is tremendous power to use that ability to promote species conservation and management via niche construction but many questions remain (Smith 2007).

Hyperkeystone species

As in our earlier discussion of humans as predators, the occurrence of humans in given locations can have impacts on the abundance and diversity of co-occurring species. By definition, such human impacts extend to so-called keystone species, the species that have disproportionate impacts on a trophic system relative to their own abundance (Paine 1969, Power et al. 1996). Thus, humans are often referred to as a hyperkeystone species because they can influence the distribution and abundance of keystone species with subsequent effects that can cascade through trophic system structure (Worm and Paine 2016). An example of this principle is represented by a collapse in sea otter (*Enhydra lutris*) populations in the Pacific Northwest of North America, in the 1990s, which in turn released sea urchin (*Strongylocentrotus polyacanthus*) populations with negative consequences on kelp (*Laminaria* spp.) distribution and abundance (Estes et al. 1998). This collapse was due to increased orca (*Orcinus orca*) predation of sea otters, which some, in turn, attribute to the catastrophic decline of great whales, the primary prey of orcas, due to industrial whaling by humans (Springer et al. 2003). Therefore, the ecological keystone effects of these non-human species (otters and orcas) were governed at a higher ('hyper') level by humans. Importantly, human behaviour at the very top of this trophic system can initiate cascading mechanisms that flow all the way to the level of primary producers at the very bottom of the food web. Furthermore, while the nuances of this particular example are debated (e.g. see Springer et al. 2008), this is but one example of many that could be cited where humans have reduced, either directly or indirectly, populations of keystone species.

Of course, it is not only via predation (e.g. over-hunting or persecution) that the hyperkeystone effects of humans can be initiated. Translocation of non-human keystone species (Figure 1.1), which is also a form of pseudomutualism discussed at length to follow, of animals for population introduction, reintroduction, or supplementation also provides a good example of humans acting as hyperkeystone species. One of the most-studied examples of this would be the wolf reintroduction to the Greater Yellowstone Ecosystem in the Rocky Mountains of the United States. Via a series of translocations starting in 1995, wolves from Canada were captured, translocated, and held in a holding pen before eventual release into the Greater Yellowstone Ecosystem (Fritts et al. 1997). Not only has this action, among many others occurring both locally and globally, led to the wide-scale restoration of wolves in the United States, but it has also triggered trophic cascades (Figure 1.1). Though the precise mechanisms are still debated, humans undoubtedly contributed to the recovery of various tree species in Yellowstone by the reduction of elk (*Cervus elaphus*) populations resulting from wolf predation, human hunting policies, and winter severity (Kimble et al. 2011, Ripple and Beschta 2012, Beschta and Ripple 2016). Thus, given the ability of humans to change species assemblages, particularly as it relates to large mammals that sit at the very top of trophic systems, they have a profound capacity to act as a hyperkeystone species.

Risk-responders

We have made a compelling case above that humans are the most voracious predator on planet Earth. That does not mean, however, that humans do not also experience risks from animals, which can be quite extreme. Take for instance, vehicle collisions with animals that lead to extensive human fatalities each year. In response to these risks, extensive efforts have been made to alert drivers to hotspots of animal movement across roads and various policies have been implemented to reduce speed limits in areas where collisions have a higher probability (Clevenger, Chruszcz, and Gunson 2001, Meisingset et al. 2014, Huijser et al. 2016). Thus, humans clearly respond to risks from animals by changing their behaviours (i.e. acting as risk-responders). Another form of risk derives from direct interaction with animals. The most virulent animal taxon globally is undoubtedly the mosquito (*Culicidae*), which can transmit a number of diseases to humans. Around 600,000 people die per year from malaria alone (World Health Organization 2023). Consequently, humans use a variety of sprays, sleep under nets, and take medications to reduce the likelihood of encountering mosquitoes. However, while vehicle collisions and mosquitoes claim more human lives annually than any other type of animal interaction combined, it is fierce large predators that are often associated with the greatest risk responses from people due to their charismatic nature.

As discussed previously, competition over prey and access to space have always defined the nature and strength of human–carnivore relationships. Interpretation of the Palaeolithic cave drawings in France reveals that early humans had both a respect for and fear of the large carnivores that they shared their habitats with (Chapter 2 , this volume). Untold numbers of humans have died over evolutionary time because of negative interactions with large carnivores. Though global estimates of human deaths from large carnivores per year are difficult to estimate, bear (Genus *Ursus*) attacks on people have been so common and so numerous that an entire book has been devoted to defining the characteristics in which these attacks have occurred, with special reference to the practices that people could implement to reduce risk (Herrero 2018). Thus, while the number of people that die each year from carnivores would not even be remotely comparable to those killed by vehicle collisions or mosquitoes, the fear of large carnivore attack is often far greater (Kushnir and Packer 2019). All humans have, at one time or another, conceptualized how they might respond when encountering a large carnivore. People travelling in habitats where they could encounter carnivores may carry weapons (e.g. spray, knives, guns), intentionally emit auditory cues (e.g. whistling, talking, and singing), and increase their vigilance, among other examples of behavioural plasticity. These examples illustrate the fact that humans recognize that a diversity of animal species carries risks, but especially predators such as large carnivores. Intriguingly, such changes in human behaviour likely do entail fitness consequences in the classical ecological sense of altered survival, but investigations connecting these ideas in an ecological framework are, as yet, rare (Moll et al. 2021).

Pseudomutualists

In an era of dramatic environmental change associated with the coupled climate and biodiversity crisis, it has become common to highlight the negative impacts of humans on animal populations. However, there are many ways in which humans have demonstrated the capacity to aid animals. The first way in which they have done so is via protection. Humans the world over pursue advanced training as scouts, rangers, and management officials to protect animals from illegal harvest and habitat destruction. There are certain species that are so rare, for example, that individuals within their populations are subject to near continuous monitoring by armed rangers, or even by unpiloted aerial systems or drones (Mulero-Pázmány et al. 2014). This form of mutualism is intensive, costly, and dangerous to humans, so the benefits must be substantial. And while such actions are often associated with the preservation of ecotourism which provides an important source of conservation revenue, human motivations to protect animals are far greater than commerce alone. The diverse ways in which humans care for wild animals has led to the growth of wildlife rehabilitation centres that are designed to nurse animals back to health in an effort to return them to the wild. Humans have also developed rapid veterinary response units that intervene to help injured animals in nature. Finally, humans have also invested in genetic data banks and captive breeding for species of conservation concern to help supplement and restore their populations (Ralls and Ballou 1986, Ryder et al. 2000).

A second way that humans have benefited animal populations is via translocation. Humans voluntarily undertake complex, costly, and challenging translocations to introduce or re-introduce species of conservation concern (Berger-Tal, Blumstein, and Swaisgood 2020). That is not to suggest that translocation is not without its risks to the animal subjects (e.g. capture myopathy, disease transmission) (Griffith et al. 1993, Cunningham 1996, Breed et al. 2019), however, the actions are principally motivated by recovering some of the biodiversity that has been lost and promoting a more nature-positive future. Finally, both protection and translocation are aided by research-informed conservation. Thus, human pseudomutualism with animals extends to science, at least that which intends to address the research-implementation gap so as to more effectively conserve and manage animal populations (Knight et al. 2008, Dubois et al. 2020, Cazalis et al. 2022). We intentionally did not use mutualism to describe these human–animal relationships because it has not yet been quantified precisely how the actions described here benefit survival and reproduction. Undoubtedly, there are strong and positive emotional impacts that are generated by animal population protection, translocation, and research-informed conservation work. However, it is presently unclear whether these emotional responses have fitness implications. It is perfectly reasonable to hypothesize that it very well could and thus, as others have done before (Moll et al. 2021), we also highlight this area as an exciting and important area of research inquiry.

Conclusion

In this chapter, we have demonstrated the diverse ways in which humans shape animal populations with special reference to large carnivores. We have charted the history of interactions between people and predators and discussed how human evolution has been defined, in part, by the nature and strength of large carnivore relationships. We have explored the coupled human and natural systems over which human–carnivore interactions occur. Within the context of trophic system ecology, we then highlighted the impact of humans when they behave as predators, niche constructors, risk-responders, pseudomutualists, and as a hyperkeystone species. From this evidence, we have shown that humans are an enigmatic species that can destroy nature, as a means of expanding human society, and in turn, implement extensive and laborious methods to restore nature. Humans envision themselves to be fundamental components of trophic systems and yet, somehow, separate from the interactions that define predators and prey within. People are prolific builders, agriculturalists, pastoralists, and predators and yet, human systems have strived to protect certain species at all costs. It has been suggested that the coupled climate and biodiversity crisis presents the greatest existential threat to human society. Humans have inhabited Earth for a fraction of the time that it has been in existence, and the planet will carry on with or without people. In the meantime, however, people have the opportunity to positively affect the integrity of coupled human and natural systems globally by living, as much as is possible, in balance with animals and recognizing the fundamental role of people within the Animal Kingdom.

References

Abrams, P. A. 2000. The evolution of predator-prey interactions: theory and evidence. *Annual Review of Ecology and Systematics* 31(1):79–105.

Andrews, P. 2020. Last common ancestor of apes and humans: morphology and environment. *Folia Primatologica*. S. Karger AG:122–148. Available at: https://doi.org/10.1159/000501557.

Antón, M., A. Galobart, and A. Turner. 2005. Coexistence of scimitar-toothed cats, lions and hominins in the European Pleistocene. Implications of the postcranial anatomy of *Homotherium latidens* (Owen) for comparative palaeoecology. *Quaternary Science Reviews* 24(10–11):1287–1301.

Baerwald, T. J., P. L. Firth, and S. L. Ruth. 2016. The dynamics of coupled natural and human systems program at the U.S. National Science Foundation: lessons learned in interdisciplinary funding program development and management. *Current Opinion in Environmental Sustainability* 19:123–133.

Bakker, E. S. and J. C. Svenning. 2018. Trophic rewilding: impact on ecosystems under global change. *Philosophical Transactions of the Royal Society B: Biological Sciences* 373(1761): 20170432.

Bateman, P. W. and P. A. Fleming. 2012. Big city life: carnivores in urban environments. *Journal of Zoology* 287(1):1–23.

Bates, L. A., et al. 2007. Elephants classify human ethnic groups by odor and garment color. *Current Biology* 17(22):1938–1942.

Beck, J. M., et al. 2019. Improving human-lion conflict research through interdisciplinarity. *Frontiers in Ecology and Evolution* 7:243.

Berger-Tal, O., D. T. Blumstein, and R. R. Swaisgood. 2020. Conservation translocations: a review of common difficulties and promising directions. *Animal Conservation* 23(2):121–131.

Beschta, R. L. and W. J. Ripple. 2016. Riparian vegetation recovery in Yellowstone: the first two decades after wolf reintroduction. *Biological Conservation* 198:93–103.

Blackwell, B. F., et al. 2016. No single solution: application of behavioural principles in mitigating human–wildlife conflict. *Animal Behaviour* 120:245–254.

Boivin, N. L., et al. 2016. Ecological consequences of human niche construction: examining long-term anthropogenic shaping of global species distributions. *Proceedings of the National Academy of Sciences of the United States of America* 113(23):6388–6396.

Bourdeau, P. E., M. T. Bach, and S. D. Peacor. 2016. Predator presence dramatically reduces copepod abundance through condition-mediated non-consumptive effects. Freshwater Biology, 61(6):1020–1031. Available at: https://doi.org/10.1111/fwb.12766.

Braczkowski, A. R., et al. 2023. The unequal burden of human-wildlife conflict. *Communications Biology* 6(1):182.

Brazier, R. E., et al. 2021. Beaver: nature's ecosystem engineers. *Wiley Interdisciplinary Reviews: Water* 8(1):e1494.

Breed, D., et al. 2019. Conserving wildlife in a changing world: understanding capture myopathy—a malignant outcome of stress during capture and translocation. *Conservation Physiology* 7(1):coz027.

Burgin, S. and N. Hardiman. 2015. Effects of non-consumptive wildlife-oriented tourism on marine species and prospects for their sustainable management. *Journal of Environmental Management* 151:210–220.

Carter, J., et al. 2008. Subtle cues of predation risk: starlings respond to a predator's direction of eye-gaze. *Proceedings of the Royal Society B: Biological Sciences* 275(1644):1709–1715.

Carvalho, S., et al. 2012. Chimpanzee carrying behaviour and the origins of human bipedality. *Current Biology* 22(6):R180–R181.

Cazalis, V., et al. 2022. Bridging the research-implementation gap in IUCN Red List assessments. *Trends in Ecology and Evolution* 37(4):359–370.

Ceballos-Mago, N., C. E. González, and D. J. Chivers. 2010. Impact of the pet trade on the Margarita capuchin monkey Cebus apella margaritae. *Endangered Species Research* 12(1):57–68.

Chaka, S. N. M., et al. 2021. Predicting the fine-scale factors that correlate with multiple carnivore depredation of livestock in their enclosures. *African Journal of Ecology* 59(1):74–87.

CIESIN, IFPRI, and CIAT 2011. Global Rural-Urban Mapping Project, Version 1 (GRUMPv1): Urban Extents Grid. NASA Socioeconomic Data and Applications Center (SEDAC).

Clevenger, A. P., B. Chruszcz, and K. E. Gunson. 2001. Highway mitigation fencing reduces wildlife-vehicle collisions. *Wildlife Society Bulletin (1973–2006)* 29(2):646–653.

Cohen, J. E. 1995. Population growth and earth's human carrying capacity. *Science* 269:341–346.

Creel, S. and J. J. Rotella. 2010. Meta-analysis of relationships between human offtake, total mortality and population dynamics of gray wolves (*Canis lupus*). *PLoS One* 5(9):e12918.

Cueto, M., et al. 2016. Under the skin of a lion: unique evidence of upper Paleolithic exploitation and use of cave lion (*Panthera spelaea*) from the lower gallery of La Garma (Spain). *PLoS One* 11(10):e0163591.

Cunningham, A. A. 1996. Disease risks of wildlife translocations. *Conservation Biology* 10(2):e0163591.

Dangerfield, J. M., T. S. McCarthy, and W. N. Ellery. 1998. The mound-building termite Macrotermes michaelseni as an ecosystem engineer. *Journal of Tropical Ecology* 14(4):507–520.

Darimont, C. T., et al. 2015. The unique ecology of human predators. *Science* 349(6250):858–860.

Dubois, N. S., et al. 2020. Bridging the research-implementation gap requires engagement from practitioners. *Conservation Science and Practice* 2(1):e134.

Eaton, L., et al. 2016. Non-consumptive effects of native and invasive predators on juvenile Caribbean parrotfish. *Environmental Biology of Fishes* 99(5):499–508.

Ellis, E. C. 2011. Anthropogenic transformation of the terrestrial biosphere. *Philosophical Transactions of the Royal Society A: Mathematical, Physical and Engineering Sciences* 369(1938):1010–1035.

Ellis, E. C., et al. 2023. People have shaped most of terrestrial nature for at least 12,000 years. *Proceedings of the National Academy of Sciences of the United States of America* 118(17):e2023483118.

Elton, C. and M. Nicholson. 1942. The ten-year cycle in numbers of the lynx in Canada. *Journal of Animal Ecology* 11(2):215–244.

Espigares, M. P., et al. 2013. *Homo* vs. *Pachycrocuta*: earliest evidence of competition for an elephant carcass between scavengers at Fuente Nueva-3 (Orce, Spain). *Quaternary International* 295:113–125.

Estes, J. A., et al. 1998. Killer whale predation on sea otters linking oceanic and nearshore ecosystems. *Science* 282(5388):473–476.

Fahrig, L. 2003. Effects of habitat fragmentation on biodiversity. *Annual Review of Ecology, Evolution, and Systematics* 34:487–515.

Felix, N., et al. 2022. Retaliatory killing negatively affects African lion (*Panthera leo*) male coalitions in the Tarangire-Manyara Ecosystem, Tanzania. *PLoS One* 17(8):e0272272.

Frid, A. and L. Dill. 2002. Human-caused disturbance stimuli as a form of predation risk. *Ecology* 6(1):11.

Fritts, S. H., et al. 1997. Planning and implementing a reintroduction of wolves to Yellowstone National Park and central Idaho. *Restoration Ecology* 5(1):7–27.

Gallup Jr, G. G. 1982. Self-awareness and the emergence of mind in primates. *American Journal of Primatology* 2(3):237–248.

Galway-Witham, J. and C. Stringer. 2018. How did Homo sapiens evolve?. *Science* 360(6395):1296–1298.

Gaynor, K. M., et al. 2024. Hunting mode and habitat selection mediate the success of human hunters. *Movement Ecology* 12(1):29.

Gilpin, M. E. 1973. Do hares eat lynx?. *The American Naturalist* 107(957):727–730.

Griffith, B., et al. 1993. Animal translocations and potential disease transmission. *Journal of Zoo and Wildlife Medicine* 24(3):231–236.

Groucutt, H. S., et al. 2015. Rethinking the dispersal of *Homo sapiens* out of Africa. *Evolutionary Anthropology* 24(4):149–164.

Grove, M. 2017. Environmental complexity, life history, and encephalisation in human evolution. *Biology and Philosophy* 32(3):395–420.

Harmand, S., et al. 2015. 3.3-million-year-old stone tools from Lomekwi 3, West Turkana, Kenya. *Nature* 521: 310–315.

Harris, J. B. C., et al. 2017. Measuring the impact of the pet trade on Indonesian birds. *Conservation Biology* 31(2):394–405.

Harrison, T. 2010. Apes among the tangled branches of human origins. *Science* 327:532–534.

Havlík, P., et al. 2013. Crop productivity and the global livestock sector: implications for land use change and greenhouse gas emissions. *American Journal of Agricultural Economics* 95(2):442–448.

Haynes, G. 2012. Elephants (and extinct relatives) as earth-movers and ecosystem engineers. *Geomorphology* 157–158:99–107.

Hebblewhite, M., et al. 2005. Human activity mediates a trophic cascade caused by wolves. *Ecology* 86(8): 2135–2144.

Hendry, A. P., K. M. Gotanda, and E. I. Svensson. 2017. Human influences on evolution, and the ecological and societal consequences. *Philosophical Transactions of the Royal Society B: Biological Sciences* 372(1712):20160028.

Herrero, S. 2018. *Bear Attacks: Their Causes and Avoidance*. Rowman & Littlefield, Lanham, MD.

Hewes, G. W. 1964. Hominid bipedalism: independent evidence for the food-carrying theory. *Science* 146(3642):416–418.

Huey, R. B. and E. R. Pianka 1981. Ecological consequences of foraging mode. *Ecology* 62(4):991–999.

Huijser, M. P., et al. 2016. Effectiveness of short sections of wildlife fencing and crossing structures along highways in reducing wildlife-vehicle collisions and providing safe crossing opportunities for large mammals. *Biological Conservation* 197:61–68.

Inskip, C., et al. 2013. Human-tiger conflict in context: risks to lives and livelihoods in the Bangladesh sundarbans. *Human Ecology* 41(2):169–186.

Johansson, M., et al. 2016. Targeting human fear of large carnivores—many ideas but few known effects. *Biological Conservation* 201:261–269.

Kimble, D. S., et al. 2011. Aspen recovery since wolf reintroduction on the Northern Yellowstone winter range. *Rangeland Ecology and Management* 64(2):119–130.

Kimmerer, R. W. 2013. *Braiding Sweetgrass*. Milkweed Editions, Minneapolis, MN.

Kissui, B. M. 2008. Livestock predation by lions, leopards, spotted hyenas, and their vulnerability to retaliatory killing in the Maasai steppe, Tanzania. *Animal Conservation* 11(5):422–432.

Knight, A. T., et al. 2008. Knowing but not doing: selecting priority conservation areas and the research-implementation gap. *Conservation Biology* 22(3):610–617.

Kushnir, H. and C. Packer. 2019. Perceptions of risk from man-eating lions in southeastern Tanzania. *Frontiers in Ecology and Evolution* 7:47.

Langergraber, K. E., et al. 2012. Generation times in wild chimpanzees and gorillas suggest earlier divergence times in great ape and human evolution. *Proceedings of the National Academy of Sciences of the United States of America* 109(39):15716–15721.

Laws, R. M. 1970. Elephants as agents of habitat and landscape change in East Africa. *Oikos* 21(1):1–15.

Lima, S. L. 1998. Nonlethal effects in the ecology of predator-prey interactions. *BioScience* 48(1):25–34.

Lima, S. L. and L. M. Dill. 1990. Behavioral decisions made under the risk of predation: a review and prospectus. *Canadian Journal of Zoology* 68(4):619–640.

Liu, J., et al. 2007. Coupled human and natural systems. *Ambio* 36(8):639–649.

Liu, J., et al. 2021. Coupled human and natural systems: the evolution and applications of an integrated framework. *Ambio* 50(10):1778–1783.

Liu, M., et al. 2023. Ungulates' behavioral responses to humans as an apex predator in a hunting-prohibited area of China. *Animals* 13(5):845.

Madin, E. M. P., et al. 2016. Human activities change marine ecosystems by altering predation risk. *Global Change Biology* 22:44–60.

Maity, T. and A. Saxena. 2024. Challenges and innovations in food and water availability for a sustainable Mars colonization. *Life Sciences in Space Research* 42:27–36.

Martinez, D. 2003. Protected areas, indigenous peoples, and the western idea of nature. *Ecological Restoration* 21(4):247–250.

McComb, K., et al. 2014. Elephants can determine ethnicity, gender, and age from acoustic cues in human voices. *Proceedings of the National Academy of Sciences of the United States of America* 111(14):5433–5438.

Mckee, M. H., et al. 1997. Importance of consumptive and non-consumptive prey mortality in a coupled predator-prey system. *Freshwater Biology* 38(1):193–201.

McMahon, J. D., et al. 2018. Covariance between predation risk and nutritional preferences confounds interpretations of giving-up density experiments. *Ecology* 99(7):1517–1522.

Meisingset, E. L., et al. 2014. Targeting mitigation efforts: the role of speed limit and road edge clearance for deer-vehicle collisions. *Journal of Wildlife Management* 78(4):679–688.

Miller, J. R. B., J. M. Ament, and O. J. Schmitz. 2014. Fear on the move: predator hunting mode predicts variation in prey mortality and plasticity in prey spatial response. *Journal of Animal Ecology* 83(1):214–222.

Milner-Gulland, E. J. 2012. Interactions between human behaviour and ecological systems. *Philosophical Transactions of the Royal Society B: Biological Sciences* 367(1586):270–278.

Moll, R. J., et al. 2021. A framework for the eltonian niche of humans. *BioScience* 71:928–941.

Montgomery, R. A., G. J. Roloff, J. J. Millspaugh, and M. Nylen-Nemetchek. 2014. Living amidst a sea of agriculture: predicting the occurrence of Canada lynx within an ecological island. *Wildlife Biology* 20:145–154.

Montgomery, R. A., et al. 2018. Discordant scales and the potential pitfalls for human-carnivore conflict mitigation. *Biological Conservation* 224:170–177.

Montgomery, R. A., et al. 2022. The hunting modes of human predation and potential nonconsumptive effects on animal populations. *Biological Conservation* 265. DOI: 10.1016/j.biocon.2021.109398.

Montgomery, R. A., D. W. Macdonald, and M. W. Hayward. 2020. The inducible defences of large mammals to human lethality. *Functional Ecology* 34(12):2426–2441.

Muhly, T. B., et al. 2013. Humans strengthen bottom-up effects and weaken trophic cascades in a terrestrial food web. *PLoS One* 8(5):e64311.

Mulero-Pázmány, M., et al. 2014. Remotely piloted aircraft systems as a rhinoceros anti-poaching tool in Africa. *PLoS One* 9(1):e83873.

Packer, C., et al. 2011. Effects of trophy hunting on lion and leopard populations in Tanzania. *Conservation Biology* 25(1):142–153.

Paine, R. T. 1969. A note on trophic complexity and community stability. *The American Naturalist* 103(929):91–93.

Panger, M. A., et al. 2002. Older than the oldowan? Rethinking the emergence of hominin tool use. *Evolutionary Anthropology* 11(6):235–245.

Pessarrodona, A., et al. 2019. Consumptive and non-consumptive effects of predators vary with the ontogeny of their prey. *Ecology* 100(5):e02649.

Power, M. E., et al. 1996. Challenges in the quest for keystones: identifying keystone species is difficult—but essential to understanding how loss of species will affect ecosystems. *BioScience* 46:609–620.

Preisser, E. L., J. L. Orrock, and O. J. Schmitz. 2007. Predator hunting mode and habitat domain alter nonconsumptive effects in predator-prey interactions. *Ecology* 88(11):2744–2751.

Raichlen, D. A. and H. Pontzer. 2021. Energetic and endurance constraints on great ape quadrupedalism and the benefits of hominin bipedalism. *Evolutionary Anthropology* 30(4):253–261.

Ralls, K. and J. Ballou. 1986. Captive breeding programs for populations with a small number of founders. *Trends in Ecology and Evolution* 1(1):19–22.

Ripple, W. J., et al. 2014. Status and ecological effects of the world's largest carnivores. *Science* 343(6167):1241484.

Ripple, W. J. and R. L. Beschta. 2012. Trophic cascades in Yellowstone: the first 15 years after wolf reintroduction. *Biological Conservation* 145(1):205–213.

Rode, J., et al. 2021. Why so negative? Exploring the socio-economic impacts of large carnivores from a European perspective. *Biological Conservation* 255:108918.

Russo, G., et al. 2023. First direct evidence of lion hunting and the early use of a lion pelt by Neanderthals. *Scientific Reports* 13(1):16405.

Ryder, O. A., et al. 2000. DNA banks for endangered animal species. *Science* 288(5464):275–277.

Schmitz, O. J. 2007. Predator diversity and trophic interactions. *Ecology* 88(10):2415–2426.

Schmitz, O. J. 2008. Effects of predator hunting mode on grassland ecosystem function. *Science* 319(5865): 952–954.

Seed, A. and R. Byrne. 2010. Animal tool-use. *Current Biology* 20(23):R1032–R1039.

Senut, B., et al. 2018. Palaeoenvironments and the origin of hominid bipedalism. *Historical Biology* 30(1–2):284–296.

Sheriff, M. J., et al. 2020. Non-consumptive predator effects on prey population size: a dearth of evidence. *Journal of Animal Ecology* 89(6):1302–1316.

Smith, B. D. 2007. The ultimate ecosystem engineers. *Science* 315(5820):1797–1798.

Smith, J. A., et al. 2017. Fear of the human 'super predator' reduces feeding time in large carnivores. *Proceedings of the Royal Society B: Biological Sciences* 284(1857):20170433.

Springer, A. M., et al. 2003. Sequential megafaunal collapse in the North Pacific Ocean: an ongoing legacy of industrial whaling?. *Proceedings of the National Academy of Sciences of the United States of America* 100(21):12223–12228.

Springer, A. M., et al. 2008. Mammal-eating killer whales, industrial whaling, and the sequential megafaunal collapse in the North Pacific Ocean: a reply to critics of Springer et al. 2003. *Marine Mammal Science* 24(2): 414–442.

Stringer, C. 2016. The origin and evolution of homo sapiens. *Philosophical Transactions of the Royal Society B: Biological Sciences* 371(1698):20150237.

Thaker, M., et al. 2011. Minimizing predation risk in a landscape of multiple predators: effects on the spatial distribution of African ungulates. *Ecology* 92(2):398–407.

Tollrian, R. and C. D. Harvell (eds). 1999. *The Ecology and Evolution of Inducible Defenses*. Princeton University Press, Princeton, NJ.

Treves, A., R. B. Wallace, and S. White. 2009. Participatory planning of interventions to mitigate human-wildlife conflicts. *Conservation Biology* 23(6):1577–1587.

Vitousek, P. M., et al. 1997. Human domination of Earth's ecosystems. *Science* 277(5325):494–499.

Voigt, M., et al. 2022. Deforestation projections imply range-wide population decline for critically endangered Bornean orangutan. *Perspectives in Ecology and Conservation* 20(3):240–248.

Wang, S., et al. 2018. Structure, function, and dynamic mechanisms of coupled human–natural systems. *Current Opinion in Environmental Sustainability* 33:87–91.

Western, D. 2001. Human-modified ecosystems and future evolution. *Proceedings of the National Academy of Sciences of the United States of America* 98(10):5458–5465.

Wikenros, C., et al. 2015. Behavioural responses of ungulates to indirect cues of an ambush predator. *Behaviour* 152(7–8):1019–1040.

Wirsenius, S., C. Azar, and G. Berndes. 2010. How much land is needed for global food production under scenarios of dietary changes and livestock productivity increases in 2030?. *Agricultural Systems* 103(9):621–638.

World Health Organization. 2023. *World Malaria Report 2023*. CC BY-NC-SA 3.0 IGO. World Health Organization, Geneva.

Worm, B. and R. T. Paine. 2016. Humans as a hyperkeystone species. *Trends in Ecology and Evolution* 31(8): 600–607.

van Zanten, H. H. E., et al. 2016. Global food supply: land use efficiency of livestock systems. *International Journal of Life Cycle Assessment* 21(5):747–758.

Zimmermann, A., B. McQuinn, and D. W. Macdonald. 2020. Levels of conflict over wildlife: understanding and addressing the right problem. *Conservation Science and Practice* 2(10):e259.

CHAPTER 2

Human–Carnivore Interactions in the Pleistocene

Matt W. Hayward, Blaire Van Valkenburgh, Cassandra K. Bugir, and Chris N. Johnson

Introduction

The Pleistocene began with a wave of invasive species crossing between the continents now known as North and South America during the Great Faunal Interchange 2.58 million years ago, and ended with another wave of invasive species entering North America just before the beginning of the Holocene, 11,700 years ago. The final connection of the Isthmus of Panama allowed North America camelids, tapirs, deer, horses, gomphotheres, felid and canid carnivorans, and multiple rodents, to move to the south; and ground sloths (*Megatherium* spp.), terror birds (Phorusrhacids), armadillos (*Glyptodon* spp.), and capybaras to move north to round out the Pliocene and welcome in the Pleistocene (Marshall 1988, Weir et al. 2009). The end of the era of glaciation phases was nigh when a highly intelligent bipedal mammal crossed the Bering Strait from Asia to enter the Americas, accompanied by dogs (*Canis lupus familiaris*). This was the culmination of the sapient human invasion of all habitable continents. Hence, the start and end of the Pleistocene had immense ramifications for biodiversity. Throughout the Pleistocene, the climate alternated between long glacial periods lasting 40–100,000 years—during which global mean temperatures were up to 10°C colder than today, sea levels were 120 m lower, and glaciers reached to 40° north and south of the equator—and short interglacial periods of 10–15,000 years, where conditions were similar to those of today.

Extinction rates of vertebrates were exceptionally high during the Pleistocene epoch, especially in the Late Pleistocene (that is, in the last glacial cycle, from the penultimate glacial maximum around 130,000 years ago to the end of the Pleistocene 11,700 years ago). Pleistocene extinctions were concentrated on large-bodied species, resulting in the extinction of at least 104 genera of animals (Barnosky et al. 2004), involving at least 168 species, weighing more than 10 kg (Pučkov 1997). This pattern of selective extinction of large-bodied species was unprecedented over the preceding 65 million years of the Cenozoic, during which the likelihood of extinction was either unrelated to body size or, at some times, was higher in small-bodied taxa (Lyons et al. 2004b, Smith et al. 2018).

Size-selective extinction in the Pleistocene had strong effects on large carnivores. At the beginning of the Pleistocene, large vertebrate carnivores were prominent in the terrestrial faunas of all the world's habitable continents; by its end, all the largest species were extinct. The extinct species included, among others, the sabre-toothed and scimitar-toothed cats (*Smilodon* and *Homotherium*), dire wolf (*Canis dirus*), and short-faced bear (*Arctodus simus*) from the Americas; the machairodontine sabre-toothed cats of Africa and Eurasia; and the Australian marsupial lion (*Thyalcoleo carnifex*). Some of the extinct species, such as the various sabre-toothed cats and long-legged hyaenas (*Chasmaporthetes*), were morphologically and functionally

Matt W. Hayward et al., *Human–Carnivore Interactions in the Pleistocene*. In: *Apex Predators in the Anthropocene*. Edited by: Robert A. Montgomery and Remington J. Moll, Oxford University Press. © Oxford University Press (2026). DOI: 10.1093/9780191996665.003.0002 unlike living carnivores and would look most unfamiliar (and alarming) if humans saw them alive today; others were closely related and functionally similar but larger than extant species such as lions (*Panthera leo*), cheetahs (*Acinonyx jubatus*), and leopards (*Panthera pardus*).

These extinctions coincided in general with the evolution and global expansion of humans, but they also overlapped periods of climate change. Consequently, two primary causes have been proposed for the Pleistocene megafaunal extinctions: climate change or human overkill, or possibly some combination of both (Stuart 1999, Barnosky et al. 2004, Johnson 2006).

The interactions between humans and large carnivores during the Pleistocene extinctions has been less studied. Carnivores may have killed humans for food, and they may have been killed by people. They may also have competed for the same prey resources. Understanding the interactions between humans and other large carnivores may illuminate the causes of their extinctions. This chapter reviews human–carnivore interactions in the Pleistocene, including the possible role of humans in the mass extinction event. Insights into this issue also have implications for species conservation today where 4% of mammals (242 species) and 10% of carnivorans (29 species) are listed on the International Union for Conservation of Nature (IUCN) Red List as being threatened by climate change, and 12% and 22% (or 690 and 66 species) by persecution and overharvesting analogues of prehistoric human impacts; (www.redlist.org accessed 12/10/2023).

Ecological relationships of humans and carnivores in the Pleistocene

Humans as predators and prey

Humans in the Pleistocene were hunter-gatherers who preyed on a wide range of animal species, including large vertebrates. Large-bodied prey species were important to people for a variety of purposes. Mammals, especially, supplied a wide range of resources (Figure 2.1). Beyond being hunted for food, their hides were used for clothing and bedding; their bones and antlers were used to construct tools, weapons, and even houses; and sinews and tendons were used as thread for sewing and lashing. In addition to their practical uses, some mammals likely held cultural and symbolic significance for Pleistocene humans. For example, cave art and other artefacts illustrate that humans have long incorporated large mammals and other animals into their spiritual and religious beliefs (Figures 2.1c and d).

That anatomically modern humans hunted even the largest species is evident from a number of locations across North America, Siberia, and Western Europe, where four to eight-tonne mammoths (*Mammuthus spp.*) and mastodons (*Mammut americanum*) were killed for their meat, hides, and bones (Agam and Barkai 2018). Four-tonne giant ground sloths (*Megatherium americanum*) and large (up to two tonnes), armoured glyptodont mammals were distributed throughout the Americas during the Pleistocene, and there is evidence of humans hunting and killing them (Bustos et al. 2018, Politis et al. 2019, Carlini et al. 2022). There is tantalizing evidence that humans hunted the largest carnivores as well. For example, human-modified *Smilodon* bones suggest that they may have been hunted by humans (Serangeli et al. 2015). Neanderthal people also hunted lions 48,000 years ago, likely using spears (Russo et al. 2023). In Australia, the evidence for hunting of megafauna is not as strong as in Eurasia and the Americas, perhaps reflecting the greater antiquity of human-megafauna interaction on that continent alongside poor conditions for preservation of animal remains in Australian archaeological sites. Nonetheless, there are clear signs that people used giant kangaroos and diprotodons, and harvested the eggs of some large flightless birds (Johnson et al. 2023).

During the early part of the Pleistocene, humans used simple tools, such as wooden spears and stone-tipped spears, to hunt large prey. However, as humans continued to evolve and develop more advanced technologies, their hunting weaponry also became more sophisticated. For example, the bow and arrow was likely invented in Africa around 70,000 years ago, and was used in Europe by 54,000 years ago when anatomically modern humans first interacted with Neanderthals (Metz et al. 2023). This was followed by the invention of the atlatl or woomera, a spear-throwing device,

Figure 2.1 (a) Replication of a Cro-Magnon hunter using an atlatl or woomera from the Dordogne region of France at Grottes du Roc de Cazelle. (b) Recreated scene of butchery and meat drying activities of Cro-Magnon people at Grottes du Roc de Cazelle. (c) Replica of a 25,000-year-old artwork depicting aurochs at Lascaux, France. (d) Matt Hayward attempting to recreate Cro-Magnon art alongside other children, with the resulting image of *Bison priscus* inset, at Le Parc du Thot, France.

Source: © Matt W. Hayward/Laugery—grottes du roc Cazellle-Dordogne.

around 30,000 years ago, which allowed for greater speed and distance when throwing spears due to increased leverage and thrust (Raymond 1986). As humans continued to develop more advanced hunting weaponry, they were able to hunt more efficiently and effectively, leading to greater success in acquiring food and other resources. The evolution of hunting weaponry was likely an important factor in the success and expansion of human populations during the Pleistocene period, and substantially increased the threat humans posed to competing carnivores. The development of novel hunting technologies presents intriguing complexities in how carnivores might have historically managed risk from humans by modifying behavioural traits, and how such induced responses might manifest today in the face of modern technologies (e.g. guns) and less threatening activities (e.g. tourism; see Chapters 7–9, this volume).

An analysis of 57 published studies reporting on 6887 kills from 24 countries spanning a period from 200,000 to 10,000 ybp revealed that anatomically modern humans living in the Pleistocene killed at least 257 prey species that ranged in body mass from 0.2 to 12,700 kg (Bugir 2022). Warthogs (*Phacochoerus aethiopicus*) and bovids were significantly preferred prey, and species weighing anywhere between 0.2 kg and 209 kg were foraged upon (Bugir 2022). Overall, carnivores as a group were significantly avoided as prey, but brown and cave bears (*Ursus arctos* and *U. spelaeus*), wild cats (*Felis silvestris*), cave lions (*Panthera leo spelaea*), Eurasian lynx (*Lynx lynx*), grey wolves (*Canis lupus*) and wolverines (*Gulo gulo*) were all killed in proportion to their relative abundance in prey communities, while spotted hyaenas (*Crocuta crocuta*) were avoided (Bugir 2022).

In Eurasia, anatomically modern humans overlapped with Neanderthal people (*Homo neanderthalensis*), who also hunted large prey. Synthesis of data from 46 sites across 12 Eurasian countries revealed that Neanderthals significantly preferred steppe bison (*Bison priscus*) and mountain gazelle (*Gazella gazelle*), while avoiding bears (*Ursus* spp.), grey wolves, aurochs (*Bos primigenius*), Iberian ibex (*Capra pyrenaica*), and Pyrenean chamois (*Rupicapra pyrenaica*) (Bugir 2022). This suggested that Neanderthals avoided the most dangerous megafauna, and notably avoided dangerous predators. It is also likely that, when Neanderthals moved into a new area, they targeted the largest prey available (e.g. larger-bodied males of sexually dimorphic species at Kebara Cave, Israel), before switching to killing prey in accordance with their availability once prey species began recognizing the threat Neanderthals posed (Speth and Tchernov 2001). At this time, Neanderthals were living in Eurasia where more archaic hominids also existed, like *Homo erectus* and Denisovans, and so most competitor and prey species had evolved mechanisms to recognize the threats and opportunities Neanderthals afforded.

For the first 200,000 years of anatomically modern human existence, humans were restricted to Africa and again, humans, competitors, and prey had co-evolved. Even when humans expanded their distribution out of Africa into Eurasia 200,000 years ago, they were entering faunal communities that likely recognized groups of bipedal primates as threats, as well as potential food sources. Yet human intelligence and communication skills enabled these individuals to be impressive predators. As human technological proficiency improved with advances in stone flaking and projectile weapon techniques (Figure 2.1a), the size of the species we killed likely increased. These may be underestimates for carnivores if people were killing them in the field and not bringing their carcasses back to butchery or living sites. Yet anatomically modern humans would also have been laying traps for smaller species, meaning that there were very few species that were free of risk from predatory humans. All this illustrates that humans were killing carnivores even though they do not appear to be actively selectively hunting them, as well as exploiting a similar prey base to almost all predatory species. The technological advances that humans have made since the Pleistocene mean that we now target all but the smallest prey species (Darimont et al. 2017, Bugir et al. 2021b). Furthermore, the remarkable consistency in the size of the prey targeted by different human groups in these studies illustrates their robustness.

These observations make it clear that Pleistocene humans killed carnivores, even though they do not appear to have selectively hunted them. Reports of tooth marks on human skeletal remains show that carnivores also hunted humans (Selvaggio 1994a), and were doing so for at least 500,000 years (Daujeard et al. 2016) and probably much longer. Palaeolithic art works also illustrate the importance that our ancestors placed on understanding carnivores. The Grotte Chauvet-Pont-d'Arc in France contains a number of Palaeolithic cave paintings, including depictions of animals that are now extinct. One painting, known as the 'Panel of the Lions', shows a group of lions attacking an auroch (Figure 2.2). The level of detail on the postures, and therefore behaviour, of the lions in the painting demonstrates close human observation of this species. Whether this was because lions posed a threat to people, or provided opportunities for scavenging, or were of spiritual interest, is unknown. Being small-toothed and soft-pawed creatures, humans are always likely to have been vulnerable to attack by large carnivores, and on that basis might have been attractive prey (Brain 1981, Berger and Clarke 1995, Hart and Sussman 2011). That would be the case for humans who were alone and unarmed when found by a predator, but cooperative defence would have made humans less vulnerable to predation than other species of potential prey. What we do not know is how often they were targeted relative to other prey or how often such predation (or scavenging events) may have occurred within a single human lifetime.

Figure 2.2 Image of the replica of the 30–33,000-year-old Panel of the Lions from Chauvet Cave, France, held at the Brno Museum Anthropos. Creative Commons licence.
Source: © HTO.

Humans as competitors

The great diversity of animals hunted by humans, including the very largest species, would have meant that humans competed for prey resources to some extent with probably all other species of large, terrestrial carnivores (Van Valkenburgh 2001). The effects of this competition would have been asymmetric: human populations would have exerted stronger competitive effects on other species of large carnivores than those species did on humans. Being omnivorous, our ancestors would have been less constrained by prey abundance than the large carnivores with whom they competed, which were specialized consumers of large vertebrates (with the major exception of ursids) (Hayward et al. 2007). Therefore, humans would have been able to persist even after driving populations of large prey to densities too low to sustain populations of obligate carnivores, by switching to use of plant material and smaller prey, including aquatic resources (e.g. fish, molluscs).

One other characteristic trait of humans would have made them especially potent competitors. Human intelligence, and especially language abilities, would have given human ancestors a special capacity for cooperative hunting, and also conferred the skill and planning needed to devise weapons and traps. This meant that humans were no longer constrained to optimally forage on a limited set of prey species, in contrast to large carnivores, such as lions and wild dogs (*Lycaon pictus*), that specialize on a narrow range of prey species (Hayward and Kerley 2005, Hayward et al. 2016, Hayward et al. 2023). So, for instance, humans could set traps that yielded small protein sources, like rodents, that would be suboptimal and therefore unsustainable for other human-sized carnivorous species

(Carbone et al. 1999). Human ingenuity gave our ancestors another asset as hunters: the ability to cause death at a distance using missile technology, especially where that involved weapon systems like atlatls (or spear throwers) that increase the velocity, accuracy, and stopping power of a lethal projectile. It is likely that this ability, which was unprecedented in vertebrate evolution, allowed human hunters to have very large impacts on prey that had not co-evolved with them and were therefore naïve to the risk posed by a person at, what for any other predator would be, a safe distance. Because of this ability to drive populations of prey to low numbers, predation by humans would have had large impacts on other carnivores with overlapping prey preferences. Ultimately, in the Holocene, humans became ecosystem engineers by domesticating food sources and creating fields to grow crops, which again emphasizes the idea of asymmetric effects of prey depletion on humans and large carnivores.

Humans may have competed with other carnivores via scavenging, or kleptoparasitism, as well as through predation. Indeed, humans may have had a physiological advantage as scavengers because diurnal activity patterns would have enabled human ancestors to more easily access carcasses in the heat of the day when larger predators and scavengers would be thermally stressed (Hayward and Slotow 2009) and more likely to give up a carcass to avoid an energetically costly interaction. Passive scavenging by human ancestors—that is, collection of carcass remains once the predator has departed—may have yielded as much as 900 kilojoules per day, while active or power scavenging by driving predators off their kills could have yielded vastly more (Blumenschine et al. 1991, Tappen 2001). However, scavenging still only produces 20% of meat in the diet of modern Hadza hunter-gatherers in Tanzania (Bunn 2001).

Humans as collaborators

Not all human–carnivore interactions during the Pleistocene were negative for the carnivores. Humans may have facilitated access to food as some carnivores would have scavenged from kills made by humans (either by kleptoparasitism or working over kill sites after humans had finished using them), although humans also likely scavenged meat from the kills of other predators (Selvaggio 1994b, Domínguez-Rodrigo 2002).

It is also possible that the differing hunting styles of humans and some other large carnivores were complementary to one another and were combined in cooperative hunting. There are examples of such interspecific cooperation between humans and other predators. Until recently, some hunting people coordinated their activities with dolphins and orcas in the hunting of fish and whales in ways that provide increased hunting success for both partners in the interaction (Cram et al. 2022). One technique used by Blackfoot people to hunt bison (*Bison bison*) in Alberta, Canada, involved driving herds over natural deadfalls known as 'buffalo jumps'. This required that bison be herded over long distances. Blackfoot traditions strongly imply that people initially learnt their skills in herding bison by imitating the behaviour of wolves, whose hunting of bison involved shadowing herds (Barsh and Marlor 2003). Cram et al. (2022) suggest that the development of the Blackfoot technique of engineering mass deaths of bison included a stage in which humans and wolves worked together to drive bison towards deadfalls; presumably, they then amicably shared the resulting surplus. They further suggest that wolves and people might have cooperated in hunting other large and dangerous prey that wolves could run down and corner but found difficult to kill, that being accomplished by humans who followed the chase and carried lethal weapons. Ultimately, we do not know to what extent Pleistocene humans refined their hunting skills by learning from other large carnivores or hunted cooperatively with other species, but the diversity of Pleistocene carnivores suggests there would have been many opportunities for such relationships to develop. And there is of course one case in which we know for certain that a long-term collaboration between humans and another carnivore did develop in the Pleistocene.

All domestic dogs are descended from the grey wolf. Genetic analysis of ancient dogs and wolves shows that dogs separated from their wolf ancestors between 14,000 and 40,000 years ago, most probably around 28,000 years ago (Bergström et al. 2022). The oldest known specimen of an animal that was clearly a dog rather than a wolf is dated to about

17,000 years ago, and comes from an archaeological site in northern Spain (Hervella et al. 2022). Although it is still unclear exactly when and where dogs originated, this evidence leaves no doubt that grey wolves were domesticated in the Late Pleistocene in Eurasia, in association with hunting and gathering people.

There are two main hypotheses on how the association between wolves and people developed, which differ according to whether they assign the initial impetus to wolves or to people. The first proposes that the process began with wolves being attracted to people for the sake of the food we unintentionally provided in the form of scraps, carrion, and faeces (Coppinger and Coppinger 2001, Zeder 2012, Morey and Jeger 2015). The association might have begun with some wolves waiting for hunters to move on from kills of large-bodied prey so that they could scavenge the remains, or visiting abandoned campsites to collect whatever food was left behind. Then, building on their growing familiarity with people, those wolves might have started loitering around camps and settlements, following people when they shifted camp, and eventually forming durable attachments to them.

The alternative hypothesis suggests that domestication resulted from people capturing and raising wolf cubs (Germonpré et al. 2021, Mech and Janssens 2022). People probably hunted wolves to some extent, for food, cultural purposes (like Maasai moran do with lion hunting), and perhaps to reduce the number of competing predators. Hunters would have been able to occasionally capture wolf cubs when the mother was killed or became separated from her offspring. The keeping and raising of infant animals captured in this way is common in traditional hunting societies, and usually it is done out of affection for and interest in the young animals. Often, the care given to them extends to women breastfeeding them (Simoons and Baldwin 1982). This was common, for example, among traditional Aboriginal people in Australia who captured dingo pups in the wild and raised them within human families (Brumm and Koungoulos 2022). Hand-raising might have been highly significant as an initial step towards domestication, especially when it involved breastfeeding, if it produced adults who were more tolerant of proximity to humans than was typical of their species. However, as Brumm et al. (2023) argue, it is likely that the genetic changes required to produce adult animals capable of calmly cohabiting with people would have taken many generations to accomplish, and at least in the early stages of the process most hand-raised adults probably lived outside of human society (as was the case for dingoes in Australia), although their reduced aversion to people may have allowed them to stay close to human groups, from which they may have continued to obtain some food.

So, although these two hypotheses on the origin of the domestic dog make contrasting claims over which species was initially responsible for the development of the bond between people and wolves, both envisage a stage in which mobile bands of people were loosely associated with collections of wolves. At that point the two species were likely to have scavenged from one another and they may have shadowed one another on hunting expeditions. They might also have hunted cooperatively as a precursor to a later stage in the process when humans intentionally kept hunting dogs to increase rates of prey capture (Perri 2020). It is possible that some of the same elements of the relationship between people and wolves also developed in relation to other species of Pleistocene carnivores. But, it was only in the case of the wolf that the relationship matured to the point of full domestication, perhaps due to fundamental similarities in human and wolf society that allowed them to merge with one another (or because in other potential cases, the carnivores went extinct because of other interactions with people before domestication became a possibility).

Chimpanzees (*Pan troglodytes*) represent our closest common ancestors, and are often used as a model species to illustrate the behaviour of our ancient ancestors (Stanford 1996, 2001). Their interactions with prey likely represent the earliest ancestral condition of hominids. Chimpanzees have been recorded killing 24 different bird, ungulate, rodent, and primate species, but significantly prefer colobus monkeys (*Piliocolobus tephrosceles*) and most preferred prey are about one-fifth of their own body mass (Bugir et al. 2021a). Males perform the majority of the hunting (79–98%) and immature individuals are targeted, although meat only contributed <5% of the total annual diet (Stanford 2001). Hence, our deep-time ancestors would have been competing with smaller predators weighing less than 10 kg, while being prey to larger predators, like leopards and raptors (Hayward et al. 2006). They would not have been direct threats to larger predators. Given that *Homo erectus* was larger than chimpanzees, they are likely to have been able to kill larger prey and drive more species of carnivores off kills.

Megafaunal extinction in the Pleistocene

What were the outcomes of interactions between humans and carnivores during the Pleistocene? Two main answers have been proposed—human-induced extinctions or climate-induced extinctions.

Human-induced extinctions of the megafauna

The extinctions of the megafauna by the end of the Pleistocene were the culmination of 50,000 years of humans invading new lands as novel predators—*Homo ergaster* and *H. erectus* left Africa for Eurasia between 1.3 and 1.8 million years ago, followed by *H. sapiens* between 50,000 and 100,000 years ago. Anatomically modern humans had colonized Australia by 53,000 to 60,000 years ago (Roberts et al. 1994), the Americas by at least 13,300 years ago and possibly 23,000 years ago (Bennett et al. 2021), Madagascar by 10,000 years ago (Hansford et al. 2021), and New Zealand by 700 years ago (Anderson and McGovern-Wilson 1990). Many have argued that the intercontinental invasion of our species ultimately led to the demise of the megafauna and many Pleistocene carnivores, either directly or indirectly. This section investigates the impact of these broad extinction drivers and the flow of effects to carnivores.

Humans may have either directly hunted and competed with animals, leading to their extinction or altered habitats in association with these factors to do so. In Africa, where humans evolved in an environment with many species of large carnivores, the evolution of human foraging behaviour throughout the Pleistocene resulted in a gradual increase in the occupation of the large carnivore niche by hominins. This niche encroachment would have placed increasing competitive pressure on large carnivores and is associated with a contraction of the functional diversity of the non-human carnivore guild due to the extinction of several specialized clades (Werdelin and Lewis 2013). Extinction rates of large but not small carnivores in East Africa increased steadily over the last three million years, in a trend that cannot be explained by changes in temperature and rainfall, but is closely related to increasing brain volume in hominins (Faurby et al. 2020). Brain volume is a likely proxy for the capacity and effectiveness of humans as hunters of large-bodied prey. The rising rate of large carnivore extinction was also broadly correlated with a declining area of forest cover between three and one million years ago, possibly indicating an effect of environmental change on extinction, independent of humans. However, Faurby et al. (2020) showed that in North America over the same period, and in the absence of humans, there was no comparable decline in the representation of large-bodied species in carnivore communities, and no effect of changes in forest cover on the diversity of large carnivores. Hence, there is a compelling case that the effects of humans on carnivore diversity began long before the origination and expansion of *H. sapiens*. The rise of hominins have also caused biodiversity declines over several million years in other groups of large animals (Anieli et al. 2022, Dembitzer et al. 2022).

The other main proposed outcome of human–carnivore interactions in the Pleistocene was that the expansion of anatomically modern humans caused a step-change in predation pressure on large animals that resulted in a global wave of extinction of megafauna, including large carnivores. This wave of extinction dramatically reduced megafaunal abundance and diversity in Eurasia, Australia, and the Americas during the Late Pleistocene; a similar reduction of megafauna (including large carnivores) occurred in the Holocene when humans reached large islands such as Madagascar, New Zealand, and the islands of the Caribbean.

The plausibility of this model of megafaunal extinction by human 'overkill' rests on two points. The first is that Pleistocene humans hunted large animals—even the very largest species—among a wide range of prey species. The evidence of hunting is patchy, and for some species, and in some places, it is absent (or at least, has not yet been uncovered). This is no doubt partly due to the incompleteness of the fossil record but could also suggest that the rates of hunting of large animals were low, possibly too low to have had significant demographic impacts. However, demographic modelling shows that the slow reproductive rates of megaherbivores could have made them vulnerable to quite low rates of offtake, which might even have been limited to killing of juveniles and young adults but could nonetheless have driven declines to extinction (Johnson 2002, Van Valkenburgh et al. 2016). The rarity of carnivores in the diets of Pleistocene humans could further suggest that human hunting did not directly cause the extinction of carnivores. On the other hand, large carnivores have naturally low population densities. This could help to explain their rarity in the archaeological record, but would have been a significant factor causing them to be susceptible to even very low rates of killing. The second point is that human hunters possess unique features, most notably intelligence and ingenuity, that made it possible for us to cause extinction of some prey species through high killing rates and prey switching.

Human overkill

Wildlife are completely naïve to the risks novel predators pose. Charles Darwin reported that the warrah or Falklands wolf *Dusicyon australis* was fearless in approaching him as it had no recognition of humans as a threat (Darwin 1859), yet the species was extinct less than 50 years later. While prey species that have evolved alongside predators rapidly overcome their naivete (Berger et al. 2001), species that have no history of evolutionary interaction with a particular predator may not develop appropriate responses to novel threats soon enough to avert extinction (e.g. the warrah). The human overkill hypothesis posits that invasive human populations colonized new continents or islands and cause rapid declines in the unsuspecting faunal communities (Martin 1967).

The further back in time we go, the harder it is to clearly see the causes of mass extinction events, and so more recent colonization events are most illuminative. The arrival of the Maori to New Zealand led to the extinction of nine species of flightless ratite moas, from the 12 kg *Eurypteryx curtus* up to giants of 250 kg. Moas occurred across the two main islands, and within 150 years of human arrival they were extinct. Butchery sites have been discovered that show the bountiful food was used wastefully, with only drumsticks carved off the carcass and the rest left to rot (Anderson 1989). Genetic evidence shows that less than 2000 people with a population density of about 0.1 person per square kilometre were responsible for this entire extinction event (Holdaway et al. 2014). There is no support for the role of climate change in driving the extinctions of the moa in New Zealand (Rawlence et al. 2012). New Zealand supported a single giant raptor at this time, Haast's eagle *Hieraaetus moorei*—whose primary prey was the giant 230 kg moa *Dinornis robustus*. Its extinction probably arose as a consequence of the extinction of the giant moa, and the fact that humans hunted its eggs. A similar pattern of human colonization and faunal extinction occurred across the Pacific.

Madagascar once supported a rich fauna including 17 species of giant lemurs, elephant birds, several hippopotami, the aardvark-like *Plesiorycteropus* sp., giant tortoises and large carnivores. However, all of the largest of Madagascar's fauna are now extinct, with only small species (less than 10 kg) persisting (Hansford et al. 2021). Humans arrived in Madagascar from Asia up to 10,000 years ago, but it wasn't until 1200–900 years ago that the megafauna became extinct (Hansford et al. 2021). While the cause of these extinctions has been most recently attributed to human-induced changes in landscapes (Hansford et al. 2021), the variable resource availability within the five main habitats on the island and their attraction to humans could also explain the regionality of extinctions there as humans inhabited each at different densities. Nonetheless, Madagascar's extinctions were clearly not a blitzkrieg like in New Zealand.

Debate continues as to the causes of the extinctions in Australia and the Americas (Cooper et al. 2015), but increasingly evidence indicates that the extinctions would not have occurred without the presence of humans (Saltré et al. 2016, Van Der Kaars et al. 2017b), and that humans were the primary driver (Bergman et al. 2023). Africa and Europe provide an experimental control for the extinction events elsewhere around the world, as humans co-evolved with wildlife here for at least 1.8 million years. Over the past 50,000 years, Europe lost nine genera of megafauna and Africa eight, while Australia lost 21, North America 33, and South America 50 (Barnosky et al. 2004). Climate change affected each continent differently and at different times, but not all faced climate change at the time of human colonization or when most species went extinct (Barnosky et al. 2004, Bergman et al. 2023).

The reduction in prey base associated with human overkill is likely to have been exacerbated by predation by carnivores. The densities of carnivores typically lag behind that of their prey (e.g. O'Donoghue et al. 1997). Hence, megaherbivores may have been driven into a predator-pit by human and carnivore hunting, from which they were unable to escape. This in turn may have led to large carnivores going extinct due to reductions in preferred prey.

The human overkill hypothesis is said to rely on a relatively short window of time from human arrival to animal extinction. This may not be absolutely necessary as optimal foraging theory suggests that once a prey item becomes too scarce, predators are unlikely to continue to hunt it and so it may persist at a low density. Contemporary species threatened by invasive species show that some can persist alongside novel predators without immediately being driven extinct, particularly in large landmasses (e.g. some critical weight range marsupials persisting alongside foxes and cats in Australia). Indeed, evidence suggests megafauna persisted alongside humans in Australia for up to 4000 years before going extinct (Van Der Kaars et al. 2017b). However, this is not a long time in an evolutionary sense given mammal species last 2–3 million years on average.

Top-down forcing hypothesis

The top-down forcing hypothesis of the Pleistocene extinctions posits that colonizing human predation impacts on megaherbivores were exacerbated by that of multiple, sympatric carnivores (Janzen 1983, Ripple and Van Valkenburgh 2010). This is based on evidence that large herbivores in the Pleistocene occurred at low densities and their populations were primarily controlled by top-down factors (Johnson 2002, Brook and Johnson 2006, Ripple and Van Valkenburgh 2010, Van Valkenburgh et al. 2016), hence any additional predation pressure is likely to have sent them over the precipice. The top-down forcing hypothesis is not rejected due to the scarcity of known human kill sites of large herbivores as these are likely to have been cleaned up by scavenging members of the Pleistocene carnivore guild that persisted alongside humans (Ripple and Van Valkenburgh 2010). Humans were likely free of competitive limitation from carnivores because of our omnivory, and had a reduced risk from intraguild predation through group-living, management of fire, mastery of projectile weaponry, presence of commensal dogs, and use of complex language for communication (Donadio and Buskirk 2006, Ripple and Van Valkenburgh 2010). Apparently, the Pleistocene carnivore guild exhibited high levels of competition, as illustrated by the three-times higher frequency of tooth breakage in Pleistocene carnivores than those of today (Van Valkenburgh 2009). This is also supported by higher levels of tooth breakage of modern African carnivores historically than more recently, where populations are much smaller; and in resource-constrained arid areas compared to more mesic areas (Mann et al. 2017). Therefore, this multi-predator theory offers further clarity to the role of humans in the Pleistocene extinctions. Again, once the megaherbivores were driven extinct, predators that could not prey switch or cope with vegetative food sources would have followed.

Climate change as a cause of Pleistocene extinctions

The main alternative hypothesis on the cause of megafaunal extinction is that it was a result of climate change. The Pleistocene was characterized by climatic variability with global temperatures fluctuating by up to 6°C. At peak glaciation, the East Antarctic Ice Sheet was 500 m thick during the glacial chills, whereas it is now only 50 m thick and has been since the end of the Pleistocene (Suganuma et al. 2014). The temperature changes around the world were not constant, but varied regionally. There were also climatic fluctuations immediately prior to interglacial periods that are known as 'flickers' that involved short-term, high amplitude changes in conditions, which may have driven rapid shifts in faunal distributions and communities (Roy et al. 1996). Species that survived the Pleistocene must have been either resilient to the degree of climatic variability they experienced or able to retreat to climatic refugia from which they recolonized broader areas thereafter (Levinsky et al. 2007, Ohlemuller et al. 2011). This variability in climatic conditions could have led to the demise of the many Pleistocene megafauna, including carnivores, that went extinct over this period.

Assessing the evidence of the cause of the Pleistocene extinctions

Studies that have investigated patterns of extinction at global scales have consistently supported the spread of anatomically modern humans as the primary driver of megafaunal extinction. These studies demonstrate what Paul Martin referred to as the 'deadly syncopation', in which arrival of our species on the continents and large islands outside Africa was followed by differential extinction of large-bodied vertebrates (Martin 2005). The duration of overlap of people and megafauna was evidently quite variable, but the sequence of events was consistent, and it recurred regardless of changes in climate. Several studies have used statistical models to compare the relative influence of humans and climate change on rates of extinction of large vertebrates, and have shown that human arrival was the most important predictor of extinction while climate had little or no influence (Sandom et al. 2014b, Bartlett et al. 2016, Araujo et al. 2017, Lemoine et al. 2023). Global range expansion of modern humans was correlated not only with the extinction of megafauna, but also with genomic markers of changed population size of surviving megafauna, which lost on average 91% of their abundance in a pattern that also cannot be explained by climate change (Bergman et al. 2023).

The importance of human arrival rather than climate change in accounting for the large-scale pattern of megafaunal extinction is exemplified by a comparison of North America and Australia. Humans reached the Americas during the transition from the Last Glacial Maximum to the Holocene. Therefore, the wave of extinctions that followed human arrival coincided with rapid and extreme climate change, which could possibly have caused or at least contributed to extinction. But, in Australia, humans arrived during a period of relative climate stability, in the middle of the last glacial cycle. Nonetheless, the arrival of humans in Australia was followed by a wave of megafaunal extinction that was as severe or more so than in America, and was uncorrelated with any change in climate (Saltré et al. 2016). A parsimonious interpretation of the two cases would discard climate change as an explanatory variable in models that attempt to explain megafaunal extinction on both continents. That conclusion is strengthened by consideration of the large islands that were occupied by people during the stable climatic conditions of the Holocene. In New Zealand, for example, there is no evidence for a pulse of extinction coinciding with the extinction of megafauna in Australia, despite the two land masses being subject to the same large-scale patterns of climate. Instead, New Zealand's megafauna persisted through the Late Pleistocene and the transition to the Holocene, only to go extinct within 200 years of the arrival of people in the late thirteenth century CE (Perry et al. 2014).

Just as studies that encompass large spatial scales agree in attributing megafaunal extinction mainly to humans, so do studies that consider long periods of time. As we have already noted, the differential extinction of large-bodied species is characteristic of the last glacial cycle, during which modern humans expanded over the globe. Such extinction did not occur in previous glacial cycles (with the exception of the loss of some large-bodied animals in Africa, also noted here), several of which were similar in their pattern of fluctuating temperature and rainfall to the last cycle (Masson-Delmotte et al. 2010). Therefore, another piece of evidence suggesting climate change was not a major factor in the Late Pleistocene extinction of megafauna is that those species had survived many previous changes of similar nature and magnitude. Davoli et al. (2024) used climate-based species distribution models to reconstruct the diversity of Europe's megafauna species in the current interglacial period (the Holocene) and the previous interglacial, around 127,000 years ago, as they would be without human influence. Their analysis showed that if species distributions were controlled by climate alone, the diversity and abundance of megafauna should have been similar in the two periods. In fact, megafaunal biomass in the Holocene was reduced by 94.5% and species richness by 70.8%, from the levels that would be expected if climate was the main determinant of the distribution of species. These reductions can be attributed to the impacts of prehistoric *H. sapiens* (Davoli et al. 2024).

Other studies of Late Pleistocene megafaunal extinction focus on smaller regions or single sites within shorter time periods and attempt to interpret the causes of extinction by detailed analysis of the chronology of extinction in relation to indicators of regional climate, environmental change, and human population dynamics. Studies of this kind are more likely to conclude that climate change contributed to extinction, and they tend to find more complexity in the extinction process. For example, O'Keefe et al. (2023) found that the extinction of the Rancho Le Brea regional megafauna of southern California was preceded by a period of climatic drying, which saw a woodland ecosystem that was characterized by high herbivore biomass and low fire activity transition to a dry chaparral system subject to frequent fires that were likely to have been ignited by people. This state shift was associated with a rapid decline to extinction of megafauna. These results reflect a complex interplay of climate and human impacts, and the authors argue that it is not possible to attribute extinction simply to one or the other (O'Keefe et al. 2023).

While the data from Rancho Le Brea suggest that regional environmental change was an important factor in creating the conditions for megafaunal extinctions, other intensive studies have come to different conclusions. For example, Gill and colleagues (Gill et al. 2009, Gill et al. 2012) studied environmental change and megafaunal extinction in north-eastern North America and while they found that changes in vegetation and increased vegetation followed megafaunal decline, there were no changes in environmental conditions that preceded megafaunal decline and might plausibly have contributed to causing it, leaving human hunting as the most likely cause. There have been many other local-scale studies of megafaunal extinction in other parts of the world, and they have similarly come to a wide range of conclusions on the involvement of climate-driven environmental change and changed fire regimes as factors that interacted in complex ways with direct human impacts to cause extinction. These varying conclusions may be difficult to reconcile with the large-scale studies which, as noted earlier, quite consistently find a dominant role for the direct impact of humans as the cause of megafaunal extinction. We suggest that it is possible to draw two general conclusions from the full range of studies. First, the extinction of megafauna was an almost inevitable result of the arrival of modern humans in new lands during the Late Pleistocene. But second, despite this inevitability, the details of extinction dynamics—elements such as the existence of lags between human arrival and megafaunal extinction, and the speed and synchroneity of declines of megafaunal species—were often influenced by features of the local environment that interacted in complex ways with direct human impacts, in much the same way as regional factors, like drought and fire, change the local extinction risk of species today.

Climate and human effects on prey

The Pleistocene was characterized by an abundance of large mammals—megaherbivores and their predators. There is a strong relationship between the richness of megaherbivore species and that of their predators (Van Valkenburgh et al. 2016). For example, when only one megaherbivore is present in a community, only 32% of sites supported three hypercarnivores (species for which meat comprises >75% of their diet), whereas 91% of sites with three or more megaherbivores supported this many hypercarnivore species (Van Valkenburgh et al. 2016). In Pleistocene communities with two to six megaherbivore species, there were between four and seven hypercarnivore species, including two or three carnivores that exceeded 100 kg (Van Valkenburgh et al. 2016).

Humans may have exacerbated the population pressures on carnivores by reducing the availability of the preferred prey of each predator. Given that 50% to 80% of the variability in extant African carnivore population densities is explained by the abundance of their preferred prey (Hayward et al. 2007), it is likely that the abundance of carnivores in the Pleistocene was also tightly linked to bottom-up factors related to their prey. Hence, the combination of slow life history traits and a reduction in carnivore carrying capacity (prey abundance) due to human hunting may have increased the chances of predators being driven extinct via interspecific (carnivore-carnivore) competition.

Climate change during the Pleistocene likely altered the distribution and abundance of prey species. Humans would probably have been more resilient to both exploitation and interference competition with carnivores because of our generalist ecology. Humans are omnivores and capable of switching food sources between animal and plant matter. Also, humans are not morphologically constrained to rely on large prey items, as is likely for species like *Smilodon* or *Homotherium*, in the same way as modern apex predators like African lions *Panthera leo* or tigers *P. tigris* (Hayward and Kerley 2005, Hayward et al. 2012). Instead, we can set traps and snares for small game or fish that can be eaten alongside vegetable matter to sustain us entirely. Furthermore, humanity's mastery of fire unlocked more nutrients from starch and protein per gram of meat during the digestion process (Carmody and Wrangham 2009), which likely enabled our ancestors to survive through resource-constrained periods when specialist carnivores could not.

During the Pleistocene, the Earth experienced several episodes of global cooling and warming, which had significant impacts on ecosystems and the species that inhabited them. Some researchers argue that climate change was the primary cause of the extinction of many large mammal species, with human hunting playing a secondary role or being a consequence of these environmental changes.

Certainly, the climate changed relatively rapidly, in an evolutionary sense, during the transition from glacial to interglacial periods, and biodiversity may have been unable to cope with this. In support of this hypothesis, many of the large mammal species that went extinct during the Pleistocene were adapted to colder environments, such as the woolly mammoth. As the Earth warmed at the end of the Pleistocene, these species may have been unable to adapt to the new environmental conditions, leading to their decline and eventual extinction. There is some evidence that the faunal community present in the terminal glaciation of the Pleistocene were unique and may have been more susceptible to increasing temperatures than species that persisted through the earlier interglacial periods (Scott 2010).

Additionally, the timing of human arrival in different parts of the world did not always coincide with the extinction of large mammal species. For example, in Australia, human populations arrived by at least 50,000 years ago, but the extinction of many large marsupial species did not occur until much later, around 45,000 years ago, when the climate began to dry out and become more arid (Wroe et al. 2004).

A key requirement of climate change as the cause of the Pleistocene extinctions is that climate change needed to occur contemporaneously with the extinctions. The further back in time we go, the harder it is to link climate change with the extinctions, but certainly not all sites of megafaunal extinction were facing a changing climate at the time of the extinction event. Hence, there is conjecture even within authorship teams as to the importance of climate change as a driver of the Pleistocene extinction event (e.g. Cooper et al. 2015, Van Der Kaars et al. 2017a).

Climate change during the Pleistocene is likely to have altered habitats (Graham and Lundelius 1984). Habitat specialist species may have been unable to track their habitat as it followed its bioclimatic envelope. In the absence of refugia, these species may have gone extinct.

However, it is important to note that climate change alone may not fully explain the Pleistocene megafaunal extinctions. Many large mammal species were able to survive previous episodes of climate change, and it is likely that the combination of environmental changes and human hunting pressure played a complex and interacting role in driving the extinctions. Overall, while the evidence for climate change as the primary cause of the Pleistocene megafaunal extinctions is not as strong as the evidence for human hunting, it is still an important factor to consider in understanding these complex ecological dynamics.

Indirect impacts of humans on Pleistocene megafauna

In addition to direct hunting of large mammals, there is evidence to suggest that humans also had indirect impacts on Pleistocene megafauna. For example, as humans migrated across the globe and expanded their populations, they brought with them a suite of domesticated animals, such as dogs, goats, and sheep, or commensal animals, like rats and mice. This group of animals could have had significant impacts on local ecosystems, competing with wild herbivores for resources such as grasses and shrubs, or competing for food and directly preying upon carnivores.

Humans also elicit fear in wildlife, including carnivores, and extant wildlife are more fearful of humans than even the most dangerous of predators (Chapters 7 and 9, this volume) (Zanette et al. 2023). This can lead to behavioural responses to humans in which carnivores avoid areas inhabited by people, and can ultimately lead to trophic cascades, where a change in the dynamics at one trophic level can influence species further along the trophic chain (see Chapters 10–12, this volume) (Schmitz et al. 2004). Thus, by creating halos of low carnivore density around areas of high human use (rock shelters or caves), humans further lowered the carrying capacity of a site. When human population densities expanded to have a relatively even distribution across the landscape, there would no longer have been areas of refuge for carnivores.

Another indirect impact of humans on Pleistocene megafauna is the modification of landscapes through fire and other land-use practices. Some researchers argue that humans used fire to clear land for hunting and foraging, leading to the transformation of ecosystems and the decline of some large mammal species. For example, in Australia, the use of fire by Aboriginal people may have contributed to the decline of the giant marsupials that once inhabited the continent (Flannery 1990, 1994). The impact of fire on carnivores during the Pleistocene is not fully understood, but it is clear that fire was a significant factor in shaping landscapes and ecosystems during this time period.

The impact of fire on carnivores would have varied depending on the species and their ecological requirements. Some carnivores, such as hyenas and jackals, are known to scavenge from fire-affected landscapes and may have benefited from the increased availability of food. Others, such as some large cats and wolves, may have been negatively impacted if their preferred prey species were negatively affected by fire.

Overall, the impact of fire on carnivores during the Pleistocene was likely complex and varied depending on the species and the specific ecological context. While fire may have had some negative impacts on carnivores, it is also possible that some species may have benefited from the increased availability of food and altered landscapes.

Finally, some researchers have proposed that humans may have indirectly impacted megafauna through the transmission of diseases (MacPhee 1997). As human populations expanded and came into contact with other species, they may have introduced new diseases to which these animals had no immunity, leading to population declines and extinctions. Using other novel diseases affecting biodiversity as a proxy for what may have occurred in the Pleistocene, Lyons et al. (2004a) found it unlikely that a disease was responsible for the extinction of the megafauna in the Pleistocene. Indeed, it seems an evolutionarily unstable strategy for a disease to send its host extinct, and even diseases like Tasmanian devil facial tumour disease have evolved reduced lethality to avert this. COVID-19 evolved in a similar way of reduced lethality but increased transmission rates.

Overall, while the evidence for these indirect impacts of humans on Pleistocene megafauna is less clear than for direct hunting, they highlight the complexity of human–carnivore interactions and the ways in which human activities can have far-reaching and unintended consequences for ecosystems and the species that inhabit them.

Morphological changes

Some species survived the Pleistocene extinction event, but exhibited substantial morphological changes in the process. Australia's larger macropods that survived the arrival of humans have declined in body mass (by up to 45%)—theoretically because humans selected the largest individuals and those species responded evolutionarily via breeding earlier at smaller body sizes (Flannery

1990). This likely led to corresponding evolution in predators that persisted, and we can see this in the jaguar *Panthera onca*, where individuals today are 15–20% smaller than individuals alive in the mid-Pleistocene, and sport proportionally shorter legs (Kurtén 1973). It seems likely that humans have altered predator-prey interactions by driving preferred prey extinct and keeping the remaining prey at lower densities, such that jaguars today no longer exhibit preferential predation on any members of the Cetartiodactyla in contrast to the other pantherine felids (Hayward et al. 2016). Thus, the impacts of humans in the Pleistocene are still being felt today.

Conclusion

Overall, the nature of human–carnivore interactions during the Pleistocene was likely complex and influenced by a range of factors, including hunting, the use of fire, and competition for resources. The loss of large herbivores due to human hunting had a significant impact on many carnivore species, and combined with their slow life histories and climate change, may have contributed to their eventual extinction. The loss of the megaherbivores fundamentally affected ecosystem structure, seed dispersal, surface albedo, and biogeochemical cycling, which still affects ecosystems today (Doughty et al. 2016, Malhi et al. 2016, Turvey and Crees 2019).

While scientists have debated the causes of the Pleistocene mass extinction event for decades (Yule et al. 2014), adequate explanations of a cause must account for the spatiotemporal patterns of extinction observed (Bartlett et al. 2016). That is, the driver must have occurred contemporaneously with the extinction event. More and more evidence is accumulating to suggest that, in the absence of humans, the rate of extinction would not have been excessive despite climate change occurring at some sites or regions alongside population declines and extinctions (Sandom et al. 2014a, Sandom et al. 2014b). Climate change might have assisted humans, but had humans not arrived, the majority of species would have persisted throughout the Pleistocene (Brook et al. 2007). This assertion is evidenced by those megafaunal species that survived on remote islands that were not colonized by humans until very recently (e.g. mammoths on Wrangel Island until ~4 k ybp (Arppe et al. 2019); Stellar's sea cows *Hydrodamalis gigas* (Turvey and Risley 2006)). Furthermore, whole genome sequencing to infer the population sizes of 142 species of the past 100,000 years has revealed that faunal biomass is now at 25% of what it was, and climate has limited power to explain these declines compared to human colonization (Bergman et al. 2023). Finally, robust, high-resolution reconstruction of climates around the world has revealed that humans were the dominant driver of megafaunal extinctions (Bartlett et al. 2016). The top-down forcing by multiple predator hypothesis may explain the various anomalies that have confused debates around the role of humans historically.

Understanding the past is critical to our ability to cope with current and future perturbations (Sandom et al. 2018). The direction and amplitude of human—carnivore interactions prehistorically are being replicated today (Ripple et al. 2014), and the implications on ecosystems will have ramifications long into the future unless we can abate the worst outcomes (Davis et al. 2018). Unfortunately, the degree of fragmentation found in the habitats relied upon by carnivores, and the lack of sufficient prey (Ripple et al. 2015) have resulted in an absence of sufficient refugia for carnivores to retreat to—and consequently global extinction of many species appears imminent (Ripple et al. 2014). As the other chapters in this volume attest, a modern world in which large carnivores and humans coexist is remarkably complex and full of dynamic interactions, making large carnivore conservation one of the great challenges of the modern era.

Acknowledgements

The authors thank the editors for their patience in receiving this chapter. We particularly thank Mary C. Stiner for taking the time to offer insightful comments on an earlier draft of this chapter.

References

Agam, A. and R. Barkai. 2018. Elephant and mammoth hunting during the Paleolithic: a review of the relevant archaeological, ethnographic and ethno-historical records. *Quaternary* 1:3.

Anderson, A. 1989. *Prodigious Birds*. Cambridge University Press, Cambridge.

Anderson, A. and R. McGovern-Wilson. 1990. The pattern of prehistoric Polynesian colonisation in New Zealand. *Journal of the Royal Society of New Zealand* 20:41–63.

Anieli, G. P., A. G., A. Antonelli, D. Silvestro, and S. Faurby. 2022. Two major extinction events in the evolutionary history of turtles: one caused by a meteorite, the other by hominins. bioRxiv:2022.2007.2020.500661.

Araujo, B. B. A., L. G. R. Oliveira-Santos, M. S. Lima-Ribeiro, J. A. F. Diniz-Filho, and F. A. S. Fernandez. 2017. Bigger kill than chill: the uneven roles of humans and climate on late Quaternary megafaunal extinctions. *Quaternary International* 431:216–222.

Arppe, L., J. A. Karhu, S. Vartanyan, D. G. Drucker, H. Etu-Sihvola, and H. Bocherens. 2019. Thriving or surviving? The isotopic record of the Wrangel Island woolly mammoth population. *Quaternary Science Reviews* 222:105884.

Barnosky, A. D., P. L. Koch, R. S. Feranec, S. L. Wing, and A. B. Shabel. 2004. Assessing the causes of late Pleistocene extinctions on the continents. *Science* 306:70–75.

Barsh, R. L. and C. Marlor. 2003. Driving bison and Blackfoot science. *Human Ecology* 31:571–593.

Bartlett, L. J., et al. 2016. Robustness despite uncertainty: regional climate data reveal the dominant role of humans in explaining global extinctions of Late Quaternary megafauna. *Ecography* 39:152–161.

Bennett, M. R., et al. 2021. Evidence of humans in North America during the last glacial maximum. *Science* 373:1528–1531.

Berger, J., J. E. Swenson, and I.-L. Persson. 2001. Recolonizing carnivores and naive prey: conservation lessons from Pleistocene extinctions. *Science* 291:1036–1039.

Berger, L. R. and R. J. Clarke. 1995. Eagle involvement in accumulation of the Taung child fauna. *Journal of Human Evolution* 29:275–299.

Bergman, J. R., et al. 2023. Worldwide Late Pleistocene and Early Holocene population declines in extant megafauna are associated with Homo sapiens expansion rather than climate change. *Nature Communications* 14:7679.

Bergström, A., et al. 2022. Grey wolf genomic history reveals a dual ancestry of dogs. *Nature* 607:313–320.

Blumenschine, R. J., A. Whiten, and K. Hawkes. 1991. Hominid carnivory and foraging strategies, and the socio-economic function of early archaeological sites [and discussion]. *Philosophical Transactions of the Royal Society of London, Series B* 334:211–221.

Brain, C. 1981. *The Hunters of the Hunted? An Introduction to African Cave Taphonomy*. University of Chicago Press, Chicago, IL.

Brook, B. W., et al. 2007. Would the Australian megafauna have become extinct if humans had never colonised the continent? Comments on 'A review of the evidence for a human role in the extinction of Australian megafauna and an alternative explanation' by S. Wroe and J. Field. *Quaternary Science Reviews* 26:560–564.

Brook, B. W. and C. N. Johnson. 2006. Selective hunting of juveniles as a cause of the imperceptible overkill of the Australian Pleistocene megafauna. *Alcheringa: An Australasian Journal of Palaeontology* 30:39–48.

Brumm, A., M. Germonpré, and L. Koungoulos. 2023. The human-initiated model of wolf domestication—an expansion based on human-dingo relations in Aboriginal Australia. *Frontiers of Psychology* 14:1082338.

Brumm, A. and L. Koungoulos. 2022. The role of socialisation in the taming and management of wild dingoes by Australian Aboriginal people. *Animals* 12:2285.

Bugir, C. 2022. Prey Preferences of Hominids. Unpublished PhD thesis. University of Newcastle, Callaghan, NSW, Australia.

Bugir, C. K., T. M. Butynski, and M. W. Hayward. 2021a. Prey preferences of the chimpanzee (*Pan troglodytes*). *Ecology and Evolution* 11:7138–7146.

Bugir, C. K., et al. 2021b. Prey preferences of modern human hunter-gatherers. *Food Webs* 26:e00183.

Bunn, H. T. 2001. Hunting, power scavenging, and butchering by Hadza foragers and by Plio-Pleistocene Homo. In *Meat-Eating and Human Evolution*, edited by C. B. Stanford and H. T. Bunn. Oxford University Press, Oxford.

Bustos, D., et al. 2018. Footprints preserve terminal Pleistocene hunt? Human-sloth interactions in North America. *Science Advances* 4:eaar7621.

Carbone, C., G. M. Mace, S. C. Roberts, and D. W. Macdonald. 1999. Energetic constraints on the diet of terrestrial carnivores. *Nature* 402:286–288.

Carlini, A. A., J. D. Carrillo-Briceño, A. Jaimes, O. Aguilera, A. E. Zurita, J. Iriarte, and M. R. Sánchez-Villagra. 2022. Damaged glyptodontid skulls from Late Pleistocene sites of northwestern Venezuela: evidence of hunting by humans? *Swiss Journal of Palaeontology* 141:1–14.

Carmody, R. N. and R. W. Wrangham. 2009. The energetic significance of cooking. *Journal of Human Evolution* 57:379–391.

Cooper, A., C. Turney, K. A. Hughen, B. W. Brook, H. G. McDonald, and C. J. Bradshaw. 2015. Abrupt warming events drove Late Pleistocene Holarctic megafaunal turnover. *Science* 349:602–606.

Coppinger, R. and L. Coppinger. 2001. *Dogs: A New Understanding of Canine Orgin, Behavior, and Evolution*. University of Chicago Press, Chicago, IL.

Cram, D. L., et al. 2022. The ecology and evolution of human-wildlife cooperation. *People and Nature* 4: 841–855.

Darimont, C. T., B. F. Codding, and K. Hawkes. 2017. Why men trophy hunt. *Biology Letters* 13:20160909.

Darwin, C. 1859. *On the Origin of Species: By Means of Natural Selection or the Preservation of Favoured Races in the Struggle for Life* (reprinted in 1964). Harvard University, Cambridge, MA.

Daujeard, C., et al. 2016. Pleistocene hominins as a resource for carnivores: A c. 500,000-year-old human femur bearing tooth-marks in North Africa (Thomas Quarry I, Morocco). *PLoS One* 11:e0152284.

Davis, M., S. Faurby, and J.-C. Svenning. 2018. Mammal diversity will take millions of years to recover from the current biodiversity crisis. *Proceedings of the National Academy of Sciences* 115:11262–11267.

Davoli, M., S. Monsarrat, R. Ø. Pedersen, P. Scussolini, D. N. Karger, S. Normand, and J.-C. Svenning. 2024. Megafauna diversity and functional declines in Europe from the Last Interglacial to the present. *Global Ecology and Biogeography* 33:34–47.

Dembitzer, J., R. Barkai, M. Ben-Dor, and S. Meiri. 2022. Levantine overkill: 1.5 million years of hunting down the body size distribution. *Quaternary Science Reviews* 276:107316.

Domínguez-Rodrigo, M. 2002. Hunting and scavenging by early humans: the state of the debate. *Journal of World Prehistory* 16:1–54.

Donadio, E. and S. W. Buskirk. 2006. Diet, morphology, and interspecific killing in Carnivora. *American Naturalist* 167:524–536.

Doughty, C. E., et al. 2016. Global nutrient transport in a world of giants. *Proceedings of the National Academy of Sciences* 113:868–873.

Faurby, S., D. Silvestro, L. Werdelin, and A. Antonelli. 2020. Brain expansion in early hominins predicts carnivore extinctions in East Africa. *Ecology Letters* 23: 537–544.

Flannery, T. F. 1990. Pleistocene faunal loss: implications of the aftershock for Australia's past and future. *Archaeology in Oceania* 25:45–67.

Flannery, T. F. 1994. *The Future Eaters: An Ecological History of the Australasian Lands and People*. Reed Books, Sydney.

Germonpré, M., M. Van den Broeck, M. Lázničková-Galetová, M. V. Sablin, and H. Bocherens. 2021. Mothering the orphaned pup: the beginning of a domestication process in the Upper Palaeolithic. *Human Ecology* 49:677–689.

Gill, J. L., J. W. Williams, S. T. Jackson, J. P. Donnelly, and G. C. Schellinger. 2012. Climatic and megaherbivory controls on late-glacial vegetation dynamics: a new, high-resolution, multi-proxy record from Silver Lake, Ohio. *Quaternary Science Reviews* 34:66–80.

Gill, J. L., J. W. Williams, S. T. Jackson, K. B. Lininger, and G. S. Robinson. 2009. Pleistocene megafaunal collapse, novel plant communities, and enhanced fire regimes in North America. *Science* 326:1100–1103.

Gowlett, J. A. 2016. The discovery of fire by humans: a long and convoluted process. *Philosophical Transactions of the Royal Society B: Biological Sciences* 371:20150164.

Graham, R. W. and E. L. J. Lundelius. 1984. Coevolutionary disequilibrium and Pleistocene extinctions. In *Quaternary Extinctions: A Prehistoric Revolution*, edited by P. S. Martin and R. Klein. University of Arizona Press, Tucson, A. Z.

Hansford, J. P., A. M. Lister, E. M. Weston, and S. T. Turvey. 2021. Simultaneous extinction of Madagascar's megaherbivores correlates with late Holocene human-caused landscape transformation. *Quaternary Science Reviews* 263:106996.

Hart, D. and R. W. Sussman. 2011. The influence of predation on primate and early human evolution: impetus for cooperation. In *Origins of Altruism and Cooperation*, edited by R. Sussman and R. Cloninger. Springer, New York.

Hayward, M. W., P. Henschel, J. O'Brien, M. Hofmeyr, G. Balme, and G. I. H. Kerley. 2006. Prey preferences of the leopard (*Panthera pardus*). Journal of Zoology 270: 298–313.

Hayward, M. W., W. Jedrzejewski, and B. Jedrzejewska. 2012. Prey preferences of the tiger *Panthera tigris*. *Journal of Zoology* 286:221–231.

Hayward, M. W., et al. 2016. Prey preferences of the jaguar *Panthera onca* reflect the post-Pleistocene demise of large prey. *Frontiers in Ecology and Evolution* 3:e148

Hayward, M. W., and G. I. H. Kerley. 2005. Prey preferences of the lion (*Panthera leo*). *Journal of Zoology* 267:309–322.

Hayward, M. W., et al. 2023. Diet selection of the coyote *Canis latrans*. *Journal of Mammalogy* 104(6): 1338–1352.

Hayward, M. W., J. O'Brien, and G. I. H. Kerley. 2007. Carrying capacity of large African predators: predictions and tests. *Biological Conservation* 139:219–229.

Hayward, M. W. and R. Slotow. 2009. Temporal partitioning of activity in large African carnivores: tests of multiple hypotheses. *South African Journal of Wildlife Research* 39:109–125.

Hervella, M., A. San-Juan-Nó, A. Aldasoro-Zabala, K. Mariezkurrena, J. Altuna, and C. De-la-rua. 2022. The domestic dog that lived ~17,000 years ago in the Lower Magdalenian of Erralla site (Basque Country): a radiometric and genetic analysis. *Journal of Archaeological Science: Reports* 46:103706.

Holdaway, R. N., M. E. Allentoft, C. Jacomb, C. L. Oskam, N. R. Beavan, and M. Bunce. 2014. An extremely low-density human population exterminated New Zealand moa. *Nature Communications* 5:5436.

Janzen, D. H. 1983. The Pleistocene hunters had help. *The American Naturalist* 121:598–599.

Johnson, C. N. 2002. Determinants of loss of mammal species during the Late Quaternary 'megafauna'extinctions: life history and ecology, but not body size. *Proceedings of the Royal Society of London. Series B: Biological Sciences* 269:2221–2227.

Johnson, C. N. 2006. *Australia's Mammal Extinctions: A 50,000 Year History*. CSIRO Publishing, Canberra, Australia.

Johnson, C. N., J. Dortch, and T. H. Worthy. 2023. Interactions with megafauna. In *The Oxford Handbook of the Archaeology of Indigenous Australia and New Guinea*, edited by I. Mcniven and B. David. Oxford University Press, Oxford.

Jones, R. 1969. Fire-stick farming. *Australian Natural History* 3:224–228.

Kurtén, B. 1973. Pleistocene jaguars in North America. *Societas Scientiarum Fennica* 62:1–23.

Lemoine, R. T., R. Buitenwerf, and J.-C. Svenning. 2023. Megafauna extinctions in the late-Quaternary are linked to human range expansion, not climate change. *Anthropocene* 44:100403.

Levinsky, I., F. Skov, J.-C. Svenning, and C. Rahbek. 2007. Potential impacts of climate change on the distributions and diversity patterns of European mammals. *Biodiversity and Conservation* 16:3803–3816.

Lyons, K. S., F. A. Smith, P. J. Wagner, E. P. White, and J. H. Brown. 2004a. Was a 'hyperdisease' responsible for the late Pleistocene megafaunal extinction? *Ecology Letters* 7:859–868.

Lyons, S. K., F. A. Smith, and J. H. Brown. 2004b. Of mice, mastodons and men: human-mediated extinctions on four continents. *Evolutionary Ecology Research* 6:339–358.

MacPhee, R. D. 1997. The 40,000-year plague: humans, hyperdisease, and first-contact extinctions. In *Natural Change and Human Impact in Madagascar*, edited by S. M. Goodman and B. D. Patterson. Springer, New York.

Malhi, Y., C. E. Doughty, M. Galetti, F. A. Smith, J.-C. Svenning, and J. W. Terborgh. 2016. Megafauna and ecosystem function from the Pleistocene to the Anthropocene. *Proceedings of the National Academy of Sciences* 113:838–846.

Mann, S. A., B. Van Valkenburgh, and M. W. Hayward. 2017. Tooth fracture within the African carnivore guild: the influence of intraguild competition and resource availability. *Journal of Zoology* 303:261–269.

Marshall, L. G. 1988. Land mammals and the great American interchange. *American Scientist* 76:380–388.

Martin, P. S. 1967. Prehistoric overkill. In *Pleistocene Extinctions: The Search for a Cause*, edited by P. S. Martin and H. E. Wright. Yale University Press, New Haven, U. S. A.

Martin, P. S. 2005. *Twilight of the Mammoths*. University of California Press, Berkeley and Los Angeles.

Masson-Delmotte, V., et al. 2010. EPICA Dome C record of glacial and interglacial intensities. *Quaternary Science Reviews* 29:113–128.

Mech, L. D. and L. A. A. Janssens. 2022. An assessment of current wolf *Canis lupus* domestication hypotheses based on wolf ecology and behaviour. *Mammal Review* 52:304–314.

Metz, L., J. E. Lewis, and L. Slimak. 2023. Bow-and-arrow, technology of the first modern humans in Europe 54,000 years ago at Mandrin, France. *Science Advances* 9:eadd4675.

Morey, D. F. and R. Jeger. 2015. Paleolithic dogs: why sustained domestication then? *Journal of Archaeological Science: Reports* 3:420–428.

O'Donoghue, M., S. Boutin, C. J. Krebs, and E. J. Hofer. 1997. Numerical responses of coyotes and lynx to the snowshoe hare cycle. *Oikos* 80:150–162.

O'Keefe, F. R., et al. 2023. Pre-younger dryas megafaunal extirpation at Rancho La Brea linked to fire-driven state shift. *Science* 381:eabo3594.

Ohlemuller, R., B. Huntley, S. Normand, and J.-C. Svenning. 2011. Potential source and sink locations for climate-driven species range shifts in Europe since the Last Glacial Maximum. *Global Ecology and Biogeography* 21(2):152–163.

Perri, A. R. 2020. Prehistoric dogs as hunting weapons: the advent of animal biotechnology. In *Dogs: Archaeology Beyond Domestication*, edited by B. Bethke and A. Burtt. University Press of Florida, Gainesville, F. L.

Perry, G. L. W., A. B. Wheeler, J. R. Wood, and J. M. Wilmshurst. 2014. A high-precision chronology for the rapid extinction of New Zealand moa (Aves, Dinornithiformes). *Quaternary Science Reviews* 105:126–135.

Politis, G. G., P. G. Messineo, T. W. Stafford Jr, and E. L. Lindsey. 2019. Campo Laborde: a Late Pleistocene giant ground sloth kill and butchering site in the Pampas. *Science Advances* 5:eaau4546.

Pučkov, P. V. 1997. *Were the Mammoths Killed by the Warming?: (testing of the climatic versions of wurm extinctions)*. Schmalhausen Inst. of Zoology, Ukraine.

Rawlence, N. J., J. L. Metcalf, J. R. Wood, T. H. Worthy, J. J. Austin, and A. Cooper. 2012. The effect of climate and environmental change on the megafaunal moa of New Zealand in the absence of humans. *Quaternary Science Reviews* 50:141–153.

Raymond, A. 1986. Experiments in the function and performance of the weighted atlatl. *World Archaeology* 18:153–177.

Ripple, W. J., et al. 2014. Status and ecological effects of the world's largest carnivores. *Science* 343:1241484.

Ripple, W. J., et al. 2015. Collapse of the world's largest herbivores. *Science Advances* 1:e1400103.

Ripple, W. J. and B. Van Valkenburgh. 2010. Linking top-down forces to the Pleistocene megafaunal extinctions. *BioScience* 60:516–526.

Roberts, R. G., R. Jones, N. A. Spooner, M. J. Head, A. S. Murray, and M. A. Smith. 1994. The human colonisation of Australia: optical dates of 53,000 and 60,000 years bracket human arrival at Deaf Adder Gorge, Northern Territory. *Quaternary Science Reviews* 13:575–583.

Roy, K., J. W. Valentine, D. Jablonski, and S. M. Kidwell. 1996. Scales of climatic variability and time averaging in Pleistocene biotas: implications for ecology and evolution. *Trends in Ecology & Evolution* 11: 458–463.

Russo, G., et al. 2023. First direct evidence of lion hunting and the early use of a lion pelt by Neanderthals. *Scientific Reports* 13:16405.

Saltré, F., et al. 2016. Climate change not to blame for late Quaternary megafauna extinctions in Australia. *Nature Communications* 7:10511.

Sandom, C., S. Faurby, B. Sandel, and J.-C. Svenning. 2014a. Global late Quaternary megafauna extinctions linked to humans, not climate change. *Proceedings of the Royal Society B: Biological Sciences* 281:20133254.

Sandom, C. J., et al. 2018. Learning from the past to prepare for the future: felids face continued threat from declining prey. *Ecography* 41:140–152.

Schmitz, O. J., V. Krivan, and O. Ovadia. 2004. Trophic cascades: the primacy of trait-mediated indirect interactions. *Ecology Letters* 7:153–163.

Scott, E. 2010. Extinctions, scenarios, and assumptions: changes in latest Pleistocene large herbivore abundance and distribution in western North America. *Quaternary International* 217:225–239.

Selvaggio, M. M. 1994a. Carnivore tooth marks and stone tool butchery marks on scavenged bones: archaeological implications. *Journal of Human Evolution* 27: 215–228.

Selvaggio, M. M. 1994b. Evidence from carnivore tooth marks and stone-tool-butchery marks for scavenging by hominids at FLKZinjanthropus Olduvai Gorge, Tanzania. Rutgers The State University of New Jersey, School of Graduate Studies.

Serangeli, J., T. Van Kolfschoten, B. M. Starkovich, and I. Verheijen. 2015. The European saber-toothed cat (*Homotherium latidens*) found in the 'Spear Horizon' at Schöningen (Germany). *Journal of Human Evolution* 89:172–180.

Simoons, F. J. and J. A. Baldwin. 1982. Breast-feeding of animals by women: its socio-cultural context and geographic occurrence. *Anthropos* 77:421–448.

Smith, F. A., R. E. Elliott Smith, S. K. Lyons, and J. L. Payne. 2018. Body size downgrading of mammals over the late Quaternary. *Science* 360:310–313.

Speth, J. D. and E. Tchernov. 2001. Neandertal hunting and meat-processing in the Near East. In *Meat-Eating and Human Evolution*, edited by C. B. Stanford and H. T. Burns. Oxford University Press, Oxford.

Stanford, C. B. 1996. The hunting ecology of wild chimpanzees: implications for the evolutionary ecology of Pliocene hominids. *American Anthropologist* 98:96–113.

Stanford, C. B. 2001. A comparison of social meat-foraging by chimpanzees and human foragers. In *Meat-Eating and Human Evolution*, edited by C. B. Stanford and H. T. Burns. Oxford University Press, Oxford.

Stuart, A. J. 1999. Late Pleistocene megafaunal extinctions: a European perspective. In *Extinctions in Near Time*, edited by R. D. E. Macphee. Kluwer Academic/Plenum Publishers, New York.

Suganuma, Y., H. Miura, A. Zondervan, and J. I. Okuno. 2014. East Antarctic deglaciation and the link to global cooling during the Quaternary: evidence from glacial geomorphology and 10Be surface exposure dating of the Sør Rondane Mountains, Dronning Maud Land. *Quaternary Science Reviews* 97:102–120.

Tappen, M. 2001. Deconstructing the Serengeti. In *Meat-Eating and Human Evolution*, edited by C. B. Stanford and H. T. Burns. Oxford University Press, Oxford.

Turvey, S. T. and J. J. Crees. 2019. Extinction in the Anthropocene. *Current Biology* 29:R982–R986.

Turvey, S. T. and C. L. Risley. 2006. Modelling the extinction of Steller's sea cow. *Biology Letters* 2:94–97.

Van Der Kaars, S., et al. 2017a. Humans rather than climate the primary cause of Pleistocene megafaunal extinction in Australia. *Nature Communications* 8:14142.

Van Der Kaars, S., et al. 2017b. Humans rather than climate the primary cause of Pleistocene megafaunal extinction in Australia. *Nature Communications* 8:1–7.

Van Valkenburgh, B. 2001. The dog-eat-dog world. *Meat-Eating and Human Evolution*, edited by C. B. Stanford and H. T. Burns. Oxford University Press, Oxford.

Van Valkenburgh, B. 2009. Costs of carnivory: tooth fracture in Pleistocene and Recent carnivorans. *Biological Journal of the Linnean Society* 96:68–81.

Van Valkenburgh, B., M. W. Hayward, W. J. Ripple, C. Meloro, and V. L. Roth. 2016. The impact of large terrestrial carnivores on Pleistocene ecosystems. *Proceedings of the National Academy of Sciences of America* 113:862–867.

Weir, J. T., E. Bermingham, and D. Schluter. 2009. The great American biotic interchange in birds. *Proceedings of the National Academy of Sciences* 106:21737–21742.

Werdelin, L. and M. E. Lewis. 2013. Temporal change in functional richness and evenness in the Eastern African Plio-Pleistocene Carnivoran guild. *PLoS One* 8:e57944.

Wroe, S., J. Field, R. Fullagar, and L. S. Jermin. 2004. Megafaunal extinction in the late Quaternary and the global overkill hypothesis. *Alcheringa* 28:291–331.

Yule, J. V., R. J. Fournier, C. X. Jensen, and J. Yang. 2014. A review and synthesis of late Pleistocene extinction modeling: progress delayed by mismatches between ecological realism, interpretation, and methodological transparency. *The Quarterly Review of Biology* 89:91–106.

Zanette, L. Y., et al. 2023. Fear of the human 'super predator' pervades the South African savanna. *Current Biology* 33(21):4689-4696.e4.

Zeder, M. A. 2012. The domestication of animals. *Journal of Anthropological Research* 68:161–190.

CHAPTER 3

Human Disturbances Define Carnivore Ecology in the Anthropocene

Christopher J. Schell, Justin Brashares, and Lauren Stanton

Introduction

Ecology as a discipline is principally concerned with the relationships of organisms to each other and to their environments. These complex relationships often form over centuries, with evolutionary processes working across time to optimize organismal fitness to local environmental conditions. Species interactions (e.g. competition, predator-prey interactions, commensalism, and mutualism) can be an especially potent driver of ecological and evolutionary change, as these interactions affect species abundances and distributions across a landscape, which can affect each species' ecological niches (Agrawal 2001). For example, apex predators and mesocarnivores will frequently partition their space use and temporal activity to avoid competing for shared resources (i.e. niche partitioning) (Sévêque et al. 2020, 2022, Curras et al. 2022, Cervantes et al. 2023). Ecosystem instability caused by a disturbance (e.g. wildfires, hurricanes, etc.) can temporarily disrupt species interactions by either directly affecting the population dynamics of either interacting species, or indirectly by augmenting resource distributions and availability. Both direct and indirect effects can increase mortality and reduce effective population sizes (Burton et al. 2020). Importantly, the spatial and temporal extent of the disturbance determines its overall impact on ecosystem structure and function (Burton et al. 2020). Disturbances over prolonged timescales with increased geographic reach can irreversibly degrade ecosystems and prevent successional growth, driving biodiversity loss that further jeopardizes ecosystem health, function, and services.

In the Anthropocene, human disturbances have become that proverbial long-term, geographically widespread ecosystem threat, spurring the sixth mass extinction that our planet has experienced (Young et al. 2016, Newman 2019, Finn et al. 2023). Certainly, ecological disturbances are not uncommon processes and ecosystems are inherently dynamic and conditional (Guiden et al. 2019). Moreover, humans have shaped the ecological and evolutionary processes of sympatric species for centuries, dictating species life histories, population sizes, and range shifts on macroecological scales (Di Marco and Santini 2015, Sullivan et al. 2017, Smith et al. 2019a). However, modern anthropogenic disturbances are especially potent, occurring under rapid timescales that can overwhelm nonhuman entities at all levels of ecological organization. These unprecedented anthropogenic pressures compel species to rapidly adapt or face extinction (Hendry et al. 2017). Unfortunately, there are more ecological 'losers' than there are 'winners' in this era (Finn et al. 2023), with nearly 1 million plant and animal species threatened with extinction by 2100 (IPBES 2019). Human societies are dependent on functioning ecosystems and their derived benefits as well (Des Roches et al. 2021a, 2021b, Lambert and Schell 2023), and continual ecosystem deterioration may place up to 5 billion people at risk of diminished ecosystem services (Chaplin-Kramer et al. 2019). Nevertheless, anthropogenic

Christopher J. Schell, Justin Brashares, and Lauren Stanton, *Human Disturbances Define Carnivore Ecology in the Anthropocene*.
In: *Apex Predators in the Anthropocene*. Edited by: Robert A. Montgomery and Remington J. Moll, Oxford University Press.
 DOI: 10.1093/9780191996665.003.0003

disturbances are expected to intensify rather than dissipate (Jaureguiberry et al. 2022a).

At a perilous crossroads in this narrative are apex predators, who are especially challenged by the pervasive and long-term disturbances humans effectuate on the globe (Stier et al. 2016, Ripple et al. 2017, McInturff et al. 2021). The need for large, connected landscapes and contention over shared food resources place apex predators in precarious situations with humans, increasing conflicts that often lead to local extinction (Treves and Karanth 2003, Treves et al. 2006). In addition, slow life-history strategies (i.e. long generation times with low reproductive output) amplify carnivore extinction vulnerability, as reproductive output from surviving individuals may be insufficient to compensate for increased human-caused mortalities. Consequently, carnivores are forced to negotiate increasingly contested spaces (e.g. agricultural lands, working lands, urban areas, etc.) where their occupation of landscapes is perceived as a threat. The peril faced by carnivores is particularly concerning because apex predators tend to be some of the most functionally important organisms in an ecosystem, providing top-down regulatory processes that help balance prey abundances (e.g. grey wolves *Canis lupus* in Yellowstone), which subsequently facilitate smaller species coexistence across trophic levels (Ostfeld and Holt 2004, Ripple et al. 2014). The widespread decline of large carnivores globally thus represents one of the most pressing issues compromising ecosystem stability and nature's contributions to people (Estes et al. 2011).

Here, we articulate how human disturbances have restructured carnivore ecology in the Anthropocene. To do this, we apply the framework developed by the Intergovernmental Science-Policy Platform on Biodiversity and Ecosystem Services (IPBES) that outlines the five most prominent direct drivers of the biodiversity crisis to apex predators (IPBES 2019, Lozano et al. 2019). These direct drivers include (1) land use and change; (2) direct exploitation; (3) climate change; (4) pollution; and (5) non-native species interactions (1). For each direct driver, we define the general impacts of the disturbance category, then provide specific examples describing how the ecology of apex predators has been affected (Figure 3.1). We provide evidence demonstrating how each direct driver shapes the organismal (e.g. boldness, foraging, physiology), population (e.g. age and reproductive structure, growth rates, survival), and community ecology (e.g., predator-prey interactions, competition) of predators (Figure 3.2). We then conclude on how addressing the indirect drivers of biodiversity loss (i.e. the underlying social, economic, cultural, and political considerations) are essential for developing resilient conservation and management strategies that can withstand the unpredictability of the Anthropocene.

Land use and change

Habitat loss has been the most impactful and disruptive threat to biodiversity worldwide (Jaureguiberry et al. 2022b). The sheer scale of urban development and agricultural production is astounding, with nearly a third (32%) of global land area in the last 60 years (Winkler et al. 2021) dominated by anthropogenic uses. Urbanization and agriculture are estimated to substantially grow by 2050 to keep pace with an exponentially growing human population (est. 2.5 billion increase). Recent estimates suggest that new urban land cover alone will be somewhere between 0.82 to 1.53 million km^2 (Simkin et al. 2022), approximately 3.6 times more area than the state of California, USA. Agriculture dwarfs these values in comparison, with almost half (44%) of the globe's habitable land currently being used for agriculture, which covers nearly 48 million km^2 of the Earth's surface (Ellis et al. 2010). Demand for agricultural production is likely to rise 70% to 100% by 2050, putting excessive strain on global markets to increase production and transform millions of new hectares into crops and pastures (Zabel et al. 2019).

Human land use dominance has manifold consequences for carnivore ecology, all of which emanate from negotiating contested geographies to acquire increasingly limited resources. In developed and developing areas, habitat fragmentation due to amplified anthropogenic land use reduces the size, connectivity, and quality of ecological niches. Carnivores thus find themselves in a small fraction of

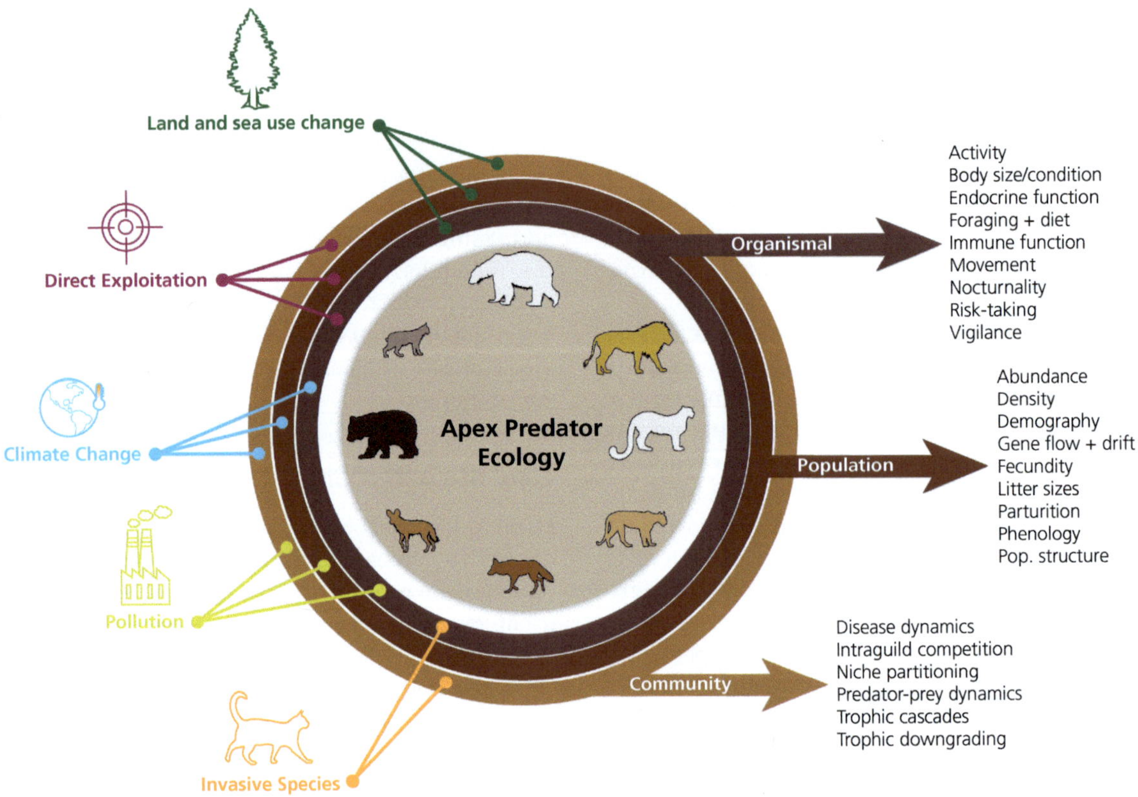

Figure 3.1 Conceptual framework illustrating how direct drivers of biodiversity loss impact apex predator ecology at various ecological scales. Evidence across a range of carnivore taxa demonstrate how land/sea use change, direct exploitation, climate change, pollution, and invasive species affect the organismal (e.g. activity levels, risk-taking, etc.), population (e.g. abundance, demography, etc.), and community (e.g. disease, niche partitioning, etc.) ecology of these species.

suitable habitat (i.e. **realized niches**), as a growing amount of potential habitat space (i.e. the **fundamental niche**) is becoming increasingly occupied by humans (Hutchinson 1957, Chase 2003, Dragonetti et al. 2023, Ma et al. 2024). The resultant network of habitat patches operates as a collection of 'islands', where habitat fragments are separated by a complex matrix of impervious surfaces (e.g. roadways, buildings, fencing, etc.) (Dunn et al. 2022). Subpopulations on smaller, isolated islands are especially vulnerable to disruptive and stochastic events (e.g. storms, disease, drought, etc.) that lead to genetic bottlenecks because of the difficulty in individuals navigating to those islands, reducing the potential for genetic rescue, and elevating the likelihood of deleterious effects from inbreeding, leading to local extinction.

This theoretical framework serves as the backdrop for carnivores navigating expanding urban-wildland interfaces and human development, with recent examples illustrating how predator behavioural strategies, mortality risk, population structure, and species interactions are all shaped by development. The most widespread and generalizable pattern observed across predators are increased avoidance behaviours. Apex predators generally try to avoid human settlements, reducing their movements and space use of areas with high human population and road densities to avoid conflicts with people (Basille et al. 2013, Schuette et al. 2013, Clinchy et al. 2016, Smith et al. 2017, 2022, Everatt et al. 2019, Suraci et al. 2019). Some evidence suggests that human presence may be more of a deterrent to carnivore space use than infrastructure (Nickel et al. 2020, Lewis et al. 2021), though roads are an especially potent deterrent (Basille et al. 2013, Evans et al. 2019, Kautz et al. 2021) (see Chapter 6). Further, carnivores increase their nocturnality (see Chapter 7) to acquire resources when anthropogenic activity is at a nadir (Ordiz et al. 2017, Gaynor et al. 2018, Lamb et al. 2020, Mills and Harris 2020, Blašković et al. 2022, Sévêque et al. 2022, Abernathy et al. 2023). Taken together, these results illustrate a potent, human-mediated **landscape of fear** that is modulating both predator and prey avoidance behaviours in response to perceived predation risk by humans (Støen et al. 2015, Gaynor et al. 2019). Such behavioural responses have the potential to reduce dispersal and gene flow, ultimately reducing effective population size and making apex predators even more susceptible to local extinction.

Despite potential persecution or mortality associated with utilizing more human-dominated habitats, large carnivores are compelled to take more risks (Martínez-Abraín et al. 2019). Evidence in mountain lions (*Puma concolor*) across the American West (Wang et al. 2017, Blecha et al. 2018, Riley et al. 2021), leopards (*Panthera pardus*) in India (Surve et al. 2022), African lions (*Panthera leo*) across sub-Saharan Africa (Mills and Harris 2020, Mills et al. 2023), brown bears (*Ursus arctos*) and black bears (*Ursus americanus*) across the North American continent (Lamb et al. 2017, Johnson et al. 2018, Evans et al. 2019), caracals (*Caracal caracal*) in South Africa (Leighton et al. 2023), and grey wolves across North America and Europe (Barker et al. 2023, Zanni et al. 2023) suggests that these species are increasingly using more developed habitats, with consequences for their survival (see Box 3.1; Figure 3.2). Large predators face trade-offs when utilizing more human-dominated habitats, as they may be able to acquire high-value food resources yet be more susceptible to vehicular collisions, poisoning, and lethal removal (Leighton et al. 2023, Mills et al. 2023). This high-risk, high-reward scenario typifies a human-mediated **ecological trap**, where the benefits from resource acquisition are neutralized by increased mortality risks (Lamb et al. 2017, Penteriani et al. 2018). This process is highlighted by evidence of decreased fitness in brown bears that utilize areas with high berry resources (i.e. benefit), which coincidentally also have higher human densities (i.e. cost) (Lamb et al. 2017, 2020, Proctor et al. 2023).

Human development can also alter large carnivore niche partitioning and overlap. Human land use concentrates species into the remaining habitats (Moll et al. 2018, Smith et al. 2018, Manlick and Pauli 2020, Curras et al. 2022), with increasing niche overlap leading to intraguild competition (Sévêque et al. 2020, Penjor et al. 2022), competitive exclusion of smaller predators from remnant habitats (Prugh et al. 2023), modified trophic interactions (Estes et al. 2011, Fischer et al. 2012, Ripple et al. 2014, Dorresteijn et al. 2015), or alterations in carnivore diets (Bateman and Fleming 2012, Moss et al. 2016a). These processes are accentuated by species differences in tolerance of and proximity to human activity, which can establish a spatiotemporal **human shield** that partitions land use patterns of apex predators, mesocarnivores, and prey (Atickem et al. 2014, Moll et al. 2018, Cervantes et al. 2023, Abernathy et al. 2023). This shielding effect leads to mesocarnivores and prey species residing closer to human settlements to avoid predation (Berger 2007); however, increasing carnivore tolerance of human disturbance has the potential to override this effect (Blecha et al. 2018, Boutin 2018). Consequently, increasing niche overlap due to urban and agricultural expansion boosts opportunities for consuming more human food subsidies, livestock, or domestic pet predation events (Bateman and Fleming 2012, Butler et al. 2015, Ciucci et al. 2020, Surve et al. 2022, Mills et al. 2023).

Direct exploitation

Humans are the ultimate predator on Earth (Darimont et al. 2015, Oriol-Cotterill et al. 2015). Recent global estimates suggest that humanity preys on approximately one-third of all vertebrates, directly harvesting more than 15,000 vertebrates across terrestrial and aquatic ecosystems (Darimont et al. 2023). In addition, humans exploit up to 300 times more species relative to other carnivorous organisms, with approximately 40% of harvested species threatened with extinction due to human use (Darimont et al. 2023). Increased efficiency and

Box 3.1 How mountain lions navigate the American West

Mountain lions across the American West have been an especially compelling model organism to demonstrate the multivariate impacts of rising human land use on large carnivore ecology. Across the West Coast and Mountain West, pumas must navigate a patchwork of heavily trafficked roadways, densely populated urban and suburban centres, and expansive working lands that make finding resources and refugia increasingly difficult. This has resulted in entire subpopulations being isolated from each other. In the Los Angeles metro area, for instance, the 101 freeway separates the Santa Monica Mountains subpopulation from the greater LA area, with little evidence to suggest successful dispersal of individuals across US-101 (Riley et al. 2014). Similarly, mountain lion populations along the Western Slope and Front Range, Colorado, saw marked differences in genetic distance and gene flow, with impervious surfaces (e.g. roadways, buildings) impeding gene flow for the Front Range (i.e. more urbanized) subpopulations, and increased genetic distance among the two subpopulations (Trumbo et al. 2019).

Like other apex predators—including black bears (Zeller et al. 2019) and wolves (Smith et al. 2022)—mountain lions often avoid habitats with intensive human activity, recreation, or presence, showing amplified avoidance behaviours in response to human disturbances (see Chapter 8) (Riley et al. 2014, Suraci et al. 2019, Yovovich et al. 2020). For example, pumas customarily avoid residential development, despite higher energetic rewards found in or near human settlements (Wang et al. 2017, Blecha et al. 2018). Similarly, heightened recreation and wildlife habitat use by people, particularly when dogs are present, significantly decreases habitat use by mountain lions (Reilly et al. 2017). Despite puma preference for avoiding human settlements, evidence suggests there is a threshold by which mountain lions will ultimately use developed habitat more frequently (Smith et al. 2019b, Riley et al. 2021). This is largely to acquire prey items or other resources (e.g., water) that are similarly contested among people and pumas. Hunger is an especially potent motivator, as lions are more likely to traverse human-dominated areas when they are hungry (Blecha et al. 2018, Boutin 2018). This problem is exacerbated by climate-induced droughts and urban expansion, which result in prey and water access being more proximal to human-dominated environments (Abrahms et al. 2023). To navigate this conundrum, pumas can both increase their nocturnality (Nickel et al. 2020, Abernathy et al. 2023) and expand their dietary niche (Moss et al. 2016a, b) to reduce the potential for human-puma conflicts when human densities are moderate to high.

(A) Mountain lion
Source: Vishal Subramanyan

(B) Mountain lion cubs
Source: 'P-48 and P-49' by Santa Monica Mountains National Recreation Area

Driver	Species	Region/ Country	Ecological Outcome(s)	Ecological Scale(s)	References
Land/sea use and change	Mountain lion	United States	Greater avoidance of high human activity areas; Increased use of human-dominated landscapes; Ingestion of human food subsidies	Organismal	Riley et al. 2021; Smith et al. 2022, 2017; Suraci et al. 2019; Wang et al. 2017
Land/sea use and change	Black bear	United States	Increased use of human-dominated landscapes to acquire resources in anthropogenic environments; heightened mortality risk	Organismal	Evans et al., 2019; Johnson et al. 2018; Lamb et al. 2017
Direct Exploitation	Brown bear	Sweden, Scandanavia	Decreased survival and fecundity in periods of high hunting pressures; increased litter sizes and adult female body mass	Organismal, Population	Gosselin et al. 2014; Leclerc et al. 2019; Zedrosser et al.2011
Direct Exploitation	Grey wolf	Italy, Poland	Decline in native prey associated with heightened livestock depredation; Harvesting pressures on moose leads to increased predation on caribou	Community	Ciucci et al. 2020; Klich et al.2021
Direct Exploitation	Orca	North Pacific Ocean and Bering Sea	Prey switching to sea otters after declines in seal and sea lion populations due to fisheries	Community	Estes et al. 1998; Springer et al. 2003
Climate Change	African wild dog	Botswana, Kenya, Zimbabwe	Delayed timing in parturition and denning periods, altered activity budgets and nocturnality due to warming temperatures	Organismal, Population	Abrahms et al. 2022; Rabaiotti et al. 2021; Woodroffe et al. 2017
Climate Change	Polar bear	Arctic Basin and Chukchi Sea	Declines in adult body condition, mean litter sizes, and cub survival; Altered space use, foraging, and diets	Population	Atwood & Wilder 2021; Hamilton et al. 2017; Laidre et al. 2020; Pagano & Williams 2021
Climate Change	Snow leopard	Nepal	Warming temperatures decouple predator-prey relationships, leading to increased livestock depredation	Community	Aryal et al. 2014a; b; Kichloo et al. 2023; Rezaei et al. 2023
Pollution	Bobcat	United States	Exposure to anticoagulant rodenticides suppressed immune and organ function, making individuals more susceptible to mange	Organismal, Community	Riley et al. 2007; Serieys et al. 2018
Pollution	Mountain lion	United States	Decreased spatial overlap with prey due to increased light pollution, reducing predation success; Modified vigilance, activity, and foraging due to noise	Organismal, Community	Ditmer et al. 2021
Invasive Species	African lion	Kenya	Reduced predation success of zebras and prey switching to African buffalo due to landscape alterations by invasive ant species	Community	Kamaru et al. 2024
Invasive Species	Iberian lynx	Spain	Transmission of feline leukemia virus (FeLV) due to contact with domestic cats leads to mortality	Community	Meli et al. 2009

Figure 3.2 Key examples of direct drivers influencing apex predator ecology in the literature. Each of the selected case studies are organized by direct driver, region or country of the associated studies, the ecological outcomes observed during the study, the ecological scale(s) at which carnivores exhibited responses to the selected driver, and the associated references.

Source: (1) 'Female Mountain Lion in Verdugos' by Santa Monica Mountains National Recreation Area; (2) 'Black bear and dumpster' by YellowstoneNPS; (3) 'A brown bear sow with two spring cubs Photo courtesy M. Whalen' by KatmaiNPS; (4) 'Black and grey wolf (female from Druid pack, "Half Black") standing in road near Lamar River bridge' by YellowstoneNPS; (5) 'Orca 12/20/14' by Mike Charest, available under a Creative Commons Attribution 2.0 Generic (CC BY 2.0) licence https://creativecommons.org/licenses/by/2.0/deed.en; (6) 'African wild dog' by Mathias Appel, available under a Creative Commons CC0 1.0 Universal licence https://creativecommons.org/publicdomain/zero/1.0/deed.en; (7) 'Polar Bear Jumping— Arturo de Frias Marques' by Bering Land Bridge National Preserve, available under a Creative Commons Attribution-ShareAlike 2.0 (CC BY-SA 2.0) Generic licence https://creativecommons.org/licenses/by-sa/2.0/deed.en; (8) 'Snow Leopard Walking in the Snow' by Eric Kilby, available under a Creative Commons Attribution-Share Alike 2.0 (CC BY-SA 2.0) Generic licence https://creativecommons.org/licenses/by-sa/2.0/deed.en; (9) 'Bobcat along the Madison River' by YellowstoneNPS; (10) 'P-30 deceased' by Santa Monica Mountains National Recreation Area; (11) 'African lion, Panthera leo at Kgalagadi Transfrontier Park, Northern Cape, South Africa' by Derek Keats, available under a Creative Commons Attribution 2.0 Generic (CC BY 2.0) licence https://creativecommons.org/licenses/by/2.0/deed.en; (12) 'Lince ibérico (Lynx pardinus), Almuradiel, Ciudad Real, España, 2021-12-19, DD 01' by Diego Delso, available under a Creative Commons Attribution-ShareAlike 4.0 International (CC BY-SA 4.0) licence https://creativecommons.org/licenses/by-sa/4.0/deed.en.

technological advances in trapping, hunting (see Chapter 4), and fishing techniques have accelerated exploitation success, precipitating tremendous growth in food production, medicine, commerce, and other industries that rely on direct species harvest (Ripple et al. 2016, Lennox et al. 2022). As a result, industry-accelerated overexploitation has become a major selective agent, driving extreme defaunation and species trait changes on scales unrivalled by other relevant ecological factors (Darimont et al. 2009, Young et al. 2016). Indeed, human societies have cemented their trophic position after the Industrial Revolution, starting an especially notable period of sustained harvesting pressure over the past 180 years (Young et al. 2016).

To contextualize direct exploitation and its effects on human-carnivore ecology, it is helpful to articulate the importance of predation as an ecological mechanism. **Predation** is the consumption of one organism (i.e. prey) by another organism (i.e. predator), resulting in the flow of energy across trophic levels. Predation ties trophic strata together, from primary producers to apex predators, by modulating organismal behaviour at lower trophic levels (e.g. anti-predator responses, movement, foraging), species abundances, demography, species interactions, and ultimately biodiversity and nutrient cycling (Ostfeld and Holt 2004). **Top-down control** exerted by apex predators has the potential to shape the structure and function of ecological communities through **trophic cascades** that connect all organisms in a food web to predation (Ripple et al. 2014, Wallach et al. 2015). The removal of the highest trophic actors in a food web through direct exploitation thus reduces or eliminates top-down control (i.e. **trophic downgrading**, see Chapter 12), which can decouple species interactions and destabilize ecosystems (Estes et al. 2011, Ripple et al. 2014, Oriol-Cotterill et al. 2015, Leighton et al. 2023). Across ecosystems globally, humans have supplanted apex predators as the *de facto* apex predator; yet anthropogenic predation pressures neither replicate the functional significance of, nor the ecosystem services provided by, other apex predators (Wallach et al. 2015, Lennox et al. 2022). As a result, the culmination of exploitation practices functionally upends millennia of coevolved predator–prey interactions with substantial consequences for apex predators and human–carnivore interactions.

Direct effects from harvesting or hunting occur when individuals in a population are the principal hunting or harvesting target (i.e. when exploitation is directed towards the apex predator specifically; see Chapter 4). Strong direct harvest often occurs for the largest individuals of reproductive age, creating an intense 'harvest selection' pressure on adults (i.e. the reproductive 'capital' of a population) relative to juveniles (i.e. the reproductive 'interest') (Darimont et al. 2015). These direct effects often lead to shifts in animal behaviour within a population (Leclerc et al. 2017, Andersen et al. 2018, Schell et al. 2021). For instance, Brown bears in Sweden that use habitats closer to roads, as well as individuals with reduced activity and movement rates during hunting hours, were more likely to be harvested. This dynamic possibly augmented individual shyness and activity in, as well as avoidance, of areas with high human footprints (Leclerc et al. 2019). Demographic factors also change in response to harvest selection. Regulated hunting for brown bears in Sweden claims 74% of males over 3 years old and 72% of females (Bischof et al. 2018), supporting the principle that the reproductive 'capital' of a population experiences some of the greatest pressures. Bears in Sweden and Scandinavia show similar results, with Scandinavian Brown bears showing decreased survival and fecundity (i.e. fitness) in periods of high hunting pressures, contributing to decreased population growth rates (Gosselin et al. 2014). Some brown bear populations respond to harvesting pressures by increasing mean litter sizes and adult female body mass (Zedrosser et al. 2011), suggesting that increased human footprint may select for faster **life-history strategies** (Figure 3.2).

Indirect effects (see Chapters 9, 10, and 11), in contrast, can occur when the prey species of an apex predator is the principal target, ultimately reducing prey abundance and limiting food resources for carnivores (Wolf and Ripple 2016). Humans and carnivores are often in direct competition for prey resources (Ripple et al. 2014, Coimbra et al. 2020). Per capita meat consumption has elevated with growing global human populations, and the combined effects of expanding agriculture, urban development, and commercial bushmeat trade have coalesced to reduce the carnivore prey base (Ripple et al. 2014, 2016). Smaller prey bases (i.e. **prey depletion**) may compel carnivores to incorporate more anthropogenic resource subsidies (e.g. livestock, waste, etc.), elevating the potential for human–wildlife conflicts (Newsome et al. 2015, Wolf and

Ripple 2016, van Eeden et al. 2018, Carter et al. 2019). For instance, evidence from grey wolves in Poland suggests that a decline in the wild boar (*Sus scrofa*) population predicted heightened livestock depredation (Klich et al. 2021). Similar evidence from wolves in the southern Apennines, Italy suggest that a large proportion of wolf diets incorporated livestock carrion subsidies (i.e. livestock whose cause of death was something other than predation) rather than wild prey (Ciucci et al. 2020). Additional evidence from Persian leopards (*Panthera pardus tulliana*) in Iran suggests that leopards frequent protected areas with higher livestock densities when wild prey abundances are low, contributing to elevated illegal killings (Soofi et al. 2018, 2022). Further, snow leopards (*Panthera uncia*) across the Trans-Himalayas and Altai regions were more likely to hunt livestock when the preferred wild ungulate abundances were low (Suryawanshi et al. 2017).

Prey-switching is not solely restricted to a transition from less abundant wild prey to livestock but may also occur when predators consume other wild organisms in the food web as an adaptive response to prey depletion. Generalist predators with wide dietary niche breadth may begin to consume alternative prey to persist, yet such switching may aggravate or accelerate population declines of keystone species or those of conservation concern (Gilljam et al. 2015). For instance, the complex interactions among climate change, human land use, and harvesting pressures contribute to exacerbating apparent competition among moose (*Alces alces*) and the threatened caribou (*Rangifer tarandus*), where wolf predation rates on caribou are disproportionately higher in Québec, Canada (Labadie et al. 2023). In a classic example, orcas (*Orcinus orca*) switched from consuming Steller sea lions (*Eumetopias jubatus*) and harbour seals (*Phoca vitulina*) to alternatively consuming sea otters (*Enhydra lutris*) in the Aleutian archipelago, which led to precipitous declines in otter population sizes and initiated a trophic cascade that witnessed the proliferation of sea urchins and the decimation of kelp forests (Estes et al. 1998, Springer et al. 2003). Importantly, this entire trophic cascade began with human-induced reductions in marine prey for the sea lion and seal populations, which constrained pinniped species abundances that drove orcas to seek alternative prey (Estes et al. 1998, Springer et al. 2003). In both examples, exploitation-driven prey-switching by carnivores became a conservation crisis, imperilling species at lower trophic positions (Figure 3.2).

Climate change

Global climate change is one of the defining crises of our time. Decades of anthropogenic greenhouse gas emissions have precipitated climate warming on a global scale, with warming projected to surpass the ominous 1.5°C threshold by this century (IPCC 2022). The consequences from this warming are manifold, including changes in the frequency, duration, and intensity of acute climate events (e.g. droughts, wildfires, heatwaves, storms, etc.), long-term directional changes in precipitation and climate, augmented carbon and nutrient cycling in the atmosphere and oceans, and increased interannual variability in climate conditions (Abrahms et al. 2023, Bellard et al. 2012). Climate change creates a unique bind for apex predators. Paired with broad landscape conversion and overexploitation, intensifying droughts, wildfires, flooding, and variable snow and sea ice conditions restructure the size and geographic parameters of previously suitable habitat conditions. As a result, carnivores seeking prey, water, or refugia must now traverse anthropogenic spaces under heightened risks (Abrahms 2021, Kichloo et al. 2023). In addition, the persistent lack of access to resources can detrimentally affect individual body condition and health, further compelling individuals to take on risk.

Climate-induced spatial overlap among predators and prey (Aryal et al. 2014a, 2014b, Baral et al. 2023), predators and people (Aryal et al. 2014a, Johnson et al. 2018, Atwood and Wilder 2021), and sympatric predators (Kichloo et al. 2023, Rezaei et al. 2023, Rafiq et al. 2023) produce variable outcomes that can augment predation success, human-carnivore conflicts, and heterospecific competition. For instance, bharal (i.e. blue sheep, *Pseudois nayaur*), one of the most common prey species of snow leopards, move to lower elevations as warming temperatures shift vegetation to lower elevations in Nepal Himalaya (Aryal et al. 2014b). Snow leopards follow their prey in response, but in doing so, increase predation of livestock, which is then precipitated by retaliatory killings (Aryal et al. 2014a). Intraguild competition and predation may also increase as severe drought and lack of access to water resources disintegrates spatial niches, often to the detriment of conservation-relevant species (Kichloo et al. 2023, Rezaei et al. 2023, Rafiq et al. 2023). Notably, not all climate-induced shifts in temperature, precipitation, and droughts have detrimental impacts on carnivores in the short-term; rather, such climate events may increase predator-prey sympatry that results in heightened hunting success (Veldhuis et al. 2020). Large-bodied herbivores are particularly water-dependent and heat sensitive, with more thermoregulatory constraints than smaller-bodied mammals (Veldhuis et al. 2019). Thus, large herbivore shifts in habitat use and increased nocturnal activity to avoid drought conditions or high daily ambient temperatures coincidentally exposes prey species to greater predation risk by increasing spatial and temporal overlap with predators (Riginos 2015, Creel et al. 2016, Brivio et al. 2024).

The timing of salient ecological events (i.e. phenology) is particularly sensitive to climate change, resulting in phenological mismatches among predators and the environmental cues they use to predict the timing of resource availability. For example, warmer temperatures have led to longer active seasons for black bears in Durango, Colorado, who search for additional resources in developed areas when resources elsewhere are depleted, increasing human-caused mortality risk (Laufenberg et al. 2018). Warmer temperatures in the same system reduced the duration of bear hibernation by leading to earlier spring emergence periods and a postponement of hibernation in the fall (Johnson et al. 2018). Mismatched timing of food resources and post-hibernation periods was also found for Asiatic bears (*Ursus thibetanus*), which increased the likelihood of human-bear attacks and retaliatory culling (Honda and Kozakai 2020). Timing in parturition (i.e., births) and denning periods are also impacted by warming temperatures in African wild dogs (*Lycaon pictus*) in Botswana (Abrahms et al. 2022) and grey wolves in North America (Mahoney et al. 2020), either by delaying parturition for 7–21 days or initiating denning earlier by approximately 14 days, respectively. Responses from both canids are efforts to avoid denning and pup-rearing during times with the highest ambient temperatures, yet there are consequential trade-offs in reproductive success (see Box 3.2; Figure 3.2).

High daytime temperatures in arid, tropical, and semi-arid environments make activities such as hunting, movement, and dispersal extremely cost-prohibitive, straining thermoregulatory processes that sustain homeostasis and thus individual fitness (Veldhuis et al. 2019, Bonebrake et al. 2020). Carnivores plastically respond by reducing the amount of time allocated to foraging and hunting during the day, and in turn amplify their nocturnality when ambient temperatures are high (Woodroffe et al. 2017, Gaynor et al. 2018, Rabaiotti and Woodroffe 2019, Rafiq et al. 2023). By boosting nocturnal activity, some evidence suggests carnivores reap potential benefits, as increased access to, and reduced visibility of, prey at night bolsters predation efficiency (Creel et al. 2016, Mills and Harris 2020). However, changes to diel patterns also produce several fitness trade-offs, as intraguild competition may increase for resources (Cozzi et al. 2012, Rafiq et al. 2023), and human–carnivore attacks at night become more prevalent (Packer et al. 2011). In Arctic ecosystems, the progressive loss of sea ice has transformed the landscape for polar bears (*Ursus maritimus*), and in response, bear presence on land has noticeably increased (Atwood and Wilder 2021). Polar bears rely on sea ice for hunting seals (Abrahms et al. 2023), and the dramatic loss of sea ice over decades has reduced spatial overlap between polar bears and seals, effectively limiting predation success (Hamilton et al. 2017) while simultaneously elevating locomotor costs for bears due to increased swimming in search of prey (Pagano and Williams 2021). The combination of reduced caloric intake and elevated energy expenditures increases starvation risks, driving polar bears to roam coastal habitats where they prey-switch to sensitive bird populations (Prop et al. 2015, Reimer et al. 2019).

Demographic responses to climate change are perhaps the most consequential for carnivore persistence in the Anthropocene. Climate change effects have substantial impacts on the demographic rates of plants and animals, with climate-induced

Box 3.2 Climate changes place African wild dogs in a complex bind

African wild dogs are a compelling ecosystem sentinel for illustrating the complex impacts of climate change on predators. Wild dogs are social, cursorial predators specialized on hunting terrestrial ungulates during the day through coordinated hunting efforts among pack members (Creel and Creel 1995, Vucetich and Creel 1999). Coincidentally, pack size dictates predation success, with reductions in group size limiting hunting efficiency that accelerate **Allee Effects** and jeopardize pack persistence (Creel and Creel 1995). Intraguild competition with lions and spotted hyenas places additional selective pressure on the species to work cooperatively in hunting and rearing offspring (Creel and Creel 1996, 1998). Across sub-Saharan Africa, wild dogs exhibit highly seasonal reproduction, with selection shaping the timing of parturition, denning, and pup-rearing to co-occur with cooling temperatures (McNutt et al. 2019). However, with rising ambient temperatures, optimal reproductive strategies are facing tremendous pressures. Wild dogs have seen accumulating delays (7–21 days) in parturition dates and denning periods over several decades in response to rising ambient temperatures (Abrahms et al. 2022). Such shifts are arguably adaptive to rapid climate changes, as individuals that den in areas with higher maximum temperatures have lower reproductive success (Abrahms et al. 2022). In addition, pups reared in areas with higher ambient temperatures are less likely to be successfully recruited into future packs (Woodroffe et al. 2017).

However, these adaptive responses are tempered by limited behavioural flexibility in hunting and nocturnality. Despite African wild dogs' ability to increase their nocturnal activity with heightened temperatures, they only appear to do so on moonlit nights outside of the pup-rearing period (Rabaiotti and Woodroffe 2019), largely to protect offspring from predation. Moreover, increased nocturnality increases their temporal overlap with lions and hyenas, increasing mortality risk from other apex predators (Cozzi et al. 2012, Rafiq et al. 2023). Maximizing hunting efforts while mitigating the risk to offspring imposes strict ecological limits on the breadth of responses African wild dogs can take. Further, the convergence of elevated heat and human activity detrimentally affect adult wild dog survivorship, leading either to increased lethal removal by humans or zoonotic disease transmission with domestic dogs (Rabaiotti et al. 2021). Taken together, climate change binds African wild dogs in a 'phenological trap' where the environmental cues that adults use to assess future conditions are decoupled from the cue's fitness outcomes, ultimately leading to maladaptive responses that lower fitness and precipitate population declines.

(A) 'African wild dog, Lycaon pictus (young) at uMkhuze Game Reserve, kwaZulu-Natal, South Africa' by Derek Keats, available under a Creative Commons Attribution 2.0 Generic (CC BY 2.0) licence https://creativecommons.org/licenses/by/2.0/deed.en.

(B) 'Fighting African wild dogs' by Tambako The Jaguar, available under a Creative Commons Attribution-NoDerivs 2.0 Generic (CC BY-ND 2.0) licence https://creativecommons.org/licenses/by-nd/2.0/deed.en

pressures being especially harmful during critical windows of a species' life cycle (e.g. reproduction, gestation, parturition, denning, pup-rearing, etc.) (Selwood et al. 2015). Notable declines in the reproductive fitness of African wild dogs (Woodroffe et al. 2017, Abrahms et al. 2022), polar bears (Atwood and Wilder 2021, Laidre et al. 2020, Rode et al. 2018), and grey wolves (Mahoney et al. 2020) have been attributed to rising daily maximum temperatures (African wild dogs) or declines in snow and sea ice conditions (wolves and polar bears). For polar bears and African wild dogs, some of the most imperilled carnivores on the planet, adult body condition, mean litter sizes, cub survival, denning success, and recruitment all decline in response to higher ambient temperatures or sea ice reductions (Woodroffe et al. 2017, Rode et al. 2018, Rabaiotti and Woodroffe 2019, Laidre et al. 2020, Abrahms et al. 2022). These issues are compounded by human development and intraguild competition, as reduced spatiotemporal partitioning among intraguild competitors and humans combine to elevate mortality risks (Towns et al. 2009, Miller et al. 2015, Wilder et al. 2017, Rabaiotti et al. 2021, Rafiq et al. 2023).

Pollution

Although environmental contamination by humans has occurred for over a millennium, the Industrial Revolution and associated urbanization during the late eighteenth century spurred the extensive patterns of pollution that we see today (Power and Worsley 2018). The burning of fossil fuels, burgeoning traffic, and lack of waste management in factories, businesses, and homes during this time marked a significant escalation in air, water, soil, and noise pollution. Subsequent decades saw an increase in human populations, land development, mass production and use of chemicals, advancements in transport, large-scale warfare, widespread agricultural pesticide application, and unregulated dumping of industrial waste, which furthered the complexity and intensity of pollutants deposited into the environment (Massard-Guilbaud and Mathis 2017). Today, historic and ongoing contamination continues to degrade the environment and pose significant public health concerns. For instance, exposure to particulate matter (Fuller et al. 2022) and lead (Larsen and Sánchez-Triana 2023) is responsible for 7 million and 5.5 million deaths per year, respectively, as well as numerous cognitive and physiological impairments. Efforts to reduce pollution in recent years via regulation of chemicals and waste, cleaner industrial processes, and renewable energy have collectively contributed to a reduction in pollution and slowed rates of greenhouse gas emissions (Heim and Schwarzbauer 2013, Awewomom et al. 2024). Nevertheless, pollution still poses a major problem for both human and non-human life.

Deposition of pollutants into the environment, from toxicants like anticoagulant rodenticides to volatile organic compounds, threatens apex predators. Terrestrial and aquatic organisms alike are exposed to pollutants in the air, water, and soil, as well as in the food they consume (Smith et al. 2007). However, for wide-ranging, long-lived apex predators, chemicals from pollutants can both **bioaccumulate** (i.e. build up) in their tissues over time, as well as become **biomagnified** (i.e. increase in concentration) as they are passed up the food chain. For this reason, apex predators are uniquely positioned as ecosystem sentinels for monitoring pollutants, spurring research on taxa like raptors and marine mammals in recent years (Gkotsis et al. 2022, Treu et al. 2022).

Pollution has pervasive lethal and sublethal impacts for carnivores across anthropogenic gradients. The lethal impacts of large-scale, acute environmental contamination events, such as oil spills, are obvious and quick to generate indignation by the media, affected groups, and the public (Whitehead 2013). **Sublethal** impacts caused by chronic pollutant exposure are less apparent, yet far more common and can be just as deleterious to predators and their role in regulating ecosystems (Peterson and Schulte 2016, Rodríguez-Jorquera et al. 2017). Sublethal effects are harmful when chemical compounds from pollutants disrupt an animal's physiology and behaviour. For example, urban carnivores, such as bobcats (*Lynx rufus*), are at risk of secondary exposure to anticoagulant rodenticide via consumption of contaminated rodents (Baudrot et al. 2020). This poisoning can suppress immune and organ function, making predators more susceptible to opportunistic infections, notably mange (Serieys et al. 2018), which can cause rapid local population declines (Riley et al. 2007, Niedringhaus et al. 2019). Chemicals in pesticides, microplastics, heavy metals, pharmaceuticals, and other pollutants can also impair cognitive function, disrupt endocrine signalling, and cause metabolic dysfunction (Saaristo et al. 2018). These changes lead to the abnormal development and altered expression of several important behaviours, including territoriality and aggression, courtship and copulation, parental care, communication, and general activity and motor control (Zala and Penn 2004). For instance, aquatic predators exposed to pollutants have decreased predation success due to disrupted sensory abilities and impaired motor control (Clance et al. 2023). As apex predators impose top-down control on many prey species, such behavioural alterations are expected to have cascading effects (Rohr et al. 2006, Clements and Rohr 2009).

Artificial light at night (ALAN) represents an emerging pollutant that has gained considerable attention due to its disruptive effects, from individuals to communities (Gaston et al. 2012, 2014, 2015, Sanders and Gaston 2018, Dominoni et al. 2020). The proliferation of ALAN with expanding built environments has drastically altered natural light and dark cycles with downstream consequences for all organisms, humans included (Davies et al. 2013, Swaddle et al. 2015, Davies and Smyth 2018). All organisms are entrained on stable and predictable light regimes, such that our circadian rhythms (i.e. biological clocks) have articulated a series of behavioural, physical, and mental sequences attuned to light and dark cycles over a 24-hour period. ALAN disrupts those cycles, leading to elevated glucocorticoid concentrations (i.e. stress hormones) that interrupt melatonin secretion, immune function, and metabolic processes, all of which elevate cancer risks (Dominoni et al. 2016, Schmid et al. 2023). ALAN also produces cascading effects on individual foraging decisions and species interactions (Seymoure et al. 2023), particularly through the amplification and dissipation of predator-prey dynamics (Sanders and Gaston 2018). Though no study to date has assessed the organismal consequences (e.g. endocrine disruption, immunosuppression) of ALAN in apex predators, a few notable studies suggest that ALAN can be detrimental to predation success (Figure 3.2). Evidence in mountain lions across the U.S. Intermountain West suggest that lions preferentially hunt mule deer (*Odocoileus hemionus*) in the darkest habitats, yet those deer use well-lit anthropogenic environments for foraging opportunities (Ditmer et al. 2021). ALAN may therefore exacerbate spatial partitioning among predator and prey, potentially reducing lion predation success.

Chronic anthropogenic noise pollution has upended natural acoustic environments, and as a result, is considered one of the most hazardous forms of pollution in aquatic and terrestrial environments (Shannon et al. 2016, Kunc and Schmidt 2019). Deafening traffic noise from roadways, expanding urbanization, motorized recreation, growing global transportation networks, and resource extraction processes have made anthropogenic noise a widespread phenomenon with a mounting number of consequences for wildlife (Barber et al. 2010, Kok et al. 2023). Anthropogenic noise alters animal communication (Berger-Tal et al. 2019), vigilance (Klett-Mingo et al. 2016, Miller et al. 2022, Collins et al. 2022), activity patterns (Bunkley et al. 2015, Dominoni et al. 2020), foraging strategies (Siemers and Schaub 2011, Luo et al. 2015, Sweet et al. 2022, Blackburn et al. 2023), stress physiology (Tennessen et al. 2014, Kleist et al. 2018), and reproductive success (Summers et al. 2011, Injaian et al. 2018), ultimately affecting species abundances and community assemblages that cascade to influence community structure and function (Barber et al. 2010, Swaddle et al. 2015, Halfwerk and Slabbekoorn 2015, Kunc et al. 2016, Wilson et al. 2021a). Most investigations assessing the impacts of anthropogenic noise exposure have focused on invertebrates, birds, bats, and marine mammals (Shannon et al. 2016, Kunc and Schmidt 2019). However, some evidence from experimental manipulations suggests that anthropogenic noises (i.e. human voices) modify vigilance, activity, and foraging behaviours in mountain lions, bobcats, badgers (*Meles meles*), spotted hyenas (*Crocuta crocuta*), and leopards (Clinchy et al. 2016, Smith et al. 2017, Suraci et al. 2019, Barker et al. 2023, Zanette et al. 2023). Though evidence directly investigating the impacts of anthropogenic noise on apex predators is sparse, carnivore responses to a growing human footprint are well-documented (Tucker et al. 2018, Gaynor et al. 2018, Suraci et al. 2019, Rutz et al. 2020) and anthropogenic noise is a prominent component of that footprint (see Chapter 7).

Invasive species

Humans have introduced non-native species for centuries, with profound impacts on local ecosystems and global biodiversity. As traders and colonizers moved across the globe, they unintentionally, and sometimes intentionally, transported plant, animal, and microbial species into new habitats (Hulme 2009). Well-known examples include rats (*Rattus rattus*), which have decimated bird populations on islands and caused major epidemics (Banks and Hughes 2012), and European starlings (*Sturnus vulgaris*), which compete with native bird species for resources and can cause significant economic damage (Linz et al. 2017). In addition, wild populations of domesticated species like pigs and cats introduced for agricultural purposes or companionship have precipitated major ecosystem disruptions (Pimentel et al. 2000, 2005). It is estimated that nearly 17,000 species have been introduced by humans, and, despite efforts to control the spread of non-native species, this number continues to grow (Seebens et al. 2017).

Such human-mediated introductions of non-native species, especially domesticated species, present unique and severe challenges for apex predators. Non-native predators may outcompete native predators for resources, alter prey availability, or even hybridize with native predators (Allendorf et al. 2001, Glen and Dickman 2005). Such disruptions may lead to native predator declines, changes within their ecological communities, or both. In one incredible example from the Laikipia region of Kenya, non-native big-headed ants (*Pheidole megacephala*) disrupted a native ant-acacia mutualism, leaving acacia trees susceptible to herbivory by elephants (*Loxodonta africana*) (Kamaru et al. 2024). The resulting thinned vegetation decreased zebra (*Equus quagga*) predation success by African lions who, in turn, began to depredate African buffalo (*Syncerus caffer*), yielding currently unknown consequences within this savannah community (Figure 3.2). Although non-native wild species are difficult to manage and can cause significant ecological and economic damage (Pyšek and Richardson 2010), their threat to native apex predators arguably pales in comparison to the myriad problems caused by non-native domesticated species (Doherty et al. 2017; Hassell et al. 2017; Hughes & Macdonald 2013; Loss et al. 2022).

Human societies have unique connections to domesticated species. Their cultural, religious, and economic significance is frequently the rationale for prioritizing the well-being of domesticated individuals, occasionally at the detriment of their wild counterparts (Torres et al. 2018). This is especially applicable for apex predators who may depredate livestock (e.g. chickens, sheep, cattle) and free-ranging pets (e.g. cats, dogs), leading to conflict with humans (Hassell et al. 2017, Craft et al. 2017, Ünal et al. 2020) (see also Chapter 5). The economic and emotional toll of these losses can often stimulate retaliatory killings and removals (Lennox et al. 2018). Perhaps the greatest example is the extirpation of North America's apex predators after the influx of livestock in the 1800s and early 1900s. Eradication campaigns led by European colonists and federal agencies decimated populations of grey wolves, mountain lions, and coyotes (*Canis latrans*) (Ripple et al. 2013). In their absence, a **mesopredator release** was observed, wherein a lack of top-down control on foxes (*Vulpes vulpes*) and raccoons (*Procyon lotor*) saw rising mesopredator abundances and decreasing native prey (Prugh et al. 2009, Brashares et al. 2010). Unlike larger carnivores, coyote populations have since recovered and expanded beyond their historic range (Hody and Kays 2018). Coyotes have assumed the *de facto* role of apex predator in many habitats, notably cities, which also harbour large populations of domestic cats (Flockhart et al. 2016, Herrera et al. 2022). Coincidentally, ideological differences in how to curb domestic cat populations and the conservation threat they pose remain an immensely contentious societal issue (Crowley et al. 2020). Advocacy for the protection of domestic cats and the removal of coyotes is a common occurrence in urban landscapes (Stanton et al. 2023), despite the widespread, negative impacts of domestic cats on the environment (e.g. depredation of native wildlife and disease transmission) (Loss et al. 2022).

Domesticated species also serve as reservoirs for several zoonotic diseases that can spillover to wildlife, causing outbreaks that threaten prey, predators, and people (Plowright et al. 2021, 2024). Zoonotic disease transmission can occur directly (e.g. bites and scratches) or indirectly (e.g. through the deposition of urine and faeces into the environment), with myriad downstream consequences for individual health, population structure, and community interactions (Keesing and Ostfeld 2021, Marrana 2022). Some diseases, like feline leukaemia virus (FeLV), are primarily spread through contact with domestic cats (e.g. predation), and pose a risk to endangered populations of wild felids, including the Iberian Lynx (*Lynx pardinus*) and the Florida Panther (*Puma concolor coryi*) (Meli et al. 2009, Chiu et al. 2019). Other diseases, such as Toxoplasmosis, affect a wide diversity of taxa, including people (Torrey and Yolken 2013). Cats serve as the definitive host for the generalist protozoan parasite *Toxoplasma gondii*, shedding oocysts in their faeces that can persist in soil and water for several months to years (Shapiro et al. 2019). Consumption of cysts embedded in infected animal tissues amplifies pathogen transmission to wild carnivores (Hollings et al. 2013). Infection by *T. gondii* both threatens the health of immunocompromised people and wildlife, triggering confusion, increased risk-taking, and other abnormal behaviours in wildlife (Milne et al. 2020, Johnson and Johnson 2021). Disease-driven behavioural modifications have been linked to increased vehicular collisions in Persian leopards (*Panthera pardus saxicolor*) (Namroodi et al. 2016), elevated boldness that led to African lion-caused mortalities in infected spotted hyena cubs (Gering et al. 2021), and increased risk-taking and aggression in individual grey wolves with implications for pack dynamics (Meyer et al. 2022).

Additional infection risks to apex predators stem from free-ranging domestic dogs that carry pathogens for deadly diseases like rabies, canine distemper virus, and parvovirus (Young et al. 2011). The high contact rates among social carnivores like wolves (Laurenson et al. 1998, Millán et al. 2016), African wild dogs (Woodroffe et al. 2012), tigers (*Panthera tigris*) (Gilbert et al. 2015), and African lions (Roelke-Parker et al. 1996) can lead to rapid pathogen transmission and population declines. Because free-ranging domestic cats and dogs typically occur at higher densities in cities, it is likely that people and predators in developed habitats may be at greatest risk of infection from domesticated species (Wilson et al. 2021b).

Conclusion—indirect drivers

Each of the direct drivers detailed here can have profound impacts on predator persistence, with interactions among them emphasizing the urgency for action. An incomplete or insufficient diagnosis of how human disturbances dictate apex predator ecology in the twenty-first century runs the risk of exacerbating calamities that hasten ecosystem collapse. However, any policies or efforts that do not sufficiently address the **indirect drivers** underlying the direct drivers will likely be ineffective in promoting coexistence. Indirect drivers comprise the human-mediated factors that underpin the frequency, strength, and directionality of direct drivers, including the social, economic, cultural, demographic, political, and institutional considerations of society (Díaz and Malhi 2022). The physical manifestation of direct drivers is predicated on the stability and structure of human societies. If our aim is to repair the harms produced by direct drivers, then improving our understanding of the connections and feedbacks between societal processes and predator ecology will be essential (Abrahms et al. 2023, Harris et al. 2023, McInturff et al. 2021, Schell et al. 2021). These connections may include: (1) the influence of socioeconomic and development patterns in shaping carnivore ecology and human-carnivore interactions (Ravenelle and Nyhus 2017, Braczkowski et al. 2023, Johnson et al. 2023); (2) the impacts of political affiliations and governance systems in facilitating or impeding carnivore coexistence legislation (Lute et al. 2020, van Eeden et al. 2021, Cusack et al. 2022); (3) how demographics and sociocultural values, perceptions, and attitudes across stakeholder groups affect human-predator interactions and policies (Zorondo-Rodríguez et al. 2020, Darimont et al.

2021, Connolly and Nelson 2023); and (4) how geopolitical conflicts and environmental injustices affect conservation efforts and human-carnivore interactions (Bouley et al. 2018, Stalmans et al. 2019, Jordan et al. 2020, McInturff et al. 2021, Bonsen et al. 2024).

Taken together, we will need more comprehensive solutions that aim to solve for both parts of the social-ecological equation: carnivores *and* people (see Chapter 15). Human disturbances are a direct result of our collective impact on ecosystems and nature, with apex predators acutely sensitive to such ecological drivers. It will therefore be imperative that future research and applied strategies both address the direct disturbances affecting carnivores, as well as the indirect drivers emanating from society, if we hope to fully understand human–carnivore ecology in the Anthropocene.

References

Abernathy, H. N., et al. 2023. Rain, recreation and risk: human activity and ecological disturbance create seasonal risk landscapes for the prey of an ambush predator. *Journal of Animal Ecology* 92:1840–1855.

Abrahms, B. 2021, 30 July. *Human-Wildlife Conflict Under Climate Change*. American Association for the Advancement of Science, Washington, D. C.

Abrahms, B., et al. 2023. Climate change as a global amplifier of human–wildlife conflict. *Nature Climate Change* 13:224–234.

Abrahms, B., K. Rafiq, N. R. Jordan, and J. W. Mcnutt. 2022. Long-term, climate-driven phenological shift in a tropical large carnivore. *Proceedings of the National Academy of Sciences of the United States of America* 119(27):e2121667119.

Agrawal, A. A. 2001. Phenotypic plasticity in the interactions and evolution of species. *Science* 294:321–326.

Allendorf, F. W., R. F. Leary, P. Spruell, and J. K. Wenburg. 2001. The problems with hybrids: setting conservation guidelines. *Trends in Ecology & Evolution* 16: 613–622.

Andersen, K. H., L. Marty, and R. Arlinghaus. 2018. Evolution of boldness and life history in response to selective harvesting. *Canadian Journal of Fisheries and Aquatic Sciences* 75:271–281.

Aryal, A., D. Brunton, W. Ji, R. K. Barraclough, and D. Raubenheimer. 2014a. Human-carnivore conflict: ecological and economical sustainability of predation on livestock by snow leopard and other carnivores in the Himalaya. *Sustainability Science* 9:321–329.

Aryal, A., D. Brunton, and D. Raubenheimer. 2014b. Impact of climate change on human-wildlife-ecosystem interactions in the Trans-Himalaya region of Nepal. *Theoretical and Applied Climatology* 115:517–529.

Atickem, A., L. E. Loe, and N. C. Stenseth. 2014. Individual heterogeneity in use of human shields by mountain nyala. *Ethology* 120:715–725.

Atwood, T. C., and J. M. Wilder. 2021. Human-polar bear interactions. In *Ethology and Behavioral Ecology of Sea Otters and Polar Bears*, edited by R. W. Davis and A. M. Pagano. Springer, New York.

Awewomom, J., et al. 2024. Addressing global environmental pollution using environmental control techniques: a focus on environmental policy and preventive environmental management. *Discover Environment* 2. DOI:10.1007/s44274-024-00033-5.

Banks, P. B., and N. K. Hughes. 2012. A review of the evidence for potential impacts of black rats (Rattus rattus) on wildlife and humans in Australia. *Wildlife Research* 39(1):78–88.

Baral, K., B. Adhikari, S. Bhandari, R. M. Kunwar, H. P. Sharma, A. Aryal, and W. Ji. 2023. Impact of climate change on distribution of common leopard (Panthera pardus) and its implication on conservation and conflict in Nepal. *Heliyon* 9:e12807.

Barber, J. R., K. R. Crooks, and K. M. Fristrup. 2010, March. The costs of chronic noise exposure for terrestrial organisms. *Trends in Ecology & Evolution* 25(3):180–189.

Barker, K. J., et al. 2023. Large carnivores avoid humans while prioritizing prey acquisition in anthropogenic areas. *Journal of Animal Ecology* 92:889–900.

Basille, M., et al. 2013. Selecting habitat to survive: the impact of road density on survival in a large carnivore. PLoS One 8:e65493.

Bateman, P. W. and P. A. Fleming. 2012. Big city life: carnivores in urban environments. *Journal of Zoology* 287:1–23.

Baudrot, V., J. Fernandez-de-Simon, M. Coeurdassier, G. Couval, P. Giraudoux, and X. Lambin. 2020. Trophic transfer of pesticides: The fine line between predator–prey regulation and pesticide–pest regulation. *Journal of Applied Ecology* 57:806–818.

Bellard, C., C. Bertelsmeier, P. Leadley, W. Thuiller, and F. Courchamp. 2012. Impacts of climate change on the future of biodiversity. *Ecology Letters* 15:365–377.

Berger, J. 2007. Fear, human shields and the redistribution of prey and predators in protected areas. *Biology Letters* 3:620–623.

Berger-Tal, O., B. B. M. Wong, U. Candolin, and J. Barber. 2019. What evidence exists on the effects of anthropogenic noise on acoustic communication in animals? A systematic map protocol. *Environmental Evidence* 8. DOI:10.1186/s13750-019-0165-3.

Bischof, R., et al. 2018. Regulated hunting re-shapes the life history of brown bears. *Nature Ecology and Evolution* 2:116–123.

Blackburn, G., B. J. Ashton, A. Thornton, S. Woodiss-Field, and A. R. Ridley. 2023. Cognition mediates response to anthropogenic noise in wild Western Australian magpies (*Gmynorhina tibicen dorsalis*). *Global Change Biology* 29:6912–6930.

Blašković, S., T. Gomerčić, I. Topličanec, and M. Sindičić. 2022. Temporal overlap of human and apex predator activity on wildlife trails and forest roads. *Journal of Vertebrate Biology* 71: 22029.1–9.

Blecha, K. A., R. B. Boone, and M. W. Alldredge. 2018. Hunger mediates apex predator's risk avoidance response in wildland-urban interface. *Journal of Animal Ecology* 87:609–622.

Bonebrake, T. C., E. L. Rezende, and F. Bozinovic. 2020. Climate change and thermoregulatory consequences of activity time in mammals. *American Naturalist* 196:45–56.

Bonsen, G. T., et al. 2024, 1 August. Navigating complex geopolitical landscapes: challenges in conserving the endangered Arabian wolf. *Biological Conservation* 296:110655.

Bouley, P., M. Poulos, R. Branco, and N. H. Carter. 2018. Post-war recovery of the African lion in response to large-scale ecosystem restoration. *Biological Conservation* 227:233–242.

Boutin, S. 2018. Hunger makes apex predators do risky things. *Journal of Animal Ecology* 87:530–532.

Braczkowski, A. R., C. J. O'Bryan, C. Lessmann, C. Rondinini, A. P. Crysell, S. Gilbert, M. Stringer, L. Gibson, and D. Biggs. 2023. The unequal burden of human-wildlife conflict. *Communications Biology* 6:182.

Brashares, J. S., L. R. Prugh, C. J. Stoner, and C. W. Epps. 2010. Ecological and conservation implications of mesopredator release. In *Trophic Cascades: Predators, Prey, and the Changing Dynamics of Nature*, edited by J. Terborgh and J. A. Estes. Island Press, Washington, D. C.

Brivio, F., et al. 2024. Seeking temporal refugia to heat stress: increasing nocturnal activity despite predation risk. *Proceedings of the Royal Society B: Biological Sciences* 291:20231587.

Bunkley, J. P., C. J. W. McClure, N. J. Kleist, C. D. Francis, and J. R. Barber. 2015. Anthropogenic noise alters bat activity levels and echolocation calls. *Global Ecology and Conservation* 3:62–71.

Burton, P. J., A. Jentsch, and L. R. Walker. 2020. The ecology of disturbance interactions. *BioScience* 70:854–870.

Butler, J. R. A. A., J. D. C. C. Linnell, D. Morrant, V. Athreya, N. Lescureux, and A. S. McKeown. 2015. Dog eat dog, cat eat dog: social-ecological dimensions of dog predation by wild carnivores. In Free-Ranging Dogs and Wildlife Conservation, edited by M. E. Gompper. Oxford University Press, Oxford.

Carter, N. H., S. A. Levin, and V. Grimm. 2019. Effects of human-induced prey depletion on large carnivores in protected areas: lessons from modeling tiger populations in stylized spatial scenarios. *Ecology and Evolution* 9:11298–11313.

Cervantes, A. M., et al. 2023. Carnivore coexistence in Chicago: niche partitioning of coyotes and red foxes. *Urban Ecosystems* 26:1293–1307.

Chaplin-Kramer, R., et al. 2019. Global modeling of nature's contributions to people. *Science* 366:255–258.

Chase, J. M. 2003. *Ecological Niches: Linking Classical and Contemporary Approaches*. The University of Chicago Press, Chicago, I. L.

Chiu, E. S., S. Kraberger, M. Cunningham, L. Cusack, M. Roelke, and S. Vandewoude. 2019. Multiple introductions of domestic cat feline leukemia virus in endangered Florida Panthers. *Emerging Infectious Diseases* 25:92–101.

Ciucci, P., S. Mancinelli, L. Boitani, O. Gallo, and L. Grottoli. 2020. Anthropogenic food subsidies hinder the ecological role of wolves: Insights for conservation of apex predators in human-modified landscapes. *Global Ecology and Conservation* 21:e00841.

Clance, L. R., S. L. Ziegler, and F. J. Fodrie. 2023. Contaminants disrupt aquatic food webs via decreased consumer efficiency. *Science of The Total Environment* 859:160245.

Clements, W. H., and J. R. Rohr. 2009. Community responses to contaminants: using basic ecological principles to predict ecotoxicological effects. *Environmental Toxicology and Chemistry* 28(9):1789–1800.

Clinchy, M., L. Y. Zanette, D. Roberts, J. P. Suraci, C. D. Buesching, C. Newman, and D. W. Macdonald. 2016. Fear of the human 'super predator' far exceeds the fear of large carnivores in a model mesocarnivore. *Behavioral Ecology* 284:arw117.

Coimbra, Z. H., L. Gomes-Jr, and F. A. S. Fernandez. 2020. Human carnivory as a major driver of vertebrate extinction. *Perspectives in Ecology and Conservation* 18:283–293.

Collins, A. C., T. W. Vickers, and F. M. Shilling. 2022. Behavioral responses to anthropogenic noise at highways vary across temporal scales. *Frontiers in Ecology and Evolution* 10:891595.

Connolly, E. and H. Nelson. 2023. Jaguars in the borderlands: multinatural conservation for coexistence in the Anthropocene. *Frontiers in Conservation Science* 4:851254.

Cozzi, G., F. Broekhuis, J. W. Mcnutt, L. A. Turnbull, D. W. Macdonald, and B. Schmid. 2012. Fear of the dark or dinner by moonlight? Reduced temporal partitioning among Africa's large carnivores. *Ecology* 93:2590–2599.

Craft, M. E., F. Vial, E. Miguel, S. Cleaveland, A. Ferdinands, and C. Packer. 2017. Interactions between domestic and wild carnivores around the greater Serengeti ecosystem. *Animal Conservation* 20:193–204.

Creel, S. and N. M. Creel. 1995. Communal hunting and pack size in African wild dogs, *Lycaon pictus*. *Animal Behaviour* 50:1325–1339.

Creel, S. and N. M. Creel. 1996. Limitation of African wild dogs by competition with larger carnivores. *Conservation Biology* 10:526–538.

Creel, S. and N. M. Creel. 1998. Six ecological factors that may limit African wild dogs, Lycaon pictus. *Animal Conservation* 1:1–9.

Creel, S., N. M. Creel, A. M. Creel, and B. M. Creel. 2016. Hunting on a hot day: effects of temperature on interactions between African wild dogs and their prey. *Ecology* 97:2910–2916.

Crowley, S. L., M. Cecchetti, and R. A. McDonald. 2020. Our wild companions: domestic cats in the Anthropocene. *Trends in Ecology & Evolution* 35(6):477–483.

Curras, M. R., E. Donadio, A. D. Middleton, and J. N. Pauli. 2022. Carnivore niche partitioning in a human landscape. *American Naturalist* 199:496–509.

Cusack, J. J., et al. 2022. Quantifying the checks and balances of collaborative governance systems for adaptive carnivore management. *Journal of Applied Ecology* 59:1038–1049.

Darimont, C. T., S. M. Carlson, M. T. Kinnison, P. C. Paquet, T. E. Reimchen, and C. C. Wilmers. 2009. Human predators outpace other agents of trait change in the wild. *Proceedings of the National Academy of Sciences of the United States of America* 106:952–954.

Darimont, C. T., et al. 2023. Humanity's diverse predatory niche and its ecological consequences. *Communications Biology* 6:609.

Darimont, C. T., C. H. Fox, H. M. Bryan, and T. E. Reimchen. 2015. The unique ecology of human predators. *Science* 349:858–860.

Darimont, C. T., et al. 2021. Large carnivore hunting and the social license to hunt. *Conservation Biology* 35: 1111–1119.

Davies, T. W., J. Bennie, R. Inger, and K. J. Gaston. 2013. Artificial light alters natural regimes of night-time sky brightness. *Scientific Reports* 3:1–6.

Davies, T. W., and T. Smyth. 2018. Why artificial light at night should be a focus for global change research in the 21st century. *Global Change Biology* 24:872–882.

Díaz, S., and Y. Malhi. 2022. Biodiversity: Concepts, Patterns, Trends, and Perspectives. Annual Review of Environment and Resources 47:31–63.

Ditmer, M. A., et al. 2021. Artificial nightlight alters the predator–prey dynamics of an apex carnivore. *Ecography* 44:149–161.

Doherty, T. S., et al. 2017. The global impacts of domestic dogs on threatened vertebrates. *Biological Conservation* 210:56–59.

Dominoni, D. M., J. C. Borniger, and R. J. Nelson. 2016. Light at night, clocks and health: From humans to wild organisms. *Biology Letters* 12:20160015.

Dominoni, D., J. A. H. Smit, M. E. Visser, and W. Halfwerk. 2020. Multisensory pollution: Artificial light at night and anthropogenic noise have interactive effects on activity patterns of great tits (Parus major). *Environmental Pollution* 256:113314.

Dorresteijn, I., J. et al. 2015. Incorporating anthropogenic effects into trophic ecology: predator–prey interactions in a human-dominated landscape. *Proceedings of the Royal Society B: Biological Sciences* 282:20151602.

Dragonetti, C., V. Y. Mendez Angarita, and M. Di Marco. 2023. Scenarios of change in the realized climatic niche of mountain carnivores and ungulates. *Conservation Biology* 37:e14035.

Dunn, R. R., et al. 2022. A theory of city biogeography and the origin of urban species. *Frontiers in Conservation Science* 3:1–18.

van Eeden, L. M., M. S. Crowther, C. R. Dickman, D. W. Macdonald, W. J. Ripple, E. G. Ritchie, and T. M. Newsome. 2018. Managing conflict between large carnivores and livestock. *Conservation Biology* 32:26–34.

van Eeden, L. M., S. S. Rabotyagov, M. Kather, C. Bogezi, A. J. Wirsing, and J. Marzluff. 2021. Political affiliation predicts public attitudes toward gray wolf (Canis lupus) conservation and management. *Conservation Science and Practice* 3:e387.

Ellis, E. C., K. K. Goldewijk, S. Siebert, D. Lightman, and N. Ramankutty. 2010. Anthropogenic transformation of the biomes, 1700 to 2000. *Global Ecology and Biogeography* 19:589–606.

Estes, J. A., et al. 2011. Trophic downgrading of planet earth. *Science* 333:301–306.

Estes, J. A., M. T. Tinker, T. M. Williams, and D. F. Doak. 1998. Killer whale predation on sea otters linking oceanic and nearshore ecosystems. *Science* 282: 473–476.

Evans, M. J., J. E. Hawley, P. W. Rego, and T. A. G. Rittenhouse. 2019. Hourly movement decisions indicate how a large carnivore inhabits developed landscapes. *Oecologia* 190:11–23.

Everatt, K. T., J. F. Moore, and G. I. H. Kerley. 2019. Africa's apex predator, the lion, is limited by interference and exploitative competition with humans. *Global Ecology and Conservation* 20:e00758.

Finn, C., F. Grattarola, and D. Pincheira-Donoso. 2023. More losers than winners: investigating Anthropocene defaunation through the diversity of population trends. *Biological Reviews* 4:1732–1748.

Fischer, J. D., S. H. Cleeton, T. P. Lyons, and J. R. Miller. 2012. Urbanization and the predation paradox: the role of trophic dynamics in structuring vertebrate communities. *BioScience* 62:809–818.

Flockhart, D. T. T., D. R. Norris, and J. B. Coe. 2016. Predicting free-roaming cat population densities in urban areas. *Animal Conservation* 19:472–483.

Fuller, R., et al. 2022. Pollution and health: a progress update. *Lancet Planet Health* 6(6):e535–e547.

Gaston, K. J., T. W. Davies, J. Bennie, and J. Hopkins. 2012. Reducing the ecological consequences of night-time light pollution: options and developments. *Journal of Applied Ecology* 49:1256–1266.

Gaston, K. J., J. P. Duffy, S. Gaston, J. Bennie, and T. W. Davies. 2014. Human alteration of natural light cycles: causes and ecological consequences. *Oecologia* 176: 917–931.

Gaston, K. J., M. E. Visser, and F. Hölker. 2015. The biological impacts of artificial light at night: the research challenge. *Philosophical Transactions of the Royal Society B: Biological Sciences* 370:20140133.

Gaynor, K. M., J. S. Brown, A. D. Middleton, M. E. Power, and J. S. Brashares. 2019. Landscapes of Fear: Spatial Patterns of Risk Perception and Response. *Trends in Ecology & Evolution* 34:355–368.

Gaynor, K. M., C. E. Hojnowski, N. H. Carter, and J. S. Brashares. 2018. The influence of human disturbance on wildlife nocturnality. *Science* 360:1232–1235.

Gering, E., et al. 2021. Toxoplasma gondii infections are associated with costly boldness toward felids in a wild host. *Nature Communications* 12:3842.

Gilbert, M., et al. 2015. Canine distemper virus as a threat to wild tigers in Russia and across their range. *Integrative Zoology* 10:329–343.

Gilljam, D., A. Curtsdotter, and B. Ebenman. 2015. Adaptive rewiring aggravates the effects of species loss in ecosystems. *Nature Communications* 6:8412.

Gkotsis, G., et al. 2022. Assessment of contaminants of emerging concern in European apex predators and their prey by LC-QToF MS wide-scope target analysis. *Environment International* 170:107623.

Glen, A. S., and C. R. Dickman. 2005. Complex interactions among mammalian carnivores in Australia, and their implications for wildlife management. *Biological Reviews of the Cambridge Philosophical Society* 80(3): 387–401.

Gosselin, J., A. Zedrosser, J. E. Swenson, and F. Pelletier. 2014. The relative importance of direct and indirect effects of hunting mortality on the population dynamics of brown bears. *Proceedings of the Royal Society B: Biological Sciences* 282:20141840.

Guiden, P. W., S. L. Bartel, N. W. Byer, A. A. Shipley, and J. L. Orrock. 2019. Predator–prey interactions in the anthropocene: reconciling multiple aspects of novelty. *Trends in Ecology and Evolution* 34:616–627.

Halfwerk, W. and H. Slabbekoorn. 2015. Pollution going multimodal: the complex impact of the human-altered sensory environment on animal perception and performance. *Biology Letters* 11(4):20141051.

Hamilton, C. D., K. M. Kovacs, R. A. Ims, J. Aars, and C. Lydersen. 2017. An Arctic predator–prey system in flux: climate change impacts on coastal space use by polar bears and ringed seals. *Journal of Animal Ecology* 86:1054–1064.

Harris, N. C., C. E. Wilkinson, G. Fleury, and Z. N. Nhleko. 2023. Responsibility, equity, justice, and inclusion in dynamic human–wildlife interactions. *Frontiers in Ecology and the Environment* 21:380–387.

Hassell, J. M., M. Begon, M. J. Ward, and E. M. Fèvre. 2017. Urbanization and disease emergence: dynamics at the wildlife–livestock–human interface. *Trends in Ecology & Evolution* 32(1):55–67.

Heim, S. and J. Schwarzbauer. 2013. Pollution history revealed by sedimentary records: a review. *Environmental Chemistry Letters* 11:255–270.

Hendry, A. P., K. M. Gotanda, and E. I. Svensson. 2017. Human influences on evolution, and the ecological and societal consequences. *Philosophical Transactions of the Royal Society B: Biological Sciences* 372:20160028.

Herrera, D. J., et al. 2022. Prey selection and predation behavior of free-roaming domestic cats (*Felis catus*) in an urban ecosystem: Implications for urban cat management. *Biological Conservation* 268:109503.

Hody, J. W. and R. Kays. 2018. Mapping the expansion of coyotes (*Canis latrans*) across North and central America. *ZooKeys* 2018:81–97.

Hollings, T., M. Jones, N. Mooney, and H. McCallum. 2013. Wildlife disease ecology in changing landscapes: mesopredator release and toxoplasmosis. *International Journal for Parasitology: Parasites and Wildlife* 2:110–118.

Honda, T. and C. Kozakai. 2020. Mechanisms of human-black bear conflicts in Japan: in preparation for climate change. *Science of the Total Environment* 739: 140028.

Hughes, J. and D. W. Macdonald. 2013. A review of the interactions between free-roaming domestic dogs and wildlife. *Biological Conservation* 157:341–351.

Hulme, P. E. 2009. Trade, transport and trouble: managing invasive species pathways in an era of globalization. *Journal of Applied Ecology* 46(1):10–18.

Hutchinson, G. E. 1957. Concluding remarks. *Cold Spring Harbor Symposia on Quantitative Biology* 22:415–427.

Injaian, A. S., L. Y. Poon, and G. L. Patricelli. 2018. Effects of experimental anthropogenic noise on avian settlement patterns and reproductive success. *Behavioral Ecology* 29:1181–1189.

IPBES. 2019. *Summary for Policymakers of the Global Assessment Report on Biodiversity and Ecosystem Services of the Intergovernmental Science-Policy Platform on Biodiversity and Ecosystem Services*, edited by E. S. Brondizio, J. Settele, S. Díaz, and H. T. Ngo. IPBES Secretariat, Bonn, Germany.

IPCC. 2022. *Climate Change 2022: Impacts, Adaptation and Vulnerability Working Group II Contribution to the Sixth Assessment Report of the Intergovernmental Panel on Climate Change*. Cambridge University Press, Cambridge, U. K., and New York, N. Y., U. S. A.

Jaureguiberry, P., et al. 2022a. The direct drivers of recent global anthropogenic biodiversity loss. *Science Advances* 8:9982.

Jaureguiberry, P., et al. 2022b. The direct drivers of recent global anthropogenic biodiversity loss. *Science Advances* 8:9982.

Johnson, H. E., et al. 2018. Human development and climate affect hibernation in a large carnivore with implications for human–carnivore conflicts. *Journal of Applied Ecology* 55:663–672.

Johnson, S. K., and P. T. J. Johnson. 2021. Toxoplasmosis: recent advances in understanding the link between infection and host behavior. *Annual Review of Animal Biosciences* 9:249–264.

Johnson, T. F., N. J. B. Isaac, A. Paviolo, and M. González-Suárez. 2023. Socioeconomic factors predict population changes of large carnivores better than climate change or habitat loss. *Nature Communications* 14:74.

Jordan, N. R., B. P. Smith, R. G. Appleby, L. M. van Eeden, and H. S. Webster. 2020. Addressing inequality and intolerance in human–wildlife coexistence. *Conservation Biology* 34:803–810.

Kamaru, D. N., et al. 2024. Disruption of an ant-plant mutualism shapes interactions between lions and their primary prey. *Science* 383:433–438.

Kautz, T. M., N. L. Fowler, T. R. Petroelje, D. E. Beyer, N. J. Svoboda, and J. L. Belant. 2021. Large carnivore response to human road use suggests a landscape of coexistence. *Global Ecology and Conservation* 30:e01772.

Keesing, F., and R. S. Ostfeld. 2021. Dilution effects in disease ecology. *Ecology Letters* 24:2490–2505.

Kichloo, M. A., K. Sharma, and N. Sharma. 2023. Climate casualties or human disturbance? Shrinking distribution of the three large carnivores in the Greater Himalaya. *Climatic Change* 176:118.

Kleist, N. J., R. P. Guralnick, A. Cruz, C. A. Lowry, and C. D. Francis. 2018. Chronic anthropogenic noise disrupts glucocorticoid signaling and has multiple effects on fitness in an avian community. *Proceedings of the National Academy of Sciences* 115:E648–E657.

Klett-Mingo, J. I., I. Pavón, and D. Gil. 2016. Great tits, Parus major, increase vigilance time and reduce feeding effort during peaks of aircraft noise. *Animal Behaviour* 115:29–34.

Klich, D., et al. 2021. Predation on livestock as an indicator of drastic prey decline? The indirect effects of an African swine fever epidemic on predator–prey relations in Poland. *Ecological Indicators* 133: 108419.

Kok, A. C. M., et al. 2023. How chronic anthropogenic noise can affect wildlife communities. *Frontiers in Ecology & Evolution* 11. DOI:10.3389/fevo.2023.1130075

Kunc, H. P., K. E. McLaughlin, and R. Schmidt. 2016. Aquatic noise pollution: Implications for individuals, populations, and ecosystems. *Proceedings of the Royal Society B: Biological Sciences* 283(1836): 20160839.

Kunc, H. P., and R. Schmidt. 2019. The effects of anthropogenic noise on animals: a meta-analysis. *Biology Letters* 15:20190649.

Labadie, G., C. Hardy, Y. Boulanger, V. Vanlandeghem, M. Hebblewhite, and D. Fortin. 2023. Global change risks a threatened species due to alteration of predator–prey dynamics. *Ecosphere* 14:e4485.

Laidre, K. L., et al. 2020. Interrelated ecological impacts of climate change on an apex predator. *Ecological Applications* 30:e02071.

Lamb, C. T., et al. 2020. The ecology of human–carnivore coexistence. *Proceedings of the National Academy of Sciences* 117:17876–17883.

Lamb, C. T., G. Mowat, B. N. McLellan, S. E. Nielsen, and S. Boutin. 2017. Forbidden fruit: human settlement and abundant fruit create an ecological trap for an apex omnivore. *Journal of Animal Ecology* 86:55–65.

Lambert, M. R. and C. J. Schell. 2023. Cities as the solution to the biodiversity crisis. In *Urban Biodiversity and Equity*, edited by M. R. Lambert and C. J. Schell. Oxford University Press, Oxford.

Larsen, B., and E. Sánchez-Triana. 2023. Global health burden and cost of lead exposure in children and adults: a health impact and economic modelling analysis. *The Lancet Planetary Health* 7:e831–e840.

Laufenberg, J. S., H. E. Johnson, P. F. Doherty, and S. W. Breck. 2018. Compounding effects of human development and a natural food shortage on a black bear population along a human development-wildland interface. *Biological Conservation* 224:188–198.

Laurenson, K., C. Sillero-Zubiri, H. Thompson, F. Shiferaw, S. Thirgood, and J. Malcolm. 1998. Disease as a threat to endangered species: Ethiopian wolves, domestic dogs and canine pathogens. *Animal Conservation* 1:273–280.

Leclerc, M., A. Zedrosser, and F. Pelletier. 2017. Harvesting as a potential selective pressure on behavioural traits. Journal of Applied Ecology 54:1941–1945.

Leclerc, M., A. Zedrosser, J. E. Swenson, and F. Pelletier. 2019. Hunters select for behavioral traits in a large carnivore. *Scientific Reports* 9:12371.

Leighton, G. R. M., W. Froneman, L. E. K. Serieys, and J. M. Bishop. 2023. Trophic downgrading of an adaptable carnivore in an urbanising landscape. *Scientific Reports* 13:21582.

Lennox, R. J., et al. 2022. The roles of humans and apex predators in sustaining ecosystem structure and function: contrast, complementarity and coexistence. *People and Nature* 4:1071–1082.

Lennox, R. J., A. J. Gallagher, E. G. Ritchie, and S. J. Cooke. 2018. Evaluating the efficacy of predator removal in a conflict-prone world. *Biological Conservation* 224: 277–289.

Lewis, J. S., et al. 2021. Human activity influences wildlife populations and activity patterns: implications for spatial and temporal refuges. *Ecosphere* 12: e03487.

Linz, G., R. Johnson, and J. Thiele. 2017. European starlings. In *Ecology and Management of Terrestrial Vertebrate Invasive Species in the United States*, edited by W. C. Pitt, J. Beasley, and G. W. Witmer. CRC Press, Boca Raton, F. L.

Loss, S. R., et al. 2022. Review and synthesis of the global literature on domestic cat impacts on wildlife. *Journal of Animal Ecology* 91(7):1361–1372.

Lozano, J., A., et al. 2019. Human-carnivore relations: a systematic review. *Biological Conservation* 237:480–492.

Luo, J., B. M. Siemers, and K. Koselj. 2015. How anthropogenic noise affects foraging. *Global Change Biology* 21:3278–3289.

Lute, M. L., N. H. Carter, J. V. López-Bao, and J. D. C. Linnell. 2020. Conservation professionals' views on governing for coexistence with large carnivores. *Biological Conservation* 248:108668.

Ma, D., B. Abrahms, J. Allgeier, T. Newbold, B. C. Weeks, and N. H. Carter. 2024. Global expansion of human-wildlife overlap in the 21st century. *Science Advances* 10:7706.

Mahoney, P. J., et al 2020. Denning phenology and reproductive success of wolves in response to climate signals. *Environmental Research Letters* 15:125001.

Manlick, P. J. and J. N. Pauli. 2020. Human disturbance increases trophic niche overlap in terrestrial carnivore communities. *Proceedings of the National Academy of Sciences* 117:26842–26848.

Di Marco, M. and L. Santini. 2015. Human pressures predict species' geographic range size better than biological traits. *Global Change Biology* 21:2169–2178.

Marrana, M. 2022. Epidemiology of disease through the interactions between humans, domestic animals, and wildlife. *One Health: Integrated Approach to 21st Century Challenges to Health*, edited by J. C. Prata, A. I. Ribeiro, and T. Rocha-Santos. Elsevier, New York.

Martínez-Abraín, A., J. Jiménez, and D. Oro. 2019. Pax Romana: 'refuge abandonment' and spread of fearless behavior in a reconciling world. *Animal Conservation* 22:3–13.

Massard-Guilbaud, G., and C. F. Mathis. 2017. A brief introduction to the history of pollution: from local to global. *Microbial Ecotoxicology*, edited by C. Cravo-Laureau, C. Cagnon, B. Lauga, and R. Duran. Springer International Publishing.

McInturff, A., C. E. B. Cannon, P. S. Alagona, and D. N. Pellow. 2021. Meeting at the crossroads: an environmental justice framework for large carnivore reintroductions and recoveries. *Elementa* 9:1–20.

McNutt, J. W., R. Groom, and R. Woodroffe. 2019. Ambient temperature provides an adaptive explanation for seasonal reproduction in a tropical mammal. *Journal of Zoology* 309:153–160.

Meli, M. L., et al. 2009. Feline leukemia virus and other pathogens as important threats to the survival of the critically endangered Iberian lynx (*Lynx pardinus*). *PLoS One* 4:e4744.

Meyer, C. J., et al. 2022. Parasitic infection increases risk-taking in a social, intermediate host carnivore. *Communications Biology* 5:1180.

Millán, J., et al. 2016. Patterns of exposure of Iberian wolves (*Canis lupus*) to canine viruses in human-dominated landscapes. *EcoHealth* 13:123–134.

Miller, P. J. O., S. Isojunno, E. Siegal, F.-P. A. Lam, P. H. Kvadsheim, and C. Curé. 2022. Behavioral responses to predatory sounds predict sensitivity of cetaceans to anthropogenic noise within a soundscape of fear. *Proceedings of the National Academy of Sciences* 119:e2114932119.

Miller, S., J. Wilder, and R. R. Wilson. 2015. Polar bear-grizzly bear interactions during the autumn open-water period in Alaska. *Journal of Mammalogy* 96: 1317–1325.

Mills, K. L., et al. 2023. Tradeoffs between resources and risks shape the responses of a large carnivore to human disturbance. *Communications Biology* 6:986.

Mills, K. L. and N. C. Harris. 2020. Humans disrupt access to prey for large African carnivores. *eLife* 9:1–16.

Milne, G., et al. 2020. Infectious causation of abnormal host behavior: toxoplasma gondii and its potential association with dopey fox syndrome. *Frontiers in Psychiatry* 11:513536.

Moll, R. J., et al. 2018. Humans and urban development mediate the sympatry of competing carnivores. *Urban Ecosystems* 21:765–778.

Moss, W. E., M. W. Alldredge, K. A. Logan, and J. N. Pauli. 2016a. Human expansion precipitates niche expansion for an opportunistic apex predator (*Puma concolor*). *Scientific Reports* 6:39639.

Moss, W. E., M. W. Alldredge, and J. N. Pauli. 2016b. Quantifying risk and resource use for a large carnivore in an expanding urban-wildland interface. *Journal of Applied Ecology* 53:371–378.

Namroodi, S., A. Gholami, and E. Shariat-Bahadori. 2016. Toxoplasmosis may lead to road kills of Persian leopards (*Panthera pardus saxicolor*) in Golestan National Park, Iran. *Journal of Wildlife Diseases* 52(2): 436–438.

Newman, E. A. 2019. Disturbance ecology in the anthropocene. *Frontiers in Ecology and Evolution* 7:75.

Newsome, T. M., et al. 2015. The ecological effects of providing resource subsidies to predators. *Global Ecology and Biogeography* 24:1–11.

Nickel, B. A., J. P. Suraci, M. L. Allen, and C. C. Wilmers. 2020. Human presence and human footprint have nonequivalent effects on wildlife spatiotemporal habitat use. *Biological Conservation* 241:108383.

Niedringhaus, K. D., J. D. Brown, K. M. Sweeley, and M. J. Yabsley. 2019. A review of sarcoptic mange in North American wildlife. *International Journal of Parasitology: Parasites & Wildlife* 9:285–297.

Ordiz, A., S. Sæbø, J. Kindberg, J. E. Swenson, and O. G. Støen. 2017. Seasonality and human disturbance alter brown bear activity patterns: implications for circumpolar carnivore conservation? *Animal Conservation* 20:51–60.

Oriol-Cotterill, A., M. Valeix, L. G. Frank, C. Riginos, and D. W. Macdonald. 2015. Landscapes of Coexistence for terrestrial carnivores: the ecological consequences of being downgraded from ultimate to penultimate predator by humans. *Oikos* 124:1263–1273.

Ostfeld, R. S. and R. D. Holt. 2004. Are predators good for your health? Evaluating evidence for top-down regulation of zoonotic disease reservoirs. *Frontiers in Ecology & Environment* 2(1):13–20.

Packer, C., A. Swanson, D. Ikanda, and H. Kushnir. 2011. Fear of darkness, the full moon and the nocturnal ecology of African lions. *PLoS One* 6:e22285.

Pagano, A. M. and T. M. Williams. 2021. Physiological consequences of Arctic sea ice loss on large marine carnivores: unique responses by polar bears and narwhals. *Journal of Experimental Biology* 224:jeb228049.

Penjor, U., C. Astaras, S. A. Cushman, Z. Kaszta, and D. W. MacDonald. 2022. Contrasting effects of human settlement on the interaction among sympatric apex carnivores. *Proceedings of the Royal Society B: Biological Sciences* 289:20212681.

Penteriani, V., et al. 2018. Evolutionary and ecological traps for brown bears *Ursus arctos* in human-modified landscapes. *Mammal Review* 48:180–193.

Peterson, E. K., and B. A. Schulte. 2016. Impacts of pollutants on beavers and otters with implications for ecosystem ramifications. *Journal of Contemporary Water Research & Education* 157:33–45.

Pimentel, D., L. Lach, R. Zuniga, and D. Morrison. 2000. Environmental and economic costs of nonindigenous species in the United States. *BioScience* 50(1): 53–65.

Pimentel, D., R. Zuniga, and D. Morrison. 2005. Update on the environmental and economic costs associated with alien-invasive species in the United States. *Ecological Economics* 52:273–288.

Plowright, R. K., et al. 2024. Ecological countermeasures to prevent pathogen spillover and subsequent pandemics. *Nature Communications* 15:2577.

Plowright, R. K., et al. 2021. Land use-induced spillover: a call to action to safeguard environmental, animal, and human health. *The Lancet Planetary Health* 5:e237–e245.

Power, A., and A. Worsley. 2018. Historical urban pollution. In *Urban Pollution: Science and Management*, edited by S. M. Charlesworth and C. A. Booth. Wiley, Hoboken, N. J.

Proctor, M. F., et al. 2023. Berries and bullets: influence of food and mortality risk on grizzly bears in British Columbia. *Wildlife Monographs* 213:1–77.

Prop, J., et al. 2015. Climate change and the increasing impact of polar bears on bird populations. *Frontiers in Ecology and Evolution* 3:16684865.

Prugh, L. R., et al. 2023. Fear of large carnivores amplifies human-caused mortality for mesopredators. *Science* 380:754–758.

Prugh, L. R., et al. 2009. The rise of the mesopredator. *BioScience* 59:779–791.

Pyšek, P., and D. M. Richardson. 2010. Invasive species, environmental change and management, and health. *Annual Review of Environment and Resources* 35:25–55.

Rabaiotti, D., R. Groom, J. W. McNutt, J. Watermeyer, H. M. K. O'Neill, and R. Woodroffe. 2021. High temperatures and human pressures interact to influence mortality in an African carnivore. *Ecology and Evolution* 11:8495–8506.

Rabaiotti, D. and R. Woodroffe. 2019. Coping with climate change: limited behavioral responses to hot weather in a tropical carnivore. *Oecologia* 189:587–599.

Rafiq, K., N. R. Jordan, K. Golabek, J. W. McNutt, A. Wilson, and B. Abrahms. 2023. Increasing ambient temperatures trigger shifts in activity patterns and temporal partitioning in a large carnivore guild. *Proceedings of the Royal Society B: Biological Sciences* 290: 20231938.

Ravenelle, J. and P. J. Nyhus. 2017. Global patterns and trends in human–wildlife conflict compensation. *Conservation Biology* 31:1247–1256.

Reilly, M. L., M. W. Tobler, D. L. Sonderegger, and P. Beier. 2017. Spatial and temporal response of wildlife to recreational activities in the San Francisco Bay ecoregion. *Biological Conservation* 207:117–126.

Reimer, J. R., H. Brown, E. Beltaos-Kerr, and G de Vries. 2019. Evidence of intraspecific prey switching: stage-structured predation of polar bears on ringed seals. *Oecologia* 189:133–148.

Rezaei, S., A. Mohammadi, S. Shadloo, M. Ranaie, and H. Y. Wan. 2023. Climate change induces habitat shifts and overlaps among carnivores in an arid and semi-arid ecosystem. *Ecological Informatics* 77:102247.

Riginos, C. 2015. Climate and the landscape of fear in an African savanna. *Journal of Animal Ecology* 84:124–133.

Riley, S. P. D., C. Bromley, R. H. Poppenga, F. A. Uzal, L. Whited, and R. M. Sauvajot. 2007. Anticoagulant exposure and notoedric mange in bobcats and mountain lions in urban Southern California. *The Journal of Wildlife Management* 71:1874–1884.

Riley, S. P. D., et al. 2014. Individual behaviors dominate the dynamics of an urban mountain lion population isolated by roads. *Current Biology* 24:1989–1994.

Riley, S. P. D., J. A. Sikich, and J. F. Benson. 2021. Big cats in the big city: spatial ecology of mountain lions in greater Los Angeles. *Journal of Wildlife Management* 85:1527–1542.

Ripple, W. J., et al. 2016. Bushmeat hunting and extinction risk to the world's mammals. *Royal Society Open Science* 3:160498.

Ripple, W. J., et al. 2014. Status and ecological effects of the world's largest carnivores. *Science* 343:1241484.

Ripple, W. J., A. J. Wirsing, C. C. Wilmers, and M. Letnic. 2013. Widespread mesopredator effects after wolf extirpation. *Biological Conservation* 160:70–79.

Ripple, W. J., C. Wolf, T. M. Newsome, M. Hoffmann, A. J. Wirsing, and D. J. McCauley. 2017. Extinction risk is most acute for the world's largest and smallest vertebrates. *Proceedings of the National Academy of Sciences of the United States of America* 114: 10678–10683.

Des Roches, S., et al. 2021a. Socio-eco-evolutionary dynamics in cities. *Evolutionary Applications* 14:248–267.

Des Roches, S., L. H. Pendleton, B. Shapiro, and E. P. Palkovacs. 2021b. Conserving intraspecific variation for nature's contributions to people. *Nature Ecology and Evolution* 5(5):574–582.

Rode, K. D., et al. 2018. Den phenology and reproductive success of polar bears in a changing climate. *Journal of Mammalogy* 99:16–26.

Rodríguez-Jorquera, I. A., et al. 2017. Contamination of the upper class: occurrence and effects of chemical pollutants in terrestrial top predators. *Current Pollution Reports* 3(3):206–219.

Roelke-Parker, M. E., et al. 1996. A canine distemper virus epidemic in Serengeti lions (*Panthera leo*). *Nature* 379:441–445.

Rohr, J. R., J. L. Kerby, and A. Sih. 2006. Community ecology as a framework for predicting contaminant effects. *Trends in Ecology & Evolution* 21:606–613.

Rutz, C., et al. 2020. COVID-19 lockdown allows researchers to quantify the effects of human activity on wildlife. *Nature Ecology & Evolution* 4:1156–1159.

Saaristo, M., et al. 2018. Direct and indirect effects of chemical contaminants on the behaviour, ecology and evolution of wildlife. *Proceedings of the Royal Society B: Biological Sciences* 285(1885):20181297.

Sanders, D. and K. J. Gaston. 2018. How ecological communities respond to artificial light at night. *Journal of Experimental Zoology Part A: Ecological and Integrative Physiology* 329:394–400.

Schell, C. J., L. A. Stanton, J. K. Young, L. M. Angeloni, J. E. Lambert, S. W. Breck, and M. H. Murray. 2021. The evolutionary consequences of human–wildlife conflict in cities. *Evolutionary Applications* 14: 178–197.

Schmid, D. W., P. Capilla-Lasheras, D. M. Dominoni, N. Müller-Klein, S. Sommer, and A. Risely. 2023. Circadian rhythms of hosts and their gut microbiomes: Implications for animal physiology and ecology. *Functional Ecology* 00:1–12.

Schuette, P., A. P. Wagner, M. E. Wagner, and S. Creel. 2013. Occupancy patterns and niche partitioning within a diverse carnivore community exposed to anthropogenic pressures. *Biological Conservation* 158:301–312.

Seebens, H., et al. 2017. No saturation in the accumulation of alien species worldwide. *Nature Communications* 8:14435.

Selwood, K. E., M. A. Mcgeoch, and R. Mac Nally. 2015. The effects of climate change and land-use change on demographic rates and population viability. *Biological Reviews* 90:837–853.

Serieys, L. E. K., et al. 2018. Urbanization and anticoagulant poisons promote immune dysfunction in bobcats. *Proceedings of the Royal Society B: Biological Sciences* 285:20172533.

Sévêque, A., L. K. Gentle, J. V. López-Bao, R. W. Yarnell, and A. Uzal. 2020. Human disturbance has contrasting effects on niche partitioning within carnivore communities. *Biological Reviews* 95:1689–1705.

Sévêque, A., L. K. Gentle, J. Vicente López-Bao, R. W. Yarnell, and A. Uzal. 2022. Impact of human disturbance on temporal partitioning within carnivore communities. *Mammal Review* 52(1):67–81.

Seymoure, B., A. Dell, F. Hölker, and G. Kalinkat. 2023. A framework for untangling the consequences of artificial light at night on species interactions. *Philosophical*

Transactions of the Royal Society B. Biological Sciences 378(1892):20220356.

Shannon, G., et al. 2016. A synthesis of two decades of research documenting the effects of noise on wildlife. *Biological Reviews* 91:982–1005.

Shapiro, K., et al. 2019. Environmental transmission of *Toxoplasma gondii*: oocysts in water, soil and food. *Food & Waterborne Parasitology* 15:e00049.

Siemers, B. M. and A. Schaub. 2011. Hunting at the highway: traffic noise reduces foraging efficiency in acoustic predators. *Proceedings of the Royal Society B: Biological Sciences* 278:1646–1652.

Simkin, R. D., K. C. Seto, R. I. McDonald, and W. Jetz. 2022. Biodiversity impacts and conservation implications of urban land expansion projected to 2050. *Proceedings of the National Academy of Sciences of the United States of America* 119:1–10.

Smith, A. F., S. Ciuti, D. Shamovich, V. Fenchuk, B. Zimmermann, and M. Heurich. 2022. Quiet islands in a world of fear: wolves seek core zones of protected areas to escape human disturbance. *Biological Conservation* 276:109811.

Smith, F. A., R. E. Elliott Smith, S. K. Lyons, J. L. Payne, and A. Villaseñor. 2019a. The accelerating influence of humans on mammalian macroecological patterns over the late Quaternary. *Quaternary Science Reviews* 211:1–16.

Smith, J. A., T. P. Duane, and C. C. Wilmers. 2019b. Moving through the matrix: promoting permeability for large carnivores in a human-dominated landscape. *Landscape and Urban Planning* 183:50–58.

Smith, J. A., et al. 2017. Fear of the human 'super predator' reduces feeding time in large carnivores. *Proceedings of the Royal Society B: Biological Sciences* 284:20170433.

Smith, J. A., A. C. Thomas, T. Levi, Y. Wang, and C. C. Wilmers. 2018. Human activity reduces niche partitioning among three widespread mesocarnivores. *Oikos* 127:890–901.

Smith, P. N., et al. 2007. Contaminant exposure in terrestrial vertebrates. *Environmental Pollution* 150(1):41–64.

Soofi, M., et al. 2018. Livestock grazing in protected areas and its effects on large mammals in the Hyrcanian forest, Iran. *Biological Conservation* 217:377–382.

Soofi, M., et al. 2022. Quantifying the relationship between prey density, livestock and illegal killing of leopards. *Journal of Applied Ecology* 59:1536–1547.

Springer, A. M., et al. 2003. Sequential megafaunal collapse in the North Pacific Ocean: An ongoing legacy of industrial whaling? *Proceedings of the National Academy of Sciences* 100:12223–12228.

Stalmans, M. E., T. J. Massad, M. J. S. Peel, C. E. Tarnita, and R. M. Pringle. 2019. War-induced collapse and asymmetric recovery of large-mammal populations in Gorongosa National Park, Mozambique. *PLoS One* 14:e0212864.

Stanton, L. A., C. E. Wilkinson, L. Angeloni, S. Benson-Amram, C. J. Schell, and J. K. Young. 2023. Animal behavior, cognition, and human–wildlife interactions in urban areas. In *Urban Biodiversity and Equity*, edited by M. R. Lambert and C. J. Schell. Oxford University Press, Oxford.

Stier, A. C., et al. 2016. Ecosystem context and historical contingency in apex predator recoveries. *Science Advances* 2(5):e1501769.

Støen, O. G., et al. 2015. Physiological evidence for a human-induced landscape of fear in brown bears (*Ursus arctos*). *Physiology and Behavior* 152:244–248.

Sullivan, A. P., D. W. Bird, and G. H. Perry. 2017. Human behaviour as a long-term ecological driver of non-human evolution. *Nature Ecology and Evolution* 1:1–11.

Summers, P. D., G. M. Cunnington, and L. Fahrig. 2011. Are the negative effects of roads on breeding birds caused by traffic noise? *Journal of Applied Ecology* 48:1527–1534.

Suraci, J. P., M. Clinchy, L. Y. Zanette, and C. C. Wilmers. 2019. Fear of humans as apex predators has landscape-scale impacts from mountain lions to mice. *Ecology Letters* 22:1578–1586.

Surve, N. S., S. Sathyakumar, K. Sankar, D. Jathanna, V. Gupta, and V. Athreya. 2022. Leopards in the city: the tale of Sanjay Gandhi National Park and Tungareshwar Wildlife Sanctuary, two protected areas in and adjacent to Mumbai, India. *Frontiers in Conservation Science* 3:787031.

Suryawanshi, K. R., et al. 2017. Impact of wild prey availability on livestock predation by snow leopards. *Royal Society Open Science* 4:170026.

Swaddle, J. P., et al. 2015. A framework to assess evolutionary responses to anthropogenic light and sound. *Trends in Ecology and Evolution* 30:550–560.

Sweet, K. A., B. P. Sweet, D. G. E. Gomes, C. D. Francis, and J. R. Barber. 2022. Natural and anthropogenic noise increase vigilance and decrease foraging behaviors in song sparrows. *Behavioral Ecology* 33:288–297.

Tennessen, J. B., S. E. Parks, and T. Langkilde. 2014. Traffic noise causes physiological stress and impairs breeding migration behaviour in frogs. *Conservation Physiology* 2:cou032.

Torres, D. F., E. S. Oliveira, and R. R. N. Alves. 2018. Understanding human–wildlife conflicts and their implications. *Ethnozoology: Animals in Our Lives*, edited by R. R. Nóbrega Alves and U. P. Albuquerque. Academic Press, Elsevier, London, U. K.

Torrey, E. F., and R. H. Yolken. 2013. Toxoplasma oocysts as a public health problem. *Trends in Parasitology* 29: 380–384.

Towns, L., A. E. Derocher, I. Stirling, N. J. Lunn, and D. Hedman. 2009. Spatial and temporal patterns of problem polar bears in Churchill, Manitoba. *Polar Biology* 32:1529–1537.

Treu, G., et al. 2022. Using environmental monitoring data from apex predators for chemicals management: towards better use of monitoring data from apex predators in support of prioritisation and risk assessment of chemicals in Europe. Environmental Sciences Europe 34:82.

Treves, A., and K. U. Karanth. 2003. Human-carnivore conflict and perspectives on carnivore management worldwide. *Conservation Biology* 17:1491–1499.

Treves, A., R. B. Wallace, L. Naughton-Treves, and A. Morales. 2006. Co-managing human–wildlife conflicts: a review. *Human Dimensions of Wildlife* 11:383–396.

Trumbo, D. R., et al. 2019. Urbanization impacts apex predator gene flow but not genetic diversity across an urban-rural divide. *Molecular Ecology* 28:4926–4940.

Tucker, M. A., et al. 2018. Moving in the Anthropocene: global reductions in terrestrial mammalian movements. *Page Science* 359(6374):466–469.

Ünal, Y., B. K. Pekin, İ. Oğurlu, H. Süel, and A. Koca. 2020. Human, domestic animal, Caracal (Caracal caracal), and other wildlife species interactions in a Mediterranean forest landscape. *European Journal of Wildlife Research* 66:5.

Veldhuis, M. P., T. R. Hofmeester, G. Balme, D. J. Druce, R. T. Pitman, and J. P. G. M. Cromsigt. 2020. Predation risk constrains herbivormes' adaptive capacity to warming. *Nature Ecology and Evolution* 4:1069–1074.

Veldhuis, M. P., et al. 2019. Large herbivore assemblages in a changing climate: incorporating water dependence and thermoregulation. *Ecology Letters* 22:1536–1546.

Vucetich, J. A., and S. Creel. 1999. Ecological interactions, social organization, and extinction risk in African wild dogs. *Conservation Biology* 13:1172–1182.

Wallach, A. D., W. J. Ripple, and S. P. Carroll. 2015. Novel trophic cascades: apex predators enable coexistence. *Trends in Ecology & Evolution* 30(3):146–153.

Wang, Y., J. A. Smith, and C. C. Wilmers. 2017. Residential development alters behavior, movement, and energetics in an apex predator, the puma. *PLoS One* 12:e0184687.

Whitehead, A. 2013. Interactions between oil-spill pollutants and natural stressors can compound ecotoxicological effects. *Integrative and Comparative Biology* 53: 635–647.

Wilder, J. M., et al. 2017. Polar bear attacks on humans: implications of a changing climate. *Wildlife Society Bulletin* 41:537–547.

Wilson, A. A., et al. 2021a. Artificial night light and anthropogenic noise interact to influence bird abundance over a continental scale. *Global Change Biology* 27: 3987–4004.

Wilson, A. G., S. Wilson, N. Alavi, and D. R. Lapen. 2021b. Human density is associated with the increased prevalence of a generalist zoonotic parasite in mammalian wildlife. *Proceedings of the Royal Society B: Biological Sciences* 288:20211724.

Winkler, K., R. Fuchs, M. Rounsevell, and M. Herold. 2021. Global land use changes are four times greater than previously estimated. *Nature Communications* 12:1–10.

Wolf, C., and W. J. Ripple. 2016. Prey depletion as a threat to the world's large carnivores. *Royal Society Open Science* 3:160252.

Woodroffe, R., R. Groom, and J. W. McNutt. 2017. Hot dogs: High ambient temperatures impact reproductive success in a tropical carnivore. *Journal of Animal Ecology* 86:1329–1338.

Woodroffe, R., K. C. Prager, L. Munson, P. A. Conrad, E. J. Dubovi, and J. A. K. Mazet. 2012. Contact with domestic dogs increases pathogen exposure in endangered African wild dogs (*Lycaon pictus*). *PLoS One* 7:e30099.

Young, H. S., D. J. McCauley, M. Galetti, and R. Dirzo. 2016. Patterns, causes, and consequences of anthropocene defaunation. *Annual Review of Ecology, Evolution, and Systematics* 47:333–358.

Young, J. K., K. A. Olson, R. P. Reading, S. Amgalanbaatar, and J. Berger. 2011. Is wildlife going to the dogs? Impacts of feral and free-roaming dogs on wildlife populations. *BioScience* 61:125–132.

Yovovich, V., M. L. Allen, L. T. Macaulay, and C. C. Wilmers. 2020. Using spatial characteristics of apex carnivore communication and reproductive behaviors to predict responses to future human development. *Biodiversity and Conservation* 29:2589–2603.

Zabel, F., R. Delzeit, J. M. Schneider, R. Seppelt, W. Mauser, and T. Václavík. 2019. Global impacts of future cropland expansion and intensification on agricultural markets and biodiversity. *Nature Communications* 10:1–10.

Zala, S. M., and D. J. Penn. 2004. Abnormal behaviours induced by chemical pollution: a review of the evidence and new challenges. *Animal Behaviour* 68(4):649–664.

Zanette, L. Y., et al. 2023. Fear of the human 'super predator' pervades the South African savanna. *Current Biology* 33:4689–4696.e4.

Zanni, M., R. Brogi, E. Merli, and M. Apollonio. 2023. The wolf and the city: insights on wolves conservation in the Anthropocene. *Animal Conservation* 26:256923215.

Zedrosser, A., S. M. J. G. Steyaert, H. Gossow, and J. E. Swenson. 2011. Brown bear conservation and the ghost of persecution past. *Biological Conservation* 144: 2163–2170.

Zeller, K. A., D. W. Wattles, L. Conlee, and S. Destefano. 2019. Black bears alter movements in response to anthropogenic features with time of day and season. *Movement Ecology* 7:19.

Zorondo-Rodríguez, F., D. Moreira-Arce, and S. Boutin. 2020. Underlying social attitudes towards conservation of threatened carnivores in human-dominated landscapes. *ORYX* 54:351–358.

SECTION TWO

The Lethal Effects of Humans on Carnivore Populations

CHAPTER 4

The Impacts of Hunting on Large Carnivore Populations

Andrés Ordiz, Jon E. Swenson, Cyril Milleret, and Richard Bischof

The regulation of wildlife populations promoting increases, decreases, or a sustainable yield is a core principle of wildlife management (Fryxell et al. 2014). Removal of individuals (e.g. via hunting or trapping) is a common procedure to achieve goals for wildlife populations, including carnivores (Lindsey et al. 2007, Packer et al. 2009, Swenson et al. 2017). However, the broad consequences and implications of carnivore removal are still poorly understood and underappreciated. Killing of carnivores, especially apex predators, can have unintended effects on demography, behaviour, evolution, and ecological systems in general, beyond the direct numerical effects (Treves 2009, Ordiz et al. 2013a, 2021).

Historically, large carnivores have been seen as a threat, rather than as a key element, in ecosystems. Predation is both the mechanism through which carnivores exert their primary ecological role and the main source of conflict with humans (Ordiz et al. 2013a). The present distribution and abundance of large carnivores, their genetic structure, and other morphological traits, such as body size, are reflections of long-term persecution aimed at extirpation (Ripple et al. 2014). Although a paradigm change towards active or passive restoration and sustainability has caused a comeback of many carnivore species in parts of their former range (Ingeman et al. 2022), lethal control continues, as does the potential for indirect effects.

We start this chapter by briefly reviewing the history of persecution that led to the eradication of many large carnivore populations and brought others to the brink of extinction. Legal persecution stopped in many places during the second half of the twentieth century, typically replaced by protection or regulated hunting. For a few species, harvest implies the selection of specific morphological traits, not necessarily intentional selection, which can have indirect effects on ecological and/or evolutionary aspects beyond the direct, numerical effect of removals.

Later on in the chapter, we aim to discuss the direct numerical effects following individual removal and population size reduction and a variety of indirect effects: (a) other demographic effects, such as changes in sex and age structure; (b) social effects, which may be different—or not—for 'solitary' or group-living species, such as territory turnover, disruption, and/or diffusion of dominance or group dissolution, which in turn also impact demographic structure; (c) evolutionary—selective effects, including the distribution and persistence of specific traits and, (d) behavioural effects that (*i*) may affect the demography of the population and (*ii*) may go beyond the affected species, in relation to the role that apex predators can play in ecosystems (Figure 4.1 provides a summary of all these points). In that sense, human harvest of prey species, not just carnivores, also affects predator-prey interactions and thus the cascading effects of carnivores in ecosystems (Terborgh et al. 2010). We will only briefly touch upon the

Andrés Ordiz et al., *The Impacts of Hunting on Large Carnivore Populations*. In: *Apex Predators in the Anthropocene*. Edited by: Robert A. Montgomery and Remington J. Moll, Oxford University Press. © Oxford University Press (2026). DOI: 10.1093/9780191996665.003.0004 latter two topics, as they are central elsewhere in this book (Chapters 10, 11, 12).

As examples we use primarily bears (brown bear and American black bear; *Ursus arctos* and *U. americanus*), African and Asiatic lions (*Panthera leo leo* and *P. leo persica*), and grey wolves (*Canis lupus*), as the effects of hunting have been studied more extensively in these apex predators than in many other carnivore species, and because they belong to different families (ursids, felids, and canids) and are still present on different continents. Nonetheless, other species have experienced long-term persecution and continue to be managed in ways that have both direct and indirect consequences for their populations and ecology (e.g. a mustelid, the wolverine *Gulo gulo*, Moqanaki et al. 2023).

History of persecution

At the global scale, some large carnivore species went extinct, but most have suffered important population declines and range contractions over hundreds of years, in particular during the past two centuries, due to direct (persecution, harvest) and indirect (habitat loss and degradation, depletion of prey) factors (Ripple et al. 2014). For instance, the extinction of the thylacine (*Thylacinus cynocephalus*) involved a fast, hunting-related population decline and other stressors caused by European settlement of their Australian range (Prowse et al. 2013).

There are marked similarities in the pattern of carnivore persecution and eradication across different continents. For instance, the African lion occupies approximately 10% of its historic range (Mweetwa et al. 2018) and the grey wolf was eradicated from most of Europe, the United States of America (U. S. A.), and Mexico, with its present global range reduced by approximately one-third. Brown bears experienced a similar decline (Ripple et al. 2014), and other bear species have faced a 50% population size reduction in the last century (Servheen et al. 1999). Historically, large carnivores have been seen as a threat to people, livestock, and as competitors for prey species. To this day, fear of perceived and actual threats still drives carnivore management on most continents (Woodroffe et al. 2005, Ripple et al. 2014, Sollund and Goyes 2021).

Persecution has been widespread and has varied in magnitude, ranging from wholesale eradication attempts, fuelled by bounties starting to be paid 2000 years ago in Europe (Linnell et al. 2005), the government-sponsored eradication campaigns targeting large carnivores across the contiguous USA and elsewhere during the nineteenth and twentieth centuries, to the occasional poisoning of individuals all over predators' ranges (Woodroffe and Ginsberg 1998, Woodroffe 2000, Linnell et al. 2005). Some studies show that range contractions were more likely in regions with high rural human population density, livestock density, or cropland (Woodroffe 2000, Wolf and Ripple 2017), yet government policies and cultural factors affecting public tolerance for predators may be more influential than human densities *per se* (Linnell et al. 2005). Both in North America and Europe, intense efforts to eradicate large carnivore populations were highly successful, driving populations to local extinction, even in areas with low human population densities. In southern Europe, however, some carnivore populations persisted in areas with high human density despite centuries of persecution, although most of those populations remain critically endangered even after protective measures were adopted (Linnell et al. 2005). Most large carnivore mortality is caused, intentionally or not, by people, a general pattern that holds for both protected and nonprotected areas (Woodroffe and Ginsberg 1998, Rauset et al. 2016). Persecution, habitat fragmentation, prey depletion, and disease are widespread drivers of carnivore declines (Woodroffe and Ginsberg 1998). This is also showcased in Europe, where, over centuries, persecution, extermination of ungulate prey and forest clearance caused the local extinction of many carnivore populations (Boitani 1995, Breitenmoser 1998, Linnell et al. 2001).

Nowadays (as of 2024), some carnivore populations are recovering in their former ranges in North America and Europe, yet the pace of recovery varies and so does the probability of success in the long term (Bruskotter and Shelby 2010, Chapron et al. 2014, Ausband and Mech 2023). Some carnivore recovery has occurred in other areas as well (Ingeman et al. 2022). For instance, Amur tigers (*Panthera tigris altaica*) in the Asian Far East dropped to historically low numbers in the 1940s, then began recovering decades after hunting stopped (Miquelle et al. 2005). Overall, these historic perspectives provide important context in which to assess the current range dynamics of recovering species (Wolf and Ripple 2017).

In the following sections we review the demographic, behavioural, and long-term evolutionary effects that persecution has had, and continues to have, on large carnivores. The motivation and implementation of carnivore harvest has changed (i.e. total eradication is not a widespread, publicly supported goal anymore), though most large carnivore mortality continues to be caused by humans and our activities.

Carnivore hunting today

Today, many carnivore populations are managed, often with some type of offtake as a core element of the management strategy. Numerical control, target population levels, and conflict resolution tend to be among the primary motivations at national and jurisdictional levels (Treves 2009, Bischof et al. 2012, Swenson et al. 2017). Occasionally, hunting is used to support carnivore conservation (Treves 2009), with the intent, successful or not, to engender greater local acceptance (Chapron and Treves 2016, Pepin et al. 2017). Lethal control of carnivores is also applied across different continents with the intention of limiting damages to wild prey populations or domesticated livestock (e.g. Inskip and Zimmermann 2009, Fernández-Gil et al. 2016). Proximate motivations for hunting large carnivores are sport, trophies, predator control, fur, and other body parts. Hunting can be legal, requiring permits that are constrained by quotas and bag limits, and also illegal, where these systems do not exist or are disregarded. Lethal control of carnivore populations can be conducted by private individuals (e.g. sport hunters and livestock owners) or agencies and their staff. Here, we use the term 'hunting' to be inclusive of a diversity of different forms, involving all the aforementioned motivations, as well as all removal methods such as shooting, trapping, and poisoning. In spite of its use to reach management goals, hunting large carnivores also raises concerns (Treves 2009, Ordiz et al. 2013a, 2021) because it can trigger both lethal and indirect effects at the individual and population level.

Direct effects of hunting large carnivores

The slow life-history characteristics that are typical of large carnivores make them particularly vulnerable to higher mortality rates caused by humans (Wolf and Ripple 2017). Brown bears, for instance, in different parts of North America have reported high human-caused mortality rates accounting for 119 of 131 known mortalities (90%) in two Canadian national parks (Benn and Herrero 2002). McLellan et al. (1999) reported that about 80% of their radio-collared bears in British Columbia were killed by people, a percentage remarkably similar in the following years (84%, McLellan 2015). From ~1975 to the early 1990s, human-caused mortality accounted for ~75% of brown bear mortality in and around Yellowstone National Park (Mattson et al. 1992, 1998), a period when that bear population was not legally hunted. Bischof et al. (2018) reported that in a hunted population of brown bears in Scandinavia, the cause of mortality for 74% of radio-collared adult (>3 yr) male bears and 72% of adult females was human hunting. Yet, the combined mortality from other causes showed greater selectivity regarding sex and age than legal hunting, so their role in population dynamics and evolution should not be underestimated (Bischof et al. 2009, 2018). In Slovenia, from 1998 to 2008, Krofel et al. (2012) reported that 97% of the mortality of ~1000 dead bears was caused by humans, with legal hunts accounting for 59% of that. In smaller, southern European bear populations that are officially protected, human-caused mortality continues to be the most important threat (Ciucci and Boitani 2008), yet in some areas, ongoing bear recovery suggests that poaching has declined to some extent (Zarzo-Arias et al. 2018). Gantchoff et al. (2020) reviewed 31 studies that monitored the fate of 2630 American black bears and found that 77% of mortality was directly caused by humans.

African lions have lost over 90% of their historical range (Montgomery et al. 2018, Mweetwa et al. 2018), and today only ten populations, four in East Africa and six in Southern Africa, qualify as lion strongholds, with ~24,000 of the ~32,000 lions inhabiting the whole continent (Riggio et al. 2013). Furthermore, lion populations in West, Central, and East Africa are likely to suffer a projected 50% decline over the next two decades, with only Southern Africa providing a more promising scenario (Bauer et al. 2015). Consistent with other large felids (Inskip and Zimmermann 2009), the primary threats to lions are anthropogenic, including habitat loss and human encroachment, conflict with humans and livestock, poaching via snares, hunting, disease, and prey depletion (Mweetwa et al. 2018, Vinks et al. 2021). The world's only remaining Asiatic lion population (*Panthera leo persica*), in Gujarat (India), is much smaller, but has been increasing in recent decades (Meena et al. 2014), from 177 lions in 1968 to ~ 411 in 2010. In 2000–2010, recorded mortality (n = 88) of Asiatic lions was caused by natural (55%) and human-related (43%) factors, including a severe poaching episode, despite a historically positive attitude towards lions by the local community (Banerjee and Jhala 2012, Meena et al. 2014).

Mortality rates and their causes vary across the range of African lions. In Laikipia County, Kenya, which has a mixed of protected areas, conservancies, and village lands, ~90% of recorded lion mortality was caused by humans in 1998–2016 (Frank 2023). The sample included 337 lion deaths, of which 202 were monitored animals, and human-caused deaths accounted for 88% of these mortalities. Among the other 125 non-monitored lions, human-caused mortality accounted for 93.6% of the cases. The intensity of human-caused mortality was lower in some national parks in both Tanzania and Namibia (Woodroffe and Frank 2005), but not in others. For instance, all recorded mortalities (n = 5) in one year in a study in Mozambique were caused by humans, in an area with a minimum population estimate of 34 lions (Everatt et al. 2014). In Hwange National Park, Zimbabwe, Loveridge et al. (2017) reported that among 206 lions that died between 1999 and 2012, 88% of male and 67% of female mortalities were caused by human factors. Specifically, male mortality was mostly caused by trophy hunting, whereas female mortality resulted from snaring, retaliatory killing, and hunting. In Tanzania, an important country for the remaining stronghold populations of lion, legal trophy hunting seems to have been the main driver of a decline in lion abundance, thus exceeding sustainable levels (Packer et al. 2010). Similar concerns have been expressed for hunted lions in other countries, such as Cameroon (Croes et al. 2011) and Zimbabwe (Groom et al. 2014). Beyond protected areas, many of which are insufficient in size to support viable populations of lions, pastoralist landscapes are essential for lion survival (Frank 2023). Lion conservation in the future will thus remain fraught with conflict. In Masailand, East Africa, home to the largest contiguous lion population in Africa, lions outside protected areas are in danger of being extirpated by pastoralists (Winterbach et al. 2012). This creates a source-sink dynamic where intense levels of lion mortality at reserve boundaries can and even lead populations to local extinction (Loveridge et al. 2010).

In grey wolves, human-caused mortality has also been found to be between 70% and 80% of their overall mortality (Ballard et al. 1987, Murray et al. 2010, Sunde et al. 2021). Hill et al. (2022) reviewed cause-specific mortality records from 21 North American studies reporting an aggregate of 1442 wolf mortalities. Overall, human causes accounted for 61% of the wolf mortality, including illegal killing (23%), legal harvest (16%), and lethal management (12%) (Hill et al. 2022). Human-caused mortality is a major limiting factor of population growth rates and geographical expansion in several grey wolf populations, in North America (e.g. Gude et al. 2012, Ausband and Mech 2023) and in Europe (e.g. Wabakken et al. 2001, Kojola et al. 2009, Quevedo et al. 2019).

Illegal killing of carnivores can turn part of a range into a sink (Sunde et al. 2021) and/or slow population growth rates and reduce population sizes. For instance, the Scandinavian wolf population could have been almost four times as large in 2009 without poaching in the previous decade (Liberg et al. 2012), and very large fluctuations in the Finnish wolf population size seemed to be explained by poaching as a driving factor (Suutarinen and Kojola 2017).

Other demographic effects (changes to sex and age structure)

Wildlife managers often inform their practices and decisions based on information on the effects of direct mortality on a population's growth rate, ignoring indirect effects that may also alter population growth (e.g. Frank et al. 2017a). Furthermore, it has been argued that human causes of large carnivore mortality are often underestimated, which reinforces the role of human-caused mortality in large carnivore population dynamics (Bergstrom 2017).

Indirect effects of hunting are, among others, hunting-induced changes in a population, including human-induced selection, that result in an additive change to the population growth rate, 'lambda', beyond what is due to direct mortality (Frank et al. 2017a; see also Chapters 7 and 9). For instance, varying hunting pressure can affect the relative importance of demographic traits. In brown bears, fecundity of adult females is more important for population growth under low hunting pressure, and survival is more important under high hunting pressure (Gosselin et al. 2015). Thus, understanding the link between harvest and risk effects on the target populations is important for the management of large carnivores.

Harvest can affect the age and sex structure of large carnivore populations, depending on target selectivity, hunting pressure and method, among other factors. For example, the age and sex structure of different wolf populations differed as a result of human-caused mortality, in periods of heavy and low hunting pressure (Jedrzejewska et al. 1996, Sidorovich et al. 2007), and similar findings have been reported for American black bears, brown bears, and polar bears (*Ursus maritimus*) (McLellan and Shackleton 1988, Derocher et al. 1997, Noyce and Garshelis 1997). In brown bears, the hunting of adult females has been found to have the greatest negative effect on population growth rate (Knight and Eberhardt 1985, Gosselin et al. 2015). The opposite case has also been found, as population growth rate increases the most when breeding females are protected (in Sweden, females with dependent cubs cannot be hunted; Van de Walle et al. 2021). Hunting can also influence the age structure of large carnivore populations. Specifically, an underrepresentation of older animals in the population can occur, even if harvest is not selective to age (Bischof et al. 2008a). Therefore, beyond the fact that human-caused mortality typically accounts for most mortality in large carnivores, it can be additive to natural mortality, because hunting affects adults rather than juveniles (Milner et al. 2007, Bischof et al. 2009, 2018, Murray et al. 2010, Banerjee and Jhala 2012).

If female bears, for instance, are generally younger, population growth may be reduced, because females breeding for the first time (primiparous females) have smaller litter sizes and greater cub mortality than multiparous females (Gosselin et al. 2017). Human-caused mortality can also change the male age structure and influence age-specific reproductive rates, which might result in lower quality cubs and potentially lower recruitment. Younger bear males start having a larger contribution to recruitment than under natural conditions, where older, prime-aged individuals would do most of the breeding (Frank et al. 2017a).

Additive mortality, as the term implies, involves multiple mortality causes that are combined without mortality due to other causes diminishing. However, the consequences of human mortality on the demography of carnivores are challenging to quantify, as the reproductive status of the individuals at the time of death can play a key role in recruitment or the survival of dependent individuals (e.g. Brainerd et al. 2008). Furthermore, the population dynamic consequences of removing individuals can be larger than the sum of individual deaths when the removal causes indirect effects (Milner et al. 2007). For instance, the removal of dominant, breeding individuals can erode the social structure of populations, and this has been documented for both social species and even species that typically maintain solitary life histories, as we will describe in the following section.

Effects of hunting on the social structure of large carnivores

Human-caused mortality can also alter large carnivore social structure, as documented for canids, ursids, and felids (e.g. Rutledge et al. 2010, Frank et al. 2017a, b, and Loveridge et al. 2016, respectively). In grey wolves, the loss of breeding individuals is especially important and can lead to abandonment of territories, dissolution of pairs and packs, and smaller pack size (Brainerd et al. 2008, Milleret et al. 2017). Pup survival was higher in larger packs (>6 wolves) than in smaller ones, and the number of adult wolves remaining after breeder loss, together with pup age, had the greatest influence on pup survival (Brainerd et al. 2008). In Scandinavia, no single case of 'divorce' or 'abandonment' was observed among 153 identified wolf pairs, and when causes of pair dissolution were identified, death of one or both wolves was always involved, leading to a median pair bond duration of only two years (Milleret et al. 2017). This high dissolution rate suggests that anthropogenic factors limit wolf recolonization, which is in line with the finding that poaching prevented a much faster increase in the wolf population (Liberg et al. 2012). In contrast, reducing human-caused mortality restores the natural social structure of kin-based wolf packs, without a marked change in density, which led Rutledge et al. (2010) to suggest that conservation policy should consider effects of wolf harvesting beyond direct effects on population size.

In a variety of species, sexually selected infanticide (SSI) can occur when the mating success of a male increases after he kills dependent young that he has not sired, and SSI has been documented in both lions and bears (Swenson et al. 1997, Whitman et al. 2004). Among lions, adult sex ratios are more skewed towards females when hunting is intense, and even if adult males are typically targeted by trophy hunters, survival of all age and sex classes has been found to be lowest when male lion removals were highest (Loveridge et al. 2016). With high hunting pressure, frequent removal of male lions increases infanticide of dependent cubs, leading to low juvenile and subadult survival and also to the premature departure of subadults, resulting in higher disperser mortality (Loveridge et al. 2016). Likewise, sex structure is different in other protected versus hunted areas (e.g. Sogbohossou et al. 2014). Removal of males can have far-reaching effects, where male replacement is associated with infanticide. As trophy hunting of large carnivores tends to increase the rate of male removals, excessive hunting could limit recruitment through the negative effects of infanticide on cub survival. In lions, it seems that such effects can be avoided by targeting males of certain ages (Whitman et al. 2004).

Hunting brown bears, a more solitary and less territorial species than gregarious wolves or lions, can create empty areas in the landscape and neighbouring animals shift their home ranges, filling those gaps (Leclerc et al. 2017, Frank et al. 2017b). When 'new' males occupy those areas, increased SSI induces lower cub survival, with subsequent negative impacts on population growth (Swenson et al. 1997, Swenson 2003, Gosselin et al. 2015, 2017). To reduce SSI risk, females employ different strategies, including multimale mating and changes in habitat selection. These types of effects have been found to present reproductive costs, as found in bear populations in Canada and Sweden (Wielgus and Bunnell 2000, Frank et al. 2017a). SSI in brown bears has been documented elsewhere, e.g. in Italy (Davoli et al. 2018), Spain (Penteriani et al. 2020), and Japan (Ito et al. 2022). However, it appears to be uncommon in other areas (McLellan 2005). Therefore, bear hunting and the potential effects of SSI can act as an ecological trap, at least in some bear populations (Penteriani et al. 2018). Hunting also influences dispersal patterns in large carnivores, and thus the distribution of individuals, as shown for mountain lions (*Puma concolor*, Robinson et al. 2008) and bears (Gilroy et al. 2015).

Among social carnivores, reproductive success is dependent on the size, stability, and social cohesion of groups, as described earlier in terms of pup survival for wolves (Brainerd et al. 2008). Breeding female lions in large prides are able to defend cubs cooperatively, which can increase reproductive success. Synchronous breeding is more likely in prides with a stable coalition of males, which is dependent upon the intensity of hunting (Loveridge et al. 2007). The intensity of hunting affected both male and female home-range sizes, which declined in periods of low hunting, corresponding to increases in adult males and male coalitions (Loveridge et al. 2007). As in brown bears (Steyaert et al. 2013), female lions adapt their movements and habitat use to avoid potentially infanticidal males, which pose a larger threat when male turnover in prides is high (Loveridge et al. 2016).

Effects of hunting on evolutionary/selective traits

The selectivity of wildlife harvest, with special reference to trophy hunting, can cause genetic changes in wild populations, with negative evolutionary effects on targeted species (Coltman et al. 2003, Festa-Bianchet 2003, Loveridge et al. 2007, Allendorf et al. 2008, Darimont et al. 2009). Adult large carnivores experience low risk of predation by other animals, whereas hunting by humans typically targets adult individuals. North American hunters tend to select large male brown bears (McLellan and Shackleton 1988). This tendency could artificially select for smaller and less reproductively successful phenotypes (Frank et al. 2017a). In Scandinavia, where trophy hunting is not common, brown bear hunting also exerts a high degree of selective force. Whereas most natural mortality affects young bears, hunting targets single, adult bears, while a specific segment of the population is protected by regulation (females and their dependent cubs, Bischof et al. 2018).

The positive relationship between hunting mortality risk and age counters the natural pattern of reduced mortality during adulthood in large mammals. In this way, hunting is additive to the natural mortality of younger individuals, and the effects can influence demographic traits, social instability, altered life histories, and evolutionary changes (Bischof et al. 2018, van de Walle et al. 2021). Hunted populations may be stable or even growing under sustainable harvest practices, but changes in vital rates and effects on the influence of intrinsic and extrinsic determinants can alter the fitness consequences of life-history strategies, potentially driving evolutionary change (Bischof et al. 2018). Van de Walle et al. (2021) quantified and compared the impact of hunting intensity, hunting regulations, and expected hunting-induced selection on female reproductive traits among populations of Scandinavian brown bears. They concluded that, to predict population consequences of harvest-induced selection, it is essential to integrate both hunting intensity and regulations, particularly if hunting selectivity targets female reproductive strategies (van de Walle et al. 2021). Indeed, human persecution may already have shaped the life histories and behaviour of hunted large mammals, including European brown bears (Zedrosser et al. 2013). Long-term, human-caused selection may explain why European brown bears are less aggressive to people and invest more in reproduction in relation to female body mass than brown bears in North America, where the history of human persecution is much shorter (Zedrosser et al. 2011).

In the 1970s there were an estimated ~750 grey wolves in the contiguous USA, in northern Minnesota and on Isle Royale, Michigan; now, more than 6000 wolves are present in 11 states (Ausband and Mech 2023). In Europe, wolves are recolonizing parts of their former range, with an estimated ~12,000 wolves over 800,000 km^2 in 28 European countries (Hindrikson et al. 2017). This recovery is likely facilitated by a mixture of protection of wildland zones, socioeconomic changes, and recovery of wild ungulates. In human-dominated landscapes, however, wolves are exposed to direct and indirect human impacts that affect their ecology, behaviour, conservation, and evolution (Ciucci et al. 2020), with past genetic bottlenecks influencing present populations.

It is important to note that even in populations that are, or have recently been, recovering after severe bottlenecks, effective population size estimates reveal the historical signatures of large-scale efforts of predator extermination, despite decades of recovery efforts (vonHoldt et al. 2024). These authors studied the genomic estimates of effective population size along with diversity, subdivision, and inbreeding from 437 grey wolf samples from North America, collected between 1986 and 2021. vonHoldt et al. (2024) highlighted the importance of effective population size estimates for grey wolf conservation, which are ~5.2–9.3% of census estimates for the species. They showed that when grey wolf populations exceed the minimum effective population sizes needed to avoid extinction due to inbreeding depression in the short term, and wolf recovery is seen as a conservation success (Ausband and Mech 2023), they are still below population sizes predicted to avoid the long-term risk of extinction (vonHoldt et al. 2024). Likewise, the effective population size of Scandinavian wolves falls well below the advised conservation goals (Laikre et al. 2016), and a similar issue has recently been shown for wolves in Spain (Salado et al. 2022).

Iberian wolves (*Canis lupus signatus*) declined in range and population size in the last few centuries due to human persecution and habitat fragmentation, consistent with the pattern described earlier for large carnivores in Europe (Boitani 1995). However, wolves were never extinct in the Iberian Peninsula (Spain and Portugal). Wolf numbers increased from a historical low around 1970 and have stabilized in recent decades (Ordiz et al. 2022), but genetic diversity has declined in this period (Salado et al. 2022). Salado et al. compared historical and present mitochondrial diversity in Iberian wolves; only 6 of 10 mitochondrial DNA haplotypes observed in historical individuals were found in modern wolves. Furthermore, a haplotype that was restricted to southern Spain has gone extinct, where wolves are now officially gone as well. The study of Salado et al. (2022) illustrates a lag between demographic and genetic diversity changes, showing that, after severe population declines, genetic diversity of large carnivore populations can continue to be lost in stable or even expanding populations. Such populations may thus be of conservation concern even after their demographic trajectory has been reversed (Salado et al. 2022).

Debates about wolf management have been particularly intense in Spain. In 2021, wolves were granted a level of protection that prevented the use of hunting as a management tool at the country scale (Ordiz et al. 2022), a decision heavily discussed by regional governments. The discussion includes a debate over the historical (larger) versus present (severely reduced) distribution of the species. This is a crucial criterion to determine the level of protection according to Spanish law (Clavero et al. 2022a, b, Nores and López-Bao 2022). Recolonization of the historical Iberian wolf range is necessary to prevent loss of genetic diversity through drift or local extinction, issues that could be prevented by an increase in population size (Salado et al. 2022). Reducing lethal would improve their conservation status and favour transboundary and genetic connectivity (Quevedo et al. 2019). In summary, long-term persecution and the subsequent demographic history of wolves in both North America and Europe modulate their present genomic diversity, which should be considered, beyond mere population sizes and their recent trends, by management agencies (Chapron 2014, Laikre et al. 2016, 2022, vonHold et al. 2024, Salado et al. 2022).

Evolution sometimes can occur rapidly, over a few generations, and changes can be hard to reverse, which should cause managers to exhibit caution (Frank et al. 2017a). In Scandinavia, the protection of female brown bears and their dependent cubs (family groups) from hunting creates a strong link between survival and reproduction, increasing the influence of management on the life history of bears (Bischof et al. 2018). Mothers initiate the family break-up (Dahle and Swenson 2003) and most bear cubs are weaned during their second spring, but some litters stay with their mother for an extra year. The proportion of litters that remain with their mother for an extra year has increased during the past three decades; 12% during the first, 14% during the second, and 34% during the most recent decade, and this increase in interlitter interval reduces the rate of population growth rate. However, this change in the length of maternal care overlaps a period of pronounced growth of the Swedish bear population, so further study is needed to tease apart the potential effects of density dependence and hunting (Bischof et al. 2018). In any case, it is well established that life-history trade-offs and age-specific variation in demographic rates are crucial components of ecology and evolution (Stearns 1992), thus the potential of hunting to cause evolutionary effects should not be ignored (Stenseth and Dunlop 2009).

For species targeted by trophy hunters, the most sought-after individuals often have well-developed secondary sexual traits, such as large antlers, horns, tusks, or mane (Whitman et al. 2004, Mysterud 2011). Lions are hunted in up to 15 African countries (Lindsey et al. 2007) and males are often selected based on mane length and colour (Whitman et al. 2004). Manes and other sexually selected traits generally grow with age, but high-quality individuals may show early physical development and may thus be shot before they have bred. This, in turn, can have negative evolutionary consequences at the population and even species level (Coltman et al. 2003, Whitman et al. 2004).

Milner et al. (2007) suggested that, in response to the direct and indirect demographic effects and the evolutionary consequences of selective harvesting, hunting regimes should try to mimic natural mortality patterns more closely. However, this is rarely the case (e.g. see Fig. 3 in Bischof et al. 2018). It should also be noted that implementing compensatory harvest to mimic natural, age-selective mortality patterns may increase overall mortality, which can impose selection (Bischof et al. 2008b) and eventually reduce population size (Mysterud and Bischof 2010). Overall, a deeper understanding of the potential evolutionary effects of human-caused mortality of large carnivores warrants further research.

Effects of hunting and human disturbance on large carnivore behaviour

Research based on fine-scale empirical data over the last two decades shows that carnivores adjust their habitat use and activity patterns to avoid hunting and other human activities. Thus, behavioural responses of large carnivores living in human-dominated landscapes resemble those of prey under predation risk (Ordiz et al. 2011).

A variety of human activities affect brown bear behaviour in Scandinavia, where bear mortality is mostly human caused (Bischof et al. 2009). Human activities affect bear habitat selection at different temporal and spatial scales (Ordiz et al. 2017). Within this context, bears are even more nocturnal after hunting seasons start (Ordiz et al. 2012), after encounters with people (Ordiz et al. 2013b), and where road density is higher (Ordiz et al. 2014). At a very fine scale, bears seek concealment cover to avoid humans (Ordiz et al. 2011), particularly after encounters with people in the forest (Salhén et al. 2015). As for the bears, human-related factors affect grey wolf habitat selection at multiple spatial scales (Wam et al. 2012, Ordiz et al. 2015, Ciucci et al. 2018) and temporal patterns of activity, which in human-dominated landscapes is reduced during daytime (Ciucci et al. 1997, Milleret et al. 2018). The selection of breeding sites by wolves is affected more by the distribution of human infrastructure than by habitat itself in an area where they were hunted heavily (Jedrzejewska et al. 1996, Theuerkauf et al. 2003). Similarly, lions adjust foraging and space use to reduce encounters with humans, especially at times when the latter are most active (Valeix et al. 2012, Oriol-Coterill et al. 2015, Everatt et al. 2019).

Behavioural responses of individuals matter by themselves, because they may imply fitness costs (Ordiz et al. 2013b). Also, because individual behavioural responses can propagate to demography and population dynamics, understanding the behaviour of large carnivores exposed to human harvest and other human activities is relevant for conservation and management (Oriol-Coterill et al. 2015, Macdonald 2016). For instance, brown bear abundance is higher in less disturbed areas than in more disturbed places (Linke et al. 2013). Hunting can also affect the expression of some behaviours in surviving individuals. For example, harvest tends to favour social reorganization in some species, promoting male turnover, leading for instance to an increase of SSI, as explained before. In this context, Gosselin et al. (2015) documented that bear hunting in Scandinavia had negative indirect effects on the population through an increase in SSI, which lowered cub survival and possibly fecundity rates. As a consequence, SSI could explain 13.6% of the variation in population growth (Gosselin et al. 2015). This study illustrates the links among the behaviour of individuals, social structure of the species, and consequences on population growth, documenting how hunting can originate additional negative effects on the population. Therefore, they concluded that all these behavioural and demographic factors should be considered when establishing harvest quotas and policies (Gosselin et al. 2015). Similar findings, linking individual behavioural responses to potential effects in terms of population growth, have been suggested for lions and for other brown bear populations (Milner et al. 2007, Proctor et al. 2023).

Concluding remarks

The intertwined nature of direct and indirect effects of large carnivore hunting

The direct and indirect effects of large carnivore hunting that we have described are interconnected (Figure 4.1). The removal of individuals can affect

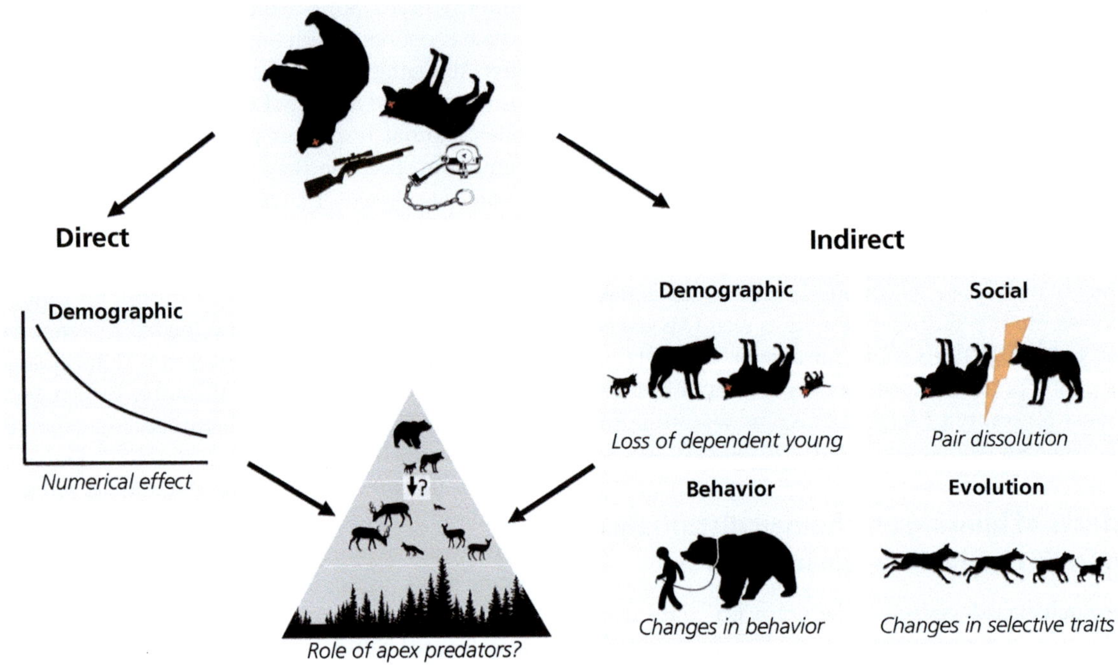

Figure 4.1 Summary of direct (lethal) and indirect, risk effects of hunting large carnivores for these species and potentially for their ecological function in nature.

Credit of the silhouette used in the Evolution box: Robert Adrian Hillman/Shutterstock.

population sizes, and the responses of surviving individuals to hunting can also lead to a complex cascade of hunting-caused changes that reverberate at the population level. For instance, hunting (selective hunting in particular, but not only) can cause demographic changes in terms of age and sex classes and growth rate, usually implying changes in the social structure and individual behaviours. This in turn can affect the demographic structure of populations. For example, the hunting of male lions and male brown bears can reduce offspring survival via infanticide, as described earlier, which can negatively affect population trends of both species. Likewise, removal of breeding wolves alters the social structure of wolf populations and pup survival. This demonstrates the ways in which direct and indirect effects can ultimately scale up to have implications at the population level for both gregarious species and for those typically considered as 'solitary'. Careful study of these effects can yield critically important information for managers, which should have a responsibility to take these complex interactions into account when attempting to keep a population at target levels that are compatible with long-term conservation goals.

Effects of large carnivore demographic and behavioural responses at the ecosystem level

The position of large carnivore species such as bears, lions, and wolves, among others, in the apex predator level of trophic systems illustrates that changes in their populations can alter multiple natural processes in lower trophic levels (Bergstrom et al. 2009). Brown bears, for example, influence both the dispersal (Ordiz et al. 2015) and predation patterns of wolves (Tallian et al. 2017, 2022), which in turn affect prey demography (Tallian et al. 2021) and behaviour (Sahlén et al. 2016). However, the demographic and behavioural responses of large carnivores to human activity, including hunting, affect those interactive effects with other species (Ordiz et al. 2013a, 2021). The most obvious human effect on apex predator function is arguably the direct reduction of carnivore numbers. A reduced predator population may not be able to numerically control populations of prey species and/or mesopredators, whose release leads to biodiversity loss. Also, the 'landscape of fear' where risk effects may be induced by predators on prey can also occur via hunting on large carnivores (Ordiz et al. 2011, 2012, Valeix et al. 2012, Oriol-Coterrill et al. 2015). Variation in the social structure and behaviour of predators in relation to human harvest and other activities can thus limit their effects on ecosystems, because their own habitat selection and temporal activity patterns are constrained in human-dominated landscapes (Tallian et al. 2023). We suggest that assessing and eventually restoring the apex predator function of large carnivores is a key topic in conservation (Brook et al. 2012, Ritchie et al. 2012, Ordiz et al. 2013a, 2021).

One of the reasons why large carnivores have historically been persecuted by humans is because they are seen as a competitor for prey, both wild and domesticated (Frank and Woodroffe 2001). The depletion of wild prey, in turn, presents a major threat to many large carnivore species worldwide (Wolf and Ripple 2016). Furthermore, the spatial and temporal avoidance of humans by large carnivores creates human shields for prey, as documented in different ecosystems (Hebblewhite et al. 2005, Berger 2007, Muhly et al. 2013, Rogala et al. 2011). On the one hand, spatial and temporal separation of predators and humans (i.e. predators avoiding human infrastructure, especially during daytime, and being more active at night) can facilitate coexistence (Oriol-Coterill et al. 2015), but it also alters natural interactions, among other effects (Gaynor et al. 2018); see also Chapters 12 and 14.

Overabundant herbivores occur in areas where they are released from predation risk (e.g. Hebblewhite et al. 2005) (i.e. ungulates become less accessible to carnivores because the latter are excluded from human settlements and their immediate surroundings). For instance, both grey wolves and brown bears prey heavily on moose (*Alces alces*) in Scandinavia and other ecosystems (Tallian et al. 2022). In turn, moose and other ungulates may modulate their predation risk by using human infrastructures as a shield, areas that predators tend to avoid (Berger 2007, Muhly et al. 2013). This is therefore an example of how large carnivore behavioural responses to avoid human activity and infrastructure affect lower trophic levels.

Large carnivores playing an apex predator role were ubiquitous across the planet for millions of years, and recent research reveals extensive cascading effects globally due to their disappearance and long-term persecution in aquatic and terrestrial ecosystems alike (Terborgh et al. 2010, Estes et al. 2011). Many species continue to be at risk of extinction (Ripple et al. 2014), but some are partially recovering portions of their former ranges (e.g. Cimatti et al. 2021, Ingeman et al. 2022, Ausband and Mech 2023). In this chapter we have reviewed some of the knowledge on the direct and indirect effects of human hunting pressure on large carnivores, focusing specifically on the implications for demography, behaviour, evolution, and ecosystem function. Such information could be useful for carnivore conservation, not only at the species level, but also to preserve carnivore guilds and the wider communities they are part of.

In this chapter we have not discussed the human-dimensions component (see Chapter 13) that, alongside ecology, is central to large carnivore conservation. Lack of communication and engagement among stakeholders in the research, conservation, and management spheres presents a major challenge to conservation (Knight et al. 2008). Management interventions can only be effective if they are actually implemented (e.g. Jonzén et al. 2013), and acceptance by interested stakeholders is important for consensus and implementation (Eklund et al. 2020). In other words, carnivore conservation and management in human-dominated landscapes require social and political inputs to be effective (Redpath et al. 2013). Nevertheless, the gap between accumulating science on the role of carnivores in ecosystems and on the effects that humans (via hunting and other activities) can have on them and management seems to be growing. At the moment, increasing pressure on several large carnivore populations is apparently more rooted in political than in ecological issues (Ripple et al. 2019, Laikre et al. 2022, von Hohenberg and Hager 2022, Ordiz et al. 2024).

References

Allendorf, F. W., P. R. England, G. Luikart, P. A. Ritchie, and N. Ryman. 2008. Genetic effects of harvest on wild animal populations. *Trends in Ecology and Evolution* 23:327–337.

Ausband, D. E and L. D. Mech. 2023. The challenges of success: future wolf conservation and management in the United States. *BioScience* 73:587–591.

Ballard, W. B., J. S. Whitman, and C. L. Gardner. 1987. Ecology of an exploited wolf population in South-Central Alaska. *Wildlife Monographs* 98:1.

Banerjee, K. and Y. V. Jhala. 2012. Demographic parameters of endangered Asiatic lions (*Panthera leo persica*) in Gir forests, India. *Journal of Mammalogy* 93:1420–1430.

Bauer, H., et al. 2015. Lion (*Panthera leo*) populations are declining rapidly across Africa, except in intensively managed areas. *Proceedings of the National Academy of Sciences* 112:14894–14899.

Benn, B. and S. Herrero. 2002. Grizzly bear mortality and human access in Banff and Yoho National Parks, 1971–98. *Ursus* 13:213–221.

Berger, J. 2007. Fear, human shields and the redistribution of prey and predators in protected areas. *Biology Letters* 3: 620–623.

Bergstrom, B. J. 2017. Carnivore conservation: shifting the paradigm from control to coexistence. *Journal of Mammalogy* 98:1–6.

Bergstrom, B. J., S. Vignieri, S. R. Sheffield, W. E. S. Sechrest, and A. A. Carlson. 2009. The Northern Rocky Mountain graywolf is not yet recovered. *BioScience* 59:991–999.

Bischof, R., A. Fujita., A. Zedrosser, A. Soderberg, and J. E. Swenson. 2008a. Hunting patterns, ban on baiting, and harvest demographics of brown bears in Sweden. *Journal of Wildlife Management* 72:79–88.

Bischof, R., A. Mysterud, and J. E. Swenson. 2008b. Should hunting mortality mimic the patterns of natural mortality? *Biology Letters* 4:307–310.

Bischof, R., J. E. Swenson, N. G. Yoccoz, A. Mysterud, and O. Gimenez. 2009. The magnitude and selectivity of natural and multiple anthropogenic mortality causes in hunted brown bears. *Journal of Animal Ecology* 78:656–665.

Bischof, R., E. B. Nilsen, H. Broseth, P. Mannil, J. Ozolin, and J. D. C. Linnell. 2012. Implementation uncertainty when using recreational hunting to manage carnivores. *Journal of Applied Ecology* 49:824–832.

Bischof, R., C. Bonenfant, I. M. Rivrud, A. Zedrosser, A. Friebe, T. Coulson, A. Mysterud and J.E. Swenson. 2018. Regulated hunting re-shapes the life history of brown bears. *Nature Ecology and Evolution* 2:116–123.

Boitani, L. 1995. Ecological and cultural diversities in the evolution of wolf human relationships. In *Ecology and Conservation of Wolves in a Changing World*, edited by L. N. Carbyn, S. H. Fritts, and D. R. Seip. Canadian Circumpolar Institute, Alberta, Canada.

Brainerd, S. M., et al. 2008. The effects of breeder loss on wolves. *The Journal of Wildlife Management* 72:89–98.

Breitenmoser, U. 1998. Large predators in the Alps: the fall and rise of man's competitors. *Biological Conservation* 83:279–289.

Brook, L., C. N. Johnson, and E. Ritchie. 2012. Effects of predator control on behaviour of an apex predator and indirect consequences for mesopredator suppression. *Journal of Applied Ecology* 49:1278–1286.

Bruskotter, J. T. and L. B. Shelby. 2010. Human dimensions of large carnivore conservation and management. *Human Dimensions of Wildlife* 15:311–314.

Cimatti, M., et al. 2021. Large carnivore expansion in Europe is associated with human population density and land cover changes. *Diversity and Distributions* 27:602–617.

Ciucci, P., L. Boitani, F. Francisci, and G. Andreoli. 1997. Home range, activity and movements of a wolf pack in Central Italy. *Journal of Zoology* 243:803–819.

Ciucci, P., S. Mancinelli, L. Boitani, O. Gallo, and L. Grottoli. 2020. Anthropogenic food subsidies hinder the ecological role of wolves: Insights for conservation of apex predators in human-modified landscapes. *Global Ecology and Conservation* 21:e00841.

Ciucci, P. and L. Boitani. 2008. The Apennine brown bear: a critical review of its status and conservation problems. *Ursus* 19:130–145.

Ciucci, P., L. Boitani, M. Falco, and L. Maiorano. 2018. Hierarchical, multi-grain rendezvous site selection by wolves in Southern Italy. *The Journal of Wildlife Management* 82:1049–1061.

Clavero, M., A. García-Reyes, A. Fernández-Gil, E. Revilla, and N. Fernández. 2022a. Where wolves were: setting historical baselines for wolf recovery in Spain. *Animal Conservation* 26:239–249.

Clavero, M., A. García-Reyes, A. Fernández-Gil, E. Revilla, and N. Fernández. 2022b. On the misuse of historical data to set conservation baselines: wolves in Spain as an example. *Biological Conservation* 276:109810.

Coltman, D., P. O'Donoghue, J. Jorgenson, J. T. Hogg, C. Strobeck, and M. Festa-Bianchet. 2003. Undesirable evolutionary consequences of trophy hunting. *Nature* 426:655–658.

Chapron, G. 2014. Challenge the abuse of science in setting policy. *Nature* 516:289.

Chapron, G., et al. 2014. Recovery of large carnivores in Europe's modern human-dominated landscapes. *Science* 346:1517–1519.

Chapron, G. and A. Treves. 2016. Blood does not buy goodwill: allowing culling increases poaching of a large carnivore. *Proceedings of the Royal Society Biological Series* 283:20152939

Croes, B. M., P. J. Funston, G. Rasmussen, R. Buij, A. Saleh, P. N. Tumenta, and H. H. De Iongh. 2011. The impact of trophy hunting on lions (*Panthera leo*) and other large carnivores in the Bénoué Complex, northern Cameroon. *Biological Conservation* 144:3064–3072.

Dahle, B. and J. E. Swenson. 2003. Family break-up in brown bears: are young forced to leave? *Journal of Mammalogy* 84:536–540.

Davoli, F., M. Cozzo, F. Angeli, C. Groff, and E. Randi. 2018. Infanticide in brown bear: a case-study in the Italian Alps—genetic identification of perpetrator and implications in small populations. *Nature Conservation* 25:55–75.

Darimont, C. T., S. M. Carlson, M. T. Kinnison, P. C. Paquet, T. E. Reimchen, and C. C. Wilmers. 2009. Human predators outpace other agents of trait change in the wild. *Proceedings of the National Academy of Sciences* 106:952–954.

Derocher, A. E., I. Stirling, and W. Calvert. 1997. Male-biased harvesting of polar bears in western Hudson Bay. *Journal of Wildlife Management* 61:1075–1082.

Eklund, A., M. Johansson, A. Flykt, H. Andrén, and J. Frank. 2020. Believed effect—a prerequisite but not a guarantee for acceptance of carnivore management interventions. *Biological Conservation* 241:108251.

Estes, J. A., et al. 2011. Trophic downgrading of planet Earth. *Science* 333:301–306.

Everatt, K. T., L. Andresen, and M. J. Somers. 2014. Trophic scaling and occupancy analysis reveals a lion population limited by top-down anthropogenic pressure in the Limpopo National Park, Mozambique. *PLoS One* 9:e99389.

Everatt, K. T., J. F. Moore, and G. I. H. Kerley. 2019. Africa's apex predator, the lion, is limited by interference and exploitative competition with humans. *Global Ecology and Conservation* 20:e00758.

Fernández-Gil, A., J. Naves, A. Ordiz, M. Quevedo, E. Revilla, and M. Delibes. 2016. Conflict misleads large carnivore management and conservation: brown bears and wolves in Spain. *PLoS One* 11:e0151541.

Festa-Bianchet, M. 2003. Exploitative wildlife management as a selective pressure for the life-history evolution of large mammals. In *Animal Behavior and Wildlife Conservation*, edited by M. Festa-Bianchet and M. Apollonio. Island Press, Washington, D. C.

Frank, L. G. 2023. Twenty years of lion conservation in a commercial rangeland. *Frontiers in Ecology and Evolution* 11:1141195.

Frank, L. G., and R. Woodroffe. 2001. Behaviour of carnivores in controlled and exploited populations. In *Carnivore Conservation*, edited by J. L. Gittleman, R. K. Wayne, D. W. Macdonald, S. M Funk. Cambridge University Press, Cambridge, U. K.

Frank, S. C., et al. 2017a. Indirect effects of bear hunting: a review from Scandinavia. *Ursus* 28:150–164.

Frank, S. C., et al. 2017b. Sociodemographic factors modulate the spatial response of brown bears to vacancies created by hunting. *Journal of Animal Ecology* 87:1–12.

Fryxell, J. M., A. R. E. Sinclair, and G. Caughley. 2014. *Wildlife Ecology, Conservation, and Management*. Third edition. John Wiley & Sons Limited, West Sussex, UK.

Gantchoff, M. G., et al. 2020. Mortality of a large wide-ranging mammal largely caused by anthropogenic activities. *Scientific Reports* 10:1–11.

Gaynor, K. M., C. E. Hojnowski, N. H. Carter, and J. S. Brashares. 2018. The influence of human disturbance on wildlife nocturnality. *Science* 1235:1232–1235.

Gilroy, J. J., A. Ordiz, and R. Bischof. 2015. Carnivore coexistence: value the wilderness. *Science* 347:382.

Gosselin, J., A. Zedrosser, and J. E. Swenson. 2015. The relative importance of direct and indirect effects of hunting mortality on the population dynamics of brown bears. *Proceedings of the Royal Society Biological Series* 282:20141840.

Gosselin, J., M. Leclerc, A. Zedrosser, S. Steyaert, J. E. Swenson, and F. Pelletier. 2017. Hunting promotes sexual conflict in brown bears. *Journal of Animal Ecology* 86:35–42.

Groom R. J., P. J. Funston, and R. Mandisodza. 2014. Surveys of lions *Panthera leo* in protected areas in Zimbabwe yield disturbing results: what is driving the population collapse? *Oryx* 48:385–393.

Gude, J. A., et al. 2012. Wolf population dynamics in the US Northern Rocky Mountains are affected by recruitment and human-caused mortality. *Journal of Wildlife Management* 76:108–118.

Hebblewhite, M., et al.2005. Human activity mediates a trophic cascade caused by wolves. *Ecology* 86:2135–2144.

Hill, J.E., et al. 2022. Quantifying anthropogenic wolf mortality in relation to hunting regulations and landscape attributes across North America. *Ecology and Evolution* 12:e8875

Hindrikson, M., et al. 2017. Wolf population genetics in Europe: a systematic review, meta-analysis and suggestions for conservation and management. *Biological Reviews* 92:1601–1629.

Ingeman, K. E., et al. 2022. Glimmers of hope in large carnivore recoveries. *Scientific Reports* 12:10005

Inskip, C. and A. Zimmermann. 2009. Human-felid conflict: a review of patterns and priorities worldwide. *Oryx* 43:18–34.

Ito, T., H. Katsushima, K. M. Tomita, and T. Matsumoto. 2022 Infanticide or predation? Cannibalism by a brown bear in Hokkaido, Japan. *Ursus* 33(13):1–5.

Jedrzejewska, B., W. Jedrzejewski, A. N. Bunevich, L. Milkowski, and H. Okarma. 1996. Population dynamics of wolves Canis lupus in Bialowieza primeval forest (Poland and Belarus) in relation to hunting by humans, 1847–1993. *Mammal Review* 26:103–126.

Jonzén, N., H. Sand, P. Wabakken, J. E. Swenson, J. Kindberg, O. Liberg, and G. Chapron. 2013. Sharing the bounty—adjusting harvest to predator return in the Scandinavian human–wolf–bear–moose system. *Ecological Modelling* 265:140–148.

Knight, R. R., and L. L. Eberhardt. 1985. Population dynamics of Yellowstone grizzly bears. *Ecology* 66: 323–334.

Knight, A.T., R.M. Cowling, M. Rouget, A. Balmford, A.T. Lombard, and B.M. Campbell. 2008. Knowing but not doing: selecting priority conservation areas and the research-implementation gap. *Conservation Biology* 22: 610–617.

Kojola, I., et al. 2009. Dispersal behavior and the connectivity between wolf populations in Northern Europe. *Journal of Wildlife Management* 73:309–313.

Krofel, M., M. Jonotovic, and K. Jerina. 2012. Demography and mortality patterns of removed brown bears in a heavily exploited population. *Ursus* 23:91–103.

Laikre, L., F. Olsson, E. Jansson, O. Hössjer, and N. Ryman. 2016. Metapopulation effective size and conservation genetic goals for the Fennoscandian wolf (*Canis lupus*) population. *Heredity* 117:279–289.

Laikre, L., et al. 2022. Planned cull endangers Swedish wolf population. *Science* 377:162.

Leclerc, M., S. Frank, A. Zedrosser, J. E. Swenson, and F. Pelletier. 2017. Hunting promotes spatial reorganization and sexually selected infanticide. *Scientific Reports* 7:45222.

Liberg, O., A. Aronson, H. Sand, P. Wabakken, E. Maartmann, L. Svensson, and M. Åkesson. 2012. Monitoring of wolves in Scandinavia. *Hystrix* 23:29–34.

Lindsey, P. A., P. A. Roulet, and S. S. Romañach. 2007. Economic and conservation significance of the trophy hunting industry in sub-Saharan Africa. *Biological Conservation* 134:455–469.

Linnell, J. D. C., J. E. Swenson, and R. Andersen. 2001. Predators and people: conservation of large carnivores is possible at high human densities if management policy is favourable. *Animal Conservation* 4: 345–349.

Linnell, J. D. C., C. Promberger, L. Boitani, J. E. Swenson, U. Breitenmoser, and R. Andersen. 2005. The linkage between conservation strategies for large carnivores and biodiversity: the view from the 'half-full' forests of Europe. In *Large Carnivores and the Conservation of Biodiversity*, edited by J. C. Ray, K. H. Redford, R. Steneck, and J. Berger. Island Press, Washington, D. C.

Linke, J., G. J. McDermid, M. J. Fortin, and G. B. Stenhouse. 2013. Relationships between grizzly bears and human disturbances in a rapidly changing multi-use forest landscape. *Biological Conservation* 166:54–63.

Loveridge, A., A. Searle, F. Murindagomo, and D. Macdonald. 2007. The impact of sport-hunting on the population dynamics of an African lion population in a protected area. *Biological Conservation* 134:548–558.

Loveridge, A. J., G. Hemson, Z. Davidson, and D. W. Macdonald. 2010. African lions on the edge: reserve boundaries as 'attractive sinks'. In *The Biology and Conservation of Wild Felids*, edited by D. W. Macdonald and A. J. Loveridge. Oxford University Press, Oxford.

Loveridge, A. J., M. Valeix, G. Chapron, Z. Davidson, G. Mtare, and D. W. Macdonald. 2016. Conservation of large predator populations: demographic and spatial responses of African lions to the intensity of trophy hunting. *Biological Conservation* 204:247–254.

Loveridge, A. J., M. Valeix, N. B. Elliot, and D. W. Macdonald. 2017. The landscape of anthropogenic mortality: how African lions respond to spatial variation in risk. *Journal of Applied Ecology* 54:815–825.

Macdonald, D. W. 2016. Animal behaviour and its role in carnivore conservation: examples of seven deadly threats. *Animal Behavior* 120:197–209.

Mattson, D. J., B. M. Blanchard, and R. R. Knight. 1992. Yellowstone grizzly bear mortality, human habituation, and whitebark pine seed crops. *Journal of Wildlife Management* 56:432.

Mattson, D. J. 1998. Changes in mortality of Yellowstone's grizzly bears. *Ursus* 10:129–138.

McLellan, B. N. 2005. Sexually selected infanticide in grizzly bears: the effects of hunting on cub survival. *Ursus* 16:141–156.

McLellan, B. N, and D. M. Shackleton. 1988. A comparison of grizzly bear harvest data fromMontana and southeastern British Columbia. *Wildlife Society Bulletin* 16:371–375.

McLellan, B. N., et al. 1999. Rates and causes of grizzly bear mortality in the interior mountains of British Columbia, Alberta, Montana, Washington, and Idaho. *Journal of Wildlife Management* 63:911–920.

McLellan, B. N. 2015. Some mechanisms underlying variation in vital rates of grizzly bears on a multiple use landscape. *Journal of Wildlife Management* 79:749–765.

Meena, V., D. W. Macdonald, and R. A. Montgomery. 2014. Managing success: Asiatic lion conservation, interface problems and peoples' perceptions in the Gir Protected Area. *Biological Conservation* 174:120–126.

Milleret, C., et al. 2017. Let's stay together? Intrinsic and extrinsic factors involved in pair bond dissolution in a recolonizing wolf population. *Journal of Animal Ecology* 86:43–54.

Milleret, C., et al. 2018. Habitat segregation between brown bears and gray wolves in a human-dominated landscape. *Ecology and Evolution* 8:1–17.

Milner, J. M., E. B. Nilsen, and H. Andreassen. 2007. Demographic side effects of selective hunting in ungulates and carnivores. *Conservation Biology* 21:36–47.

Miquelle, D., et al. 2005. Tigers and wolves in the Russian Far East: competitive exclusion, functional redundancy, and conservation implications. In *Large Carnivores and the Conservation of Biodiversity*, edited by J. C. Ray, K. H. Redford, R. Steneck, and J. Berger. Island Press, Washington, D. C.

Moqanaki, E., C. Milleret, P. Dupont, H. Broseth, and R. Bischof. 2023. Wolverine density distribution reflects past persecution and current management in Scandinavia. *Ecography* 2023:e06689

Montgomery, R. A., et al. 2018. Examining evident interdisciplinarity among prides of lion researchers. *Frontiers in Ecology and Evolution* 6:2018.

Muhly, T. B., M. Hebblewhite, D. Paton, J. A. Pitt, M. S. Boyce, and M. Musiani. 2013. Humans strengthen bottom-up effects and weaken trophic cascades in a terrestrial food web. *PLoS One* 8:e64311.

Murray, D. L., et al. 2010. Death from anthropogenic causes is partially compensatory in recovering wolf populations. *Biological Conservation* 143:2514–2524.

Mweetwa, T., et al. 2018. Quantifying lion (*Panthera leo*) demographic response following a three-year moratorium on trophy hunting. *PLoS One* 13:e0197030.

Mysterud, A. 2011. Selective harvesting of large mammals: how often does it result in directional selection? *Journal of Applied Ecology* 48:827–834.

Mysterud, A. and R. Bischof. 2010. Can compensatory culling offset undesirable evolutionary consequences of trophy hunting? *Journal of Animal Ecology* 79:148–160.

Nores, C. and J. V. López-Bao. 2022. Historical data to inform the legal status of species in Europe: an example with wolves. *Biological Conservation* 272:109639.

Noyce, K. V. and D. L. Garshelis. 1997. Influence of natural food abundance on black bear harvests in Minnesota. *Journal of Wildlife Management* 61:1067–1074.

Ordiz, A., M. Delibes, O.-G. Støen, and J. E. Swenson. 2011. Predators or prey? Spatio-temporal discrimination of human-derived risk by brown bears. *Oecologia* 166:59–67.

Ordiz, A., O.-G. Støen, S. Sæbø, J. Kindberg, M. Delibes, and J. E. Swenson. 2012. Do bears know they are being hunted? *Biological Conservation* 152:21–28.

Ordiz, A., R. Bischof, and J. E. Swenson. 2013a. Saving large carnivores, but losing the apex predator? *Biological Conservation* 168:128–133.

Ordiz, A., et al. 2013b. Lasting behavioural responses of brown bears to experimental encounters with humans. *Journal of Applied Ecology* 50:306–314.

Ordiz, A., J. Kindberg, S. Sæbø, J. E. Swenson, and O.-G. Støen. 2014. Brown bear circadian behavior reveals human environmental encroachment. *Biological Conservation* 173:1–9.

Ordiz, A., C. Milleret, J. Kindberg, J. Mansson, P. Wabakken, J. E. Swenson, and H. Sand. 2015 Wolves, people, and brown bears influence the expansion of the recolonizing wolf population in Scandinavia. *Ecosphere* 6:1–14.

Ordiz, A., O.-G. Støen, M. Delibes, and J. E. Swenson. 2017. Staying cool or staying safe in a human-dominated landscape: which is more relevant for brown bears? *Oecologia* 185:191–194.

Ordiz, A., M. Aronsson, J. Persson, O.-G. Støen, J. E. Swenson, and J. Kindberg. 2021. Effects of human disturbance on terrestrial apex predators. *Diversity* 13:68.

Ordiz, A., D. Canestrari, and J. Echegaray. 2022. Wolf conservation and management in Spain, an open debate. *Frontiers in Environmental Sciences* 10:2022.

Ordiz, A., D. Canestrari, and J. Echegaray. 2024. Large carnivore management at odds: Science or prejudice? *Global Ecology and Conservation* 54:e03202.

Oriol-Cotterill, A., M. Valeix, L. G. Frank, C. Riginos, and D. W. Macdonald. 2015. Landscapes of coexistence for terrestrial carnivores: the ecological consequences of being downgraded from ultimate to penultimate predator by humans. *Oikos* 124:1263–1273.

Packer, C., et al. 2009. Sport hunting, predator control and conservation of large carnivores. *PLoS One* 4: e5941

Packer, C., H. Brink, B. M. Kissui, H. Maliti, H. Kushnir, and T. Caro. 2010. Effects of trophy hunting on lion and leopard populations in Tanzania. *Conservation Biology* 25:142–153.

Penteriani, V., et al. 2018. Evolutionary and ecological traps for brown bears Ursus arctos in human-modified landscapes. *Mammal Review* 48:180–193.

Penteriani, V., et al. 2020. Female brown bears use areas with infanticide risk in a spatially confined population. *Ursus* 31:1–9.

Pepin, K. M., S. L. Kay, and A. J. Davis. 2017. Comment on: 'Blood does not buy goodwill: allowing culling increases poaching of a large carnivore'. *Proceedings of the Royal Society Biological Series* 284:20161459.

Proctor, M., et al. 2023. Berries and bullets: influence of food and mortality risk on grizzly bears in British Columbia. *Wildlife Monographs* 2023:e1078.

Prowse, T. A., et al. 2013. No need for disease: testing extinction hypotheses for the thylacine using multi-species metamodels. *Journal of Animal Ecology* 82:355–364.

Quevedo, M., et al. 2019. Lethal management may hinder population recovery in Iberian wolves. *Biodiversity and Conservation* 28:415–432.

Rauset, G. R., et al. 2016. National parks in northern Sweden as refuges for illegal killing of large carnivores. *Conservation Letters* 9:334–341.

Redpath, S. M., et al. 2013. Understanding and managing conservation conflicts. *Trends in Ecology and Evolution* 28:100–109.

Riggio, J., et al. 2013. The size of savannah Africa: a lion's (Panthera leo) view. *Biodiversity and Conservation* 22: 17–35.

Ripple, W. J., et al. 2014. Status and ecological effects of the world's largest carnivores. *Science* 343:1241484.

Ripple, W. J., S. D. Miller, J. W. Schoen, and S. P. Rabinowitch. 2019. Large carnivores under assault in Alaska. *PLoS Biology* 17:e3000090.

Ritchie, E. G., B. Elmhagen, A. S. Glen, M. Letnic, G. Ludwig, and D. A. McDonald. 2012. Ecosystem restoration with teeth: what role for predators? *Trends in Ecology and Evolution* 27:265–271.

Rogala, J. K., M. Hebblewhite, J. Whittington, C. A. White, J. Coleshill, and M. Musiani. 2011. Human activity differentially redistributes large mammals in the Canadian Rockies national parks. *Ecology and Society* 16:16.

Robinson, H. S., R. B. Wielgus, H. S. Cooley, and S. W. Cooley. 2008. Sink populations in carnivore management: cougar demography and immigration in a hunted population. *Ecological Applications* 18:1028–1037.

Rutledge, L. Y., B. R. Patterson, K. J. Mills, K. M. Loveless, D. L. Murray, and B. N. White. 2010. Protection from harvesting restores the natural social structure of eastern wolf packs. *Biological Conservation* 143:332–339.

Sahlén, V., A. Ordiz, J. E. Swenson, and O.-G. Støen. 2015. Behavioural differences between single Scandinavian brown bears (*Ursus arctos*) and females with dependent young when experimentally approached by humans. *PLoS One* 10:e0121576.

Sahlén, E., S. Noell, C. S. Deperno, J. Kindberg, G. Spong, and J. P. Cromsigt. 2016. Phantoms of the forest: legacy risk effects of a regionally extinct large carnivore. *Ecology and Evolution* 6:791–799.

Salado, I., M. Preick, N. Lupiáñez-Corpas, A. Fernández-Gil, C. Vilà, M. Hofreiter, and J. A. Leonard. 2022. Loss of mitochondrial genetic diversity despite population growth: the legacy of pastwolf population declines. *Genes* 14:75.

Servheen, C., S. Herrero, and B. Peyton. 1999. *Bears: Status Survey and Conservation Action Plan*. IUCN/SSC Bear and Polar Bear Specialists Group; IUCN Publications Unit, Cambridge, U. K.

Sidorovich, V. E., V. P. Stolyarov, N. N. Vorobei, N. V. Ivanova, and B. Jedrzejewska. 2007. Litter size, sex ratio, and age structure of gray wolves, *Canis lupus*, in relation to population fluctuations in northern Belarus. *Canadian Journal of Zoology* 85:295–300.

Sogbohossou, E. A., et al. 2014. Social structure of lions (*Panthera leo*) is affected by management in Pendjari Biosphere Reserve, Benin. *PLoS One* 9:e84674.

Sollund, R. and D. R. Goyes. 2021. State-organized crime and the killing of wolves in Norway. *Trends in Organized Crime* 24:467–484

Stearns, S. C. 1992. *The Evolution of Life Histories*. Oxford University Press, Oxford, U. K.

Stenseth, N. C. and E. S. Dunlop. 2009. Unnatural selection. *Nature* 457:803–804.

Steyaert, S. M. J. G., J. Kindberg, J. E. Swenson, and A. Zedrosser. 2013. Male reproductive strategy explains spatiotemporal segregation in brown bears. *Journal of Animal Ecology* 82:836e845.

Sunde, P., et al. 2021. Where have all the young wolves gone? Traffic and cryptic mortality create a wolf population sink in Denmark and northernmost Germany. *Conservation Letters* 14:1–10.

Suutarinen, J. and I. Kojola. 2017. Poaching regulates the legally hunted wolf population in Finland. *Biological Conservation* 215:11–18.

Swenson, J. E. 2003. Implications of sexually selected infanticide for the hunting of large carnivores. In *Animal Behavior and Wildlife Conservation*, edited by M. Festa-Bianchet and M. Apollonio. Island Press, Washington, D. C.

Swenson, J. E., F. Sandegren, A. Söderberg, A. Bjärvall, R. Franzén, and P. Wabakken. 1997. Infanticide caused by hunting of male bears. *Nature* 386:450–451.

Swenson, J. E., M. Schneider, A. Zedrosser, A. Söderberg, R. Franzén, R., and J. Kindberg. 2017. Challenges of managing a European brown bear population; lessons from Sweden, 1943–2013. *Wildlife Biology* 2017: 1943–2013.

Tallian, A., et al. 2017. Competition between apex predators? Brown bears decrease wolf kill rate on two continents. *Proceedings of the Royal Society Biological Sciences* 284:20162368.

Tallian, A., et al. 2021. The return of large carnivores: Using hunter observation data to understand the role of predators on ungulate populations. *Global Ecology and Conservation* 27:e01587.

Tallian, A., et al. 2022. Of wolves and bears: Seasonal drivers of interference and exploitation competition between apex predators. *Ecological Monographs* 92:E1498.

Tallian, A., et al. 2023. Wolves in a Human World: Social Dynamics of the Northern Hemisphere's Most Iconic Social Carnivore. In: Srinivasan, M., Würsig, B. (eds) *Social Strategies of Carnivorous Mammalian Predators*. Fascinating Life Sciences. Springer, Cham.

Terborgh, J., and J. A. Estes. 2010. *Trophic Cascades: Predators, Prey, and the Changing Dynamics of Nature*. Island Press, Washington, D. C.

Theuerkauf, J., S. Rouys, and W. Jedrzejewski. 2003. Selection of den, rendezvous, and resting sites by wolves in the Bialowieza Forest, Poland. *Canadian Journal of Zoology* 81:163–167.

Treves, A. 2009. Hunting for large carnivore conservation. *Journal of Applied Ecology* 46:1350–1356.

Valeix, M., G. Hemson, A. J. Loveridge, G. Mills, and D. W. Macdonald. 2012. Behavioural adjustments of a large carnivore to access secondary prey in a human-dominated landscape. *Journal of Applied Ecology* 49: 73–81.

Van de Walle, J., F. Pelletier, A. Zedrosser, J. E. Swenson, S. Jenouvrier, and R. Bischof. 2021. The interplay between hunting rate, hunting selectivity, and reproductive strategies shapes population dynamics of a large carnivore. *Evolutionary Applications* 14:2414–2432.

Vinks, M.A., et al. 2021. Response of lion demography and dynamics to the loss of preferred larger prey. *Ecological Applications* 31:1–12.

von Hohenberg, B. C. and A. Hager. 2022. Wolf attacks predict far-right voting. *Proceedings of the National Academy of Sciences* 119:e2202224119.

vonHoldt, B., et al. 2024. Demographic history shapes North American gray wolf genomic diversity and informs species' conservation. *Molecular Ecology* 33:e17231.

Wabakken, P., H. Sand, O. Liberg, and A. Bjärvall. 2001. The recovery, distribution, and population dynamics of wolves on the Scandinavian peninsula 1978–1998. *Canadian Journal of Zoology* 79:710–725.

Wam, H. K., K. Eldegard, and O. Hjeljord. 2012. From overlooking to concealed: predator avoidance in an apex carnivore. *European Journal of Wildlife Research* 58: 1001–1003.

Whitman, K., H. Quadling, A. Starfield, and C. Packer. 2004. Sustainable trophy hunting in African lions. *Nature* 428:175–178.

Wielgus, R. B., and F. L. Bunnell. 2000. Possible negative effects of adult male mortality on female grizzly bear reproduction. *Biological Conservation* 93: 145–154.

Winterbach, H. E. K., C. W. Winterbach, M. J. Somers, and M. W. Hayward. 2012. Key factors and related principles in the conservation of large African carnivores. *Mammal Review* 43:89–110.

Woodroffe, R. 2000. Predators and people: using human densities to interpret declines of large carnivores. *Animal Conservation* 3:165–173.

Woodroffe, R., and J. R. Ginsberg. 1998. Edge effects and the extinction of populations inside protected areas. *Science* 280:10–13.

Woodroffe, R. and L. G. Frank. 2005. Lethal control of African lions (Panthera leo): local and regional population impacts. *Animal Conservation* 8: 91–98.

Wolf, C., and W. J. Ripple. 2016. Prey depletion as a threat to the world's large carnivores. *Royal Society Open Sciences* 3:160252.

Wolf, C., and W. J. Ripple. 2017. Range contractions of the world's large carnivores. *Royal Society Open Sciences* 4:170052.

Zarzo-Arias, A., et al. 2018. Brown bear behaviour in human-modified landscapes: the case of the endangered Cantabrian population, NW Spain. *Global Ecology and Conservation* 16:e00499.

Zedrosser, A., S. Steyaert, H. Gossow, and J. E. Swenson. 2011. Brown bear conservation and the ghost of persecution past. *Biological Conservation* 144: 2163–2170.

Zedrosser, A., F. Pelletier, R. Bischof, M. Festa-Bianchet, and J. E. Swenson. 2013. Determinants of lifetime reproduction in female brown bears: early body mass, longevity, and hunting regulations. *Ecology* 94: 231–240.

CHAPTER 5

How Conflict with Humans Changes Carnivores

Anna Rouviere and Robert A. Montgomery

Introduction

Humans and carnivores have shared landscapes for tens of thousands of years (Malhi et al. 2016). Across that time, human–carnivore interactions have been commonplace. Most often, these interactions are benign. A spotted hyena (*Crocuta crocuta*) walks through a human village under the cover of darkness without incident. In another case, a person passes through a forest and spots a jaguar (*Panthera onca*) peacefully resting in a tree overhead. In these cases, the humans and the carnivores have had an interaction and yet, there has been no tangible effect. There are also many instances in which humans and carnivores interact with positive outcomes that emerge (Lozano et al. 2019). Ecotourism is a prime example of a beneficial effect that people often speak about and remember for years to come. Consider the throngs of visitors to Yellowstone National Park, USA, hoping to catch even the smallest glimpse of a wolf (*Canis lupus*). From the perspective of the carnivore, there are a number of positive outcomes that can be derived from interactions with people. Consider the ways in which coyotes (*Canis latrans*) benefit from anthropogenic food subsidies that they access within the built environment. Unfortunately, however, conflict between humans and carnivores can also occur. Conflict emanates from human–carnivore interactions that generate negative outcomes for one or both parties. Typically, these are instances in which carnivores present threats to the security or private property of people, or occasions in which people either intentionally or unintentionally reduce the survival and reproduction of carnivores. Let's imagine a coupled human and natural system in sub-Saharan Africa, for example, where spotted hyenas depredate livestock and affected livestock owners retaliate by poisoning carcasses in the hopes that the offending hyenas will die. The killing of carnivores perceived to be responsible for that conflict provides an example of the ways in which humans can shape carnivore populations directly. However, emergent research has begun to quantify how people can also change carnivores via indirect pathways.

The behavioural plasticity of carnivores seeking to avoid human interaction can generate potential non-consumptive consequences. Over time, the trade-off decisions that carnivores make to avoid people can accumulate to the point where feeding times are reduced or stress levels are increased, which can come at a detriment to their survival and reproduction (Slovikosky and Montgomery 2024). We are only just beginning to understand how these nonconsumptive effects may scale to have demographic consequences at the carnivore population level that are equivalent, or even greater, to consumptive effects. Within this chapter, we explore the diversity of ways in which humans change the ecological systems inhabited by carnivores with direct and indirect effects that shape populations. With carnivore conservation presenting a thorny challenge globally, we link the outcomes of this research effort to discussions of the implications for carnivore population persistence. By exploring

Anna Rouviere and Robert A. Montgomery, *How Conflict with Humans Changes Carnivores*. In: *Apex Predators in the Anthropocene*. Edited by: Robert A. Montgomery and Remington J. Moll, Oxford University Press. © Oxford University Press (2026). DOI: 10.1093/9780191996665.003.0005 the historical contexts and consequences of human–carnivore conflict, both for humans and carnivores, we chart a path forward for the ways in which human–carnivore coexistence may be achievable.

Human–carnivore conflict over time

Conflict in the past

For as long as humans and carnivores have shared landscapes, their interactions have been characterised by conflict (Mata-González et al. 2023). The nature and strength of this conflict, however, has varied across spatial and temporal dimensions. Palaeolithic cave drawings provide a 12,000-year-old record of the preoccupation that people have always had with carnivores (Sigari et al. 2024). Charcoal drawings, rendered in exquisite detail, provide clear evidence of how early humans observed carnivores. Humans have feared, revered, and respected carnivores. This preoccupation is likely attributable to two factors. Firstly, humans and carnivores have always competed over shared prey and shared habitat. Though humans have the most diverse dietary niche of any predator, there is considerable overlap in preferred prey when compared to that of many large carnivores (Bugir et al. 2021). Secondly, despite the fact that humans are relatively unique in being able to turn apex predators into prey (Darimont et al. 2015), carnivores can be injurious, or even lethal, to people (Bombieri et al. 2023). Early human hunting strategies involved the tracking and displacement of carnivores from their kill sites (Blumenschine et al. 1994). These effects, in combination with direct killing and rapidly shifting climate conditions, led to extensive reductions of carnivore species richness in the last 30,000 years (Meltzer 2020). Within the last 3,000 years, humans' unique use of tools at scale facilitated widespread landscape conversion for natural resource extraction (e.g. clearcutting of forests for timber) and agricultural use (Gignoux et al. 2011). Agricultural development facilitated the widescale expansion of human populations into new niches around the world (Jones et al. 2013). Then, over the last 300 years, further mechanisation of tools via the Industrial Revolution enabled the global human population to dramatically increase and, in the process, to change the structure and function of the biotic and abiotic characteristics of planet Earth (Meyer and Turner 1992). Across each of these three phases of human evolution, carnivore populations have been compromised. This follows a general rule in ecology that where human populations increase, biodiversity tends to decrease (Forester and Machlis 1996, Cardillo et al. 2004). And the decrease of carnivore populations has been accelerated by conflict with humans.

Conflict in the present

In the twenty-first century, conflict with carnivores persists and can be triggered by a number of different pathways that have substantial variation in severity. At the lower end of the spectrum, for instance, raccoons (*Procyon lotor*) may cause conflict when they choose to den inside people's homes. At the upper end of the severity spectrum, grizzly bears (*Ursus arctos horribilis*) have a propensity to attack and kill people. Both of these examples are emblematic of human–carnivore conflict where people, as a direct result of these interactions, have experienced negative effects. Clearly, however, the severity of these interactions generating the conflict significantly varies. Despite this variation, one of the most common mechanisms leading to human–carnivore conflict globally is carnivore depredation of livestock. Livestock depredation occurs when carnivores hunt domestic livestock. Thus, depredation includes instances in which fossa (*Cryptoprocta ferox*) kill the chickens of Malagasy villagers as well as those in which Eurasian lynx (*Lynx lynx*) hunt semi-domestic reindeer in northern Scandinavia. Oftentimes, affected livestock owners will retaliate against the carnivores that they perceive to be responsible for this depredation (Kissui et al. 2019). Thus, given this conflict, coupled with factors including land conversion and prey depletion, over three-quarters of the 31 species of large carnivores remaining on the planet have declining population trends (Ripple et al. 2014). Given the severity of this conflict for both humans and carnivores, extensive research has focused on the interventions that may reduce the likelihood of carnivores killing livestock or decrease the probability of people retaliating against carnivores (Rouviere and Montgomery,

2025). Importantly then, interventions seeking to reduce human–carnivore conflict need to be tailored to the actions that are aimed at preventing carnivores from killing people's livestock versus those that are designed to stop people from killing carnivores.

To increase the efficacy of these interventions, it is necessary to recognise the complexities that are inherent to conflict. Despite the naming convention, one cannot expect to reduce conflict by just focusing on humans or carnivores. In actuality, there are five dimensions that are relevant to human–carnivore conflict (Montgomery et al. 2018a). These include the: (*i*) human dimension—represented by the ways in which people maintain their lifestyles; (*ii*) carnivore dimension—associated with the life history of carnivores; (*iii*) livestock dimension—how many livestock there are and where they are located on the grazing landscape or in their enclosures; (*iv*) wild prey dimension—the distribution, abundance, and behaviour of wild prey; and (*v*) environmental dimension—including the biotic and abiotic conditions of the landscapes in which humans and carnivores may interact. Thus, the development of effective, sustainable, and suitable solutions for human–carnivore conflict is dependent upon deducing the complexity of the coupled human and natural systems across these five dimensions.

The human consequences of human–carnivore conflict

Carnivore depredation of livestock has a variety of impacts on livestock owners, from financial loss to threats to safety, or even lower productivity of livestock production. These effects, along with other factors, can lead humans to act in a variety of ways that can be detrimental to carnivore conservation.

Effects of livestock depredation on humans

Although livestock losses to carnivores can have important financial, social, and cultural consequences for livestock owners, even unsuccessful depredation attempts by carnivores can generate insecurity with ramifications that are not commonly discussed.

Effects of realised loss of livestock

The financial impacts of the livestock losses that emerge from depredation vary along a wide range, and can represent a loss of as little as <1% to as much as 70% of the annual income of affected livestock owners (Thorn et al. 2012, Gebresenbet et al. 2018). Such disparities may first simply be due to variation in the amount of depredation experienced, which then translates to proportional differences in financial loss. Factors influencing the number of livestock taken by carnivores can include the extent of preventive measures put into place by communities, the propensity of carnivores in a given area to attack livestock or to perform surplus killing, or the location of affected communities relative to carnivore distribution and density. It is thus unsurprising that communities living in or on the edge of protected areas, such as wildlife reserves or national parks, tend to experience severe financial losses. For example, the communities living in India's Kibber Wildlife sanctuary lose on average 52% of their annual income to livestock depredation (Mishra 1997), and those in Kaffa Biosphere Reserve, Ethiopia, have been found to lose between 66% and 70% of theirs (Gebresenbet et al. 2018).

However, it is also possible for communities having lost the same number of livestock to experience vastly different financial repercussions. Firstly, the type of livestock taken can be an important factor. Cattle (*Bos taurus*), for example, tend to have higher economic value when compared to other livestock (Rust and Marker 2014, Khattak et al. 2022). In sub-Saharan Africa, lions (*Panthera leo*) tend to attack cattle, while leopards (*Panthera pardus*) target smaller stock such as goats (*Capra hircus*) or sheep (*Ovis aries*) (Kissui 2008, Petracca et al. 2019, Yirga et al. 2021, Deneke et al. 2022). The same intensity of depredation carried out by lions and leopards would therefore have different financial repercussions for the affected livestock owners (Mkonyi et al. 2017). Additionally, the financial impact of depredation is highly dependent on how reliant communities are on livestock production for their income. Communities whose income is primarily derived from agricultural activities will suffer relatively greater financial impacts from damage to livestock or crops than communities with more varied sources of revenue (Rao et al. 2002, Dickman et al. 2014). As such, alternative income sources have been introduced in some areas (Samelius et al. 2021), and are otherwise often suggested as a means of reducing depredation-induced financial losses and providing increased stability of income (Bagchi and Mishra 2006, Aryal et al. 2014, Bhatia et al. 2021, Uduman et al. 2022). Finally, financial impacts of livestock depredation are reportedly higher in countries with developing economies (Braczkowski et al. 2023). In many parts of the Global South, livestock depredation can decrease existing herd sizes—which may have already been constricted due to climatic instability and disease—to below the threshold of economic viability (Lybbert et al. 2004, Ayanlade et al. 2022). Moreover, pastoralists in developing countries produce less meat per individual cattle (Braczkowski et al. 2023). Together, these factors contribute to the fact that developing economies are estimated to be two to eight times more economically vulnerable to livestock depredation than developed ones (Braczkowski et al. 2023).

In addition to financial value, livestock also have, in many areas, important social and cultural value. In certain parts of the world, cattle can be an indicator of the social status of the family that owns it (Mitchell et al. 2019), and can be used in cultural practices such as the payment of dowry (Ngowi et al. 2008). The loss of such livestock to carnivores can thus have ramifications for local communities that go beyond measurable financial loss.

Threats to safety

Carnivore attacks on livestock can constitute serious threats to human personal safety due to, among other things, the spatial proximity of livestock to human presence. This proximity increases the likelihood of human–carnivore encounters when the latter are in search of livestock. For example, tigers (*Panthera tigris*) have been reported to regularly enter villages in Bangladesh, leading to insecurity for, and attacks on, humans (Inskip et al. 2014). Spotted hyenas, which are known to visit villages in search of food, have been seen to enter homes and attack residents in their sleep (Dickman et al. 2014). Not only are humans spatially close to livestock, but those who own livestock also have a vested interest in keeping these animals alive and well. Therefore, many carnivore attacks on humans have occurred during human efforts to protect livestock or chase carnivores away from a herd or a corral (Packer et al. 2005, Dickman et al. 2014, Farhadinia et al. 2017). Even depredation attempts that do not lead to livestock death may thus constitute a danger to human safety, where the successful deterrence of offending carnivores could occur at the cost of human life or serious injury.

Risk effects on livestock

Although livestock have been reported to display considerably lower antipredator behaviour than wild prey due to a long history of domestication (Sivertsen et al. 2016), they are not entirely indifferent to predator presence, nor to perceived depredation risk. Not only have cattle been found to increase their vigilance behaviour by forming larger groups in areas of high depredation risk (Beck et al. 2021), but the psychological impacts of past depredation attempts can also have physiological consequences. Past experiences of livestock depredation within a herd can lead to an increase in stress-induced diseases (Sommers et al. 2010, Cooke et al. 2013), lower birth rates (Widman et al. 2019), and lower calf weights (Ramler et al. 2014). Such effects can in turn alter the quality and quantity of livestock-derived products (Linkowski et al. 2017, Scasta et al. 2018), and lower the economic viability of a herd. These indirect costs can at times be greater than the costs of direct loss of livestock due to depredation (Ramler et al. 2014).

While the effects of livestock depredation on humans are often quantified in relation to the number of livestock killed by carnivores, it is therefore important to note that even unsuccessful depredation attempts can have lasting consequences on human health, on livestock health, and on the economic returns of livestock production.

Human attitudes towards depredating carnivores

While it would be reasonable to assume that human attitudes towards depredating carnivores are directly related to the amount of loss that these species can inflict, this is far from the reality (Grey et al. 2017, Harvey et al. 2017, Eshete et al. 2018b,

Gebresenbet et al. 2018). Attitudes are instead often shaped by perceived depredation risk (which can differ from actual depredation risk), depredation-related characteristics of the species, and factors that are independent of depredation altogether.

Perceived risk of depredation

The risk of livestock depredation is often quantified based on the likelihood and intensity of livestock death in an area (Beattie et al. 2020, Balbuena-Serrano et al. 2021, Kachel et al. 2022). However, being able to accurately assess this risk as a member of an affected community is a difficult task (Knopff et al. 2016, Grey et al. 2017). Villagers in Tanzania, for example, estimated that they were at a similar risk of being attacked by a lion as they were of contracting HIV/AIDS or malaria. The true risk of contracting these diseases, however, was approximately 20 times higher than that of falling victim to a lion attack (Kushnir and Packer 2019).

Therefore, in addition to actual rates of depredation, a variety of factors can shape perceived depredation risk. Firstly, risk perception can be correlated with socioeconomic factors such as a person's age, gender, level of education, level of literacy, occupation, or knowledge of carnivores (Amit and Jacobson 2017, Fort et al. 2018, Eshete et al. 2018a, Kushnir and Packer 2019, Torres et al. 2020, Meena et al. 2021, Gebo et al. 2022). Secondly, risk perception can, for a given person, differ between carnivores depending on the amount of fear that the species instils. If a species is regarded as being more frightening than another, its perceived risk of depredation can be higher regardless of how much depredation it is in fact responsible for (Torres et al. 2020). In Mexico, people perceived pumas (*Puma concolor*) and jaguars to be responsible for most attacks on livestock, even though the biggest perpetrators of reported depredation events were feral dogs (*Canis lupus familiaris*), which are more familiar to humans and induce correspondingly less fear (Zarco-González et al. 2018). Finally, perceived risk can differ from actual risk when a species is mistaken for another. Cheetahs (*Acinonyx jubatus*), which attack livestock very little in Iran, are thought to be important depredators due to their physical resemblance to leopards (Farhadinia et al. 2017). Thus, the perceived intensity of depredation can depend on human socioeconomic factors, on the fear-inducing nature of carnivore species, and on the confusion of one species for another.

Depredation-related characteristics

In addition to risk perception, characteristics of a carnivore species during a depredation event may influence human attitudes. Carnivores that tend to target livestock considered to be more valuable may be regarded in a more negative light (Hazzah et al. 2009, Rust and Marker 2014). Even when depredation of high-value livestock is rare, these events can lead to more negative attitudes than many attacks on lower-value livestock. Moreover, livestock owners may hold more negative opinions of species that perform surplus killing than of species that could cause more livestock deaths overall (Muhly and Musiani 2009). Additionally, a perceived feeling of control over a species can influence attitudes. In Kenya, villagers had positive attitudes towards leopards, despite a high depredation rate, as they considered the species to be more respondent to deterrents that they had implemented than other carnivores (Mitchell et al. 2019). Thus, the ecology and life history characteristics of the carnivore species can affect the nature of human–carnivore conflict.

Factors independent of depredation

Human attitudes towards depredating carnivores can also be defined by factors entirely independent of the act of killing livestock. Firstly, people can have inherent predispositions that are informed and influenced by existing beliefs associated with a species. The negative attitudes of Malagasy people towards fossa are made stronger by cultural characterisations of this species as consuming the remains of their ancestors (Merson et al. 2019). In the East Hararghe Region of Ethiopia, the presence of spotted hyenas is, conversely, considered to have some cultural value as the species is believed to be capable of detecting and removing malevolent spirits (Baynes-Rock 2013). Secondly, the aesthetics or charisma of a carnivore can also be a contributing factor to human attitudes to depredation (Drouilly et al. 2021, Hoffmann and Montgomery 2022). Despite commonly being responsible for livestock depredation, lions are often seen in a more positive light than spotted hyenas because the former appear majestic and evoke a sense of pride, while the latter are perceived to be ugly and unclean (Mitchell et al. 2019). Similarly, in Mexico, the most frequent justification for positive attitudes towards pumas and jaguars was the aesthetics of these species (Piña-Covarrubias et al. 2022). Finally, the economic value of a species for a community can also affect opinions. For example, if a community financially profits from tourism brought on by the local presence of a carnivore, attitudes towards this species may be improved (Mitchell et al., 2019).

Human interventions to reduce conflict

There are a range of ways in which humans around the world act in response to the livestock depredation that they experience. These actions can vary along a spectrum from least to most harmful to carnivores. The magnitude and severity of these human responses depend both on the attitudes held towards a carnivore (and all contributing factors therein), and on the propensity and willingness of individual people to execute the act. For example, two people with equally negative attitudes towards a carnivore species may not have the same willingness to respond with lethal force. Socioeconomic factors and social norms (Inskip et al. 2014), as well as beliefs (both religious and cultural), all factor in. Thereby, variation in teachings regarding nature and animals can lead to significantly different rates of retaliatory killing of carnivores (Hazzah et al. 2009, Harvey et al. 2017, Merson et al. 2019). For instance, among a range of religious communities, members of the Kenyan Assemblies of God (a religion which does not include in its Bible passages discussions of respect towards nature) were found to be statistically most likely to kill a lion in Kenya (Hazzah et al. 2009). In South Africa, Afrikaans-speaking farmers were 19 times more likely to kill depredating carnivores than Setswana-speaking farmers (Thorn et al. 2012). Beliefs can also be rooted in folklore. In Samburu County, Kenya, families were reluctant to kill African wild dogs (*Lycaon pictus*) due to the local belief that such an act would prevent them from bearing a son (Mitchell et al. 2019).

Lethal interventions against depredating carnivores can be carried out directly by affected livestock owners, or by local authorities or government entities. However, interventions can also be non-lethal, ranging from highly localised sensory deterrents to larger-scale methods such as sterilisation and translocation.

Unregulated lethal interventions by affected livestock owners

Livestock owners who experience depredation can feel compelled to kill carnivores in prevention of, or in retaliation for, such attacks. This often occurs at the discretion of the owners themselves, in an unregulated fashion (Mateo-Tomás et al. 2012, Gálvez et al. 2018, Gebresenbet et al. 2018). While local retaliatory killing is often illegal, it is in some areas permitted for the defence of human or livestock life (Ogada et al. 2003, Rust 2017, Dellinger et al. 2021, Parchizadeh and Belant 2021, Nisi et al. 2022).

Carnivores are killed either directly, when the individual is within sight of and specifically targeted by the person killing it, or indirectly. Direct killing can occur by shooting (Breck et al. 2011, Kerth et al. 2013, Nelson et al. 2016, Moss et al. 2016, Minnie et al. 2018, Pereira et al. 2020, Yirga et al. 2021) or with arrows, spears, axes, or sticks (Kissui 2008, Hazzah et al. 2009, Inskip et al. 2014). Some community members also employ more unusual methods, such as electrocuting, using dynamite, or setting the animals on fire (Mishra 1997, Ramesh et al. 2020, Sebsibe 2022). Indirect killing occurs by setting a trap (Aryal et al. 2014, Santiago-Avila et al. 2018, Pereira et al. 2020, Augugliaro et al. 2020) or poisoning bait (Mateo-Tomás et al. 2012, Baynes-Rock 2013, Singh et al. 2015, Braczkowski et al. 2020) in hopes that the targeted species will be caught or killed. Species with scavenging life history traits, such as spotted hyenas, are more often killed indirectly, especially by poisoning, whereas non-scavenging species tend to be specifically targeted (Kissui 2008, Atickem et al. 2010).

While indirect methods are by definition indiscriminate, direct methods can either be employed to kill specific individuals that have depredated, or random individuals within the species. In the former case, killing most often occurs at the site of depredation, soon after or during the attack (Kolowski and Holekamp 2006, Rashid et al. 2020). In the latter case, pastoralists will directly but indiscriminately kill members of an offending species, regardless of whether these individuals have depredated or not. For example, some farmers shoot lion cubs, which are too young to attack any domestic livestock (Woodroffe and Frank 2005), or destroy wolf dens to kill any pups therein (Werhahn et al. 2019). These actions can be seen as preventive killing rather than retaliatory killing.

Regulated lethal interventions by governments

As human–carnivore conflicts became a persistent problem for livestock owners, over time many governments began establishing carnivore culling programmes. These took the form of hunts organised as state-approved operations or bounty systems. In France, King Charlemagne established a wolf-killing elite task force called the *Louveterie* in the ninth century (Moriceau 2011). In Finland, a bounty system was established in the seventeenth century to reduce the populations of several carnivores, including wolves, brown bears, red foxes (*Vulpes vulpes*), Eurasian lynx, and pine martens (*Martes martes*) (Pohja-Mykrä et al. 2005). At the same time, in 1630, the Massachusetts Bay colony was setting up the first American bounty on wolves shortly after the British colonisation of North America (Barclay 2002). Over the next three centuries, these activities spread to many states and regions of the United States. For instance, in 1929, just 40 years after the establishment of a bounty in Montana, rewards had been paid for the reported slaughter of over 111,000 wolves and 886,000 coyotes within the state (Wise 2013).

Though bounty systems tend to no longer exist, the lethal control of carnivores is still enforced by local or national governments in countries such as the United States (Hiller et al. 2015, Moss et al. 2016, Bergstrom 2017, Hill et al. 2022), Sweden (Sivertsen et al. 2016, Aronsson and Persson 2017), Argentina (Pereira et al. 2020), Norway (Tveraa et al. 2014), and India (Puri et al. 2020). Practices including 'proactive culling' usually consist of setting annual quotas for the number of carnivores to be killed, and using organised hunting, trapping, or aerial shooting to achieve this goal (Creel and Rotella 2010). Although the aim of this effort is to reduce rates of livestock depredation, the efficacy of these methods is strongly debated (Fernández-Gil et al. 2016, Treves et al. 2016, Moreira-Arce et al. 2018, Lorand et al. 2022).

Another form of lethal control is 'reactive culling', wherein only carnivore individuals that are perceived to be responsible for depredation are removed. This selective form of lethal control is used as an alternative to proactive culling and intends to optimise removal efforts and save both time and money (Baker et al. 2008). The method relies on the assumption that livestock depredation is not carried out uniformly among a carnivore population but, rather, by select individuals. Reactive culling is believed by some to prevent the spread of livestock-killing behaviour from problematic individuals to the rest of the population (Woodroffe and Frank 2005). However, the correct identification of responsible individuals can be difficult and is often based on generalisations, such as the idea that only males are depredators (Kerth et al. 2013).

Non-lethal interventions

Non-lethal means of managing human–carnivore conflict by reducing rates of livestock depredation have grown as culling methods have declined. Interventions implemented at the household scale can include both 'disruptive-stimulus' or 'aversive-stimulus' approaches (Shivik 2006). Disruptive-stimulus approaches include sensory methods that aim to perturb carnivores upon their approach to livestock, such as loud noises, shouting, lights, alarms, fires burning, or music playing (Woodroffe et al. 2007, Petracca et al. 2019, Naha et al. 2020, Behnam and Nader 2021, Sijapati et al. 2021, Filla et al. 2022). Fladry, which involves the placement of coloured pieces of fabric onto a fence or rope, is also often used to deter wolves, though it is not as effective on other species (Davidson-Nelson and Gehring 2010). Another creative use of disruptive stimuli is the painting of eyespots on the rumps of cattle to perturb the hunting behaviour of carnivores (Radford et al. 2020). Biofences are also a specific type of disruptive-stimulus approach involving human-deployed scent marks in the form of urine and scat, to simulate conspecific territory boundaries and dissuade movement across the created boundary (Ausband et al. 2013). However, as disruptive-stimulus approaches have no adverse consequences on carnivores (i.e. injury or death), the effectiveness of these types of methods is usually limited in time due to an eventual habituation by the carnivore (Miller et al. 2016, Khorozyan and Waltert 2019, Petracca et al. 2019). Efforts to limit habituation include alternating or combining methods (Lorand et al. 2022), using stimuli that appear intermittently at random times (Shivik 2006, Ohrens et al. 2019, Naha et al. 2020), or restricting the deployment of these methods to times when livestock are most vulnerable to depredation, such as during calving seasons (Shivik 2006, Miller et al. 2016).

Aversive-stimulus approaches, on the other hand, use techniques causing physical discomfort or pain in an effort to negatively condition carnivores into learning not to approach livestock. These can include electric fences (Miller et al. 2016, Wilson et al. 2017, Samelius et al. 2021), rubber bullets shot with paintball-type weapons (Shivik 2006), shock collars (Eklund et al. 2017), or capsaicin projectiles (Beckmann and Lackey 2018). Once again, these techniques have been criticised for being ineffective or conducted in such a way that carnivores can become habituated to their application.

Some non-lethal interventions, such as the use of dogs or herders, have the potential to become aversive, or even lethal, if these lead to injury or death of approaching carnivores rather than simple deterrence (Beckmann and Lackey 2018, Whitehouse-Tedd et al. 2020). Others, however, could be classed as 'neutral' methods as they do not provide disruptive nor aversive stimuli but simply prevent carnivores from reaching the livestock. These include corrals, pens, or other enclosures that offer physical barriers to depredation. These barriers require regular maintenance to be effective (Weise et al. 2018), and the most effective designs or materials depend on the carnivore species (Kolowski and Holekamp 2006, Broekhuis et al. 2017, Eklund et al. 2017).

Finally, some interventions are not implemented at the scale of individual households, but at community, landscape, or regional scales depending upon the specific management tools (Montgomery et al. 2018b). Several of the broader-scale efforts involve translocation and sterilisation. Translocation consists of moving problematic individuals from an area of high depredation to a distant site where their potential to depredate will be lower. Once the individual carnivores have been transported to their release site, they are either let go immediately ('hard release') or held in temporary enclosures during an adjustment period before being released ('soft release') (Bradley et al. 2005, Weise et al. 2015). However, translocations are expensive and time-consuming, and do not always result in decreased depredation pressures (Lorand et al. 2022). Sterilisation is another non-lethal measure used to ameliorate depredation. This intervention, achieved by preventing reproduction surgically or chemically (Gese and Terletzky 2021), is a method of population control, but is also based on the theory that carnivores with dependent offspring require more food and are thus led to a higher propensity for depredation (Shivik 2006, Eklund et al. 2017).

The carnivore consequences of human–carnivore conflict

Human interventions in response to the depredation of livestock by carnivores can have both population-level and behavioural consequences on the affected carnivore species. However, the impact is not limited to the species in question, as it can also reach the wider ecosystem in which they occur (Moll et al. 2021).

Population effects

Lethal interventions

In the absence of human-induced mortality, individual carnivores within a population can die of other causes (i.e. 'natural mortality'). These causes can include disease, predation, senescence, or density-dependent factors such as competition for resources. The introduction of human-induced mortality in addition to natural mortality can have several varying effects on the carnivore population.

In some cases, the deaths caused by human-induced mortality are simply added to the deaths caused by natural mortality, and thus lead to a directly proportional decline in the population's survival rate; this is called 'additive mortality'

(Figure 5.1, Figure 5.2) (Creel and Rotella 2010). This has been found to occur in some cases following lethal removals of carnivores in response to livestock depredation (Creel and Rotella 2010, Benson et al. 2023), even when carried out within the current hunting quotas set by some governments (Woodroffe and Frank 2005). Resulting population declines have sparked not only calls for caution when defining lethal control strategies and setting annual mortality quotas (Chapron et al. 2003), but also wider concern about the conservation implications of lethal interventions. Thus, human–carnivore conflict is among the main drivers of carnivore declines (Ripple et al. 2014),

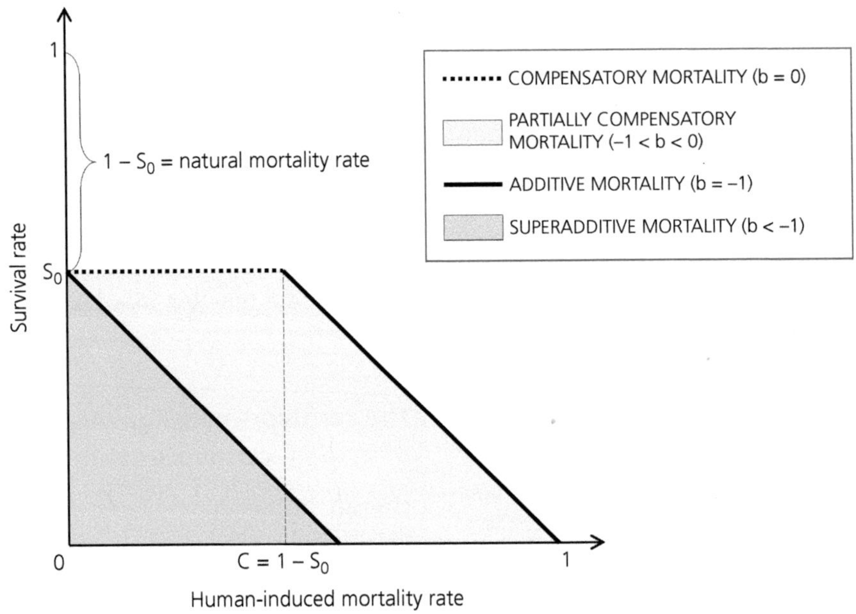

Figure 5.1 The possible effects of human-induced mortality on carnivore survival rates. Under null human-induced mortality pressure, the mortality rate of the population ($1 - S_0$) is non-zero due to natural causes. Additive mortality occurs when the survival rate is directly inversely proportional to the human-induced mortality rate (slope $b = -1$). In the case of superadditive mortality, the survival rate declines faster than the human-induced mortality rate increases ($b < -1$). Compensatory mortality occurs when the survival rate remains stable under increasing human-induced mortality pressure ($b = 1$). Mortality cannot be compensatory past point $C = 1 - S_0$, at which it would be compensating for the entirety of the natural mortality. Beyond this point, mortality becomes additive (or superadditive). Mortality can also be partially compensatory if the initial decline in the survival rate is slower than the increase in the human-induced mortality rate ($-1 < b < 0$).

LETHAL INTERVENTIONS CAN

Decrease local population size due to:

- Decreased survival rate
  - Partially compensatory mortality
  - Additive mortality
  - Superadditive mortality

Have no effect on local population size due to:

- Stable survival rate (compensatory mortality)
- Increased fecundity (compensatory natality)
- Immigration from source populations

Figure 5.2 Summary of the possible effects of human lethal interventions on carnivore population size.

and has led to many local extinctions of carnivore species (Woodroffe 2000, Wolf and Ripple 2017).

At times, the declining survival rate resulting from lethal interventions can be even greater than the rate at which carnivores are being killed, due to additional indirect deaths caused by the ramifications of human-induced mortality. This is referred to as 'superadditive mortality' (Figure 5.1, Figure 5.2) (Creel and Rotella 2010, Sandercock et al. 2011). For example, conflict can have an indirect impact on juveniles (Ausband et al. 2015). Indeed, lion cubs that are born to depredating mothers have less than half the chances of surviving to one year of age than cubs born to non-depredating mothers (Woodroffe and Frank 2005). Additional deaths can also occur when humans kill dominant individuals in populations of carnivores with hierarchical social structures. In lion prides, for example, the death of the dominant male will cause new males to compete for dominance. This competition can be lethal, and can also lead to the eviction of subadults as well as infanticide (Cooley et al. 2009, Loveridge et al. 2017, Petracca et al. 2019). Thus, the combination of direct human-induced mortality, lower juvenile survival, and lethal competition for dominance can all contribute to the superadditive effects of carnivore mortality due to conflict.

In other cases, however, no visible population declines following lethal control can be detected (Figure 5.2). Such results are perhaps counterintuitive, but can occur for several reasons. Firstly, the introduction of human-induced mortality can replace the effects of natural mortality, thus leading to a stable survival rate. In a population that has reached its environment's carrying capacity, for example, a high proportion of natural deaths occur due to density-dependent factors such as competition for resources. In these cases, human-induced mortality can replace natural mortality, as the individuals that are killed by humans would have died by competition otherwise. This phenomenon is called 'compensatory mortality' (Figure 5.1, Figure 5.2) (Boyce et al. 1999, Haight et al. 2002).

Secondly, stable population sizes can also be a result of compensatory natality, when reproductive rates increase in response to higher levels of mortality (Boyce et al. 1999). In the United States, for example, coyote litter sizes have been found to be higher among heavily exploited populations (Knowlton 1972). Finally, when mortality occurs unevenly across a landscape, a population may only appear stable due to new individuals immigrating from source populations to occupy the newly vacant space (Novaro et al. 2005, Robinson et al. 2008, Minnie et al. 2016, 2018, Pereira et al. 2020). These individuals are drawn to this area because it appears to have low competition and to be ecologically suitable for survival and reproduction. Depending on the site, they may not be able to recognise cues that indicate a high risk of human-induced mortality. These population sinks are thus called 'ecological traps', or 'attractive sinks' (Delibes et al. 2001). Because of this phenomenon, a seemingly constant local population size in response to lethal pressure may in fact mask wider declines (Robinson et al. 2008, Cooley et al. 2009). Often, the immigrating individuals are young dispersing subadults (Woodroffe and Frank 2005, Petracca et al. 2019). A population at a heavily hunted site may therefore become younger over time due to these source-sink dynamics (Robinson et al. 2008, Cooley et al. 2009, Minnie et al. 2016). In many carnivore species, younger adults tend to be bolder and more prone to depredation than older individuals (Petracca et al. 2019). Thus, heavy local hunting pressure may not only not lead to the desired reduction in carnivore numbers, but could instead lead to a counterintuitive increase in depredation rates. Additionally, young male immigration can result in increased infanticide as the new males look for breeding opportunities. In the United States, heavily hunted areas had an increase in puma kitten mortality due to the immigration of unrelated males into sink populations (Cooley et al. 2009).

Non-lethal interventions

Despite being a non-lethal intervention, translocation can also yield population-level consequences. Firstly, this intervention can lead to carnivore mortality via injury in the translocation process, as well as capture myopathy. The process of translocation itself can, if not carried out properly, cause injury or death to the animals during transportation (Linnell et al. 1997). Mortality risks persist even if the release is successful; some translocated individuals travel long distances to return to their site of capture, which can lead to high rates of road mortality (Linnell et al. 1997). Moreover, it can be difficult for translocated individuals to acclimate to their release sites (Stuparyk et al. 2018, Berger-Tal et al. 2020). In Montana, translocated wolves had significantly lower rates of survival than non-translocated wolves and failed to form stable pack structures (Bradley et al. 2005). In Kenya, every monitored lion relocated because of perceived depredation died following translocation to a national park (Woodroffe and Frank 2005). Translocation can be particularly deadly for more vulnerable individuals such as juveniles (Rogers 1986). In addition to survival rates, translocation can affect fecundity. The reproductive rates of female bears in Virginia, United States, decreased after their translocation (Comly-Gericke and Vaughan 1997).

Sterilisation is another non-lethal intervention with the potential to have effects on carnivore populations. Firstly, sterilisation by definition affects recruitment, thus potentially causing population declines (Massei and Cowan 2014). Conversely, however, it may also increase survival rates through changes in energetic allocation; individuals who do not need to expend any energy on reproduction can instead focus on survival, thus potentially leading to improved body condition and longevity. Though not much research on this topic has been carried out on carnivores, this effect has been shown in other taxa such as rabbits, deer, or equids (Williams et al. 2007, Kirkpatrick and Turner 2007, Gionfriddo et al. 2011). Some methods of sterilisation may also have effects on social structure and dynamics. For example, it has been hypothesised that methods removing gonadal hormones (such as tubal ligations and vasectomies, rather than spaying and neutering) could affect social dominance or establishment and maintenance of territories, which are thought to be hormonally controlled. However, social disruptions resulting from sterilisation have yet to be convincingly detected among these managed carnivore populations (Saunders et al. 2002, Gese and Terletzky 2021).

Behavioural effects

Lethal interventions

In the wild, prey animals assess their predation risk by evaluating spatio-temporal factors indicative of predator presence (Moll et al. 2017). Although many of the carnivore species that tend to conflict with humans are large-bodied and positioned higher up in trophic systems, empirical evidence has shown that some have learned to recognise humans as a source of mortality and can assess their own risk of human-induced mortality via cues of human presence (Boydston et al. 2003, Moss et al. 2016, Nisi et al. 2022). If a carnivore recognises a cue that it associates with high perceived mortality risk, it is very likely to adapt its behaviour to lower this perceived risk (Montgomery et al. 2020). Therefore, carnivores' use of an area can be influenced in space and time by their fear of human-induced mortality, creating what is known in predator-prey ecology as a 'landscape of fear' (Laundré et al. 2001, 2010). This behavioural plasticity can affect various aspects of behaviour, such as space use, vigilance, movement, or foraging (Green et al. 2018, Abade et al. 2018, Montgomery et al. 2020). For example, leopards in China reduce their temporal overlap with cattle when humans are present, but seek to depredate when humans are absent (Liu et al. 2023). In Botswana, both spatial and temporal avoidance of humans has been detected as lions generally alter their space use to avoid cattle posts, but when they do visit only do so during periods of human inactivity and increase their travel speed (Valeix et al. 2012). In the United States, fear of humans can cut by half the time that pumas spend at kills (Smith et al. 2015, 2017).

While proxies of human presence can sometimes reliably indicate the level of mortality risk to carnivores, patterns of human killing of carnivores are complex and difficult to predict as they are influenced by a range of factors. This unpredictability signifies that some cues can be misleading, since mortality risk is not always proportional to human density (Loveridge et al. 2017). For example, while human density and activity in California, United States, are highest in highly developed areas during the day, human-induced puma mortality is highest at intermediate levels of development and at night. Because mortality risk does not map onto cues of human presence, a mismatch arises between risk and fear-induced behavioural responses from pumas (Nisi et al. 2022). This can thus lead to the phenomenon of ecological traps introduced earlier (Loveridge et al. 2017). Similarly, the landscape inside a protected area may look similar to that on the other side of its border despite a very different mortality risk. The lack of reliable cues associated with this change can also create ecological traps and lead to a sustained high mortality rate due to a lack of appropriate behavioural modifications (Balme et al. 2010).

Non-lethal interventions

Many non-lethal interventions are structured to induce a fear response in the offending carnivore, leading to an avoidance of the area in which the intervention has been implemented. However, this fear is not caused by realised mortality, like it is for lethal interventions. Instead, it is caused by startling or disturbing stimuli in the case of disruptive-stimulus approaches, and pain or discomfort in the case of aversive-stimulus approaches. The fact that no mortality is associated with these stimuli can cause behavioural modifications to decrease in intensity over time as carnivores habituate to the lack of lethal consequences (Eklund et al. 2017). This is especially true for sensory deterrents, which have no adverse repercussions on the carnivores whatsoever; while they may be temporarily effective (Filla et al. 2022), their effectiveness is usually limited in time (Ausband et al. 2013, Petracca et al. 2019).

Ecosystem effects

Lethal interventions

Human interventions in response to livestock depredation can either directly or indirectly affect the ecosystem in which the human–carnivore interactions occur. For example, lethal interventions can cause accidental deaths of non-target species, especially in the case of indirect interventions such as bait poisoning. In Spain, 112 poisoning events led to the deaths of an estimated 228 individual animals from 25 different species, only 38% of which were mammalian carnivores (Mateo-Tomás et al. 2012). Additionally, since 2000, the American Wildlife Services has unintentionally killed vertebrates among some 150 species while conducting lethal control of predators (Bergstrom et al. 2014).

However, even in instances in which only the target species is killed by these interventions, the deaths of carnivores in aggregate can trigger trophic ramifications with implications for ecosystem structure and resilience (Bergstrom 2017). Firstly, the loss of dominant predators, commonly known as apex or top predators, can lead to an increase in the abundance of smaller predators with lower trophic roles, called mesopredators, due to their release from the top-down control exerted by top predators (Sovada et al. 1995, Crooks and Soule 1999). This phenomenon is known as 'mesopredator release' (Prugh et al. 2009). An unchecked increase in mesopredator abundance can have dramatic effects on their prey populations. Take, for example, the decline of dingoes (*Canis lupus dingo*) in Australia, which triggered an increase in red fox and feral cat (*Felis catus*) populations, leading to large-scale declines and extinctions of endemic marsupial species (Johnson et al. 2007). Furthermore, mesopredator release can have negative consequences on livestock and other domestic animals. For instance, the phenomenon can lead to increased disease transmission. Black-backed jackals (*Canis mesomelas*) carry tick species which are able to spread pathogens to dogs and potentially to sheep (Viljoen et al. 2021). Mesopredators are also, in some cases, even more prolific depredators than the top predators killed by humans. Therefore, the lethal control of large carnivores in response to livestock depredation may result in an undesired increase in depredation rates due to mesopredator release (Thorn et al. 2012). Moreover, some large carnivore species—which are not commonly considered to be mesopredators—are nonetheless under the competitive control of more dominant species, as is the case with leopards and spotted hyenas, which can be competitively excluded by lions (Packer et al. 2009, Kissui et al. 2019). In such cases, the mortality of the dominant carnivore can affect the abundance and/or distribution of these species as well.

Non-lethal interventions

Even non-lethal interventions have the potential to impact non-target species. Livestock guarding dogs, for example, have been known to harass, injure, or even kill non-depredating wildlife (Potgieter et al. 2016, Whitehouse-Tedd et al. 2020). Disruptive-stimulus approaches are likely to engender fear-based responses in non-target species as well as target species (Blackwell et al. 2016). The translocations of carnivores into new areas can potentially lead to disease transmission and to the emergence of competition between the translocated individuals and resident animals (Griffith et al. 1993, Cunningham 1996). Finally, some methods of sterilisation can lead to an adverse exposure risk for predators and scavengers of the treated animal (Massei and Cowan 2014).

The future of human–carnivore conflict

What will conflict look like in the future?

The same factors that have accelerated the decline of carnivore populations (e.g. habitat conversion, prey depletion, and human population growth) are expected to continue in future. In fact, it is a reasonable prediction to presume that human–carnivore conflict will intensify. The human population is predicted to grow to around 11 billion people until it reaches a global asymptote around the year 2100 (United Nations 2022). It goes without saying that this human population will require resources and sustenance for survival. Consequently, feeding the future given finite land globally is a particularly important concern for the persistence of human societies and for biodiversity conservation (Laurance et al. 2014). Despite efforts to rapidly expand the footprint of protected areas around the world (i.e. the 30 by 30 commitments), agricultural and infrastructure development in support of future human societies will likely increase the rates at which people and carnivores come into contact. Given that conflict is a likely effect to emerge from these interactions, interventions to reduce the probability of carnivores killing people's livestock and people retaliating against carnivores will be even more valuable in future.

One of the techniques that has been proposed to manage human–carnivore conflict, and human–wildlife conflict more broadly, at scale is fencing (Packer et al. 2013). In this case, complete fencing of protected areas to keep the wildlife in and the people, at least those that do not pay reserve entrance fees, out has been proposed. Herein, fences intend to reduce human–carnivore interaction and prevent conflict from being able to emerge. A simple enough premise that is positively complex in application. Advocates for fencing highlight that African lions living in fenced reserves are among the only populations with positive growth rates. Those opposed will highlight that fenced ecosystems are more akin to outdoor zoos rather than functional ecosystems. At the heart of this issue are two factors. Firstly, the implementation and maintenance of fenced ecosystems, particularly at a scale associated with the predicted protected area expansion, is extremely time-consuming and expensive (Pekor et al. 2019). This may be feasible in countries with well-developed economies, but options are comparatively limited in the Global South. Secondly, in addition to breaking the mystique of the wilderness, fences also lead to future scenarios in which wildlife populations, particularly large carnivores, will need to be intensively managed. Take, for instance, several reserves in Southern Africa where large carnivores, such as lions, have had to be intensively culled to avoid substantial prey depletion and associated ecosystem effects (Miller and Funston 2014). Given range restrictions, it is very difficult to achieve natural population regulation of interacting predator-prey species in fully fenced reserves.

Thereby, while it is challenging to predict what the future will hold for human–carnivore conflict, it is hard to be optimistic. We now have thousands of years of evidence identifying that human population growth is associated with declining biodiversity, with carnivores being at the top of the hit list because of their tendency to compete directly with humans. Furthermore, the one technique that is associated with a positive population growth rate (fencing) is precisely the one that will be both complex and morally multifaceted to engineer. What this context does make clear, however, is that interventions will be even more in demand in future than they are in the present.

Steps forward

We have three specific recommendations for researchers and practitioners of human–carnivore conflict to consider. First, make research more actionable. Human–carnivore conflict research has tended not to be structured in actionable ways (Gray et al. 2020). Not only that, but research effort in this area has shown considerable taxonomic bias (Hoffmann and Montgomery 2022). Consequently, species that receive considerable conservation attention are often those that are charismatic rather than conflict-inducing. Herein, we recommend that research is structured and designed in ways that enable the outcomes to directly inform policy and on-the-ground decision-making. In doing so, direct collaboration with the agencies and organisations that are responsible for developing policies is vital. These types of interdisciplinary collaborations can ensure that research better translates into the practices designed to reduce conflict. Secondly, we recommend that research fully embrace the complexity that is inherent to conflict. Though humans and carnivores are important dimensions to predicting and intervening against human–carnivore conflict, they are not the only ones. The severity of conflict also depends upon livestock, wild prey, and environmental dimensions. Thus, factors associated with each of these dimensions must be quantified to create a holistic picture of conflict so as to inform effective management interventions. Finally, the direct and indirect ways in which humans have managed conflict have fundamentally changed carnivores. However, recent recovery of many carnivore populations globally has demonstrated that carnivore populations can come back even without the direct assistance of people. In the absence of direct persecution, carnivore populations can recover. Thus, even in future population growth scenarios, human–carnivore coexistence may be achievable via effective interventions to reduce the likelihood that carnivores can depredate livestock and a willingness of people to prevent carnivore persecution.

Conclusion

In this chapter, we have presented the historical context in which conflict between humans and carnivores has occurred. We have discussed the consequences of that conflict, typically accelerated by carnivore depredation of livestock, for humans. Transitioning to carnivore consequences, we then examined the population-level, behavioural, and ecosystem-wide impacts that can emerge. Concluding the chapter with a look at future scenarios for human–carnivore conflict, we emphasise the urgency to action in the development of effective interventions. To make these interventions most effective, we highlight that the research teams must focus on interdisciplinarity and the actionability of that research. With human population growth and expansion predicted across the remainder of this century, trialling interventions that are more effective at reducing human–carnivore conflict will be essential for the betterment of livestock owners and the conservation of carnivores.

References

Abade, L., et al. 2018. Spatial variation in leopard (*Panthera pardus*) site use across a gradient of anthropogenic pressure in Tanzania's Ruaha landscape. *PLoS One* 13:e0204370.

Amit, R. and S. K. Jacobson. 2017. Stakeholder barriers and benefits associated with improving livestock husbandry to prevent jaguar and puma depredation. *Human Dimensions of Wildlife* 22:246–266.

Aronsson, M. and J. Persson. 2017. Mismatch between goals and the scale of actions constrains adaptive carnivore management: the case of the wolverine in Sweden. *Animal Conservation* 20:261–269.

Aryal, A., D. Brunton, W. Ji, R. K. Barraclough, and D. Raubenheimer. 2014. Human-carnivore conflict: ecological and economical sustainability of predation on livestock by snow leopard and other carnivores in the Himalaya. *Sustainability Science* 9:321–329.

Atickem, A., S. Williams, A. Bekele, and S. Thirgood. 2010. Livestock predation in the Bale Mountains, Ethiopia. *African Journal of Ecology* 48:1076–1082.

Augugliaro, C., P. Christe, C. Janchivlamdan, H. Baymanday, and F. Zimmermann. 2020. Patterns of human interaction with snow leopard and co-predators in the Mongolian western Altai: current issues and perspectives. *Global Ecology and Conservation* 24:e01378.

Ausband, D. E., M. S. Mitchell, S. B. Bassing, and C. White. 2013. No trespassing: using a biofence to manipulate wolf movements. *Wildlife Research* 40:207–216.

Ausband, D. E., C. R. Stansbury, J. L. Stenglein, J. L. Struthers, and L. P. Waits. 2015. Recruitment in a social carnivore before and after harvest. *Animal Conservation* 18:415–423.

Ayanlade, A., et al. 2022. Extreme climate events in sub-Saharan Africa: a call for improving agricultural technology transfer to enhance adaptive capacity. *Climate Services* 27:100311.

Bagchi, S. and C. Mishra. 2006. Living with large carnivores: predation on livestock by the snow leopard (*Uncia uncia*). *Journal of Zoology* 268:217–224.

Baker, P. J., L. Boitani, S. Harris, G. Saunders, and P. C. L. White. 2008. Terrestrial carnivores and human food production: impact and management. *Mammal Review* 38:123–166.

Balbuena-Serrano, M. M. Zarco-González, O. Monroy-Vilchis, R. G. Morato, and R. C. De Paula. 2021. Hotspots of livestock depredation by pumas and jaguars in Brazil: a biome-scale analysis. *Animal Conservation* 24:181–193.

Balme, G. A., R. Slotow, and L. T. B. Hunter. 2010. Edge effects and the impact of non-protected areas in carnivore conservation: Leopards in the Phinda-Mkhuze Complex, South Africa. *Animal Conservation* 13:315–323.

Barclay, P. D. 2002. A 'curious and grim testimony to a persistent human blindness': wolf bounties in North America, 1630–1752. *Ethics, Place and Environment* 5: 25–34.

Baynes-Rock, M. 2013. Local tolerance of hyena attacks in east Hararge region, Ethiopia. *Anthrozoos* 26:421–433.

Beattie, K., E. R. Olson, B. Kissui, A. Kirschbaum, and C. Kiffner. 2020. Predicting livestock depredation risk by African lions (*Panthera leo*) in a multi-use area of northern Tanzania. *European Journal of Wildlife Research* 66(1):11.

Beck, J. M., R. J. Moll, B. M. Kissui, and R. A. Montgomery. 2021. Do pastoralist cattle fear African lions? *Oikos* 130:422–430.

Beckmann, J. P., and C. W. Lackey. 2018. Lessons learned from a 20-year collaborative study on American black bears. *Human-Wildlife Interactions* 12:396–404.

Behnam, N.-A. and H. Nader. 2021. Analysis of conflict reduction strategies in Iran; case study: Kharvana district, East Azerbaijan province. *Journal of Wildlife and Biodiversity* 5:44–55.

Benson, J. F., et al. 2023. The ecology of human-caused mortality for a protected large carnivore. *Proceedings of the National Academy of Sciences of the United States of America* 120:e2220030120.

Berger-Tal, O., D. T. Blumstein, and R. R. Swaisgood. 2020. Conservation translocations: a review of common difficulties and promising directions. *Animal Conservation* 23:121–131.

Bergstrom, B. J. 2017. Carnivore conservation: shifting the paradigm from control to coexistence. *Journal of Mammalogy* 98:1–6.

Bergstrom, B. J., L. C. Arias, A. D. Davidson, A. W. Ferguson, L. A. Randa, and S. R. Sheffield. 2014. License to kill: reforming federal wildlife control to restore biodiversity and ecosystem function. *Conservation Letters* 7:131–142.

Bhatia, S., K. Suryawanshi, S. M. Redpath, and C. Mishra. 2021. Understanding people's responses toward predators in the Indian Himalaya. *Animal Conservation* 24:424–431.

Blackwell, B. F., T. L. DeVault, E. Fernández-Juricic, E. M. Gese, L. Gilbert-Norton, and S. W. Breck. 2016. No single solution: application of behavioural principles in mitigating human–wildlife conflict. *Animal Behaviour* 120:245–254.

Blumenschine, R. J., J. A. Cavallo, and S. D. Capaldo. 1994. Competition for carcasses and early hominid behavioral ecology—a case study and conceptual framework. *Journal of Human Evolution* 27:197–213.

Bombieri, G., et al. 2023. A worldwide perspective on large carnivore attacks on humans. *PLoS Biology* 21:e3001946.

Boyce, M. S., et al. 1999. Seasonal compensation of predation and harvesting. *Oikos* 87:419–426.

Boydston, E. E., K. M. Kapheim, H. E. Watts, M. Szykman, and K. E. Holekamp. 2003. Altered behaviour in spotted hyenas associated with increased human activity. *Animal Conservation* 6:207–219.

Braczkowski, A., J. Fattebert, R. Schenk, C. O'Bryan, D. Biggs, and M. Maron. 2020. Evidence for increasing human-wildlife conflict despite a financial compensation scheme on the edge of a Ugandan National Park. *Conservation Science and Practice* 2(12):e309.

Braczkowski, A. R., et al. 2023. The unequal burden of human-wildlife conflict. *Communications Biology* 6:182.

Bradley, E. H., et al. 2005. Evaluating wolf translocation as a nonlethal method to reduce livestock conflicts in the northwestern United States. *Conservation Biology* 19:1498–1508.

Breck, S. W., et al. 2011. Domestic calf mortality and producer detection rates in the Mexican wolf recovery area: implications for livestock management and carnivore compensation schemes. *Biological Conservation* 144: 930–936.

Broekhuis, F., S. A. Cushman, and N. B. Elliot. 2017. Identification of human–carnivore conflict hotspots to prioritize mitigation efforts. *Ecology and Evolution* 7:10630–10639.

Bugir, C. K., et al. 2021. Prey preferences of modern human hunter-gatherers. *Food Webs* 26:e00183.

Cardillo, M., A. Purvis, W. Sechrest, J. L. Gittleman, J. Bielby, and G. M. Mace. 2004. Human population density and extinction risk in the world's carnivores. *PLoS Biology* 2:e197.

Chapron, G., S. Legendre, R. Ferrière, J. Clobert, and R. G. Haight. 2003. Straťgies de conservation et de contrôle du loup (*Canis lupus*) en Europe occidentale basées sur des modèles démographiques. *Comptes Rendus—Biologies* 326:575–587.

Comly-Gericke, L. M. and M. R. Vaughan. 1997. Survival and reproduction of translocated Virginia black bears. *Bears: Their Biology and Management* 9:113–117.

Cooke, R. F., D. W. Bohnert, M. M. Reis, and B. I. Cappellozza. 2013. Wolf presence in the ranch of origin: Impacts on temperament and physiological responses of beef cattle following a simulated wolf encounter. *Journal of Animal Science* 91:5905–5911.

Cooley, H. S., R. B. Wielgus, G. M. Koehler, H. S. Robinson, and B. T. Maletzke. 2009. Does hunting regulate cougar populations? A test of the compensatory mortality hypothesis. *Ecology* 90:2913–2921.

Creel, S. and J. J. Rotella. 2010. Meta-analysis of relationships between human offtake, total mortality and population dynamics of gray wolves (*Canis lupus*). *PLoS One* 5:e12918.

Crooks, K. R. and M. E. Soule. 1999. Mesopredator release and avifaunal extinctions in a fragmented system. *Nature* 400:563–566.

Cunningham, A. A. 1996. Disease risks of wildlife translocations. *Biology* 10:349–353.

Darimont, C. T., C. H. Fox, H. M. Bryan, and T. E. Reimchen. 2015. The unique ecology of human predators. *Science* 349:858–860.

Davidson-Nelson, S. J. and T. M. Gehring. 2010. Testing fladry as a nonlethal management tool for wolves and coyotes in Michigan. *Human-Wildlife Interactions* 4:87–94.

Delibes, M., P. Ferreras, and P. Gaona. 2001. Attractive sinks, or how individual behavioural decisions determine source-sink dynamics. *Ecology Letters* 4:401–403.

Dellinger, J. A., D. K. Macon, J. L. Rudd, D. L. Clifford, and S. G. Torres. 2021. Temporal trends and drivers of mountain lion depredation in California, USA. *Human-Wildlife Interactions* 15:162–177.

Deneke, Y., A. Megaze, W. Tekalign, T. Dobamo, and H. Leirs. 2022. Livestock depredation by wild carnivores in the highlands of Wolaita zone, southern Ethiopia. *Wildlife Research* 50:301–309.

Dickman, A. J., L. Hazzah, C. Carbone, and S. M. Durant. 2014. Carnivores, culture and 'contagious conflict': multiple factors influence perceived problems with carnivores in Tanzania's Ruaha landscape. *Biological Conservation* 178:19–27.

Drouilly, M., N. Nattrass, and M. J. O'Riain. 2021. Beauty or beast? Farmers' dualistic views and the influence of aesthetic appreciation on tolerance towards black-backed jackal and caracal. *PLoS One* 16:e0248977.

Eklund, A., J. V. López-Bao, M. Tourani, G. Chapron, and J. Frank. 2017. Limited evidence on the effectiveness of interventions to reduce livestock predation by large carnivores. *Scientific Reports* 7:2097.

Eshete, G., J. Marino, and C. Sillero-Zubiri. 2018a. Ethiopian wolves conflict with pastoralists in small Afroalpine relicts. *African Journal of Ecology* 56:368–374.

Eshete, G., et al. 2018b. Does livestock predation reflect in negative local perceptions of Ethiopian wolves in South Wollo? *Tropical Ecology* 59:11–19.

Farhadinia, M. S., P. J. Johnson, L. T. B. Hunter, and D. W. Macdonald. 2017. Wolves can suppress goodwill for leopards: patterns of human-predator coexistence in northeastern Iran. *Biological Conservation* 213:210–217.

Fernández-Gil, A., J. Naves, A. Ordiz, M. Quevedo, E. Revilla, and M. Delibes. 2016. Conflict misleads large carnivore management and conservation: Brown bears and wolves in Spain. *PLoS One* 11:e0151541.

Filla, M., et al. 2022. Patterns of livestock depredation by snow leopards and effects of intervention strategies: lessons from the Nepalese Himalaya. *Wildlife Research* 49(8):719–737.

Forester, D. J. and G. E. Machlis. 1996. Modeling human factors that affect the loss of biodiversity. *Conservation Biology* 10:1253–1263.

Fort, J. L., C. K. Nielsen, A. D. Carver, R. Moreno, and N. F. V. Meyer. 2018. Factors influencing local attitudes and perceptions regarding jaguars Panthera onca and National Park conservation in Panama. *ORYX* 52: 282–291.

Gálvez, N., G. Guillera-Arroita, F. A. V. St. John, E. Schüttler, D. W. Macdonald, and Z. G. Davies. 2018. A spatially integrated framework for assessing socioecological drivers of carnivore decline. *Journal of Applied Ecology* 55:1393–1405.

Gebo, B., S. Takele, and S. Shibru. 2022. Knowledge, attitude and practice of the local people towards human–carnivore coexistence in Faragosa–Fura Landscape, Gamo Zone, southern Ethiopia. *Wildlife Biology* 2022:e01018.

Gebresenbet, F., B. Baraki, G. Yirga, C. Sillero-Zubiri, and H. Bauer. 2018. A culture of tolerance: Coexisting with large carnivores in the Kafa Highlands, Ethiopia. *ORYX* 52:751–760.

Gese, E. M. and P. A. Terletzky. 2021. Space use in free-ranging canids: Are gonadal hormones required for territory maintenance? *Canadian Journal of Zoology* 99:939–947.

Gignoux, C. R., B. M. Henn, and J. L. Mountain. 2011. Rapid, global demographic expansions after the origins of agriculture. *Proceedings of the National Academy of Sciences of the United States of America* 108:6044–6049.

Gionfriddo, J. P., A. J. Denicola, L. A. Miller, and K. A. Fagerstone. 2011. Health effects of GnRH immunocontraception of wild white-tailed deer in New Jersey. *Wildlife Society Bulletin* 35:149–160.

Gray, S. M., et al. 2020. Research-implementation gap limits the actionability of human-carnivore conflict studies in East Africa. *Animal Conservation* 23(1):7–17.

Green, D. S., L. Johnson-Ulrich, H. E. Couraud, and K. E. Holekamp. 2018. Anthropogenic disturbance induces opposing population trends in spotted hyenas and African lions. *Biodiversity and Conservation* 27:871–889.

Grey, J. N. C., S. Bell, and R. A. Hill. 2017. Leopard diets and landowner perceptions of human wildlife conflict in the Soutpansberg Mountains, South Africa. *Journal for Nature Conservation* 37:56–65.

Griffith, B., J. M. Scott, J. W. Carpenter, and C. Reed. 1993. Animal translocations and potential disease transmission. *Journal of Zoo and Wildlife Medicine* 24:231–236.

Haight, R. G., L. E. Travis, K. Nimerfro, and L. D. Mech. 2002. Computer simulation of wolf-removal strategies for animal damage control. *Wildlife Society Bulletin (1973–2006)* 30:844–852.

Harvey, R. G., V. Briggs-Gonzalez, and F. J. Mazzotti. 2017. Conservation payments in a social context: determinants of tolerance and behavioural intentions towards wild cats in northern Belize. *ORYX* 51:730–741.

Hazzah, L., M. Borgerhoff Mulder, and L. Frank. 2009. Lions and warriors: social factors underlying declining African lion populations and the effect of incentive-based management in Kenya. *Biological Conservation* 142:2428–2437.

Hill, J. E., et al. 2022. Quantifying anthropogenic wolf mortality in relation to hunting regulations and landscape attributes across North America. *Ecology and Evolution* 12:e8875.

Hiller, T. L., J. E. McFadden-Hiller, S. R. Jenkins, J. L. Belant, and A. J. Tyre. 2015. Demography, prey abundance, and management affect number of cougar mortalities associated with livestock conflicts. *Journal of Wildlife Management* 79:978–988.

Hoffmann, C. F. and R. A. Montgomery. 2022. Implications of taxonomic bias for human-carnivore conflict mitigation. *ORYX* 56:917–926.

Inskip, C., Z. Fahad, R. Tully, T. Roberts, and D. MacMillan. 2014. Understanding carnivore killing behaviour: exploring the motivations for tiger killing in the Sundarbans, Bangladesh. *Biological Conservation* 180:42–50.

Johnson, C. N., J. L. Isaac, and D. O. Fisher. 2007. Rarity of a top predator triggers continent-wide collapse of mammal prey: dingoes and marsupials in Australia. *Proceedings of the Royal Society B: Biological Sciences* 274:341–346.

Jones, B. A., et al. 2013. Zoonosis emergence linked to agricultural intensification and environmental change. *Proceedings of the National Academy of Sciences of the United States of America* 110:8399–8404.

Kachel, S., K. Anderson, and Q. Shokirov. 2022. Predicting carnivore habitat use and livestock depredation risk with false-positive multi-state occupancy models. *Biological Conservation* 271:109588.

Kerth, G., M. Gusset, J. Garbely, B. König, T. Gabanapelo, and M. Schiess-Meier. 2013. Genetic sexing of stock-raiding leopards: not only males to blame. *Conservation Genetics Resources* 5:1101–1105.

Khattak, R. H., L. Teng, T. Mehmood, S. Ahmad, E. U. Rehman, S. M. Basak, and Z. Liu. 2022. A perspective of the human–grey wolf (*Canis lupus*) conflicts in Kumrat Valley, Northern Pakistan. *Diversity* 14:887.

Khorozyan, I. and M. Waltert. 2019. How long do anti-predator interventions remain effective? Patterns, thresholds and uncertainty. *Royal Society Open Science* 6:190826.

Kirkpatrick, J. F. and A. Turner. 2007. Immunocontraception and increased longevity in equids. *Zoo Biology* 26:237–244.

Kissui, B. M. 2008. Livestock predation by lions, leopards, spotted hyenas, and their vulnerability to retaliatory killing in the Maasai steppe, Tanzania. *Animal Conservation* 11:422–432.

Kissui, B. M., C. Kiffner, H. J. König, and R. A. Montgomery. 2019. Patterns of livestock depredation and cost-effectiveness of fortified livestock enclosures in northern Tanzania. *Ecology and Evolution* 9:11420–11433.

Knopff, A. A., K. H. Knopff, and C. C. St. Clair. 2016. Tolerance for cougars diminished by high perception of risk. *Ecology and Society* 21:33.

Knowlton, F. F. 1972. Preliminary interpretations of coyote population mechanics with some management implications. *The Journal of Wildlife Management* 36:369–382.

Kolowski, J. M. and K. E. Holekamp. 2006. Spatial, temporal, and physical characteristics of livestock depredations by large carnivores along a Kenyan reserve border. *Biological Conservation* 128:529–541.

Kushnir, H. and C. Packer. 2019. Perceptions of risk from man-eating lions in southeastern Tanzania. *Frontiers in Ecology and Evolution* 7:47.

Laundré, J. W., L. Hernández, and K. B. Altendorf. 2001. Wolves, elk, and bison: reestablishing the 'landscape of fear' in Yellowstone National Park, U.S.A. *Canadian Journal of Zoology* 79:1401–1409.

Laundré, J. W., L. Hernández, and W. J. Ripple. 2010. The landscape of fear: ecological implications of being afraid. *The Open Ecology Journal* 3:1–7.

Laurance, W. F., J. Sayer, and K. G. Cassman. 2014. Agricultural expansion and its impacts on tropical nature. *Trends in Ecology & Evolution* 29(2):107–116.

Linkowski, W. A., M. Kvarnström, A. Westin, J. Moen, and L. Östlund. 2017. Wolf and bear depredation on livestock in northern Sweden 1827–2014: combining history, ecology and interviews. *Land* 6:63.

Linnell, J. D. C., R. Aanes, J. E. Swenson, J. Odden, and M. E. Smith. 1997. Translocation of carnivores as a method for managing problem animals: a review. *Biodiversity and Conservation* 6:1245–1257.

Liu, M., Y. Wang, F. Xia, B. U. Hongliang, Y. Liu, X. Shen, and L. I. Sheng. 2023. Free-ranging livestock altered the spatiotemporal behavior of the endangered North Chinese leopard (*Panthera pardus japonensis*) and its prey and intensified human–leopard conflicts. *Integrative Zoology* 18:143–156.

Lorand, C., A. Robert, A. Gastineau, J. B. Mihoub, and C. Bessa-Gomes. 2022. Effectiveness of interventions for managing human-large carnivore conflicts worldwide: scare them off, don't remove them. *Science of the Total Environment* 838:156195.

Loveridge, A. J., M. Valeix, N. B. Elliot, and D. W. Macdonald. 2017. The landscape of anthropogenic mortality: how African lions respond to spatial variation in risk. *Journal of Applied Ecology* 54:815–825.

Lozano, J., et al. 2019. Human-carnivore relations: a systematic review. *Biological Conservation* 237:480–492.

Lybbert, T. J., C. B. Barrett, S. Desta, and D. L. Coppock. 2004. Stochastic wealth dynamics and risk management among a poor population. *The Economic Journal* 114: 750–777.

Malhi, Y., C. E. Doughty, M. Galetti, F. A. Smith, J. C. Svenning, and J. W. Terborgh. 2016. Megafauna and ecosystem function from the Pleistocene to the Anthropocene. *Proceedings of the National Academy of Sciences of the United States of America* 113(4):838–846.

Massei, G. and D. Cowan. 2014. Fertility control to mitigate human-wildlife conflicts: a review. *Wildlife Research* 41:1–21.

Mata-González, M., B. M. Starkovich, M. Zeidi, and N. J. Conard. 2023. Evidence of diverse animal exploitation during the Middle Paleolithic at Ghar-e Boof (southern Zagros). *Scientific Reports* 13:19006.

Mateo-Tomás, P., P. P. Olea, I. S. Sánchez-Barbudo, and R. Mateo. 2012. Alleviating human-wildlife conflicts: identifying the causes and mapping the risk of illegal poisoning of wild fauna. *Journal of Applied Ecology* 49:376–385.

Meena, V., P. J. Johnson, A. Zimmermann, R. A. Montgomery, and D. W. Macdonald. 2021. Evaluation of human attitudes and factors conducive to promoting human-lion coexistence in the Greater Gir landscape, India. *ORYX* 55:589–598.

Meltzer, D. J. 2020. Overkill, glacial history, and the extinction of North America's Ice Age megafauna. *PNAS* 117:28555–28563.

Merson, S. D., L. J. Dollar, P. J. Johnson, and D. W. Macdonald. 2019. Retaliatory killing and human perceptions of Madagascar's largest carnivore and livestock predator, the fosa (*Cryptoprocta ferox*). *PLoS One* 14:e0213341.

Meyer, W. B. and B. L. Turner. 1992. Human population growth and global land-use/cover change. *Annual Review of Ecology and Systematics* 23:39–61.

Miller, J. R. B., K. J. Stoner, M. R. Cejtin, T. K. Meyer, A. D. Middleton, and O. J. Schmitz. 2016. Effectiveness of contemporary techniques for reducing livestock depredations by large carnivores. *Wildlife Society Bulletin* 40:806–815.

Miller, S. M. and P. J. Funston. 2014. Rapid growth rates of lion (*Panthera leo*) populations in small, fenced reserves in South Africa: a management dilemma. *South African Journal of Wildlife Research* 44:43–55.

Minnie, L., A. Gaylard, and G. I. H. Kerley. 2016. Compensatory life-history responses of a mesopredator may undermine carnivore management efforts. *Journal of Applied Ecology* 53:1891.

Minnie, L., A. Zalewski, H. Zalewska, and G. I. H. Kerley. 2018. Spatial variation in anthropogenic mortality induces a source–sink system in a hunted mesopredator. *Oecologia* 186:939–951.

Mishra, C. 1997. Livestock depredation by large carnivores in the Indian trans-Himalaya: conflict perceptions and conservation prospects. *Environmental Conservation* 24:338–343.

Mitchell, A. M., B. L. Bruyere, T. O. Otieno, S. Bhalla, and T. L. Teel. 2019. A comparison between human-carnivore conflicts and local community attitudes toward carnivores in Westgate Community Conservancy, Samburu, Kenya. *Human Dimensions of Wildlife* 24:168–179.

Mkonyi, F. J., A. B. Estes, M. J. Msuha, L. L. Lichtenfeld, and S. M. Durant. 2017. Fortified bomas and vigilant herding are perceived to reduce livestock depredation by large carnivores in the Tarangire-Simanjiro ecosystem, Tanzania. *Human Ecology* 45:513–523.

Moll, R. J., A. K. Killion, M. W. Hayward, and R. A. Montgomery. 2021. A framework for the eltonian niche of humans. *BioScience* 71:928–941.

Moll, R. J., et al. 2017. The many faces of fear: a synthesis of the methodological variation in characterizing predation risk. *Journal of Animal Ecology* 86:749–765.

Montgomery, R. A., et al. 2018a. Examining evident interdisciplinarity among prides of lion researchers. *Frontiers in Ecology and Evolution* 6:49.

Montgomery, R. A., C. F. Hoffmann, E. D. Tans, and B. Kissui. 2018b. Discordant scales and the potential pitfalls for human-carnivore conflict mitigation. *Biological Conservation* 224:170–177.

Montgomery, R. A., D. W. Macdonald, and M. W. Hayward. 2020. The inducible defences of large mammals to human lethality. *Functional Ecology* 34:2426–2441.

Moreira-Arce, D., C. S. Ugarte, F. Zorondo-Rodríguez, and J. A. Simonetti. 2018. Management tools to reduce carnivore-livestock conflicts: current gap and future challenges. *Rangeland Ecology and Management* 71: 389–394.

Moriceau, J.-M. 2011. *L'Homme contre le loup: Une guerre de deux mille ans*. Fayard, Paris.

Moss, W. E., M. W. Alldredge, and J. N. Pauli. 2016. Quantifying risk and resource use for a large carnivore in an expanding urban-wildland interface. *Journal of Applied Ecology* 53:371–378.

Muhly, T. B. and M. Musiani. 2009. Livestock depredation by wolves and the ranching economy in the Northwestern U.S. *Ecological Economics* 68:2439–2450.

Naha, D., P. Chaudhary, G. Sonker, and S. Sathyakumar. 2020. Effectiveness of non-lethal predator deterrents to reduce livestock losses to leopard attacks within a multiple-use landscape of the Himalayan region. PeerJ 8:e9544.

Nelson, A. A., M. J. Kauffman, A. D. Middleton, M. D. Jimenez, D. E. Mcwhirter, and K. Gerow. 2016. Native prey distribution and migration mediates wolf (*Canis lupus*) predation on domestic livestock in the Greater Yellowstone Ecosystem. *Canadian Journal of Zoology* 94:291–299.

Ngowi, E. E., S. W. Chenyambuga, and P. S. Gwakisa. 2008. Socio-economic values and traditional management practices of Tarime zebu cattle in Tanzania. *Livestock Research for Rural Development* 20:94.

Nisi, A. C., J. F. Benson, and C. C. Wilmers. 2022. Puma responses to unreliable human cues suggest an ecological trap in a fragmented landscape. *Oikos* 2022:e09051.

Novaro, A. J., M. C. Funes, and R. S. Walker. 2005. An empirical test of source-sink dynamics induced by hunting. *Journal of Applied Ecology* 42:910–920.

Ogada, M. O., R. Woodroffe, N. O. Oguge, and L. G. Frank. 2003. Limiting depredation by African carnivores: the role of livestock husbandry. *Conservation Biology* 17(6):1521–1530.

Ohrens, O., C. Bonacic, and A. Treves. 2019. Non-lethal defense of livestock against predators: flashing lights deter puma attacks in Chile. *Frontiers in Ecology and the Environment* 17:32–38.

Packer, C., D. Ikanda, B. Kissui, and H. Kushnir. 2005. Lion attacks on humans in Tanzania. *Nature* 436:927–928.

Packer, C., et al. 2009. Sport hunting, predator control and conservation of large carnivores. *PLoS One* 4(6): e5941.

Packer, C., et al. 2013. Conserving large carnivores: dollars and fence. *Ecology Letters* 16:635–641.

Parchizadeh, J. and J. L. Belant. 2021. Human-caused mortality of large carnivores in Iran during 1980–2021. *Global Ecology and Conservation* 27:e01618.

Pekor, A., et al. 2019. Fencing Africa's protected areas: costs, benefits, and management issues. *Biological Conservation* 229:67–75.

Pereira, J. A., et al. 2020. A small protected area facilitates persistence of a large carnivore in a ranching landscape. *Journal for Nature Conservation* 56:125846.

Petracca, L. S., et al. 2019. The effectiveness of hazing African lions as a conflict mitigation tool: implications for carnivore management. *Ecosphere* 10:e02967.

Piña-Covarrubias, E., C. Chávez, and C. P. Doncaster. 2022. Knowledge of wildlife, hunting, and human-felid interactions in Maya forest communities of the Northern Yucatán Peninsula, Mexico. *Human Ecology* 50: 1035–1045.

Pohja-Mykrä, M., T. Vuorisalo, and S. Mykrä. 2005. Hunting bounties as a key measure of historical wildlife management and game conservation: Finnish bounty schemes 1647–1975. *ORYX* 39:284–291.

Potgieter, G. C., G. I. H. Kerley, and L. L. Marker. 2016. More bark than bite? the role of livestock guarding dogs in predator control on Namibian farmlands. *ORYX* 50:514–522.

Prugh, L. R., et al. 2009. The rise of the mesopredator. *BioScience* 59:779–791.

Puri, M., A. Srivathsa, K. K. Karanth, I. Patel, and N. S. Kumar. 2020. The balancing act: Maintaining leopard-wild prey equilibrium could offer economic benefits to people in a shared forest landscape of central India. *Ecological Indicators* 110:105931.

Radford, C., J. W. McNutt, T. Rogers, B. Maslen, and N. Jordan. 2020. Artificial eyespots on cattle reduce predation by large carnivores. *Communications Biology* 3(1):430.

Ramesh, T., R. Kalle, D. Milda, V. Gayathri, M. Thanikodi, K. Ashish, and A. J. Giordano. 2020. Patterns of livestock predation risk by large carnivores in India's Eastern and Western Ghats. *Global Ecology and Conservation* 24:e01366.

Ramler, J. P., M. Hebblewhite, D. Kellenberg, and C. Sime. 2014. Crying wolf a spatial analysis of wolf location and depredations on calf weight. *American Journal of Agricultural Economics* 96:631–656.

Rao, K. S., R. K. Maikhuri, S. Nautiyal, and K. G. Saxena. 2002. Crop damage and livestock depredation by wildlife: a case study from Nanda Devi Biosphere Reserve, India. *Journal of Environmental Management* 66:317–327.

Rashid, W., J. Shi, I. U. Rahim, S. Dong, and H. Sultan. 2020. Issues and opportunities associated with trophy hunting and tourism in Khunjerab national park, Northern Pakistan. *Animals* 10:597.

Ripple, W. J., et al. 2014. Status and ecological effects of the world's largest carnivores. *Science* 343:1241484.

Robinson, H. S., R. B. Wielgus, H. S. Cooley, and S. W. Cooley. 2008. Sink populations in carnivore management: cougar demography and immigration in a hunted population. *Ecological Applications* 18:1028–1037.

Rogers, L. L. 1986. Effects of translocation distance on frequency of return by adult black bears. *Wildlife Society Bulletin (1973–2006)* 14:76–80.

Rouviere, A., and R. A. Montgomery. 2025. From encounter to death: Which stages of predation are considered within livestock depredation research? *Biological Conservation* 309:111330.

Rust, N. A. 2017. Can stakeholders agree on how to reduce human-carnivore conflict on Namibian livestock farms? A novel Q-methodology and Delphi exercise. *ORYX* 51:339–346.

Rust, N. A. and L. L. Marker. 2014. Cost of carnivore coexistence on communal and resettled land in Namibia. *Environmental Conservation* 41:45–53.

Samelius, G., et al 2021. Keeping predators out: testing fences to reduce livestock depredation at night-time corrals. *ORYX* 55:466–472.

Sandercock, B. K., E. B. Nilsen, H. Brøseth, and H. C. Pedersen. 2011. Is hunting mortality additive or compensatory to natural mortality? Effects of experimental harvest on the survival and cause-specific mortality of willow ptarmigan. *Journal of Animal Ecology* 80:244–258.

Santiago-Avila, F. J., A. M. Cornman, and A. Treves. 2018. Killing wolves to prevent predation on livestock may protect one farm but harm neighbors. *PLoS One* 13:e0189729.

Saunders, G., et al. 2002. The effects of induced sterility on the territorial behaviour and survival of foxes. *Journal of Applied Ecology* 39:56–66.

Scasta, J. D., J. L. Windh, and B. Stam. 2018. Modeling large carnivore and ranch attribute effects on livestock predation and nonlethal losses. *Rangeland Ecology and Management* 71:815–826.

Sebsibe, I. 2022. Humans-livestock predators conflict in the Central and Eastern Part of Bale Mountains National Park, Ethiopia. *BMC Ecology and Evolution* 22:113.

Shivik, J. A. 2006. Tools for the edge: what's new for conserving carnivores. *BioScience* 56:253–259.

Sigari, D., et al. 2024. The last cave lion of the late Upper Palaeolithic: the engraved feline of Grotta Romanelli (southern Italy). *Quaternary Science Reviews* 334:108670.

Sijapati, R. K., H. P. Sharma, S. Sharma, J. R. Subedi, and J. L. Belant. 2021. Livestock depredation by leopards and tigers near Bardia National Park, Nepal. *Animals* 11:1896.

Singh, R., et al. 2015. Characterizing human–tiger conflict in and around Ranthambhore Tiger Reserve, western India. *European Journal of Wildlife Research* 61:255–261.

Sivertsen, T. R., et al. 2016. Reindeer habitat selection under the risk of brown bear predation during calving season. *Ecosphere* 7(11):e01583.

Slovikosky, S. A. and R. A. Montgomery. 2024. Large mammal behavioral defenses induced by the cues of human predation. *PNAS Nexus* 3(9):382.

Smith, J. A., et al. 2017. Fear of the human 'super predator' reduces feeding time in large carnivores. *Proceedings of the Royal Society B: Biological Sciences* 284:20170433.

Smith, J. A., Y. Wang, and C. C. Wilmers. 2015. Top carnivores increase their kill rates on prey as a response to human-induced fear. *Proceedings of the Royal Society B: Biological Sciences* 282:20142711.

Sommers, A. P., C. C. Price, C. D. Urbigkit, and E. M. Peterson. 2010. Quantifying Economic Impacts of Large-Carnivore Depredation on Bovine Calves. *Journal of Wildlife Management* 74:1425–1434.

Sovada, M. A., A. B. Sargeant, and J. W. Grier. 1995. Differential effects of coyotes and red foxes on duck nest success. *The Journal of Wildlife Management* 59:1–9.

Stuparyk, B., C. J. Horn, S. Karabatsos, and J. D. Arteaga-Torres. 2018. A meta-analysis of animal survival following translocation: comparisons between conflict and conservation efforts. *Canadian Wildlife Biology and Management* 7(1):3–17.

Thorn, M., M. Green, F. Dalerum, P. W. Bateman, and D. M. Scott. 2012. What drives human-carnivore conflict in the North West Province of South Africa? *Biological Conservation* 150:23–32.

Torres, R. T., D. Lopes, C. Fonseca, and L. M. Rosalino. 2020. One rule does not fit it all: patterns and drivers of stakeholders perspectives of the endangered Iberian wolf. *Journal for Nature Conservation* 55:125822.

Treves, A., M. Krofel, and J. McManus. 2016. Predator control should not be a shot in the dark. *Frontiers in Ecology and the Environment* 14:380–388.

Tveraa, T., A. Stien, H. Brøseth, and N. G. Yoccoz. 2014. The role of predation and food limitation on claims for compensation, reindeer demography and population dynamics. *Journal of Applied Ecology* 51:1264–1272.

Uduman, A., S. Hagerman, E. Kroc, A. Watson, A. Kittle, and A. C. Burton. 2022. Attitudes towards the Sri Lankan leopard Panthera pardus kotiya in two rural communities. *ORYX* 56:528–536.

United Nations. 2022. *World Population Prospects 2022: Summary of Results*. Available at: https://www.un.org/development/desa/pd/sites/www.un.org.development.desa.pd/files/wpp2022_summary_of_results.pdf

Valeix, M., G. Hemson, A. J. Loveridge, G. Mills, and D. W. Macdonald. 2012. Behavioural adjustments of a large carnivore to access secondary prey in a human-dominated landscape. *Journal of Applied Ecology* 49: 73–81.

Viljoen, S., M. J. O'Riain, B. L. Penzhorn, M. Drouilly, I. Vorster, and J. M. Bishop. 2021. Black-backed jackals (Canis mesomelas) from semi-arid rangelands in South Africa harbour Hepatozoon canis and a Theileria species but apparently not Babesia rossi. *Veterinary Parasitology: Regional Studies and Reports* 24:100559.

Weise, F. J., et al. 2018. Size, shape and maintenance matter: a critical appraisal of a global carnivore conflict mitigation strategy—livestock protection kraals in northern Botswana. *Biological Conservation* 225:88–97.

Weise, F. J., et al. 2015. A home away from home: insights from successful leopard (*Panthera pardus*) translocations. *Biodiversity and Conservation* 24:1755–1774.

Werhahn, G., N. Kusi, C. Sillero-Zubiri, and D. W. MacDonald. 2019. Conservation implications for the Himalayan wolf *Canis (lupus) himalayensis* based on observations of packs and home sites in Nepal. *ORYX* 53:663–669.

Whitehouse-Tedd, K., R. Wilkes, C. Stannard, D. Wettlaufer, and D. Cilliers. 2020. Reported livestock guarding dog-wildlife interactions: implications for conservation and animal welfare. *Biological Conservation* 241:108249.

Widman, M., M. Steen, and K. Elofsson. 2019. Indirect costs of sheep depredation by large carnivores in Sweden. *Wildlife Society Bulletin* 43:53–61.

Williams, C. K., et al. 2007. Population responses to sterility imposed on female European rabbits. *Journal of Applied Ecology* 44:291–301.

Wilson, S. M., E. H. Bradley, and G. A. Neudecker. 2017. Case study learning to live with wolves: community-based conservation in the Blackfoot Valley of Montana. *Human-Wildlife Interactions* 11:245–257.

Wise, M. 2013. Killing Montana's wolves: stockgrowers, bounty bills, and the uncertain distinction between predators and producers. *Montana The Magazine of Western History* 63:51–67.

Wolf, C. and W. J. Ripple. 2017. Range contractions of the world's large carnivores. *Royal Society Open Science* 4:170052.

Woodroffe, R. 2000. Predators and people: using human densities to interpret declines of large carnivores. *Animal Conservation* 3:165–173.

Woodroffe, R., and L. G. Frank. 2005. Lethal control of African lions (*Panthera leo*): local and regional population impacts. *Animal Conservation* 8:91–98.

Woodroffe, R., L. G. Frank, P. A. Lindsey, S. M. K. Ole Ranah, and S. Romañach. 2007. Livestock husbandry as a tool for carnivore conservation in Africa's community rangelands: a case-control study. *Biodiversity and Conservation* 16:1245–1260.

Yirga, G., et al. 2021. Lion (*Panthera leo*) ecology and survival in protected areas of Ethiopia. *Mammalian Biology* 101:791–801.

Zarco-González, M. M., O. Monroy-Vilchis, D. Sima, A. López, and A. García-Martínez. 2018. Why some management practices determine the risk of livestock predation by felids in the Selva Maya, Mexico? Conservation strategies. *Perspectives in Ecology and Conservation* 16:146–150.

CHAPTER 6

Lethal Effects of Roads on Large Carnivores

Mairi K. P. Poisson, Andrew R. Butler, Clara R. Dawson, Adam T. Ford, Taylor S. Readyhough, Jenna K. Scherger, and Remington J. Moll

Introduction

Roads have been a crucial structure for human society for millennia, facilitating relational, cultural, commercial, scientific, and recreational activities. While such structures have likely always affected wildlife generally and large carnivores specifically, modern transportation and human population expansion have greatly intensified such effects in recent times. Globally, there are currently at least 21.6 million km of roads, ranging from major highways to small local roads (Meijer et al. 2018, International Road Federation 2023). By 2050, roads could increase up to 23%, corresponding with the addition of nearly five million kilometres of new roadways, with most of this expansion likely to occur in Africa, South Asia, East Asia, and South America (Meijer et al. 2018). Perhaps not coincidentally, these regions of the world are home to many species of conservation concern and heightened risk of extinction due to roads (Grilo et al. 2021). While the scientific literature on road impacts on wildlife has generally increased over the past two decades, studies situated in Africa, South America, and Southeast Asia are still severely lacking (Taylor and Goldingay 2010, Bennett 2017). Similarly, significant funding has been devoted to developing wildlife crossings such as bridges and underpasses in North America and Europe, but far fewer projects have been funded in Africa, Asia, and South America (Wildlife Crossings Project 2023). Thus, while much scientific and practical progress is being made to understand and mitigate road effects on large carnivores, there is more work ahead. This work must address such geographical variation.

Roads impact wildlife communities through lethal and risk-related pathways (i.e. inducing trait changes in living animals or 'risk effects'). Roads increase habitat fragmentation, which can have deleterious effects on wildlife populations, including increasing genetic isolation (Holderegger and Di Giulio 2010), and reducing species richness (Eigenbrod et al. 2008) and fitness (Bennett 2017). Roads also affect wildlife by increasing human presence in an area (Suring and Del Frate 2002) or increasing hunting pressure (Penteriani et al. 2018), an indirect lethal effect (see also Chapter 4, this volume). Roads can also have positive effects on wildlife; for example, African wild dogs (*Lycaon pictus*) use roads as travel corridors (Abrahms et al. 2016), grey wolves (*Canis lupus*) hunt along tertiary roads (Newton et al. 2017), and grey foxes (*Urocyon cinereoargenteus*) scavenge carcasses along linear features like roads and power line access clearings (Hill et al. 2018). However, research on such positive effects is scarce, and the few examples of positive effects are species or site-specific (see Figure 6.1). Research on effects of roads on wildlife populations typically examines wildlife–vehicle collisions (Shilling and Waetjen 2015), the negative impact of roads on survival (Roger et al. 2011, Snow et al. 2012), and mitigation strategies employed to reduce collisions or support wildlife

Mairi K. P. Poisson et al., *Lethal Effects of Roads on Large Carnivores*. In: *Apex Predators in the Anthropocene*. Edited by: Robert A. Montgomery and Remington J. Moll, Oxford University Press. © Oxford University Press (2026). DOI: 10.1093/9780191996665.003.0006

Figure 6.1 Camera trap images of wolves using linear structures in various locations throughout Canada, USA. (a) Wolf chases a moose (*Alces alces*) under a wildlife crossing structure ©Adam Ford/Banff Wildlife Crossings Project. (b) Two wolves travelling on an active gravel road ©Parks Canada. (c) Wolf pausing on an active gravel road ©Parks Canada. (d) Wolf travelling on an active gravel road ©Parks Canada.

populations (Huijser et al. 2008, Mccollister and Van Manen 2010). Though lethal road impacts are seemingly well-understood, they are not as frequently studied as the behavioural effects of roads (Bennett 2017).

As highly mobile, large-bodied species with slow reproductive rates, large carnivores are highly susceptible to both lethal and risk effects of roads (Rytwinski and Fahrig 2012). Large carnivores typically persist at low population levels due to food and space requirements and territoriality (Wolf and Ripple 2016, O'Neil et al. 2020). Various types of human development throughout the world generally decrease wildlife habitat availability, which also depletes prey resources for predators (Ripple et al. 2014). This depletion leads to an indirect effect of roads on large carnivores: as roads reduce prey availability, this also results in fewer large carnivores due to a lack of food resources. It is critical to understand how roads affect large carnivores, as many are already threatened by the expansion of human development and will likely experience the largest impacts with the expansion of roads worldwide.

In this chapter, we use a formal literature review and a case study focused on grey wolves to evaluate the direct, lethal effects of roads and terrestrial transportation structures on large mammalian carnivores. This work complements and expands upon other recent reviews that focused on the impacts of roads globally on vertebrate species (Rytwinski and Fahrig 2012), mammals (Grilo et al. 2021), and large predators. This review provides important insights into global trends affecting large carnivores and future research needs.

In this chapter, we also discuss how roads often have negative effects on many large carnivores around the world (Figure 6.1). However, for wolves—one of the most common large carnivores globally—the effects of roads and linear infrastructure are more nuanced. Here, we outline the benefits and costs of road use for wolves while considering how different outcomes may be dependent on the characteristics of the road itself, including specific road density, traffic volume, and the location and type of habitat the road transects.

Methods

In October 2023, we conducted a formal literature review using the Web of Science Core Collection to conduct a topic search for studies published in English at any time on the 27 large carnivore species as defined by Ripple et al. (2014). Because we sought to review all potentially relevant literature on roads and large carnivores, we implemented a broad search strategy that was

Case Study: Wolves and Roads

Wolves are a highly vagile species that occur at low densities across much of their range. Habitat features that facilitate wolf movement are thus expected to have a net positive effect on wolf fitness. For example, linear features like resource (logging) roads, trails, pipelines, and seismic lines have been shown to increase the speed and distance at which wolves travel (Dickie et al. 2017), with faster speeds being correlated to increased kill rates (Vander Vennen et al. 2016). Wolves have been found to move 1.8 to 3 times faster when travelling on roads, 1.25 times faster on trails, and 2 times faster on pipelines and seismic lines (Zimmermann et al. 2014, Dickie et al. 2017). When combined, road and seismic line density can reduce wolf territory size in support of the Exploitation Efficiency Hypothesis (Dickie et al. 2022), whereas road density alone may increase wolf territory size (Mancinelli et al. 2018). Indeed, management targets for some species of wolf prey—like endangered woodland caribou (*Rangifer tarandus caribou*) in western Canada—have sought to reduce linear feature density to slow rates of wolf predation (Dickie et al. 2023). Whereas some organisms appear to avoid many types of roads irrespective of traffic volume (Ford and Fahrig 2008) or not avoid roads at all (Bouchard et al. 2009), the positive effects of roads on wolves appear to be linked to road characteristics (Whittington et al. 2022). Broadly speaking, as road width and traffic volume increase, the net benefits of roads diminish for wolves.

For busier, more developed roads, wolves are likely to experience lethal interactions with people. Mortality from vehicle collisions is the primary cause of death in wolves (Šver et al. 2016, Garcês et al. 2020) and is most likely to occur on roads with high-speed traffic (Dennehy et al. 2021). These effects are further compounded by the social structure of wolves. For example, the death of a breeding adult can hinder pack persistence and can be ecologically disruptive when compared to the death of a subordinate individual in the pack (Cassidy et al. 2023). Landscape context—not just an individual road—is important when considering the effects of roads on wolves. Higher road density ($\leq$0.9 km/km^2) has been shown to increase the probability of wolf mortality from hunter and trapper harvest (Person and Russell 2008). In protected areas with continuous wolf habitat and fewer roads, wolves temporally increase their road-use behaviour to match human activity fluctuations by increasing their activity at night during winter and decreasing their activity when humans are more active (e.g. during the day in summer) (Smith et al. 2022).

Mitigation efforts may help to reduce the lethal effects of roads on wildlife. Wildlife is highly responsive to road closures when human activity is removed. For example, wolf detections more than doubled near roads that were temporarily closed compared to when the road was open to human access (Whittington et al. 2019). Wildlife crossing structures (e.g. overpasses and underpasses) can be an effective mitigation strategy to reduce vehicle–wildlife collisions while promoting wildlife movement and improving population connectivity (Smith et al. 2015). Wolves show a preference towards overpasses to facilitate their movement across high-speed and high-traffic highways (Mysłajek et al., 2018; Plaschke et al., 2021) and are 10 times more active at wildlife crossing structures with low human use (<50 humans crossing/hour) (Barrueto et al. 2014). Similarly, wildlife exclusion fencing near highways aims to reduce vehicle–wildlife collisions but when used alone are not effective for carnivores (Ford et al. 2022), likely because carnivores are capable of climbing over or digging under fences (Hebblewhite et al. 2003). Further, fencing may pose a barrier to wildlife movement and can support the Prey-Trap Hypothesis, which states predators use fencing to facilitate prey detection and capture (Little et al. 2002). However, in systems where fencing was installed in combination with wildlife crossing structures, wolf predation did not increase near highways because of the mitigation measures (Ford and Clevenger 2010). Ideally, a combination of reduced human activity and road mitigation strategies should be used to prioritize fewer vehicle–wildlife collisions while facilitating wildlife movement and maintaining predator–prey interactions.

informed by consultation with life sciences librarian M. Bresnahan at the University of New Hampshire. Our primary objective was to review studies documenting lethal effects. Accordingly, we included the following search terms in addition to the Latin and common species names (see Table 6.1): Surviv* OR mortalit* OR death* OR injur* OR roadkill* OR road-kill* AND Road* OR vehic** OR highway* OR motorway* AND 'large carnivore*'. This initial search resulted in 707 papers. We then classified studies according to three different research methods: (1) carcass studies examining

Table 6.1 The 27 large carnivore species included in the literature review as described by Ripple et al. (2014). The species are organized in order of IUCN Red List status, from critically endangered to least concern. The common names for each species were used as additional search terms in our literature review. We further detailed the total number of papers that included each species, as well as the number of papers by method type. For some species, the methods do not add up to the total number of papers, in which case a specific study used multiple methods and were categorized in both tracking and carcass papers

Species	Common name(s) and search terms	IUCN Red List status	# Papers	Tracking papers	Carcass papers	Review papers
Canis rufus	'red wolf' OR 'red wolves'	critically endangered	1	0	1	0
Canis simensis	'Simien jackal*' OR 'Simien fox*' OR 'Ethiopian wolf' OR 'Ethiopian wolves'	endangered	0	0	0	0
Cuon alpinus	dhole* OR 'Indian dhole*' OR 'Asiatic wild dog*' OR 'red dog*'	endangered	1	0	0	1
Lycaon pictus	'African hunting dog*' OR 'African wild dog*' OR 'painted hunting dog*'	endangered	4	0	3	1
Panthera tigris	tiger*	endangered	4	0	1	3
Acinonyx jubatus	cheetah*	vulnerable	1	0	1	0
Canis dingo	dingo*	vulnerable†	0	0	0	0
Helarctos malayanus	'sun bear*' OR 'Malayan sun bear*'	vulnerable	0	0	0	0
Melursus ursinus	'sloth bear*'	vulnerable	3	0	1	2
Neofelis diardi	'Diardi's clouded leopard*'	vulnerable	0	0	0	0
Neofelis nebulosa	'clouded leopard*'	vulnerable	0	0	0	0
Panthera leo	lion*	vulnerable	4	0	3	1
Panthera pardus	leopard*	vulnerable	10	0	6	4
Panthera uncia	'snow leopard*'	vulnerable	1	0	0	1
Tremarctos ornatus	'spectacled bear*' OR 'Andean bear*'	vulnerable	0	0	0	0
Ursus maritimus	'white bear*' OR 'polar bear*'	vulnerable	2	1	0	1
Ursus thibetanus	'Asiatic black bear*' OR 'Asian black bear*'	vulnerable	1	1	0	0
Chrysocyon brachyurus	'maned wolf' OR 'maned wolves'	near threatened	3	0	2	1
Hyaena hyaena	'striped hyena*'	near threatened	5	0	3	2
Panthera onca	jaguar*	near threatened	0	0	0	0
Parahyaena brunnea	'brown hyena*'	near threatened	1	0	0	1
Canis lupus	wolf OR wolves OR 'grey wolf' OR 'grey wolves'	least concern	19	6	9	5
Crocuta crocuta	'spotted hyena*'	least concern	3	0	3	0
Lynx lynx	'Eurasian lynx*'	least concern	7	2	1	4
Puma concolor	cougar* OR puma* OR 'mountain lion*'	least concern	17	9	8	3
Ursus americanus	'black bear*' OR 'American black bear*'	least concern	17	8	6	3
Ursus arctos	'brown bear*' OR 'grizzly bear*'	least concern	25	6	10	9

† *Canis dingo* is not included as a species by the IUCN Red List, the status listed here was retrieved from the Canid Specialist Group (Allen et al. 2017).

roadkill specifically; (2) tracking studies examining survival utilizing radio telemetry or GPS collars; and (3) review papers or meta-analyses. We retained a study for more detailed analysis if it included at least one large carnivore species and fell into one of these three research method categories. This procedure resulted in 87 relevant papers. These papers included 41 carcass studies, 27 tracking studies, four studies that included both carcass surveys and tracking in their methods, and 15 review papers. A supplementary table is available at https://doi.org/10.6084/m9.figshare.25913530.

For each paper, we recorded the large carnivore species studied, the years of the study, the countries and continents in which the study took place, the study type (carcass, tracking, or review), and the general results of the study. For review papers, we determined if the study was formal, informal, or a meta-analysis, and if positive effects were documented. We considered a review formal if it included search terms, year range (if applicable), database, and criteria for inclusion; informal reviews did not provide detailed information in a replicable manner. For carcass studies and tracking studies, we recorded the proportion of mortality due to roads, railways, human take, or other sources. We defined human take as a direct lethal activity carried out by humans, including lethal management practices, legal hunting or trapping, illegal hunting or poaching, and accidental take. We also recorded demographic mortality data (sex and stage), road type (secondary roads, highway, railway, or not specified), road density, and methods of data collection. Specifically, we noted the manner in which roadkill data was collected, such as opportunistic reporting from the public or roadway monitoring by technicians. We also recorded further information for tracking studies: positive or negative effects of roads on survival, and if roads had a significant effect on survival.

Results

The 87 relevant studies included results for 20 large carnivore species. The four most studied species were grizzly bears (*Ursus arctos;* n = 25 studies), grey wolves (n = 19 studies), American black bears (*Ursus americanus;* n = 17 studies), and cougars (*Puma concolor;* n = 17 studies; Table 6.1). Of the studies in our review, tracking studies included seven different species of large carnivores, and carcass studies included 15. Two of the tracking studies examined multiple species, one examining the hybridization of coyotes (*Canis latrans*) and wolves (Benson et al. 2014), and the other studying the survival of three species of rehabilitated bear cubs (Beecham et al. 2015). In comparison, 15 of the 45 carcass studies included data for multiple species, though not all were large carnivores.

Research study areas were primarily located in North America (n = 46) and Europe (n = 22) (Table 6.2), with the United States (n = 37) and Canada (n = 16) representing the most common study countries. Each of the four most common study species had ranges entirely or predominantly within North America and had a conservation status of 'least concern' (IUCN 2023) (IUCN Red List). There were six large carnivore species with a status of 'least concern' in our literature review, and they were included 88 times across our review, whether as the sole focal species or in multispecies studies. Meanwhile, there were 21 large carnivore species listed as 'near threatened', 'vulnerable', 'endangered', or 'critically endangered', yet they were only included 41 times in the studies in our review. Seven species of large carnivores listed as endangered (n = 1), vulnerable (n = 5), or near threatened (n = 1) did not appear in any study in our review (Table 6.1). A full list of the studies in our review can be found in the supplemental materials online.

Table 6.2 The number of papers, and the number of papers by method type, with study sites in each continent. Some papers had study sites in multiple continents, so the grand total is not equal to the 87 papers in our review nor the number of papers by each method type

Continents	# papers	Tracking papers	Carcass papers	Review papers
Africa	4	0	4	0
Asia	18	1	9	6
Europe	22	5	12	7
North America	46	27	15	7
South America	5	0	5	0
Global	3	0	0	3

Carcass studies

Of the 45 studies using carcass data, grizzly bear (n = 10), grey wolf (n = 9), and cougar (n = 8) were the most studied species, with most studies conducted in North America (n = 15) or Europe (n = 12). The four studies conducted in Africa occurred within national parks or protected areas (Gandiwa et al. 2020), with three of these in the research and biodiversity hotspot of Eastern Africa (Drews 1995, Lala et al. 2021, Naciri et al. 2023). The five South American studies were conducted in Brazil, with four focusing on highways in regions of significant deforestation within the southeast portion of the country (Miotto et al. 2012, Srbek-Araujo et al. 2015, Saranholi et al. 2016, Costa et al. 2022). Research conducted in Asia, Europe, and North America had a larger geographic spread and variation in study site type, including both protected and non-protected areas and a variety of road types.

Large carnivore mortality due to collisions increased with road speeds (Drews 1995, Colino-Rabanal et al. 2011) and higher traffic volume (Find'o et al. 2019). One study reported that grizzly bear mortality increased with road density due to the influx of people into grizzly bear habitat, in deaths reported as 'defense of life or property', rather than the direct collisions with vehicles (Suring and Del Frate 2002). Seasonal variation in roadkill mortalities was reported in some studies, but with contrasting seasonality: grey wolf roadkill was higher in summer months in Wisconsin, USA (Stenglein et al. 2015) and grizzly bear mortality peaked in June in Nova Scotia (Fudge et al. 2007), whereas Florida panthers were mostly killed between October and April (Taylor et al. 2002) and winter had higher mortality for grey wolves in Pakistan (Tahir et al. 2013). Roadkill studies rarely reported demographic data on age class and sex of carcasses. Eight studies (17.8%) reported data on the sex and/or stage of the study species. Thompson et al. (2014) found that most cougars killed on roads in South Dakota were transient subadults, and Naderi et al. (2018) determined that male leopards were 2.5 times more likely to die on roads than females in Iran.

While most of these 45 carcass papers focused on roadkill data, some collected mortality data from various sources, such as hunting and poaching. Fifteen studies reported enough information to review data on multiple mortality sources. Several studies indicated that roads were the largest mortality source for some species, such as Florida panthers (Taylor et al. 2002) and black bears in Nevada, USA (Wynn-Grant et al. 2018). When combined, roads contributed to an average of 31.5% of large carnivore mortalities (sd = 24.0%, range = 0–99.5%), while direct human activities such as hunting, trapping, poaching, and culling contributed to an average of 51.4% of mortalities (sd = 29.5%, range = 0–99.5%).

Tracking survival papers

Thirty-one studies utilized VHF radio telemetry or GPS tracking data to quantify road effects. These studies included only seven species of large carnivores, with cougar (n = 9) and American black bear (n = 8) being the most studied. Twenty-six of the studies occurred in North America, four in Europe, and one study examined bears (Family Ursidae) in Asia, Europe, and North America (Beecham et al. 2015). Most studies examined survival and causes of mortality for their study species, while five studies specifically investigated the lethal impacts of roads. There was major variation in road type and road density in the different study areas, ranging from the gravel roads of rural Alberta, Canada (Boulanger and Stenhouse 2014) to the dense highway systems of southern California, USA (Benson et al. 2023). Mortality rates due to roads also varied extensively. Eleven studies reported an effect of roads, with nine of these being statistically significant (as determined by the study authors) being negative effects. Johnson et al. (2004) found a nonlinear effect of roads, with road mortality risk for grizzly bears increasing as road density increased but decreasing at the highest road densities. Some studies using collar data reported no mortalities due to roads (n = 4), but even among these, roads impacted survival. Similar to the Suring and Del Frate (2002) black bear study, researchers in Alaska, USA, found that road density had a significant negative effect on the probability of wolf harvest (Person and Russell 2008). Basille et al. (2013) found that Eurasian lynx (*Lynx lynx*) were at higher hunting risk as road density increased, but were able to mitigate this risk by altering their habitat selection

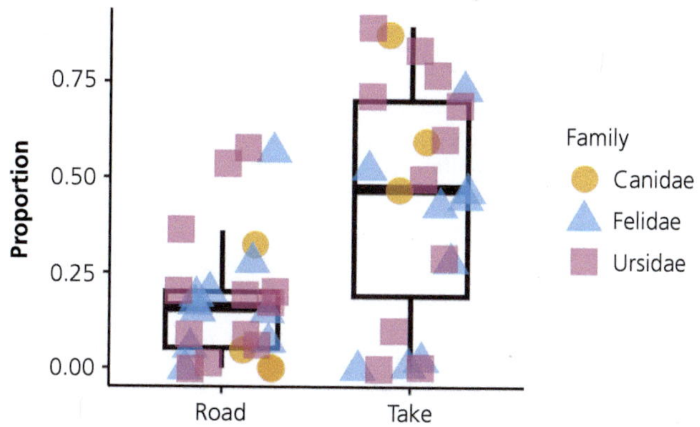

Figure 6.2 Among studies utilizing radio telemetry or GPS tracking methods, human take such as hunting, trapping, and poaching were a much higher contributor to large carnivore mortality than roads. In some studies and for some species, roads were a significant contributor but were typically not comparable to the lethality of human take.

in areas of higher road density. Overall, these studies demonstrated that roads have a negative effect on large carnivores, even if that effect is not direct mortality due to vehicle collisions.

Twenty-six studies quantified the proportion of large carnivore mortalities due to vehicle collisions, reporting an average of 18.1% (sd = 17%, range = 0–58%) (Figure 6.2). In central Florida, USA, 58% of black bear mortalities were due to vehicle collisions, the highest among the papers in our literature review. Among the 26 studies that detailed road mortalities, 23 quantified mortalities due to human take as well. Roads accounted for an average of 17.1% (sd = 0.16%, range = 0–54%) of mortalities in these 23 studies, whereas human take accounted for 44.3% (sd = 31%, range = 0–90%) of mortalities (Figure 6.2).

Fourteen of the tracking papers reported annual road mortality rates with an average proportion of 5% (sd = 6%, range = 0–17%). Total annual mortality (i.e. from all sources, including roads) in these studies ranged from 20% to 65% ($\bar{x}$ = 31%, sd = 12%). The highest annual road mortality (17%) was attributed to mining activity in British Columbia, where road mortality comprised 57% of total cougar mortality in the study (Spreadbury et al. 1996). Another study found that black bear cubs in Florida had an annual mortality of 65% in an area of higher development and road densities (Garrison et al. 2007). The annual road mortality in this study was also quite high at 13%, and the proportional road mortality for black bears in this study was 20%.

Similar to the carcass studies, there was limited demographic data detailing the sex and age of individuals among the tracking papers. Among the 11 papers that did include demographic data, mortality rates were inconsistent between males and females, juveniles, and adults (Table 6.3). Two studies showed that juveniles had a higher mortality rate than adults (Andrén et al. 2006, Tri et al. 2017), while two others found the opposite (Spreadbury et al. 1996, Hostetler et al. 2009). There was very little consistency between studies to determine any significance for mortality rates between male and female animals.

Reviews

The 15 review papers examined 17 large carnivore species, in addition to several multispecies reviews across taxa. Grizzly bears were the most common focal species among these reviews, featuring as the sole focal species in four papers and included in nine of the 15 papers overall. Grey wolves, striped hyenas (*Hyaena hyaena*), and Persian leopards (*Panthera pardus*) also featured as the focal species for one paper each, and the remaining seven papers were general reviews focusing on vertebrates (n = 5), mammals (n = 2), and northern hemisphere bears (n = 1). Eurasian lynx (*Lynx lynx*)

Table 6.3 Reported demographic data of large carnivore mortalities by the studies using radio telemetry or GPS tracking methods. There were 11 papers that included demographic data across six species. Ten of these papers were single-species analyses, while one paper examined the three North American bear species

Focal species	Road mortalities	Adult female road mortalities	Adult male road mortalities	Juvenile female road mortalities	Juvenile male road mortalities
Canis lupus	0.00	0.00	0.00	0.00	0.00
Lynx lynx	0.00	0.00	0.00	0.00	0.00
Lynx lynx	0.05	0.00	0.14	0.29	0.57
Puma concolor	0.57	0.00	0.75	0.25	0.00
Puma concolor	0.07	0.05	0.00	NA	NA
Puma concolor	0.15	0.31	0.31	0.00	0.38
Ursus americanus	0.20	NA	NA	NA	0.20
Ursus americanus	0.19	0.25	0.11	0.67	0.16
Ursus americanus	0.58	0.43	0.57	NA	NA
Ursus americanus	0.20	0.67	NA	0.33	NA
Ursus americanus, Ursus arctos, Ursus thibetanus	0.09	NA	NA	0.50	0.50

were also included in four of the 15 papers, though lacking a singular focus in any one.

Most papers (n = 10) were literature reviews regarding impacts of roads or linear features on the focal species, and five of these provided reproducible search methods. Five of the 15 papers were meta-analyses ranging from a single species to a broad taxon: Parchizadeh and Adibi (2019) examined road mortality of the Persian leopard, Theuerkauf (2009) calculated the various impacts on wolf nocturnal activity, Hill et al. (2020) analysed cause-specific mortality for mammals in relation to the Human Footprint Index, and Rytwinski and Fahrig (2012) studied the impact of roads on life history traits of vertebrates.

Overwhelmingly, roads had significant negative effects on populations of large carnivores. Many reviews identified roads as one of the leading causes of large carnivore mortality, as more mobile species seem to be more at risk of vehicle collision (Rytwinski and Fahrig 2012). In Turkey, roads are the largest threat to striped hyenas, according to a literature and grey literature review conducted by Çoğal et al. (2021). Roadkill contributed to 24% of total mortality for Persian leopards in Iran, but, similar to the results from the tracking studies noted here, this proportion was less than that of poaching and intentional poisoning (76%; Parchizadeh and Adibi 2019). Aside from direct mortality, roads affected the behaviour, survival, reproduction, and population viability of animals living in these fragmented habitats. In Theuerkauf's (2009) meta-analysis of wolf nocturnal activity, road density was one of the leading explanatory variables for increased nocturnal activity. Linnell et al. (2000) found that denning bear species, including American black bears, grizzly bears, and polar bears (*Ursus maritimus*), were more disturbed when denning within 1 km of a road, with possible den abandonment within that radius. Morales-González et al. (2020) found that grizzly bear carrying capacity was strongly reduced by roads. Proctor et al. (2020) showed that roads reduced grizzly bear reproductive rates in areas of high fragmentation, but roads also provided valuable food resources that could increase reproductive success. Further, roads increased motorized access into grizzly bear habitat, increasing the human-caused mortality rate due to hunting or poaching (Penteriani et al. 2018, Proctor et al. 2020).

According to the IUCN Red List, grizzly bear populations are of 'least concern', but grizzly bears are not considered fully recovered to their pre-colonial range in much of the United States and are thus considered 'threatened' by the US Fish and Wildlife Service (McLellan et al. 2017, U.S. Fish and Wildlife Service 2022). In a review assessing the legal battle over grizzly bear recovery in the United States, Greenwald (2023) emphasized the continued impact that roads have on this species. Notably, the author highlighted that the USFWS itself determined that limiting road density in grizzly bear recovery zones is the 'most effective way to protect grizzly bear habitat' (Greenwald 2023). In his review, Greenwald identified three court cases specifically related to road development that might threaten grizzly bear populations, all of which were lost in court. There were 23 cases related to 'roads and logging' or 'logging and roads'. Of these, 13 gained further protections for grizzly bears, often by preventing the expansion of development, eight did not gain protections, and two were still ongoing at the time of publication.

In a meta-analysis of North American mammals, Hill et al. (2020) found that 42% of mortalities were attributed to anthropogenic sources, with 7% of total mortalities the result of vehicle collisions and with higher mortality as the Human Footprint Index increased. In a formal review of vertebrates in Europe, Garcês et al. (2020) summarized the leading causes of mortality, with roadkill being the highest for Eurasian lynx and grey wolves, and among the highest for grizzly bears. Fahrig and Rytwinski's (2009) review of vertebrates in the northern hemisphere included eight large carnivore species, all experiencing negative effects of roads on abundance or distribution. Similarly, Rytwinski and Fahrig's (2012) meta-analysis of vertebrates worldwide showed that roads and/or traffic had a significant negative effect on carnivore species, and a significant negative effect on the mobility and body mass of mammals. Other transportation sectors also contribute to large carnivore mortalities, with railways being a major contributor. Dorsey et al. (2015) noted that large-bodied species are more at risk for train collisions due to their large home ranges and, for some, a preference for utilizing railways as a thoroughfare.

Discussion

Our literature review of 87 studies found that roads have the potential to be a substantial contributor to large carnivore mortality worldwide, although roads notably appeared to have a lower, and sometimes much lower, lethal impact than direct human take (see Chapter 5, this volume). The 15 reviews and meta-analyses revealed that roads had significant negative effects on carnivore species, both lethal and non-lethal. These reviews emphasized how large carnivores, as large-bodied, highly mobile species with slow reproductive rates, are at a high risk of population-level road impacts. The 45 roadkill carcass studies further underscored this point, highlighting that increases in road density, traffic density, and vehicle speed typically result in more wildlife–vehicle collisions. The 31 studies utilizing radio telemetry and GPS tracking methods found that 18% of mortalities were due to roads on average, with an annual average mortality rate of 5%. These rates varied substantially by region and by species but highlighted the considerable impact that roads can have on large carnivores, although this rate was more than two times lower than that of human take. Our review also revealed several research gaps and needs including better information on road effects across demographic categories, additional studies in the Global South, and consistent reporting of study designs to facilitate results comparability and synthesis.

We found that roads were a substantial contributor to large carnivore mortality, but human take, including legal and illegal activity, was a far larger contributor (Figure 6.2; see also Chapters 4 and 5, this volume). In a meta-analysis of North American mammals, harvest was attributed to 31% of mortalities (Hill et al. 2020). While harvest was certainly a large proportion of mammal mortality, harvest activity is typically targeted towards trophy individuals and healthy animals and is often restricted to certain seasons and species (Coe et al. 1980, Leclerc et al. 2016). Other illegal human take is also a substantial contributor and can also be targeted towards certain species or individuals. For example, despite being listed as endangered, poaching and intentional poisoning contributed to 76% of Persian leopard mortality in Iran (Parchizadeh and Adibi 2019). Meanwhile, road mortality was markedly lower than human take, but less targeted. Any animal can die on a road, regardless of conservation status, body size, or economic value. Thus, roads are likely more indiscriminate in their lethal effects than other forms of mortality, such as hunting, poaching, and intraspecific aggression, although additional research on vehicle collision variation by demographic characteristics is required to evaluate this hypothesis. Understanding the demographics of roadkill data for various species is critical for determining population-scale impacts, and our review suggests this is a major research gap.

Specifics regarding the sex and age class of study animals were lacking from the results of many empirical studies in our review. Eleven tracking studies (35.5%) and eight carcass studies (17.8%) reported either sex or age-based mortality statistics. The data reported varied substantially, such that drawing any species-specific conclusions was not possible (Table 6.3). It is believed that dispersing individuals are at higher risk of vehicle collisions (Laliberté and St-Laurent 2020, Raymond et al. 2021), but most of these studies did not include sufficient information to confirm or refute this. To develop better mitigation strategies that can address the species-specific risk of road mortality, more research studying the sex- and age-specific rates of mortality is necessary.

Road type descriptions and details were often not supplied, and when supplied the terms used were often variable and sometimes vague. For example, 26 of the carcass studies did not report road type. Among those that did report, terms like 'highway', 'motorway', 'main roads', and 'wide-lane roads' were all used. Standardization for defining road type is useful in understanding the impacts of roads, but is challenging given location-specific terminological norms. Even more useful would be to include detailed information for study areas in terms of paved and non-paved road densities, traffic speeds, and traffic volume. Studies in our review rarely provided consistent information on road densities throughout their study areas, thus we could not directly compare studies in this regard. Standardized reporting of such data could facilitate meta-analyses and enable more generalizable conclusions regarding road effects on large carnivores. In turn, such information could guide mitigation efforts such as signage, speed reductions, or passage construction. As many studies reported, mortality rates tended to increase with these variables, but it is difficult to draw comparisons across study areas and species when this information is not reported.

In survival models, roads tended to have negative effects among our sample of studies, but other research shows evidence for positive effects. Our review was not intended to capture behavioural or positive effects of roads, but it is worth noting that many studies have documented examples for certain species (e.g. *Lycaon pictus*, Abrahms et al. 2016; *Canis lupus*, Newton et al. 2017; *Urocyon cinereoargenteus*, Hill et al. 2018). Our search revealed a lack of research on the positive effects of roads, which may reveal that there are few positive effects, and those few examples are highly specific to a study area or a sample of a population. However, there might also be a bias in the literature towards what we expect, i.e. that roads harm wildlife. It is possible, for example, that populations which are already more negatively affected by roads are being studied more frequently (Fournier et al. 2019), or that a 'file-drawer' effect influences the literature, where significant negative effects are more likely to be reported than non-significant or positive effects (Jennions and Moeller 2002). Of course, any such positive road effects would be behaviourally-mediated or indirect (e.g. providing travel corridors or increasing prey availability, which in turn could influence large carnivore survival or reproduction), and thus might be difficult to detect or transient in nature. Survey methods involved in carcass roadkill studies may also lead us to certain results that reveal more about what species are killed, as opposed to what species survive. The methods for gathering these data varied, but were biased towards reports of dead animals from the general public or agencies, and less input from hunting and trapping surveys. The bias in data collection methods for the carcass studies could contribute to misunderstanding roads as a larger source of large carnivore mortality than it truly is, as this review and others show that human take tends to have a much stronger impact.

Roadkill studies are typically much more cost-effective than tracking studies and can rely on survey crews, road crews from transportation departments, or input from citizen/participatory science data (Behera and Borah 2010, Smith 2012, Taiwan Biodiversity Research Institute 2023). In comparison, tracking studies require high-cost equipment

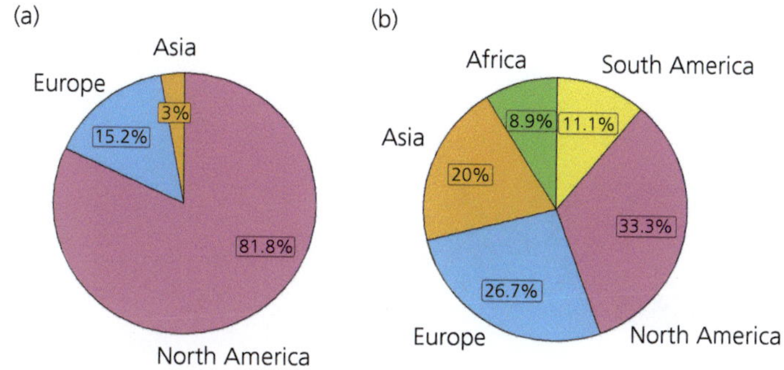

Figure 6.3 Studies using radio telemetry or GPS tracking methods (n = 31) were conducted in only three continents and predominantly in North America (81.8%). Carcass studies relying on roadkill data were conducted in five continents and were more evenly distributed across these study regions, with North America (33.3%), Europe (26.7%), and Asia (20%) representing the highest quantity of studies.

and trained, paid technicians to fit collars and conduct surveys. Tracking studies are also biased towards focal species of interest, whereas carcass studies utilizing roadkill data may consider many species in their analyses. These observations again highlight potential biases of these two methods: tracking studies are expensive and require forethought into which species to study, while carcass studies are inexpensive and study species can be determined post hoc from mortality data. Stenglein et al. (2015) analysed variation in mortality rates and detection bias of grey wolves in Wisconsin, USA, using data from both collared and non-collared individuals. Researchers approached this study with an understanding that the mortality of collared individuals would be more spatially varied and exhibit higher temporal specificity, whereas the mortality of non-collared wolves would be more opportunistic and spatially specific to areas frequented by people. These trends are apparent in their reporting of the five different mortality causes and the annual proportion within each category. Among collared individuals, illegal harvest was the highest cause of mortality, but ranked in the middle for non-collared individuals. The non-collared data likely did not accurately represent mortality sources, as the animals had to be discovered for the data to be recorded. In contrast, non-collared individuals had a much higher proportion of mortalities due to vehicles as compared to collared individuals, again suggesting the potential for vehicle-collision bias in carcass data collection methods. Therefore, caution is warranted when extrapolating mortality trends from carcass data, and tracking studies should be conducted when possible to quantify mortality sources.

Countries and organizations with better-funded programmes will have greater opportunities to implement tracking studies. The 31 tracking studies were conducted in only three continents and nine countries, with the United States featuring in 22 of the studies. North America was the study area for 81.8% of tracking papers, while there were no tracking papers in South America or Africa, and only one in Asia (Table 6.2). Meanwhile, there was a more even spread of carcass papers across the five continents where these studies occurred, though North America still featured most heavily (33.3%), and large carnivores in South America (11.1%) and Africa (8.9%) were less frequently studied (Figure 6.3). This might be an indicator of where wildlife and large carnivore research funding is awarded most, with a heavy bias towards North America. However, we note that our review was not exhaustive. We restricted our search to the scientific literature and did not capture grey literature or scientific articles published in non-English journals. This approach likely limited some of our findings about regional disparities. Still, we suspect that even despite these considerations, there is a disproportionately low amount of road research being conducted in the Global South. Such underrepresentation is important because the Global South contains many of the world's most endangered large

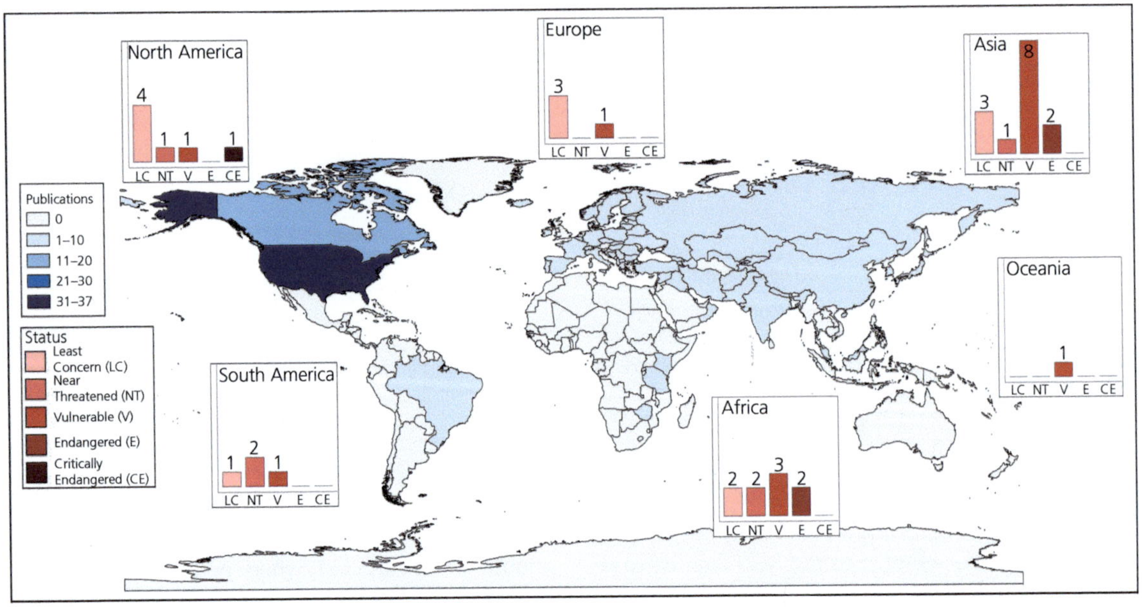

Figure 6.4 The number of publications by country for the 87 papers in our review of the lethal impact of roads on large carnivore species. The United States had the highest number of publications, with 37 total papers. The conservation status of each continent's large carnivore species is included in the bar graphs, with range data and status from the IUCN Red List. For example, in Asia there are three species with a status of 'least concern' and eight species listed as 'vulnerable'.

carnivore species (Figure 6.4), and road construction is expected to increase more in these regions than in Europe or North America (Meijer et al. 2018, IUCN 2023).

In a meta-analysis addressing the conservation risk that roads pose towards large carnivores, Quintana et al. (2022) found that all of the 36 species in their analysis are exposed to road risk, though certain species were more heavily threatened, such as the sloth bear (*Melursus ursinus*), tiger (*Panthera tigris*), and dhole (*Cuon alpinus*). Our review found that these species were rarely included in road mortality studies: tigers in four papers, sloth bears in three, and dholes in one. Quintana et al. (2022) proposed a method for determining future road risk on the 36 large carnivore species included in their study, focusing on three regions of heaviest impact: the Brazilian Amazon, sub-Saharan Africa, and Nepal. Africa and Asia have many large carnivore species listed as vulnerable or endangered, but only four papers on African large carnivores appeared in our results, and only 18 papers on Asian species. Currently, most of Africa's road density consists of smaller tertiary roads, but significant increases in construction are anticipated in the next 30 years (Meijer et al. 2018). Portions of South Asia and East Asia have very high road densities and are expected to have among the highest increases in road density by 2050 (Meijer et al. 2018). In a review of the impacts of roads on China's wildlife species, Wang et al. (2022) highlighted the lack of research on road ecology in the country. The authors suggested directing more resources towards tracking studies while also harnessing the power of participatory science data, specifically by establishing a national monitoring platform for the public to report roadkill. Future research on large carnivores should seek to address concerns about these vulnerable and endangered species in areas where road development is rapidly increasing. There is clearly a disparity between species with higher conservation need and the lack of papers studying the lethal impacts of roads, especially in regions where these species may become increasingly threatened by expanding road systems in the years to come.

Where roads are determined to have significant negative effects on wildlife, whether lethal or non-lethal, recommendations for further protections must be heeded and those actions must be studied as well. Research addressing the mitigation strategies for reducing roadkill and impacts of roads is sparse. Some research examines the effectiveness of mitigation strategies, such as Gilhooly et al.'s (2019) study examining the mitigation efforts along the TransCanada Highway as compared to the unmitigated Canadian Pacific Railway. Our review revealed few other examples, but rather highlighted a growing call for more mitigation tactics. Other research suggests improvements that could prevent animals from crossing road surfaces, such as habitat restoration near crossing points, animal proof fencing to funnel wildlife through designated passages (Ng et al. 2004), and modifications to already existing structures that cross underneath road systems (i.e. drainage culverts). In North America, funding directed towards improving or building crossing structures has increased (Huijser et al. 2008, U.S. DOT Federal Highway Administration 2023). Research into the effectiveness of these strategies must accompany this development to ensure that we are funding projects that are successful.

Conclusions

Road development represents one of many current threats to large carnivores globally, particularly as a mortality source for large-bodied species with large home ranges and slow reproductive rates. While roads are not often the leading cause of carnivore mortality, they can contribute to population declines, reduced reproductive rates, and altered home ranges. Further, they can contribute to increased lethal take as roads bring people further into carnivore habitat, thus increasing the contribution of human-caused mortality. Conversely, roads could have positive effects on some species if they lead to greater travel or hunting efficiency. Loss of large carnivore populations has had and will have significant top-down ecosystem-level effects such as trophic downgrading and mesopredator release (Estes et al. 2011, Jachowski et al. 2020, Chapters 10–12, this volume). Research addressing the ways roads affect large carnivores has been primarily focused in regions where species' conservation status is of lower concern, i.e. North America and Europe. Understanding how roads impact large carnivores in other regions of the world is critical if we are to develop mitigation strategies specific to the life history traits of species which are at the highest threat of anthropogenic-driven declines.

Acknowledgements

Partial funding support was provided by the New Hampshire Department of Transportation through Project 42372P, the National Science Foundation Graduate Research Fellowship Programme, and the New Hampshire Agricultural Experiment Station through the USDA National Institute of Food and Agriculture McIntire-Stennis Project NH 00105-M-7003422. This is Scientific Contribution number 3016.

References

Abrahms, B., N. R. Jordan, K. A. Golabek, J. W. McNutt, A. M. Wilson, and J. S. Brashares. 2016. Lessons from integrating behaviour and resource selection: activity-specific responses of African wild dogs to roads. *Animal Conservation* 19:247–255.

Allen, B. L., L. R. Allen, G. Ballard, S. M. Jackson, and P. J. Fleming. 2017. A roadmap to meaningful dingo conservation. *Canid Biology & Conservation* 20:45–56.

Andrén, H., et al. 2006. Survival rates and causes of mortality in Eurasian lynx (*Lynx lynx*) in multi-use landscapes. *Biological Conservation* 131:23–32.

Barrueto, M., A. T. Ford, and A. P. Clevenger. 2014. Anthropogenic effects on activity patterns of wildlife at crossing structures. *Ecosphere* 5:1–19.

Basille, M., et al. 2013. Selecting habitat to survive: the impact of road density on survival in a large carnivore. *PLoS One* 8:e65493.

Beecham, J. J., et al. 2015. Management implications for releasing orphaned, captive-reared bears back to the wild. *The Journal of Wildlife Management* 79:1327–1336.

Behera, S., and J. Borah. 2010. Mammal mortality due to road vehicles in Nagarjunasagar-Srisailam Tiger Reserve, Andhra Pradesh, India. *Mammalia* 74:427–430.

Bennett, V. J. 2017. Effects of road density and pattern on the conservation of species and biodiversity. *Current Landscape Ecology Reports* 2:1–11.

Benson, J. F., et al. 2023. The ecology of human-caused mortality for a protected large carnivore. *Proceedings of the National Academy of Sciences* 120:e2220030120.

Benson, J. F., B. R. Patterson, and P. J. Mahoney. 2014. A protected area influences genotype-specific survival and the structure of a Canis hybrid zone. *Ecology* 95: 254–264.

Bouchard, J., A. T. Ford, F. E. Eigenbrod, and L. Fahrig. 2009. Behavioral responses of Northern leopard frogs (*Rana pipiens*) to roads and traffic: Implications for population persistence. *Ecology and Society* 14:1–10.

Boulanger, J., and G. B. Stenhouse. 2014. The impact of roads on the demography of grizzly bears in Alberta. *PLoS One* 9:e115535.

Cassidy, K. A., et al. 2023. Human-caused mortality triggers pack instability in gray wolves. *Frontiers in Ecology and the Environment* 21:356–362.

Coe, R. J., R. L. Downing, and B. S. McGinnes. 1980. Sex and age bias in hunter-killed white-tailed deer. *The Journal of Wildlife Management* 44:245–249.

Çoğal, M., Y. İlemin, and M. Sözen. 2021. Status and distribution of the Striped Hyaena, Hyaena hyaena, in Turkey: an updated assessment (Carnivora: Mammalia). *Turkish Journal of Zoology* 45:131–141.

Colino-Rabanal, V. J., M. Lizana, and S. J. Peris. 2011. Factors influencing wolf *Canis lupus* roadkills in Northwest Spain. *European Journal of Wildlife Research* 57: 399–409.

Costa, I. M. da C., M. S. Ferreira, C. L. B. Mourão, and C. Bueno. 2022. Spatial patterns of carnivore roadkill in a high-traffic-volume highway in the endangered Brazilian Atlantic Forest. *Mammalian Biology* 102: 477–487.

Dennehy, E., L. Llaneza, and J. V. López-Bao. 2021. Contrasting wolf responses to different paved roads and traffic volume levels. *Biodiversity and Conservation* 30:3133–3150.

Dickie, M., N. Love, R. Steenweg, C. T. Lamb, J. Polfus, and A. T. Ford. 2023. In search of evidence-based management targets: a synthesis of the effects of linear features on woodland caribou. *Ecological Indicators* 154:110559.

Dickie, M., et al. 2022. Resource exploitation efficiency collapses the home range of an apex predator. *Ecology* 103:e3642.

Dickie, M., R. Serrouya, R. S. McNay, and S. Boutin. 2017. Faster and farther: wolf movement on linear features and implications for hunting behaviour. *Journal of Applied Ecology* 54:253–263.

Dorsey, B., M. Olsson, and L. Rew. 2015. Ecological effects of railways on wildlife. In *Handbook of Road Ecology*, edited by R. van der Ree, D. J. Smith, and C. Grilo. Wiley, Hoboken, N. J.

Drews, C. 1995. Road kills of animals by public traffic in Mikumi National Park, Tanzania, with notes on baboon mortality. *African Journal of Ecology* 33:89–100.

Eigenbrod, F., S. J. Hecnar, and L. Fahrig. 2008. Accessible habitat: an improved measure of the effects of habitat loss and roads on wildlife populations. *Landscape Ecology* 23:159–168.

Estes, J. A., et al. 2011. Trophic downgrading of planet Earth. *Science* 333:301–306.

Fahrig, L. and T. Rytwinski. 2009. Effects of roads on animal abundance: an empirical review and synthesis. *Ecology and Society* 14:21.

Find'o, S., M. Skuban, M. Kajba, J. Chalmers, and M. Kalaš. 2019. Identifying attributes associated with brown bear (*Ursus arctos*) road-crossing and roadkill sites. *Canadian Journal of Zoology* 97:156–164.

Ford, A. T. and A. P. Clevenger. 2010. Validity of the prey-trap hypothesis for carnivore-ungulate interactions at wildlife-crossing structures. *Conservation Biology* 24:1679–1685.

Ford, A. T., B. Dorsey, T. S. Lee, and A. P. Clevenger. 2022. A before-after-control-impact study of wildlife fencing along a highway in the Canadian Rocky Mountains. *Frontiers in Conservation Science* 3:935420.

Ford, A. T. and L. Fahrig. 2008. Movement patterns of Eastern chipmunks (*Tamias striatus*) near roads. *Journal of Mammalogy* 89:895–903.

Fournier, A. M. V., E. R. White, and S. B. Heard. 2019. Site-selection bias and apparent population declines in long-term studies. *Conservation Biology* 33: 1370–1379.

Fudge, D., B. Freedman, M. Crowell, T. Nette, and V. Power. 2007. Road-kill of mammals in Nova Scotia. *The Canadian Field-Naturalist* 121:265–273.

Gandiwa, E., C. Mashapa, N. Muboko, A. Chemura, P. Kuvaoga, and C. T. Mabika. 2020. Wildlife-vehicle collisions in Hurungwe Safari Area, northern Zimbabwe. *Scientific African* 9:e00518.

Garcês, A., F. Queiroga, J. Prada, and I. Pires. 2020. A review of the mortality of wild fauna in Europe in the last century, the consequences of human activity. *Journal of Wildlife and Biodiversity* 4:34–55.

Garrison, E. P., J. W. Mccown, and M. K. Oli. 2007. Reproductive ecology and cub survival of Florida black bears. *The Journal of Wildlife Management* 71:720–727.

Gilhooly, P. S., S. E. Nielsen, J. Whittington, and C. C. St. Clair. 2019. Wildlife mortality on roads and railways following highway mitigation. *Ecosphere* 10: e02597.

Greenwald, N. 2023. Recovery of the grizzly bear at the intersection of law and science. *Frontiers in Conservation Science* 4:10.

Grilo, C., et al. 2021. Conservation threats from roadkill in the global road network. *Global Ecology and Biogeography* 30:2200–2210.

Hebblewhite, M., M. Percy, and R. Serrouya. 2003. Black bear (*Ursus americanus*) survival and demography in the Bow Valley of Banff National Park, Alberta. *Biological Conservation* 112:415–425.

Hill, J. E., T. L. DeVault, J. C. Beasley, O. E. Rhodes, and J. L. Belant. 2018. Roads do not increase carrion use by a vertebrate scavenging community. *Scientific Reports* 8:16331.

Hill, J. E., T. L. DeVault, G. Wang, and J. L. Belant. 2020. Anthropogenic mortality in mammals increases with the human footprint. *Frontiers in Ecology and the Environment* 18:13–18.

Holderegger, R. and M. Di Giulio. 2010. The genetic effects of roads: A review of empirical evidence. *Basic and Applied Ecology* 11:522–531.

Hostetler, J. A., et al. 2009. Demographic consequences of anthropogenic influences: Florida black bears in north-central Florida. *Biological Conservation* 142:2456–2463.

Huijser, M. P., et al. 2008. *Wildlife-Vehicle Collision Reduction Study: Report to Congress*. Available at: https://www.fhwa.dot.gov/publications/research/safety/08034/

International Road Federation. 2023. *IRF World Road Statistics 2023 Edition* (Data 2016–2021).

IUCN. 2023. *The IUCN Red List of Threatened Species*.

Jachowski, D. S., A. Butler, R. Y. Y. Eng, L. Gigliotti, S. Harris, and A. Williams. 2020. Identifying mesopredator release in multi-predator systems: a review of evidence from North America. *Mammal Review* 50:367–381.

Jennions, M. D. and A. P. Moeller. 2002. Publication bias in ecology and evolution: An empirical assessment using the 'trim and fill' method. *Biological Reviews* 77: 211–222.

Johnson, C. J., M. S. Boyce, C. C. Schwartz, and M. A. Haroldson. 2004. Modeling survival: application of the Andersen-Gill model to Yellowstone grizzly bears. *Journal of Wildlife Management* 68:966–978.

Lala, F., et al. 2021. Wildlife roadkill in the Tsavo ecosystem, Kenya: identifying hotspots, potential drivers, and affected species. *Heliyon* 7:e06364.

Laliberté, J. and M.-H. St-Laurent. 2020. In the wrong place at the wrong time: moose and deer movement patterns influence wildlife-vehicle collision risk. *Accident Analysis & Prevention* 135:105365.

Leclerc, M., J. Van de Walle, A. Zedrosser, J. E. Swenson, and F. Pelletier. 2016. Can hunting data be used to estimate unbiased population parameters? A case study on brown bears. *Biology Letters* 12:20160197.

Linnell, J. D. C., J. E. Swenson, R. Andersen, and B. Barnes. 2000. How vulnerable are denning bears to disturbance? *Wildlife Society Bulletin (1973–2006)* 28:400–413.

Little, S. J., R. G. Harcourt, and A. P. Clevenger. 2002. Do wildlife passages act as prey-traps? *Biological Conservation* 107:135–145.

Mancinelli, S., L. Boitani, and P. Ciucci. 2018. Determinants of home range size and space use patterns in a protected wolf (*Canis lupus*) population in the central Apennines, Italy. *Canadian Journal of Zoology* 96:828–838.

Mccollister, M. F., and F. T. Van Manen. 2010. Effectiveness of wildlife underpasses and fencing to reduce wildlife-vehicle collisions. *The Journal of Wildlife Management* 74:1722–1731.

McLellan, B. N., M. F. Proctor, D. Huber, and S. Michel. 2017. *Ursus arctos* (amended version of 2017 assessment). The IUCN Red List of Threatened Species.

Meijer, J. R., M. A. J. Huijbregts, K. C. G. J. Schotten, and A. M. Schipper. 2018. Global patterns of current and future road infrastructure. *Environmental Research Letters* 13:064006.

Miotto, R. A., M. Cervini, R. A. Begotti, and P. M. Galetti Jr. 2012. Monitoring a puma (*Puma concolor*) population in a fragmented landscape in Southeast Brazil. *Biotropica* 44:98–104.

Morales-González, A., H. Ruiz-Villar, A. Ordiz, and V. Penteriani. 2020. Large carnivores living alongside humans: brown bears in human-modified landscapes. *Global Ecology and Conservation* 22:e00937.

Mysłajek, R. W., et al. 2018. Spatial organization in wolves *Canis lupus* recolonizing north-west Poland: large territories at low population density. *Mammalian Biology* 92:37–44.

Naciri, M., et al. 2023. Three decades of wildlife-vehicle collisions in a protected area: Main roads and long-distance commuting trips to migratory prey increase spotted hyena roadkills in the Serengeti. *Biological Conservation* 279:109950.

Naderi, M., A. Farashi, and M. A. Erdi. 2018. Persian leopard's (*Panthera pardus saxicolor*) unnatural mortality factors analysis in Iran. *PLoS One* 13:e0195387.

Newton, E. J., B. R. Patterson, M. L. Anderson, A. R. Rodgers, L. M. V. Vennen, and J. M. Fryxell. 2017. Compensatory selection for roads over natural linear features by wolves in northern Ontario: implications for caribou conservation. *PLoS One* 12:e0186525.

Ng, S. J., J. W. Dole, R. M. Sauvajot, S. P. D. Riley, and T. J. Valone. 2004. Use of highway undercrossings by wildlife in southern California. *Biological Conservation* 115:499–507.

O'Neil, S. T., J. A. Vucetich, D. E. Beyer Jr., S. R. Hoy, and J. K. Bump. 2020. Territoriality drives preemptive habitat selection in recovering wolves: implications for carnivore conservation. *Journal of Animal Ecology* 89:1433–1447.

Parchizadeh, J. and M. A. Adibi. 2019. Distribution and human-caused mortality of Persian leopards *Panthera pardus saxicolor* in Iran, based on unpublished data and Farsi gray literature. *Ecology and Evolution* 9: 11972–11978.

Penteriani, V., et al. 2018. Evolutionary and ecological traps for brown bears *Ursus arctos* in human-modified landscapes. *Mammal Review* 48:180–193.

Person, D. K. and A. L. Russell. 2008. Correlates of mortality in an exploited wolf population. *The Journal of Wildlife Management* 72:1540–1549.

Plaschke, M., M. Bhardwaj, H. J. König, E. Wenz, K. Dobiáš, and A. T. Ford. 2021. Green bridges in a re-colonizing landscape: wolves (*Canis lupus*) in Brandenburg, Germany. *Conservation Science and Practice* 3(3):364.

Proctor, M. F., B. N. McLellan, G. B. Stenhouse, G. Mowat, C. T. Lamb, and M. S. Boyce. 2020. Effects of roads and motorized human access on grizzly bear populations in British Columbia and Alberta, Canada. *Ursus* 2019: 16–39.

Quintana, I., et al. 2022. Severe conservation risks of roads on apex predators. *Scientific Reports* 12:2902.

Raymond, S., A. L. W. Schwartz, R. J. Thomas, E. Chadwick, and S. E. Perkins. 2021. Temporal patterns of wildlife roadkill in the UK. *PLoS One* 16:e0258083.

Ripple, W. J., et al. 2014. Status and ecological effects of the world's largest carnivores. *Science* 343:1241484.

Roger, E., S. W. Laffan, and D. Ramp. 2011. Road impacts a tipping point for wildlife populations in threatened landscapes. *Population Ecology* 53:215–227.

Rytwinski, T. and L. Fahrig. 2012. Do species life history traits explain population responses to roads? A meta-analysis. *Biological Conservation* 147:87–98.

Saranholi, B. H., et al. 2016. Roadkill hotspots in a protected area of Cerrado in Brazil: Planning actions to conservation. *Revista MVZ Córdoba* 21: 5441–5448.

Shilling, F. M. and D. P. Waetjen. 2015. Wildlife-vehicle collision hotspots at US highway extents: scale and data source effects. *Nature Conservation* 11:41–60.

Smith, A. F., S. Ciuti, D. Shamovich, V. Fenchuk, B. Zimmermann, and M. Heurich. 2022. Quiet islands in a world of fear: wolves seek core zones of protected areas to escape human disturbance. *Biological Conservation* 276:109811.

Smith, D. J. 2012. Determining location and design of cost-effective wildlife crossing structures along US-64 in North Carolina. *Transportation Research Record* 2270: 31–38.

Smith, D. J., R. Van Der Ree, and C. Rosell. 2015. Wildlife crossing structures: an effective strategy to restore or maintain wildlife connectivity across roads. In *Handbook of Road Ecology*, edited by R. Van Der Ree, D. J. Smith, and C. Grilo. Wiley, Hoboken, N. J.

Snow, N. P., W. F. Andelt, T. R. Stanley, J. R. Resnik, and L. Munson. 2012. Effects of roads on survival of San Clemente Island foxes. *The Journal of Wildlife Management* 76:243–252.

Spreadbury, B. R., R. R. K. Musil, J. Musil, C. Kaisner, and J. Kovak. 1996. Cougar population characteristics in Southeastern British Columbia. *The Journal of Wildlife Management* 60:962–969.

Srbek-Araujo, A. C., S. L. Mendes, and A. G. Chiarello. 2015. Jaguar (*Panthera onca* Linnaeus, 1758) roadkill in Brazilian Atlantic Forest and implications for species conservation. *Brazilian Journal of Biology* 75:581–586.

Stenglein, J. L., et al. 2015. Mortality patterns and detection bias from carcass data: an example from wolf recovery in Wisconsin. *The Journal of Wildlife Management* 79: 1173–1184.

Suring, L. H. and G. Del Frate. 2002. Spatial analysis of locations of brown bears killed in defense of life or property on the Kenai Peninsula, Alaska, USA. *Ursus* 13:237–245.

Šver, L., A. Bielen, J. Križan, and G. Gužvica. 2016. Camera traps on wildlife crossing ctructures as a tool in gray wolf (*Canis lupus*) management—five-years monitoring of wolf abundance trends in Croatia. *PLoS One* 11:e0156748.

Tahir, H. M., S. Y. Khan, and M. Arshad. 2013. Motorway (M 2): a threat to the wild animals. *Pakistan Journal of Zoology* 45:1468–1470.

Taiwan Biodiversity Research Institute. 2023. Taiwan Roadkill Observation Network.

Taylor, B. D. and R. L. Goldingay. 2010. Roads and wildlife: Impacts, mitigation and implications for wildlife management in Australia. *Wildlife Research* 37:320.

Taylor, S. K., C. D. Buergelt, M. E. Roelke-Parker, B. L. Homer, and D. S. Rotstein. 2002. Causes of mortality of free-ranging Florida panthers. *Journal of Wildlife Diseases* 38:107–114.

Theuerkauf, J. 2009. What drives wolves: Fear or hunger? Humans, diet, climate and wolf activity. *Ethology* 115:649–657.

Thompson, D. J., J. A. Jenks, and D. M. Fecske. 2014. Prevalence of human-caused mortality in an unhunted cougar population and potential impacts to management. *Wildlife Society Bulletin* 38:341–347.

Tri, A. N., et al. 2017. Harvest rates and cause-specific mortality of American black bears in the wildland–urban interface of the Mid-Atlantic region, USA. *Ursus* 28:195–207.

U.S. DOT Federal Highway Administration. 2023, 4 April. Notice of Funding Opportunity (NOFO) for Fiscal Years (FY) 2022–2023 Wildlife Crossings Pilot Program.

U.S. Fish and Wildlife Service. 2022. *Species Status Assessment for the grizzly bear (Ursus arctos horribilis) in the Lower-48 States.* Available at: https://www.fws.gov/media/species-status-assessment-grizzly-bear-lower-48-states-v22

Vander Vennen, L. M., B. R. Patterson, A. R. Rodgers, S. Moffatt, M. L. Anderson, and J. M. Fryxell. 2016. Diel movement patterns influence daily variation in wolf kill rates on moose. *Functional Ecology* 30:1568–1573.

Wang, Y., et al. 2022. Impacts of linear transport infrastructure on terrestrial vertebrate species and conservation in China. *Global Ecology and Conservation* 38: e02207.

Whittington, J., M. Hebblewhite, R. W. Baron, A. T. Ford, and J. Paczkowski. 2022. Towns and trails drive carnivore movement behaviour, resource selection, and connectivity. *Movement Ecology* 10:17.

Whittington, J., P. Low, and B. Hunt. 2019. Temporal road closures improve habitat quality for wildlife. *Scientific Reports* 9:3772.

Wildlife Crossings Project. 2023. Wildlife Crossings Project.

Wolf, C., and W. J. Ripple. 2016. Prey depletion as a threat to the world's large carnivores. *Royal Society Open Science* 3:160252.

Wynn-Grant, R., J. R. Ginsberg, C. W. Lackey, E. J. Sterling, and J. P. Beckmann. 2018. Risky business: modeling mortality risk near the urban-wildland interface for a large carnivore. *Global Ecology and Conservation* 16:e00443.

Zimmermann, B., L. Nelson, P. Wabakken, H. Sand, and O. Liberg. 2014. Behavioral responses of wolves to roads: scale-dependent ambivalence. *Behavioral Ecology* 25:1353–1364.

SECTION THREE

The Risk Effects of Humans on Carnivore Populations

CHAPTER 7

Cumulative Costs of Carnivore Behavioural Responses to Humans

Nikita Frizzelle, Justin P. Suraci, Michael Clinchy, and Liana Y. Zanette

Introduction

It is well established that the regulation of ecosystems is largely mediated by the presence of carnivores (Schmitz et al. 2000, Estes et al. 2011), in part because these predators often play the vital role of managing wildlife populations by hunting and instilling fear in those of lower trophic groups (Cunningham et al. 2018, Zanette and Clinchy 2020). However, recent research is revealing that humans are actively altering these delicate predator–prey dynamics on a global scale by excessively hunting and destroying the natural habitats of carnivore populations and their prey in nearly every biome on Earth (Ripple et al. 2014, Wolf and Ripple 2016). Darimont et al. (2015) first introduced the term "super-predator" to the field of behavioural ecology to distinguish how humans' hunting behaviour has surpassed that of the typical apex predator. The unique ecology of the human "super-predator" is characterized by humans' global superiority to all other predators within their respective food webs, as well as humans' unmatched lethality, especially towards carnivore species. Humans kill terrestrial large carnivores at nine times the rate of any other predator and exploit large carnivores at nearly four times the rate that they kill herbivores (Darimont et al. 2015). Furthermore, humans exploit up to 300 times more species than any non-human predator on the planet, preying on more than a third of all vertebrate species (Darimont et al. 2023).

More recently, an extreme and pervasive shift in mammal behaviour has been observed in response to human presence and footprint on the global scale (Sazatornil et al. 2016, Gaynor et al. 2018, Tucker et al. 2018, Montgomery et al. 2020, O'Donnell and del Barco-Trillo 2020, Doherty et al. 2021, Zhang et al. 2023), alluding to the notion that prey and predators alike are avoiding human presence out of fear. However, the experimental evidence necessary to validate that these behavioural changes in free-living large carnivores are unequivocally due to a fear of humans as a predator is limited to specific species, such as Californian pumas (*Puma concolor*), as such field experiments are notoriously difficult to employ on fierce and elusive carnivore predators (Montgomery et al. 2020). Thus, in this chapter we will emphasize vital distinctions between (1) correlative patterns, and their purported mechanistic pathways of the ecological processes underlying behavioural responses, and (2) behavioural data collected from novel experimental designs that have utilized emerging technologies to quantify the costs of exhibiting particular behaviours. Specifically, we will highlight data that measure vigilance, movement, feeding, and foraging, as these behaviours are considered the strongest behavioural determinants of population growth in wild populations (Zanette and Clinchy 2020).

Fearing humans as a "super-predator"

For millennia, humans have considered large carnivores as our most worthy competitors and predators among the animal kingdom (Serangeli et al.

Nikita Frizzelle et al., *Cumulative Costs of Carnivore Behavioural Responses to Humans*. In: *Apex Predators in the Anthropocene*. Edited by: Robert A. Montgomery and Remington J. Moll, Oxford University Press. © Oxford University Press (2026). DOI: 10.1093/9780191996665.003.0007

2015). Humankind fears their strength, power, elusiveness, and skilled hunting capabilities, and with good reason: globally, out of the 5440 documented large carnivore attacks on humans since 1950, almost one-third (32%) were fatal (Bombieri et al. 2023). In essence, our personal survivorship becomes jeopardized when a large carnivore is nearby, and the probability of these human–wildlife encounters is predicted to increase in the coming years as the global human population grows and continues encroachment into large carnivore ranges (Tilman et al. 2011, Penteriani et al. 2016, Doherty and Driscoll 2018, Bombieri et al. 2023). And yet, 77% of large terrestrial mammalian carnivore species are currently facing population declines, with at least half of these species threatened with extinction directly caused by human interference (Ripple et al. 2014, Ingeman et al. 2022). Technological advancements have motivated humans as a species to regularly pursue and target large carnivores for non-consumptive purposes, inflating the exploitation rate of hunting large carnivores to 3.7 times that of herbivore species (Darimont et al. 2015). Thus, it is pondered as to whether carnivores hold a similar perception of humans as significant lethal threats and consequentially experience predator-induced fear in our presence.

Meta-analyses demonstrate pervasiveness of effect by human predators

Numerous studies have been conducted to examine how mammals, especially large carnivores, alter their behaviour in response to human presence, human infrastructure such as roadways, and urbanization. More recently, researchers have been compiling these observational data into large-scale global meta-analyses, which repeatedly highlight that humans' impact on mammal behaviour is uniquely pervasive and profound. Altered behaviours such as foraging, habitat selection, movement, social behaviours, and vigilance have been observed in mammals attributed to global human disturbance, with more prominent behavioural changes occurring in areas with greater intensities of human disturbance (Zhang et al. 2023). Fifty-seven mammal species have been observed to reduce their movements by up to a factor of three when living in areas of high human footprint compared to areas of relatively lower human footprint (Tucker et al. 2018). Extensively built human infrastructure within urban areas is likely the main driver of habitat fragmentation, restricting the movements of individuals and reducing opportunities for long-distance migrations (Tucker et al. 2018, O'Donnell and del Barco-Trillo 2020). Roadways also impact mammal species globally, as more specialized habitat selection has been observed in mammals living in areas with high levels of road traffic (Zhang et al. 2023) (see also Chapter 6, this volume). Hence, it is unsurprising that urban mammals possess smaller home ranges in comparison to their non-urban counterparts (O'Donnell and del Barco-Trillo 2020). Additionally, habitat degradation in the form of excessive grazing by livestock in agricultural fields has been observed to significantly reduce the time spent foraging by resident mammal species (Zhang et al. 2023). However, another global meta-analysis on 77 mammal species found that while the home range sizes of mammals decreased in response to urbanization and agriculture, mammals had instead increased their movement distances in response to human disturbance, with herbivores responding slightly greater than omnivores and carnivores (Doherty et al. 2021). One possible explanation for the contrasting results may be that—in addition to the human footprint—Doherty et al. (2021) included activities involving direct human presence such as hunting, recreation, and aircrafts as forms of human disturbance. These factors independently demonstrated stronger effects on mammal movement behaviours compared to human-induced habitat modifications. In support of this hypothesis, carnivore species have been observed to be especially sensitive to tourism, with human presence altering their movement patterns, increasing their time spent vigilant, and decreasing their time spent foraging (Zhang et al. 2023). Also, breeding sites of wolves have been observed at greater distances from human settlements and villages in comparison to home sites and rendezvous sites, perhaps suggesting that these large carnivores are actively choosing to further minimize their risk of exposure to humans when they are considered to be at their most vulnerable to persecution (Sazatornil et al. 2016).

When interpreting these meta-collections of data, it is worthwhile to recognize that although there are undoubtedly pervasive and consistent patterns of altered behaviour occurring in mammal populations across the globe, it is difficult to gauge whether these adjustments are indeed indicative of a unique ultra-threatening human predator or simply a typical response to any local large carnivore. In short, a reliable control treatment is missing. Gaynor et al.'s (2018) global meta-analysis on activity patterns in mammals is a notable exception because the naturally contrasting diurnal and nocturnal habits of humans and large carnivores, respectively, introduce a trade-off to mammal species living in close proximity to areas of high human activity and/or footprint: (1) remain most active throughout daylight hours and risk a human encounter; or (2) shift to a more nocturnal lifestyle to avoid humans and minimize the risk of human encounters but increase the risk of non-human predator encounters at night (Bonnot et al. 2020, Haswell et al. 2020). Therefore, a behavioural shift towards nocturnality would suggest that mammals perceive humans to be a greater predation threat in comparison to non-human predators. Such were the main conclusions of the study, as mammals across continents, habitats, and taxa—including apex large carnivores—increased their nocturnality by an average of 36% in response to human disturbance (Gaynor et al. 2018).

Experiments demonstrate magnitude of effect by human predators

Meta-analyses present a global pattern of mammals altering their behaviour in response to human disturbance. However, while we can presume that large carnivores alter their behaviour to minimize human encounters, these data do not provide a quantitative measure of the degree to which predation risk is attributed to humans by these predators. To systematically gauge a fear of humans in large carnivores, researchers observe exhibitions of anti-predator behaviours with particular attention to the strength or onset of the response towards human presence: the more extreme, frequent, or quick the response, the greater the perceived predation risk—and thus fear—that is likely attributed to humans (Creel et al. 2019). Given the difficulty and glaring danger of studying free-living large carnivores near human observers (Suraci et al. 2017), there is a very limited number of experimental studies that have been successfully conducted to investigate their human-induced behavioural responses. However, researchers have recently circumvented this problem by using auditory playbacks of vocalizations to mimic the nearby presence of humans and other potential competitors of large carnivores (Suraci et al. 2017, Smith et al. 2017). Various mammal species have been shown to display strong anti-predator behaviours in response to the vocalizations of their local predators, suggesting that the visualization of a predator is not a requirement for prey to perceive a threat and modify their behaviours under predation risk (Hettena et al. 2014). Thus, if humans are indeed perceived as predators by large carnivores, then we would expect large carnivores to exhibit stronger—or more frequent—anti-predator behaviours in response to broadcasts of human voices compared to non-threatening vocalizations or stimuli, such as birdsong or frog mating calls. To further conclude that carnivores perceive the presence of a "super-predator", their behavioural responses would be expected to exceed those in response to positively threatening vocalizations, such as those belonging to another local carnivorous competitor. To date, auditory playbacks of human voices have been used in six experiments to investigate a fear of humans in a free-living carnivore and each study independently drew an identical data-based conclusion: carnivores fear human presence (Clinchy et al. 2016, Smith et al. 2017, Suraci et al. 2019a, Suraci et al. 2019b, Reilly et al. 2022, Zanette et al. 2023). When discussing the findings of these studies, anti-predator behaviour of mesocarnivores is generally reported separate to those of large carnivores, as it can be anticipated that the two groups may respond to predation threats differently; mesocarnivores regularly display fear towards large carnivores as non-human predators in their communities whereas large carnivores belong to the apex trophic level and typically do not fear other members of the community barring humans (Suraci et al. 2016).

The first experiment demonstrating carnivore anti-predator behavioural responses to human voice playbacks was conducted on the European badger (*Meles meles*) in Britain, where the species is heavily managed and persecuted as an agricultural pest. The badgers spent less time feeding, decreased the frequency of foraging patch visits, and displayed increased levels of vigilance compared to vocalizations of sheep, which were used as a non-threatening negative control treatment—all of which demonstrate that badgers perceive humans as predators (Clinchy et al. 2016). To fully gauge whether humans were eliciting responses akin to that of a "super-predator", additional vocalizations of one extant predator, domestic dogs (*Canis familiaris*), and two locally extinct predators, wolves (*Canis lupus*) and brown bears (*Ursus arctos*) were broadcasted to act as positive controls. In comparison to these predator treatments, badgers not only took longer to leave their dens after sunset to begin foraging, but they waited for a further 20 minutes after the playback of the human voices had ceased to finally emerge from their dens—indicating an immensely elevated fear of human presence (Clinchy et al. 2016).

In addition to the European badger, human-voice-induced anti-predator behaviours of three mesocarnivore species within a multipredator system were studied in the Santa Cruz Mountains of California: bobcats (*Lynx rufus*), skunks (*Mephitis mephitis*), and Virginia opossums (*Didelphis virginiana*) (Suraci et al. 2019a, Reilly et al. 2022). In these experiments, auditory broadcasts of human voices and frog calls—as a negative control—were positioned across a large landscape with varying degrees of human development present to investigate the effects on activity patterns and movement. Suraci et al. (2019a) found that mesocarnivore activity levels were greatly reduced when human voices were broadcast, reporting the diurnal activity of bobcats and overall activity of skunks as the most impacted behaviours. Furthermore, the strength of these responses in bobcats and skunks was found to remain constant across the human footprint gradient (Reilly et al. 2022). In contrast, opossums displayed a more expected activity pattern in response to perceived predation threats by inversely adapting their activity levels relative to the area's degree of human footprint (Reilly et al. 2022), resulting in a collective 66% reduction in foraging efficiency (Suraci et al. 2019a). These experiments demonstrate that mesocarnivores become more elusive and reduce foraging activities when nearby human presence is perceived (as illustrated in Figure 7.1).

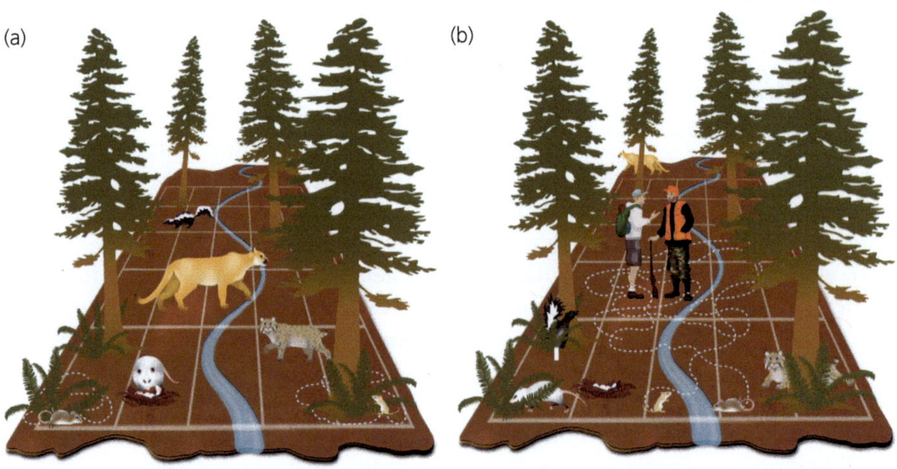

Figure 7.1 Conceptual illustrations of a human-induced landscape of fear on carnivores and their prey. Where humans are absent (a), larger carnivore species (e.g. pumas and bobcats) exhibit greater movement, activity, and/or foraging, while small mammals exhibit reduced space use. Where humans are present (b), fear of humans suppresses the activity, foraging, and/or habitat use of large carnivores and mesocarnivores, while small mammals increase their total space use and foraging intensity.

Source: Image reproduced from Suraci et al. 2019 with permission. Original artwork by Corlis Schneider.

Zanette et al. (2023) investigated a fear of human presence and hunting in a whole mammal community: comprising of four large carnivores, multiple mesocarnivore species, elephants (*Loxodonta africana*), hippopotamuses (*Hippopotamus amphibius*), rhinoceroses (*Ceratotherium simum*), and 11 ungulates. Similar to Clinchy et al. (2016)'s experiment, multiple playback treatments were conducted to gauge the community's strength of fear responses to the human voice and hunting stimuli (including gunshots and domestic dog barks) in comparison to non-threatening birdsong and threatening vocalizations of the area's top predator—the lion (*Panthera leo*). Zanette et al. (2023) found that the community as a whole were twice as likely to flee and 40% faster to abandon waterholes in response to human voices than lion vocalizations, with both predator treatments eliciting greater antipredator responses than hunting stimuli. These findings demonstrate the pervasive and profound effect of human presence on mammal behaviour across multiple trophic levels. However, when considering the carnivores exclusively, only hyenas (*Crocuta crocuta*) and leopards (*Panthera pardus*) fled significantly more to human voices than lion vocalizations; lions, African wild dogs (Lycaon pictus), and mesocarnivores did not flee in response to any playback treatment in more than 80% of trials (Zanette et al. 2023).

Globally, pumas are the most studied species in terms of investigating how wildlife alter their behaviour in response to human presence and footprint and thus, act as a model species for understanding how large carnivores perceive and fear humans as predators. Therefore, rather than discuss the experimental findings of the three remaining playback studies here (Smith et al. 2017, Suraci et al. 2019a, Suraci et al. 2019b), we have dedicated the following section of this chapter to highlight and summarize current behavioural research that explores a fear of humans in pumas.

The puma as a model species

Pumas are exposed to a plethora of human disturbances as the most widespread terrestrial mammal species in the Americas (Culver et al. 2000). Given this distribution and their vital role as a keystone species (LaBarge et al. 2022), researchers have taken special interest in understanding how urbanization and human presence affect puma ecology and behaviour.

Habitat selection and activity

Recent studies have observed that alterations in puma behaviour induced by nearby urban development are greatly related to the type of urban spaces present, or more specifically, the degree of human presence found among types of urban spaces (Wilmers et al. 2013). Heavily disturbed areas, such as residential areas or those experiencing residential development growth, have been repeatedly reported to be generally avoided by pumas (Gray et al. 2016, Blecha et al. 2018, Burdett et al. 2010, Wang et al. 2015, Kertson and Keren 2022, Nisi et al. 2022a). The short-term movement costs of navigating more rugged physical terrain to avoid areas of human-induced risk result in energetic costs, ultimately reducing daily travel distances and total home range areas by 29% and 68%, respectively (Nickel et al. 2021). However, moderately disturbed areas such as hiking trails, roadways, or agricultural land remain readily used, albeit with reduced occupancy and habitat use (Muhly et al. 2011, Kertson et al. 2013, Gray et al. 2016, Lewis et al. 2021). Moreover, Kertson et al. (2013) found that individualistic factors that may influence behaviour, such as sex and residency status, did not influence pumas' use of urban areas, further corroborating that external features of the environment are the more likely primary factors governing these behavioural adjustments.

Yet, akin to the gradient of urban development, there is also a gradient in the degree to which pumas adjust their movement and activity behaviour in response to human disturbance. In some cases, pumas decrease their avoidance of certain anthropogenic features as these features become more prevalent within their range over time (Knopff et al. 2014). In Colorado, a fine-scaled analysis showed that pumas increased their movement speed to pass through developed areas (Buderman et al. 2018). However, puma movement analysis at a coarser temporal scale—that is, over days rather than over minutes—revealed that pumas slowed their movements in more developed areas (Nisi et al. 2022b). These findings indicate two distinct strategies for avoiding human detection. Firstly, pumas may minimize the probability of encountering humans by moving quickly through finer-scaled areas where human-induced risk is high and then resuming a more relaxed pace when risk of a human encounter is lower (Buderman et al. 2018). However, when human disturbance is more widespread across developed areas, pumas as a solitary species may be able to avoid human detection more effectively by exhibiting slower, cryptic movements (Nisi et al. 2022b).

Pumas also become more active during times of day when human presence is reduced. In developed areas or areas with high human activity in the daytime, pumas primarily engaged in nocturnal behaviour (Wang et al. 2015, Azevedo et al. 2018, Alldredge et al. 2019, Figel et al. 2021). At night, pumas travelled further distances throughout areas of human development and sometimes even selected for developed areas that they would typically avoid in the daytime (Wang et al. 2017, Nisi et al. 2022a). Increased nocturnality was also reported in wildlife reserves, where pumas exhibited more nocturnal activity in a hunting reserve in comparison to a protected reserve where pumas more evenly spread their activity throughout the day (Zanon-Martinez et al. 2016). In particular, the activity of pumas often peaks at crepuscular hours (Wang et al. 2015, Zanon-Martinez et al. 2016, Azevedo et al. 2018 (males only)). This bimodal activity distribution may be the result of pumas optimizing their opportunities to hunt both large diurnal (e.g. black-tailed deer (*Odocoileus hemionus*) (Wang et al. 2017)) and nocturnal (e.g. capybara (*Hydrochoerus hydrochaeris*) (Azevedo et al. 2018)) prey while simultaneously avoiding the peak activity time of humans in the mid-day (Wang et al. 2015), although a strong bimodal pattern of activity in pumas has been reported in areas with low human disturbance as well (Frey et al. 2020).

Foraging and feeding behaviours

Initial work suggested that observed shifts in feeding behaviours of pumas meant that these large carnivores benefited from more developed areas, as greater numbers of deer were killed by pumas (especially females; Smith et al. 2015) in high housing density areas compared to areas with minimal residential development (Smith et al. 2016). Further, females selected for feeding sites near sparse developed areas within their home range, presumably due to higher abundances of mule deer (Benson et al. 2016). However, while kill rates increased, the time spent by pumas consuming prey at each kill site declined with increasing housing density (Smith et al. 2015, Smith et al. 2016). Given that pumas have also reduced kill site fidelity in more urbanized areas (Smith et al. 2015), there are two probable, yet not mutually exclusive, biological mechanisms for this shift in behaviour. The first hypothesizes that pumas are pre-emptively limiting their time to feed to minimize their chances of human detection in areas where human presence is more prominent, similar to Buderman et al.'s (2018) findings that showed pumas increasing their movement speed in human-induced 'high-risk' areas. On the other hand, the reduction in time spent feeding may be a reactive response to increased instances of pumas detecting nearby human presence at kill sites located in more developed areas, resulting in more frequent abandonment of carcass caches (Smith et al. 2017). Thus, an increase in kills over time is unlikely to be an indication of growing appetites in puma populations, but rather a necessary strategy for individuals to mitigate the negative effects of limited available feeding time per kill when hunting near human development.

Urbanization also impacts dietary preferences, with pumas shifting their selection of prey species from near complete specialization on native herbivores in wildlands to greater use of exotic and invasive synanthropes (Moss et al. 2016a, Robins et al. 2019). One study reported that individuals hunting in highly developed areas obtained approximately 20% more of their diet from alternative prey (Moss et al. 2016b). In addition to urban wildlife, Alldredge et al. (2019) found that 39% of cougars sampled during feeding site investigations had consumed domestic prey items. Despite increases in additional prey groups, however, puma diets still primarily consisted of native ungulate species, such as black-tailed deer and elk (*Cervis canadensis*), with the odds of ungulate predation remaining consistent across the wildland-urban gradient (Robins et al. 2019). As such, these observations of increased and more diversified diets among puma populations, especially towards smaller-sized urban prey, further support an apparent compensatory action by pumas to ensure adequate energy intake when feeding time is perceived to be limited for each individual kill.

Reproductive behaviours

Lastly, human disturbance across a landscape can affect the reproductive behaviours of pumas. Wilmers et al. (2013) examined the spatial occurrence of scent markings and dens in relation to the movement and feeding behaviours of 20 pumas in the Santa Cruz Mountains of California. They found that these reproductive behaviours (scent marking and denning) required at least a four times larger spatial buffer from human development areas than non-reproductive behaviours, such as movement and feeding. Furthermore, roadways with traffic speeds over 56 kph also deterred pumas, but only when they were scent-marking and denning. Residential neighbourhoods functioned as an overall deterrent to pumas, with the engagement of any type of behaviour (scent marking, denning, movement, feeding) occurring at greater distances from residential neighbourhoods relative to roadways—suggesting that greater avoidance is granted to more consistent sources of human presence or interference (Wilmers et al. 2013).

Manipulating the landscape of fear

Beyond the more observational approaches highlighted earlier, researchers have also demonstrated the direct impact of human disturbance on puma behaviour by altering the landscape of fear via the introduction—or removal—of cues that mimic human presence. When playbacks of human voices were broadcast at known kill sites, Smith et al. (2017) reported that pumas would quickly abandon carcasses and flee 83% of the time. Furthermore, pumas took longer to return to those kills in comparison to non-threatening frog calls, thereby reducing their overall feeding time by half (Smith et al. 2017). However, when vocalizations of domestic dogs were broadcast, there was minimal to no effect on puma feeding time at kill sites (Suraci et al. 2019b), suggesting that dogs do not elicit a threat response despite their regular accompaniments to human hikers and recreationists. Pumas have also been shown to avoid areas of human voice broadcasts by shifting movement pathways to create greater distances between themselves and the speakers (Suraci et al. 2019a). When travelling through these areas of perceived human presence, pumas displayed increased cautiousness—signified by decreased movement speeds (Suraci et al. 2019a). When the COVID-19 pandemic reduced human mobility by 50% in the Santa Cruz Mountains, Wilmers et al. (2021) reported that pumas responded with a relaxed aversion to urban spaces. This natural experiment corroborates that pumas avoid humans in general and distinguishes that a fear of an impending human interaction is the primary driving factor for the avoidance of human development in Californian pumas (Wilmers et al. 2021).

Fear of human footprint and infrastructure

We have already briefly introduced the effects of human footprint, infrastructure, as well as recreation and ecotourism on carnivore behaviour in pumas. The current and following sections will further elaborate on these findings by dissecting the behavioural responses of additional carnivore species by disturbance type.

Behavioural effects of human settlements and residential areas

Living alongside pumas, large carnivore species such as brown bears, black bears (*Ursus americanus*), and wolves have also exhibited human-induced changes in spatial and temporal patterns of activity to presumably avoid human presence (Ordiz et al. 2011, Zeller et al. 2019, Frey et al. 2020). As levels of human disturbance increased across a landscape, wolves increasingly undertook nocturnal activities (Frey et al. 2020). When selecting for a resting site, brown bears chose beds with higher horizontal and canopy cover during daytime hours, especially when these sites were located closer to human settlements (Ordiz et al. 2011). Individuals also demonstrated increased avoidance when greater levels of human-induced risk were introduced during the summer and autumn months, resting further from human settlements during the day as human activity became more intensive and dispersed in comparison to the springtime (Ordiz et al. 2011). Although, sometimes areas of human development can act as an attractant to black bears during years of natural food scarcity (Baruch-Mordo et al. 2012, Johnson et al. 2015) as well as seasonally—such as springtime when natural food availability is low or during autumn when black bears are in a state of hyperphagia and are preparing for an over-wintering hibernation (Johnson et al. 2015, Evans et al. 2019). Nonetheless, black bears preferentially moved through these areas of human development during the night, likely as a strategy to mitigate the risk of human encounters throughout the day (Beckmann and Berger 2003, Baruch-Mordo et al. 2012, Zeller et al. 2019).

Camera traps deployed over broad scales have been used to study the behaviour of the mesocarnivore guild of Canada and the USA (Wang et al. 2015, Stewart et al. 2016, Murray and St. Clair 2017, Wait et al. 2018, Frey et al. 2020). Like pumas (Smith et al. 2017), wolverines (*Gulo gulo*) were observed to spend minimal time at feeding sites in areas of high human footprint, whereas the variation among individuals in time spent at these sites in areas of low human footprint was significantly greater (Stewart et al. 2016). The shift to increased nocturnal activities observed in pumas was also noted in the larger-bodied bobcats and coyotes (*Canis latrans*), with coyote activity exhibiting cathemerality (Wang et al. 2015, Frey et al. 2020). In contrast, typically nocturnal martens (*Martes americana*) were observed to increase diurnal activities in response to high human disturbance, perhaps as a strategy to avoid predation from the aforementioned predators increasing their nocturnality (Frey et al. 2020). Coyotes were more likely to leave sites that were surrounded by urban landcover, while bobcats were not sensitive to urban landcover and were more likely to colonize grasslands and agricultural sites (Wait et al. 2018). Murray and St. Clair (2017) also found that healthy coyotes generally avoided residential areas in the daytime; however, diseased individuals did not exhibit a similar nocturnal shift in activity or avoid residential areas and instead selectively visited backyards with anthropogenic food regardless of time of day.

In Africa, leopards, cheetahs (*Acinonyx jubatus*), and spotted hyenas are large carnivore species that also alter their behaviour in response to nearby human inhabitants (Strampelli et al. 2018, Klaassen and Broekhuis 2018, Turner et al. 2020). Leopards generally avoided human settlements; however, the greatest contributing factor to habitat selection was the presence of bushmeat poachers and lions (Strampelli et al. 2018). Similarly, cheetahs avoided areas of high human presence, instead selecting wildlife areas with semi-closed habitat and edge density (Klaassen and Broekhuis 2018). To compare contrasting levels of human disturbance on large carnivore behaviour, Turner et al. (2020) conducted an experiment using a life-size model hyena to compare the behaviour of two hyena populations: a disturbed clan that lived adjacent to human settlements and an undisturbed clan further away from settlements. Hyenas in nearby settlements exhibited fewer risk-taking behaviours—keeping a greater distance between themselves and the model and overall spending less time near the model—than their undisturbed counterparts, perhaps indicating an increased baseline of vigilance towards conspecific and human 'intruders' alike when exposure to humans is more prevalent (Turner et al. 2020).

Behavioural effects of roadways

Paved roadways yield both benefits and unfortunately mortality to mammal populations worldwide (see Chapter 6, this volume). For example, in Botswana, African wild dogs selected and utilized roads as a resource for convenient and energy-efficient travel, especially in denser vegetative environments, only avoiding roads for high-speed running or when resting (Abrahams et al. 2016). The most heavily viewed wolf packs in Yellowstone National Park have also exhibited less road avoidance despite a twofold increase in the number of visitors over the last 40 years, suggesting an increased tolerance of vehicles (Anton et al. 2020). However, other work suggests that carnivores are not using roadways ceaselessly but are rather strategically timing their movements on these roadways to avoid high volumes of traffic and thus higher risk of mortality (Muhly et al. 2011, Murray and St. Clair 2015, Anton et al. 2020, Blaskovic et al. 2022). Wolves used areas farther from roads during the day and peak visitation in Yellowstone (Anton et al. 2020); Eurasian lynx (*Lynx lynx*), wolves, and brown bears exhibited minimal overlap in temporal activity with humans on forest roads of Croatia by avoiding road use during daylight hours; and coyotes, wolves, brown bears, and black bears avoided high-volume trails and roads despite a threefold increase in the abundance of prey species on these paths (Muhly et al. 2011). These temporal behavioural adjustments may indeed be successful in reducing occurrences of wildlife-vehicle collisions, as Murray and St. Clair (2015) reported that coyotes that were more active at dusk and crossed roads during rush hour were killed more often than coyotes that, contrastingly, exhibited less avoidance of roadways but crossed roads near midnight.

Behavioural effects of agricultural land use

Current research on the effects of agriculture on large carnivore behaviour is predominantly studied in Kenya, as pastoralists and their livestock live alongside over 21 carnivore species within many of the country's wildlife reserves and rangelands (Boydston et al. 2003, Pangle and Holekamp 2010, Schuette et al. 2013). For example, one Maasai community practices seasonal movements of their settlements and grazing areas across 8000km^2 of semi-arid rangeland, a site home to the permanent Ewaso Nyiro River (Schuette et al. 2013). When the Maasai inhabited the east side of the river, radio-collared lions occupied the entirety of area on the west side of the river; however, when the Maasai travelled to settle on the west side, the local lion population more frequently utilized densely covered habitats and were five times more likely to occupy a significantly smaller conservation area south of the Maasai settlement (Schuette et al. 2013). Moreover, lions took refuge in the conservation area during both day and night despite human activity only occurring in the daytime (Schuette et al. 2013). Although this observed shift in habitat selection likely decreases the probability of close encounters with humans, lions were 72% farther away from the river, reducing their access to the area's only permanent water source during the dry season (Schuette et al. 2013).

Hyena behaviour has also been observed to be affected by increased reliance on game reserves for livestock grazing. Boydston et al. (2003) compared female hyena behaviour from 1988–90 and 1996–8 in the Masai Mara National Reserve and found that hyenas obtained larger home ranges, ventured farther from den sites, congregated in smaller groups, and became more nocturnal over time. Additionally, hyenas gained a stronger preference for dense vegetative areas and avoided areas with the most intensive livestock grazing despite these areas containing the greatest abundance of prey (Boyston et al. 2003). Pangle and Holekamp (2010) also reported that hyenas living near grazing livestock preferred to nurse their young closer to bushes and displayed twice as much vigilance behaviour when resting than individuals living in the centre of the reserve without disturbance from livestock. To further support their observations, Pangle and Holekamp (2010) conducted a playback experiment, using cowbells to represent the presence of pastoralism and church bells as a negative control treatment, and found that the disturbed clan of hyenas responded to both treatments with increased vigilance in comparison to the undisturbed clan, corroborating their observational results and perhaps suggesting an overall heightened responsiveness to all anthropogenic stimuli.

Fear of ecotourism and outdoor recreation

National Parks and similar recreation areas around the world provide people with the opportunity to view wildlife up-close in their natural habitats; however, the presence of thousands to millions of visitors each year has been shown to alter the natural behaviours of large carnivores—along with numerous other mammal and bird species—residing in these parks (Marion et al. 2020, Dertien et al. 2021).

Behaviour of canids

In areas of significant recreational use by humans, wolves can become more nocturnal and increase their speed of travel on high-volume trails to avoid peak times and places of human presence (Haswell et al. 2020, Whittington et al. 2022). Rasmussen and MacDonald (2012) compared how African wild dogs allocated their hunting activity in two reserves: one with high human presence and persecution and one with low human presence and persecution. Individuals in the high disturbance reserve hunted at significantly different times of day compared to their undisturbed counterparts: primarily utilizing evenings and moonlight (Rasmussen and MacDonald 2012). This adjustment reduced the probability of African wild dogs encountering a human by 64%; however, it simultaneously increased their potential to encounter a hyena or lion by 70% and 37%, respectively (Rasmussen and MacDonald 2012).

Like wolves, coyotes have been temporally displaced by human recreation, displaying more nocturnal activity in recreational areas and using recreational trails primarily at night (George and Crooks 2006, Barrueto et al. 2014, Reilly et al. 2017, Lewis et al. 2021). In contrast to wolves, coyotes appear to benefit from human recreation in the form of snowmobiling, as coyotes use snowmobile trails for ease of travel and to access ungulate kills in the winter months (Kolbe et al. 2007, Gese et al. 2013). Moreover, Kolbe et al. (2007) reported that the overall influence of these trails on the foraging success and movement of coyotes was minimal.

Behaviour of felids

Current research on the impacts of tourists and recreationists on felids has been limited to lions, leopards, and ocelots (*Leopardus pardalis*). In a Peruvian National Park, ocelots were observed to temporally adjust their activity, reducing diurnal usage on tourist trails (Barcelos et al. 2022). By utilizing an unexpected flooding event, researchers in Thailand's Kaeng Krachan National Park evaluated whether leopards responded to the temporary 6-month closure of the park (Ngonrasert et al. 2017). In the absence of tourist activity, leopards moved more frequently, utilized the park's space to a greater extent, and shifted their activity towards diurnality (Ngonrasert et al. 2017). Hayward and Hayward (2009) also investigated the effects of tourist presence and absence on felid behaviour by recording stress behaviours in South African lions. Lions were more likely to exhibit stress-indicating behaviours when tourists were present in the park. Meanwhile, the display of relaxation behaviours, such as grooming and rolling, was more likely when tourists were absent (Hayward and Hayward 2009).

Behaviour of bears

In contrast to felids, the behaviour of brown bears and black bears in response to human recreation has been more extensively studied in the Northern Hemisphere. Spatially, male and solitary female bears tended to avoid both human recreational areas (Smith et al. 2012, Cristescu et al. 2013, Salvatori et al. 2023) and park trails where motorized recreation was commonplace (Ladle et al. 2019); as such, remote sites with the least human activity had the most bear activity (Coltrane and Sinnott 2015, Salvatori et al. 2023). Female brown bears with cubs, however, did not avoid motorized recreational trails but instead increased their movement rates when travelling along the trails (Ladle et al. 2019). More rapid movement patterns have also been observed in adult bears that utilize high-volume trails and roadways, suggesting that perhaps bears implement this strategy to avoid or minimize their encounter rate with recreationalists (Lewis et al. 2021, Whittington et al. 2022). When approached by humans, bears reacted non-aggressively and fled (Ordiz et al. 2013), regardless of whether the approacher was on foot or engaged in a non-motorized mode of transport, such as a bicycle (Smith et al. 2012). Additionally, bears also adopted a crepuscular and nocturnal activity pattern in recreational sites with intense human use (Coltrane and Sinnott 2015), resting and bedding throughout the daytime in contrast to resting primarily at night in areas with low human access (Cristescu et al. 2013). Bears also utilized recreational trails more often during the night with faster movement speeds than in the daytime, resulting in decreased overlaps of bear activity and human activity on park trails overall (Ordiz et al. 2013, Lewis et al. 2021).

Behaviour of other mesocarnivores

Mesocarnivores, such as lynx, bobcats, red foxes (*Vulpes vulpes*), European badgers, and skunks, have exhibited contrasting behavioural responses to human recreation and activity. Bobcats and lynx appear to be fearful of human presence, demonstrating similar temporal and spatial shifts in behaviour to those observed in pumas: individuals inhabiting recreational areas shifted their activity towards the night (George and Crooks 2006, Haswell et al. 2020, Thorsen et al. 2022) as well as spatially avoided high human activity areas and trails on a local scale (George and Crooks 2006, Lewis et al. 2021, Thorsen et al. 2022), especially avoiding areas frequently used by motorized winter recreationists (Squires et al. 2019). However, lynx did not avoid skiing areas or recreational activity at a home range scale (Squires et al. 2019, Thorsen et al. 2022), while bobcats tended to occupy sites with low human activity levels (George and Crooks 2006). Furthermore, Haswell et al. (2020) hypothesized that human presence may be dampening the top-down ecological effects of local large carnivores in mesocarnivore species, as foxes that would typically temporally avoid predators by engaging in more diurnal activity have become more nocturnal in an area of high human activity.

Contrary to Haswell et al.'s (2020) proposed notion, some studies suggest that the predation threat of large carnivores is greater than that assigned to humans in mesocarnivores (Reilly et al. 2017, Lewis et al. 2021, Salvatori et al. 2023), as many smaller carnivore species are largely nocturnal and thus may not be directly affected by recreational activities in the daytime (Lewis et al. 2021). Lewis et al. (2021) found that skunks did not demonstrate a response to human activity, while Reilly et al. (2017) observed that skunks slightly shifted their activity towards morning hours in areas of human recreation. Furthermore, site use by badgers and foxes was observed to be positively related to human activity, often using sites at closer proximity to settlements as a potential strategy to avoid large carnivore presence (Salvatori et al. 2023).

Concluding remarks

This chapter showcased how large carnivores significantly alter their behaviours in response to human presence, infrastructure, and urbanization. Global meta-analyses revealed the pervasive impact of areas with high human footprint on carnivore behaviour, with mammals exhibiting more prominent behavioural changes in areas with greater human disturbance—such as reduced movements (Tucker et al. 2018, O'Donnell and del Barco-Trillo 2020), shifts in habitat selection (Sazatornil et al. 2016, Zhang et al. 2023), and increased nocturnality (Gaynor et al. 2018). In addition to urbanized or residential areas, national parks and similar recreation areas have also been shown to significantly alter carnivore behaviours, leading to increased nocturnality and avoidance of high-traffic areas to minimize human encounters, as seen in canids, felids, bears, and mesocarnivores (Marion et al. 2020, Dertien et al. 2021). However, variations in study results suggest that specific factors like direct human presence versus habitat modifications influence these behavioural shifts differently. To test the effects of human presence on carnivore behaviour, separate from human footprint, recent experimental studies have used auditory playbacks of human voices paired with automated camera traps. These experiments have demonstrated that various carnivore species exhibit strong anti-predator behaviours in response to perceived human presence—such as reduced foraging (Clinchy et al. 2016, Smith et al. 2017, Suraci et al. 2019a, Reilly et al. 2022), increased vigilance (Clinchy et al. 2016), fleeing (Smith et al. 2017, Zanette et al. 2023), increased avoidance behaviours (Clinchy et al. 2016, Suraci et al. 2019a), and reduced activity levels (Suraci et al. 2019a, Reilly et al. 2022)—indicating that these individuals perceived humans as a significant predation threat. Given that both global observations and fine-scale experiments consistently show that carnivores significantly alter their behaviour to avoid human encounters, it is evident that large carnivores are profoundly and pervasively impacted by human presence, revealing a deep-seated fear of humans that is unmatched by any other predator. In conclusion, these behavioural observations indicate that the type of human disturbance in addition to the degree of perceived human presence serve as strong predictors to how carnivores respond to humans; thus, understanding the extent by which large carnivores fear the human "super-predator" is essential for accurately identifying potential areas of human–wildlife conflict and mitigating the negative impacts of the ongoing expansion of human activities into the native ranges of diminishing carnivore populations worldwide.

References

Abrahms, B., N. R. Jordan, K. A. Golabek, J. W. McNutt, A. M. Wilson, and J. S. Brashares. 2016. Lessons from integrating behaviour and resource selection: Activity-specific responses of African wild dogs to roads: integrating behaviour and resource selection. *Animal Conservation* 19(3):247–255.

Alldredge, M. W., F. E. Buderman, and K. A. Blecha. 2019. Human–cougar interactions in the wildland–urban interface of Colorado's front range. *Ecology and Evolution* 9(18):10415–

Anton, C. B., D. W. Smith, J. P. Suraci, D. R. Stahler, T. P. Duane, and C. C. Wilmers. 2020. Gray wolf habitat use in response to visitor activity along roadways in Yellowstone National Park. *Ecosphere* 11(6):03164.

Azevedo, F. C., F. G. Lemos, M. C. Freitas-Junior, D. G. Rocha, and F. C. C. Azevedo. 2018. Puma activity patterns and temporal overlap with prey in a human-modified landscape at Southeastern Brazil. *Journal of Zoology* 305(4):246–255.

Barcelos, D., E. M. Vieira, M. S. Pinheiro, and G. B. Ferreira. 2022. A before–after assessment of the response of mammals to tourism in a Brazilian national park. *Oryx* 56(6):854–863.

Barrueto, M., A. T. Ford, and A. P. Clevenger. 2014. Anthropogenic effects on activity patterns of wildlife at crossing structures. *Ecosphere* 5(3):art27.

Baruch-Mordo, S., K. R. Wilson, D. L. Lewis, J. Broderick, J. S. Mao, and S. W. Breck. 2012. Stochasticity in natural forage production affects use of urban areas by black bears: implications to management of human-bear conflicts. *PLoS One* 9:e85122.

Beckmann, J. P. and J. Berger. 2003. Rapid ecological and behavioural changes in carnivores: the responses of black bears (*Ursus americanus*) to altered food. *Journal of Zoology* 261:207–212.

Benson, J. F., J. A. Sikich, and S. P. D. Riley. 2016. Individual and population level resource selection patterns of mountain lions preying on mule deer along an urban-wildland gradient. *PLoS One* 11(7):e0158006.

Blašković, S., T. Gomerčić, I. Topličanec, and M. Sindičić. 2022. Temporal overlap of human and apex predator activity on wildlife trails and forest roads. *Journal of Vertebrate Biology* 71(22029):22029.

Blecha, K. A., R. B. Boone, and M. W. Alldredge. 2018. Hunger mediates apex predator's risk avoidance response in wildland-urban interface. *Journal of Animal Ecology* 87(3):609–622.

Bombieri, G., et al. 2023. A worldwide perspective on large carnivore attacks on humans. *PLOS Biology* 21(1):e3001946.

Bonnot, N. C., et al. 2020. Fear of the dark? Contrasting impacts of humans versus lynx on diel activity of roe deer across Europe. *Journal of Animal Ecology* 89(1): 132–145.

Boydston, E. E., K. M. Kapheim, H. E. Watts, M. Szykman, and K. E. Holekamp. 2003. Altered behaviour in spotted hyenas associated with increased human activity. *Animal Conservation* 6(3):207–219.

Buderman, F. E., M. B. Hooten, M. W. Alldredge, E. M. Hanks, and J. S. Ivan. 2018. Time-varying predatory behavior is primary predictor of fine-scale movement of wildland-urban cougars. *Movement Ecology* 6(1):22.

Burdett, C. L., et al. 2010. Interfacing models of wildlife habitat and human development to predict the future distribution of puma habitat. *Ecosphere* 1(1):art4.

Clinchy, M., et al. 2016. Fear of the human "super predator" far exceeds the fear of large carnivores in a model mesocarnivore. *Behavioral Ecology* 27(6):1826–1832.

Coltrane, J. and R. Sinnott. 2015. Brown bear and human recreational use of trails in Anchorage, Alaska. *Human–Wildlife Interactions* 9(1):132–147.

Creel, S., et al. 2019. What explains variation in the strength of behavioral responses to predation risk? A standardized test with large carnivore and ungulate guilds in three ecosystems. *Biological Conservation* 232:164–172.

Cristescu, B., G. B. Stenhouse, and M. S. Boyce. 2013. Perception of human-derived risk influences choice at top of the food chain. *PLoS One* 8(12):e82738.

Culver, M., W. E. Johnson, J. Pecon-Slattery, and S. J. O'Brien. 2000. Genomic ancestry of the American puma (*Puma concolor*). *Journal of Heredity* 91(3):186–197.

Cunningham, C. X., C. N. Johnson, L. A. Barmuta, T. Hollings, E. J. Woehler, and M. E. Jones. 2018. Top carnivore decline has cascading effects on scavengers and carrion persistence. *Proceedings of the Royal Society B: Biological Sciences* 285(1892):20181582.

Darimont, C. T., C. H. Fox, H. M. Bryan, and T. E. Reimchen. 2015. The unique ecology of human predators. *Science* 349(6250):858–860.

Darimont, C. T., et al. 2023. Humanity's diverse predatory niche and its ecological consequences. *Communications Biology* 6(1):609.

Dertien, J. S., C. L. Larson, and S. E. Reed. 2021. Recreation effects on wildlife: A review of potential quantitative thresholds. *Nature Conservation* 44:51–68.

Doherty, T. S. and D. A. Driscoll. 2018. Coupling movement and landscape ecology for animal conservation in production landscapes. *Proceedings of the Royal Society B: Biological Sciences* 285(1870):20172272.

Doherty, T. S., G. C. Hays, and D. A. Driscoll. 2021. Human disturbance causes widespread disruption of animal movement. *Nature Ecology & Evolution* 5(4): 513–519.

Estes, J. A., et al. (2011). Trophic downgrading of planet Earth. *Science* 333(6040):301–306.

Evans, M. J., J. E. Hawley, P. W. Rego, and T. A. G. Rittenhouse. 2019. Hourly movement decisions indicate how a large carnivore inhabits developed landscapes. *Oecologia* 190:11–23.

Figel, J. J., S. Botero-Cañola, J. D. Sánchez-Londoño, and J. Racero-Casarrubia. 2021. Jaguars and pumas exhibit distinct spatiotemporal responses to human disturbances in Colombia's most imperiled ecoregion. *Journal of Mammalogy* 102(1):333–345.

Frey, S., J. P. Volpe, N. A. Heim, J. Paczkowski, and J. T. Fisher. 2020. Move to nocturnality not a universal trend in carnivore species on disturbed landscapes. *Oikos* 129(8):1128–1140.

Gaynor, K. M., C. E. Hojnowski, N. H. Carter, and J. S. Brashares. 2018. The influence of human disturbance on wildlife nocturnality. *Science* 360(6394):1232–1235.

George, S. L. and K. R. Crooks. 2006. Recreation and large mammal activity in an urban nature reserve. *Biological Conservation* 133(1):107–117.

Gese, E. M., J. L. B. Dowd, and L. M. Aubry. 2013. The influence of snowmobile trails on coyote movements during winter in high-elevation landscapes. *PLoS One* 8(12):e82862.

Gray, M., C. C. Wilmers, S. E. Reed, and A. M. Merenlender. 2016. Landscape feature-based permeability models relate to puma occurrence. *Landscape and Urban Planning* 147:50–58.

Haswell, P. M., J. Kusak, K. A. Jones, and M. W. Hayward. 2020. Fear of the dark? A mesopredator mitigates large carnivore risk through nocturnality, but humans moderate the interaction. *Behavioral Ecology and Sociobiology* 74(5):62.

Hayward, M. W. and G. J. Hayward. 2009. The impact of tourists on lion *Panthera leo* behaviour, stress and energetics. *Acta Theriologica* 54(3):219–224.

Hettena, A. M., N. Munoz, and D. T. Blumstein. 2014. Prey responses to predator's sounds: a review and empirical study. *Ethology* 120(5):427–452.

Ingeman, K. E., et al. 2022. Glimmers of hope in large carnivore recoveries. *Scientific Reports* 12(1):10005.

Johnson, H. E., et al. 2015. Shifting perceptions of risk and reward: dynamic selection for human development by black bears in the western United States. *Biological Conservation* 187:164–172.

Kertson, B. N. and I. N. Keren. 2022. Cougar use of residential areas and interactions with people in periods of population stability and growth. *Journal of Mammalogy* 103(2):347–360.

Kertson, B. N., R. D. Spencer, and C. E. Grue. 2013. Demographic influences on cougar residential use and interactions with people in western Washington. *Journal of Mammalogy* 94(2):269–281.

Klaassen, B. and F. Broekhuis. 2018. Living on the edge: multiscale habitat selection by cheetahs in a human-wildlife landscape. *Ecology and Evolution* 8(15):7611–7623.

Knopff, A. A., K. H. Knopff, M. S. Boyce, and C. C. St. Clair. 2014. Flexible habitat selection by cougars in response to anthropogenic development. *Biological Conservation* 178:136–145.

Kolbe, J. A., J. R. Squires, D. H. Pletscher, and L. F. Ruggiero. 2007. The effect of snowmobile trails on coyote movements within lynx home ranges. *Journal of Wildlife Management* 71(5):1409–1418.

LaBarge, L. R., M. J. Evans, J. R. Miller, G. Cannataro, C. Hunt, and L. M. Elbroch. 2022. Pumas *Puma concolor* as ecological brokers: a review of their biotic relationships. *Mammal Review* 52(3):360–376.

Ladle, A., T. Avgar, M. Wheatley, G. B. Stenhouse, S. E. Nielsen, and M. S. Boyce. 2019. Grizzly bear response to spatio-temporal variability in human recreational activity. *Journal of Applied Ecology* 56(2):375–386.

Lewis, J. S., et al. 2021. Human activity influences wildlife populations and activity patterns: Implications for spatial and temporal refuges. *Ecosphere* 12(5): e03487.

Marion, S., A. Davies, U. Demšar, R. J. Irvine, P. A. Stephens, and J. Long. 2020. A systematic review of methods for studying the impacts of outdoor recreation on terrestrial wildlife. *Global Ecology and Conservation* 22:e00917.

Montgomery, R. A., D. W. Macdonald, and M. W. Hayward. 2020. The inducible defences of large mammals to human lethality. *Functional Ecology* 34(12): 2426–2441.

Moss, W. E., M. W. Alldredge, K. A. Logan, and J. N. Pauli. 2016a. Human expansion precipitates niche expansion for an opportunistic apex predator (*Puma concolor*). *Scientific Reports* 6(1):39639.

Moss, W. E., M. W. Alldredge, and J. N. Pauli. 2016b. Quantifying risk and resource use for a large carnivore in an expanding urban-wildland interface. *Journal of Applied Ecology* 53(2):371–378.

Muhly, T. B., C. Semeniuk, A. Massolo, L. Hickman, and M. Musiani. 2011. Human activity helps prey win the predator-prey space race. *PLoS One* 6(3):e17050.

Murray, M. H. and C. C. St. Clair. 2015. Individual flexibility in nocturnal activity reduces risk of road mortality for an urban carnivore. *Behavioral Ecology* 26(6): 1520–1527.

Murray, M. H. and C. C. St. Clair. 2017. Predictable features attract urban coyotes to residential yards: selection for residential areas by coyotes. *The Journal of Wildlife Management* 81(4):593–600.

Nickel, B. A., J. P. Suraci, A. C. Nisi, and C. C. Wilmers. 2021. Energetics and fear of humans constrain the spatial ecology of pumas. *Proceedings of the National Academy of Sciences* 118(5):e2004592118.

Nisi, A. C., J. F. Benson, and C. C. Wilmers. 2022a. Puma responses to unreliable human cues suggest an ecological trap in a fragmented landscape. *Oikos* 2022(5):e09051.

Nisi, A. C., et al. 2022b. Temporal scale of habitat selection for large carnivores: Balancing energetics, risk and finding prey. *Journal of Animal Ecology* 91(1):182–195.

Ngoprasert, D., A. J. Lynam, and G. A. Gale. 2017. Effects of temporary closure of a national park on leopard movement and behaviour in tropical Asia. *Mammalian Biology* 82:65–73.

O'Donnell, K. and J. Del Barco-trillo. 2020. Changes in the home range sizes of terrestrial vertebrates in response to urban disturbance: a meta-analysis. *Journal of Urban Ecology* 6(1):juaa014.

Ordiz, A., O.-G. Støen, M. Delibes, and J. E. Swenson. 2011. Predators or prey? Spatio-temporal discrimination of human-derived risk by brown bears. *Oecologia* 166(1):59–67.

Ordiz, A., et al. 2013. Lasting behavioural responses of brown bears to experimental encounters with humans. *Journal of Applied Ecology* 50(2):306–314.

Penteriani, V., et al. 2016. Human behaviour can trigger large carnivore attacks in developed countries. *Scientific Reports* 6(1):20552.

Pangle, W. M. and K. E. Holekamp. 2010. Lethal and nonlethal anthropogenic effects on spotted hyenas in the Masai Mara National Reserve. *Journal of Mammalogy* 91(1):154–164.

Rasmussen, G. S. A. and D. W. Macdonald. 2012. Masking of the zeitgeber: African wild dogs mitigate persecution by balancing time. *Journal of Zoology* 286(3):232–242.

Reilly, C. M., J. P. Suraci, J. A. Smith, Y. Wang, and C. C. Wilmers. 2022. Mesopredators retain their fear of humans across a development gradient. *Behavioral Ecology* 33(2):428–435.

Reilly, M. L., M. W. Tobler, D. L. Sonderegger, and P. Beier. 2017. Spatial and temporal response of wildlife to recreational activities in the San Francisco Bay ecoregion. *Biological Conservation* 207:117–126.

Ripple, W. J., et al. 2014. Status and ecological effects of the world's largest carnivores. *Science* 343(6167):1241484.

Robins, C. W., B. N. Kertson, J. R. Faulkner, and A. J. Wirsing. 2019. Effects of urbanization on cougar foraging ecology along the wildland–urban gradient of western Washington. *Ecosphere* 10(3):e02605.

Salvatori, M., et al. 2023. Crowded mountains: Long-term effects of human outdoor recreation on a community of wild mammals monitored with systematic camera trapping. *Ambio* 52(6):1085–1097.

Sazatornil, V., et al. 2016. The role of human-related risk in breeding site selection by wolves. *Biological Conservation* 201:103–110.

Schmitz, O. J., P. A. Hambäck, and A. P. Beckerman. 2000. Trophic cascades in terrestrial systems: a review of the effects of carnivore removals on plants. *The American Naturalist* 155(2):141–153.

Schuette, P., S. Creel, and D. Christianson. 2013. Coexistence of African lions, livestock, and people in a landscape with variable human land use and seasonal movements. *Biological Conservation* 157:148–154.

Serangeli, J., T. Van Kolfschoten, B. M. Starkovich, and I. Verheijen. 2015. The European saber-toothed cat (*Homotherium latidens*) found in the 'Spear Horizon' at Schöningen (Germany). *Journal of Human Evolution* 89:172–180.

Smith, J. A., et al. 2017. Fear of the human 'super predator' reduces feeding time in large carnivores. *Proceedings of the Royal Society B: Biological Sciences* 284(1857): 20170433.

Smith, J. A., Y. Wang, and C. C. Wilmers. 2015. Top carnivores increase their kill rates on prey as a response to human-induced fear. *Proceedings of the Royal Society B: Biological Sciences* 282(1802):20142711.

Smith, J. A., Y. Wang, and C. C. Wilmers. 2016. Spatial characteristics of residential development shift large carnivore prey habits. *The Journal of Wildlife Management* 80(6):1040–1048.

Smith, T. S., J. Oyster, S. D. Partridge, I. E. Martin, and A. Sisson. 2012. Assessing American black bear response to human activity at Kenai Fjords National Park, Alaska. *Ursus* 23(2):179–191.

Squires, J. R., L. E. Olson, E. K. Roberts, J. S. Ivan, and M. Hebblewhite. 2019. Winter recreation and Canada lynx: Reducing conflict through niche partitioning. *Ecosphere* 10(10):e02876.

Stewart, F. E. C., N. A. Heim, A. P. Clevenger, J. Paczkowski, J. P. Volpe, and J. T. Fisher. 2016. Wolverine behavior varies spatially with anthropogenic footprint: implications for conservation and inferences about declines. *Ecology and Evolution* 6(5):1493–1503.

Strampelli, P., L. Andresen, K. T. Everatt, M. J. Somers, and J. M. Rowcliffe. 2018. Habitat use responses of the African leopard in a human-disturbed region of rural Mozambique. *Mammalian Biology* 89:14–20.

Suraci, J. P., M. Clinchy, L. M. Dill, D. Roberts, and L. Y. Zanette. 2016. Fear of large carnivores causes a trophic cascade. *Nature Communications* 7(1):10698.

Suraci, J. P., et al. 2017. A new automated behavioural response system to integrate playback experiments into camera trap studies. *Methods in Ecology and Evolution* 8(8):957–964.

Suraci, J. P., M. Clinchy, L. Y. Zanette, and C. C. Wilmers. 2019a. Fear of humans as apex predators has landscape-scale impacts from mountain lions to mice. *Ecology Letters* 22(10):1578–1586.

Suraci, J. P., J. A. Smith, M. Clinchy, L. Y. Zanette, and C. C. Wilmers. 2019b. Humans, but not their dogs, displace pumas from their kills: an experimental approach. *Scientific Reports* 9(1):12214.

Thorsen, N. H., R. Bischof, J. Mattisson, T. R. Hofmeester, J. D. C. Linnell, and J. Odden. 2022. Smartphone app reveals that lynx avoid human recreationists on local scale, but not home range scale. *Scientific Reports* 12(1):4787.

Tilman, D., C. Balzer, J. Hill, and B. L. Befort. 2011. Global food demand and the sustainable intensification of agriculture. *Proceedings of the National Academy of Sciences* 108(50):20260–20264.

Tucker, M. A., et al. 2018. Moving in the Anthropocene: global reductions in terrestrial mammalian movements. *Science* 359(6374):466–469.

Turner, J. W., R. M. LaFleur, A. T. Richardson, and K. E. Holekamp. 2020. Risk-taking in free-living spotted hyenas is associated with anthropogenic disturbance, predicts survivorship, and is consistent across experimental contexts. *Ethology* 126(1):97–110.

Wait, K. R., A. M. Ricketts, and A. A. Ahlers. 2018. Land-use change structures carnivore communities in remaining tallgrass prairie: urbanization and carnivore communities. *The Journal of Wildlife Management* 82(7):1491–1502.

Wang, Y., M. L. Allen, and C. C. Wilmers. 2015. Mesopredator spatial and temporal responses to large predators and human development in the Santa Cruz Mountains of California. *Biological Conservation* 190: 23–33.

Wang, Y., J. A. Smith, and C. C. Wilmers. 2017. Residential development alters behavior, movement, and energetics in an apex predator, the puma. *PLoS One* 12(10):e0184687.

Whittington, J., M. Hebblewhite, R. W. Baron, A. T. Ford, and J. Paczkowski. 2022. Towns and trails drive carnivore movement behaviour, resource selection, and connectivity. *Movement Ecology* 10(1):17.

Wilmers C. C., A. C. Nisi, and N. Ranc. 2021. COVID-19 suppression of human mobility releases mountain lions from a landscape of fear. *Current Biology* 31(17):3952–3955.e3.

Wilmers, C. C., et al. 2013. Scale dependent behavioral responses to human development by a large predator, the puma. *PLoS One* 8(4):e60590.

Wolf, C. and W. J. Ripple. 2016. Prey depletion as a threat to the world's large carnivores. *Royal Society Open Science* 3(8):160252.

Zanette, L. Y. and M. Clinchy. 2020. Ecology and neurobiology of fear in free-living wildlife. *Annual Review of Ecology, Evolution, and Systematics* 51(1):297–318.

Zanette, L. Y., et al. 2023. Fear of the human "super predator" pervades the South African savanna. *Current Biology* 33(21):4689–4696.

Zanón-Martínez, J. I., M. J. Kelly, J. B. Mesa-Cruz, J. H. Sarasola, C. DeHart, and A. Travaini. 2016. Density and activity patterns of pumas in hunted and non-hunted areas in central Argentina. *Wildlife Research* 43(6):449.

Zeller, K. A., D. W. Wattles, L. Conlee, and S. DeStefano. 2019. Black bears alter movements in response to anthropogenic features with time of day and season. *Movement Ecology* 7(1):19.

Zhang, C., et al. 2023. Human-induced behavioural changes of global threatened terrestrial mammals. *Global Ecology and Biogeography* 00:1–15.

CHAPTER 8

Quantifying Anthropogenic Risk Effects on Carnivores

Justine A. Smith and Justin P. Suraci

Risk effects of humans on apex carnivores

Despite extensive extirpation of apex carnivores globally, many populations occur in landscapes shared with humans. While some populations have persisted in areas modified by humans, others have recolonized or been actively reintroduced to areas in which they had previously been lost. In other cases, extant populations of carnivores are currently being challenged by new anthropogenic developments and activities as humans expand their global footprint. In most places where apex carnivores and humans coexist, the carnivore modifies its behaviour to avoid the real or perceived risk of human activities. In some cases these antipredator behaviours are adaptive, as they reduce the risk of human-caused mortality. However, many human activities do not pose an immediate lethal threat to carnivores, yet do stimulate risk-averse behaviours. While both scenarios can cause risk effects in carnivores, the implications of such effects vary with the lethality of humans in the system and the intensity of carnivore response. Furthermore, humans can exert additional nonlethal effects on carnivores indirectly by altering the distribution and availability of resources, fragmenting the landscape, and increasing temperature anomalies and extreme weather events. Measuring behavioural responses to human activities and their effects on populations is thus not only essential for understanding carnivore ecology in the modern world, but also for effectively managing, conserving, and coexisting with these species.

Risk effects of humans on carnivores begin with behavioural or physiological responses (i.e. risk-induced trait responses; Peacor et al. 2020) to perceived anthropogenic threats (Figure 8.1). These include modifying habitat selection or utilization, movement behaviour, fleeing, activity budgets (foraging and vigilance), diet, activity patterns, social interactions, body condition, or stress responses. These responses may be related or sequential; when carnivores exhibit a behavioural response to human activities, they may sustain sublethal consequences, including physiological stress and energetic costs. Behavioural and physiological phenotypic responses to anthropogenic activities can result in non-consumptive effects (NCE) on populations (Smith et al. 2024). NCEs occur when phenotypic responses are strong enough to exert significant demographic costs, including reductions in reproductive success, survival, abundance, density, distribution, or population growth (Peacor et al. 2020). Phenotypic responses can also affect demographic parameters in a second species, such as competitors or resources of the apex carnivore. These community responses, called trait-mediated indirect effects (TMIE), can extend the ecological impact of human activities (Wilson et al. 2020, Smith et al. 2024). In addition to causing NCEs and TMIEs, phenotypic responses can change relationships between carnivores and humans with implications

Justine A. Smith and Justin P. Suraci, *Quantifying Anthropogenic Risk Effects on Carnivores*. In: *Apex Predators in the Anthropocene*. Edited by: Robert A. Montgomery and Remington J. Moll, Oxford University Press. © Oxford University Press (2026). DOI: 10.1093/9780191996665.003.0008 for human–wildlife conflict (e.g. Steyaert et al. 2016, Chapter 13, this volume).

Although risk effects have clear implications for populations, communities, and human–wildlife interactions, measuring such effects in apex carnivores is challenging due to their often nocturnal activity patterns, low densities, elusive nature, and history of antagonism with humans. Carnivores can be challenging to observe directly, necessitating the employment of remote technologies and passive sensors to measure their behaviour. Rapid advances in technology and statistical approaches to analyse remotely collected data have the potential to inform many outstanding questions about risk effects in carnivores. In this chapter, we will detail advances in tools and techniques used to measure risk effects in apex carnivores. We then discuss how to apply novel approaches to address modern ecological questions and conservation challenges regarding risk effects in the Anthropocene.

Early innovations in carnivore ecology field and statistical methods

Before the golden age of biologging (Wilmers et al. 2015) and molecular biology (DeYoung and Honeycutt 2005), many techniques emerged that allowed for the investigation of carnivore behaviour and physiology. Advances in field techniques and technologies gradually improved quality, quantity, and inferential value of data on carnivores. These included tracking, behavioural observations, VHF telemetry, and diet analysis. Tracking was often facilitated by snow and allowed for the investigation of carnivore presence/absence, relative abundance, reproductive behaviour, diet, and habitat use. For more conspicuous species, behavioural observations could inform the analysis of ethograms, time budgets, social behaviours, and species interactions (e.g. Ashenafi et al. 2005). Such observations continue to be used today; for example, they are the foundation of the Yellowstone winter wolf (*Canis lupus*) study (Metz et al. 2020). VHF telemetry collars date back to the 1960s and paved the way for a revolution in animal tracking (Craighead et al. 1971, Mech 1980). VHF collars allowed for the relocation of individual animals, improving the ability to locate kill and den sites, promoting the development of early home range estimation methods, and facilitating early habitat selection analyses. Scat collection occurred opportunistically, on transects, from behavioural observations, or by tracking (using animal tracks or VHF telemetry). Before the advent of modern diet analyses, scats could be dissected to identify bones and hairs to estimate diet. Scat dissection remains a popular technique today, as many more advanced methods take significant expertise in laboratory methods or are expensive at scale. While each of these methods are still in use to some degree, contemporary approaches typically allow for more data of greater accuracy or precision to be collected. As the crises of apex carnivore declines and conflict with humans have gained attention over the last few decades, the application of both foundational and contemporary methods has shifted towards measuring the risk effects of humans on carnivore behaviour and physiology.

Overview of advances in methods to study apex carnivores

Field advances

The elusiveness and low densities of apex carnivores have historically made it challenging to collect large amounts of behavioural data. However, advances in technology and sampling techniques that expand upon the early methods mentioned here have substantially increased the ability to study behavioural responses of apex carnivores to anthropogenic disturbances. Here we discuss both the technological advances allowing increased data collection from cryptic carnivores and the innovative statistical approaches designed to handle these new data streams.

Innovations in biologging

Animal biologgers often enable the collection of high-resolution location data, providing detailed information on space use, movement, and activity (Kays et al. 2015, Wilmers et al. 2015, Beltran et al. 2025). As discussed previously, VHF collars have been widely used for many decades to record coarse location information and monitor survival. However, even this long-used technology has seen advances in recent years. As tags have miniaturized, VHF transmitters have been affixed to expandable collars that can be deployed on neonates (Engebretsen et al. 2007). VHF transmitters can also be surgically implanted into the abdominal cavity of neonates when expandable collars are not effective or practical (Crawshaw et al. 2007). While the data from these tags are often not of high enough resolution to study the behaviour of neonates, they allow for tracking of offspring survival (Benson et al. 2013), which is a key metric for examining human-induced NCEs in apex carnivores.

Widespread adoption of GPS collars has revolutionized tracking of apex carnivores. Most such species are of sufficient body size to carry batteries large enough to power satellite collars (Kays et al. 2015), which transmit collected GPS locations on a fixed schedule to a server using low-orbit satellites such as the Iridium network. These collars can leverage their connection to low-orbit satellites for two-way communication, which allows the researcher to communicate with the collar remotely and change its settings. Satellite collars are thus the preferred tracking device for most apex carnivores, despite the extra battery cost to transmit locations via satellite. Alternatively, GSM (i.e. global system for mobile communication) collars use cell phone towers to download and transmit GPS locations to the researcher. These are effective in areas with good cell phone coverage and could be useful for examining apex predator behaviour in exurban to urban environments. Ultra-high frequency (UHF) collars require that the animal be found by researchers (via ground or aerial telemetry) to download the locations remotely, or that the animal's space use overlaps with a permanent radio tower network. While regularly retrieving location data from active collars can be labour-intensive, UHF technology allows for high-resolution GPS data to be retrieved from smaller carnivores (e.g. civets, *Viverra tangalunga*; Evans et al. 2016) and can also be used to reduce satellite fees or extend battery life on the collars of easily located large carnivores (e.g. African lions, *Panthera leo*; Rafiq et al. 2023a). Most apex carnivores are primarily nocturnal or live in areas with significant vegetative cover, therefore solar satellite tags are not popular with this taxon. For more diurnal carnivore populations, however, solar-powered GPS devices could eliminate the need for regular recaptures and facilitate data collection on animals without sensitizing them to trapping methods or causing additional capture stress. Additionally, satellite ear tags have begun to be deployed on carnivores (e.g. Wiig et al. 2017).

In the marine realm, the deployment of multiple types of biologging devices has been popular for decades (Watanabe and Papastamatiou 2023). Only recently has this approach been applied to apex carnivores, and advances in this area are rapidly emerging. Biologgers that can be deployed on terrestrial animals, in addition to GPS telemetry units, include those that measure an animal's acceleration, orientation, ambient temperature, and proximity to another animal, as well as those that record animal-borne photos or videos (Wilmers et al. 2015). Accelerometers have become one of the most valuable modern tools for investigating the behaviour of elusive animals (Rutishauser et al. 2011, Brown et al. 2013, Nickel et al. 2021). When processed, accelerometer data can reveal high-resolution behavioural profiles of individual animals (Wilmers et al. 2017). This tool has been implemented to quantify behaviour change and energetic costs of apex carnivore antipredator responses to nonlethal human disturbance (e.g. Figure 8.2). Although most biologging devices for apex carnivores are deployed on collars, promising studies on domestic carnivores show the potential for glue-on tags to measure behaviours that cannot be identified from collars, such as scent-marking (Bidder et al. 2020). Other types of sensors show potential in informing our understanding of apex predator behaviour. Temperature loggers can be used to examine interactions between human avoidance and thermoregulation as development and climate change continue to synergistically affect apex carnivores. Proximity sensors could indicate if social behaviour among conspecifics is affected by anthropogenic disturbances. Light-level geolocation has been used as an indicator of denning behaviour (Merkel et al. 2023). Animal-borne cameras have great potential to examine carnivore vigilance, foraging rates, diet, kill success, social interactions, movement behaviour, microhabitat selection, reproductive behaviour, and direct responses to human activities (Brockman et al. 2017, Pagano et al. 2018,

Andersen et al. 2020, Petroelje et al. 2020, Naganuma et al. 2021, Tezuka et al. 2023). Finally, animal-borne sound recorders show promise in identifying kill events (Studd et al. 2021) and recording the acoustic environment experienced by individual animals. When combined, these biologging devices have the potential to paint a much richer picture of the types of cues experienced by apex carnivores and the diversity of behavioural responses and trade-offs that subsequently emerge. Recovering these sensors can also be facilitated by advances in remote detaching technologies (Rafiq et al. 2023b). In some cases, more invasive biologgers can provide additional information about animal physiology and stress. For example, implanted heart rate monitors are compatible with some species (Laske et al. 2018, Blanchet et al. 2019).

Deployment of biologgers can be coupled with field investigations of biologging data. Field investigations allow for additional data collected from animal tracking or behavioural validation, strengthening inferences that can be made from remotely collected data. For example, accelerometer measurements can be linked to specific behaviours by directly observing tagged animals in the field or associating accelerometer data with other simultaneously collected biologging data streams such as acoustic recordings (e.g. Studd et al. 2021). Energetic expenditure can also be calibrated for animal-borne accelerometers by first measuring respiration rates in a captive setting (Williams et al. 2014). In addition, field investigations of animal location clusters can provide detailed information about apex predator diet, feeding behaviour, and resting sites (Elbroch et al. 2018, Clapp et al. 2021, 2022).

Innovations in non-invasive sampling

Despite the wealth of information that can be collected from biologgers, non-invasive sampling methods are often more appropriate given the research question, logistical or funding constraints, and concerns regarding animal welfare. Although non-invasive methods are often less suitable for measuring high-resolution space use and movement behaviour, many other behavioural and physiological responses to human disturbances can be identified using these techniques. Carnivore habitat use, activity patterns, feeding and vigilance rates, den attentiveness, fleeing behaviour, movement behaviour, group structure, diet, stress, and species interactions can all be observed with non-invasive methods. Furthermore, tying behaviour change to NCEs can also be facilitated by non-invasive estimates of distribution, reproductive success, density, and abundance.

Camera traps are a widespread and popular tool for monitoring carnivore populations. Camera traps deployed with the intention of passive observation using still images can, with the appropriate statistical techniques (see 'Statistical advances'), provide estimates of phenotypic (e.g. habitat selection, activity patterns, vigilance rates) and demographic (e.g. abundance, density, distribution) responses to anthropogenic disturbance, in addition to measuring the magnitude of human activity. However, the utility of camera traps extends well beyond data that can be collected passively. Leveraging camera traps in experimental designs provides innovative ways to investigate carnivore risk-avoidance behaviours (Smith et al. 2020b). When paired with audio playbacks (e.g. automated behavioural response systems, or ABRs; Suraci et al. 2017, Palmer et al. 2022), these integrated units can be used to examine the effects of anthropogenic auditory cues, including human voices or sounds associated with hunting (dogs, gunshots) on apex carnivore fleeing, vigilance, and feeding (Smith et al. 2017, Suraci et al. 2019a, Zanette et al. 2023), and has been used to demonstrate human-induced changes in space use at relatively large scales (Suraci et al. 2019b). Other possible experimental approaches that have not yet been adopted include exposing carnivores to human visual or olfactory cues and recording their responses with camera traps. Leveraging these manipulative approaches enables formal experimental study designs (e.g. comparison between human and control cues) into examinations of human risk effects on carnivores, a field dominated by observational studies. Like camera traps, passive acoustic monitors can estimate locations of the calls of apex carnivores and specific types of human activities, such as gunshots (Wijers et al. 2021). This novel technology could be used to estimate changes in habitat use or social behaviour of carnivores in response to anthropogenic disturbances.

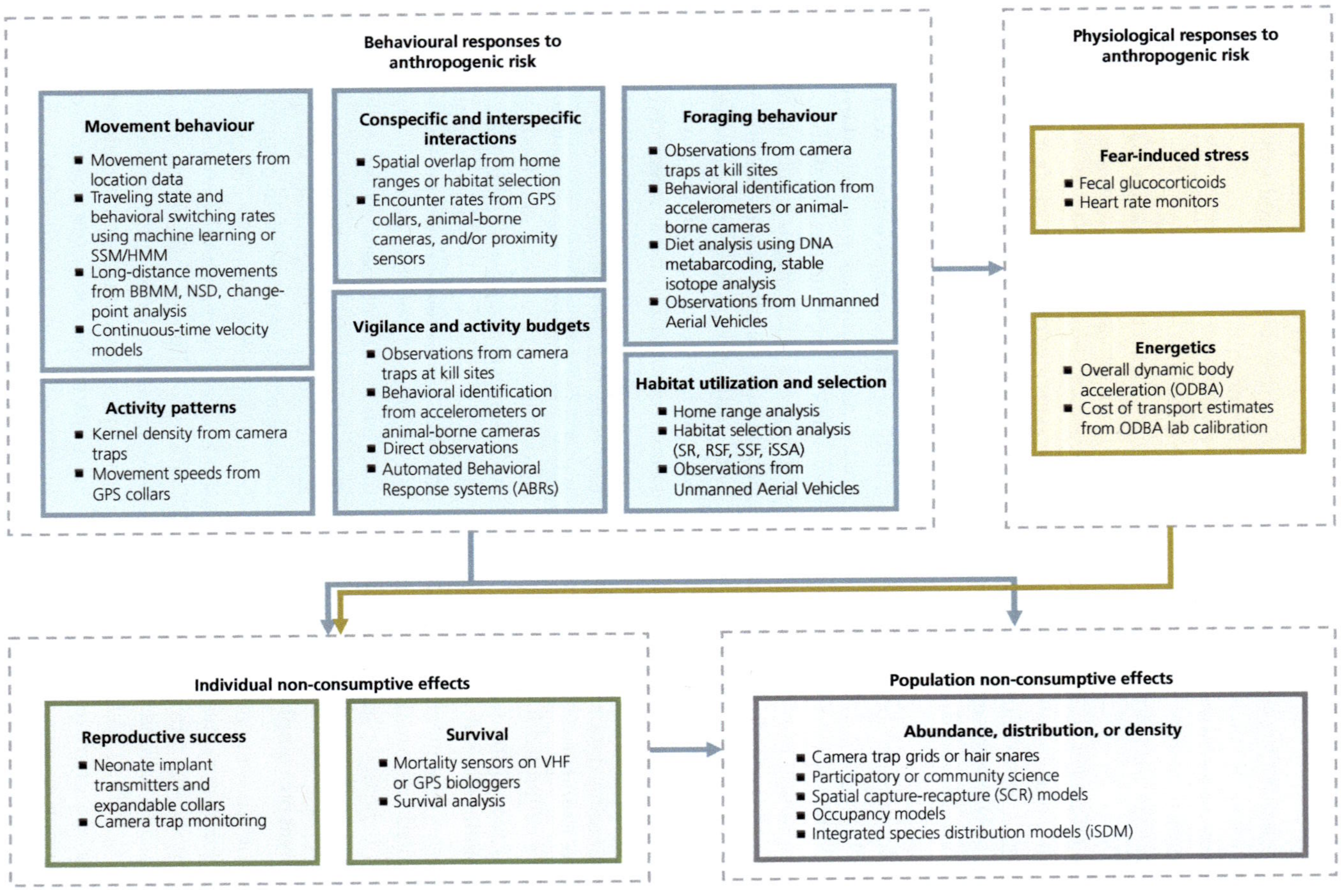

Figure 8.1 Example methods for studying risk effects in apex carnivores. Carnivore behavioural responses to human activity can affect stress and energetics, individual fitness and survival, and population abundance and density. Robust tests of anthropogenic risk effects on carnivores examine multiple steps in this process and the relationships between them. Examples provided are not exhaustive and include both data collection and analysis techniques.

Unmanned aerial vehicles (UAV), or drones, are increasingly used to measure occurrence (Platonov et al. 2013), relative density (Bushaw et al. 2019), or behaviour (Pollock et al. 2022) of carnivores, particularly those that live in relatively open landscapes and can be paired with additional data streams such as physiological measurements (Ditmer et al. 2019). Calibration against traditional ground transect surveys may be important to test for discrepancies between density estimates and the utility of UAVs (Povlsen et al. 2023). Object-based image analysis (OBIA) can increase the efficiency of detecting carnivores from UAV imagery by implementing algorithmic classification (Chrétien et al. 2015, Corcoran et al. 2021). Data derived from UAV imagery may facilitate noninvasive investigation of anthropogenic risk effects on carnivores over larger spatial scales than are possible with ground observations.

Collection of apex carnivore biological samples historically required extensive field tracking. Increasing use of scat detection dogs has promoted more efficient fecal sampling with implications for risk responses to human activities (Long et al. 2007, Mumma et al. 2017). In areas with significant human presence, scats can also be collected efficiently with the help of community scientists (Smith et al. 2018). Broad scat sampling and other tracking methods (e.g. kill site investigation) enable the testing of human impacts on diet, which can occur due to altered prey community composition (Creel et al. 2018), carnivore activity patterns (Smith et al. 2018), or anthropogenic subsidies and human–wildlife conflict (Shehzad et al. 2015). Molecular techniques such as DNA metabarcoding allow for more efficient and detailed estimation of carnivore diets (Shehzad et al. 2012). Scat collection can also be used in assays of stress hormones (e.g. glucocorticoids). Although these assays might not reliably indicate stress (Linklater et al. 2010, Linklater and Gedir 2011, McDonald et al. 2018), some evidence suggests that stress hormones in carnivores are influenced by human land use and population density (Van Meter et al. 2009, Creel et al. 2013a). Finally, non-invasively collected scat or hair samples (e.g. from hair snares) can provide genetic information about individuals. Individual genetic identification is used to provide insights into human impacts on carnivore connectivity and dispersal behaviour, survival, and population density (Carroll et al. 2024).

Non-invasive data collection can also be achieved through community or participatory science. Working with local communities to assemble data on carnivore behaviour and space use has recently gained traction as a way to leverage local knowledge and involve stakeholders in carnivore research. Participatory mapping (i.e. working with local people to convey spatial information with cartography) can provide valuable information about the space use, movement, and density of carnivores, in addition to their interactions with people and livestock (McInturff et al. 2020, Wilkinson et al. 2021). Experienced local guides can also identify animals individually, facilitating abundance estimates in critical conservation areas (Elbroch et al. 2023). At larger scales, participatory science records from web-based portals and apps (e.g. iNaturalist, Global Biodiversity Information Facility [GBIF]) can inform estimates of carnivore distributions and relative abundances. Such databases have been used to assess apex carnivore avoidance of infrastructure (Srivathsa et al. 2020) and reclamation of urban habitats during COVID-19 lockdowns (Vardi et al. 2021), but the distribution of observations can be strongly correlated with human activity (Mueller et al. 2019).

Statistical advances

Analysis of telemetry and biologging data

Home range analysis is often a first step in the evaluation of human impacts on carnivore space use and habitat utilization or selection. Traditional home range estimators include geometric (minimum convex polygon, MCP; local convex hull, LoCoH) or probabilistic (kernel density estimation, KDE) approaches (Fieberg and Börder 2012). However, MCP, KDE, and LoCoH estimators all assume that the location data are independent and identically distributed (IID; Noonan et al. 2019). Particularly as battery technology improves and higher fix rates are implemented, this assumption is often not valid. Datasets that do not meet IID assumptions are likely to produce home ranges that cannot predict future use by the animal. In the past decade, autocorrelated-KDE (AKDE) estimators, which do not require location data to be IID, have gained traction to confront this problem (Fleming et al. 2015,

Calabrese et al. 2021, Noonan et al. 2019, Silva et al. 2022). They generally produce larger home ranges than other estimators and are less sensitive to study duration (Noonan et al. 2019). Home ranges can also be estimated by producing a utilization distribution informed by the habitat selection process of the animal, which addresses issues of overestimating home range size (Signer et al. 2017). Ultimately, the movement behaviour of the animal, autocorrelation and homogeneity of the location data, environmental constraints of the system, study duration, fix rate, effective home range size (Silva et al. 2022), and perhaps most importantly, the purpose of generating a home range (Fieberg and Börger 2012), should all be considered when selecting a home range estimator. Occurrence estimators, which interpolate where the animal went but are not effective at predicting future space use (e.g. Brownian bridge density estimators, Horne et al. 2007; time-dependent LoCoH, Lyons et a. 2013), can also be useful when used in a descriptive manner (Fleming et al. 2015, Noonan et al. 2019). Regardless of which method is used to delineate home range boundaries, considering the three-dimensional structure of the landscape may be crucial to deriving accurate estimates of home range size, particularly in rugged mountainous areas (Farhadinia et al. 2020).

Habitat utilization or selection is another metric often used to describe the impact of humans on carnivore space use. One can examine how human activity affects an animal's access to or use of essential resources, or instead directly quantify avoidance of anthropogenic landscape features or disturbances. Such behavioural responses are often evaluated by using an animal's known locations and a sample of random locations available to the animal to calculate habitat selection. Specific habitat selection analysis (HSA) approaches include selection ratios (SR; Manly et al. 1993), resource selection functions (RSF; Boyce et al. 1999), step selection functions (SSF; Fortin et al. 2005), or integrated step selection analysis (iSSA; Avgar et al. 2017). SRs and RSFs assume data meet IID assumptions, which is often not the case with high-resolution GPS location data (but see autocorrelation-informed weighting proposed by Alston et al. 2022), thereby requiring location data to be rarefied. While RSFs at the scale of an individual's home range remain a valuable tool for analysing courser scales of selection, SSFs, which sample available locations at the movement step, are increasingly favoured when researchers are focused on more localized selection at the scale of individual movement decisions and habitat use within the home range. A limitation of SSF analyses is that habitat selection is often implicitly assumed to be independent of the movement process when in fact the two are inherently linked. iSSAs thus build on SSF approaches by incorporating step-level movement parameters directly into the model formulation, allowing researchers to explicitly model the movement process while estimating resource selection at the scale of the individual step. Accounting for individual variation when estimating population-level habitat selection with SSFs and iSSAs can be achieved using a two-step modelling approach (Craiu et al. 2011, 2018, Avgar et al. 2016, Prokopenko et al. 2017, Dickie et al. 2020) or by fitting a mixed Poisson regression with random intercepts for each case–control set of used and available locations and random slopes for individual animals (Muff et al. 2020). Importantly, the use of individual-level random slopes on covariates of interest facilitates the examination of functional responses in habitat selection (i.e. individual-level changes in selection across the range of covariate values driven by changes in relative availability). When conducting any HSA, estimating functional responses can be crucial to account for variability in habitat availability between animals and associated selection strengths, particularly in carnivore populations exposed to gradients of human disturbance (Heinemeyer et al. 2019, Muhly et al. 2019, Thorsen et al. 2022a). Sampling design and sample sizes for any HSA should also consider the relationship between habitat distribution and selection strength (Street et al. 2021).

Short- and long-distance movements are likely to be affected by human activities. Fear of humans can alter movement parameters like step lengths and turn angles as animals try to either move quickly through or stealthily traverse risky areas. Changes in long-distance movement patterns may be distinguished by quantifying dispersal paths with Brownian bridge movement models (BBMM; Horne et al. 2007, Sarkar et al. 2021), net-squared displacement (NSD; Börger and Fryxell 2012), daily displacement (Byrne et al. 2014), likelihood-based change point detection (Gurarie et al. 2009), and mechanistic range shift analysis (Gurarie et al. 2017). Effects of humans on movement can be assessed by comparing distributions of movement parameters (e.g. step lengths, turn angles, residence time, or time-to-return; Abrahms et al. 2016, Thorsen et al. 2022a), conducting behavioural change point analysis (Gurarie et al. 2009, Morelle et al. 2017), or using movement parameters to distinguish behavioural states (e.g. directed travel or resting) by fitting hidden Markov models (HMM; Blakey et al. 2022, Thorsen et al. 2022a). Methods that distinguish behavioural states allow for the investigation of behaviour-specific habitat selection (Abrahms et al. 2016, Suraci et al. 2019c), which can also be achieved by fitting integrated HMM-SSF models (Klappstein et al. 2023). Spatial dynamics of animal movement can also be linked directly to anthropogenic disturbances through the implementation of hierarchical continuous time velocity models (Paun et al. 2022).

Conspecific and interspecific interactions can increase or decrease depending on species- or individual-specific responses to anthropogenic disturbances (Van Scoyoc et al. 2023). One way to assess interactions is by estimating overlap between home ranges or utilization distributions (Fieberg and Kochanny 2005, Winner et al. 2018, Tilberg and Dixon 2022). Attraction or avoidance can be further assessed by testing an animal's selection for areas of high use or occurrence by another individual (Schlägel et al. 2019). Beyond spatiotemporal attraction or avoidance, encounter locations can be estimated explicitly. GPS location data alone can be used to estimate encounter rates between apex predators and their prey by setting a distance threshold between locations (Middleton et al. 2013, Smith et al. 2020a) or using locations where predator and prey tracks crossed within a time threshold (Courbin et al. 2013), although care should be taken to avoid bias due to GPS fix rates or uncollared individuals (Creel et al. 2013b). Proximity loggers have been primarily used in the investigation of conspecific social interactions and disease transmission (Prange et al. 2011, Cross et al. 2013, Rutz et al. 2015), yet show promise in quantifying encounter rates between apex carnivores and interacting species if a sufficient proportion of the population is tagged (Wilmers et al. 2015, Prugh et al. 2019). In some cases, estimating the spatially explicit probability of encounter is more informative than quantifying encounter rates, particularly when investigating the influence of human activities on prey space use as mediated by predator space use. In such scenarios, the conditional distribution of encounters (CDE) can be estimated from movement data alone (Noonan et al. 2021). Each of these approaches can contribute to the essential goal of understanding the indirect effects of humans on apex carnivores and the species with which they interact.

Perceived risk from humans can impact apex predator time allocation to crucial behaviours (e.g. hunting, feeding, territorial patrol) as well as where on the landscape predators are engaging in those behaviours, ultimately affecting the amount of habitat available for fitness-enhancing activities and thus the carrying capacity of the environment. Biologging data can provide an invaluable window into the spatially explicit behavioural patterns of free-living carnivores but must first be segmented or classified into discrete behavioural states. As noted already, HMMs based on GPS movement data alone can be used for this purpose, though the level of behavioural detail obtained from such approaches is often relatively coarse (e.g. identification of broad movement categories such as 'meandering' or 'directed travel') and dependent on GPS fix rate. Accelerometer data, either alone or in combination with other data streams, can potentially provide substantially greater behavioural detail. However, the number of unique behavioural states that can be identified will depend on whether validated training data are available (e.g. video recordings or direct observations of animals engaging in multiple relevant behaviours and time-synced to accelerometer traces; e.g. Pagano et al. 2017, Giese et al. 2021) or whether the researcher is relying on inherent structure in the accelerometry data themselves to identify behavioural states (e.g. unsupervised classification). Where labelled training data on the accelerometer signatures associated with specific behaviours are available, common machine learning approaches (including Random Forest [RF], Support Vector Machines, and Neural Networks; Nathan et al. 2012, Resheff et al. 2014, Ladds et al. 2017) have been successfully employed to identify numerous behavioural states and apply these classifications to unlabelled data from free-living individuals. For instance, Pagano et al. (2017) fit RF models to data from collar-mounted accelerometers and water conductivity sensors deployed on captive polar bears (*Ursos maritimus*) to identify 10 distinct behavioural states (including resting, swimming, and feeding), which they suggest could then be used to quantify the impacts of a changing climate on polar bear activity and energy budgets. When training data from captive animals are unavailable, researchers, have used additional collar-mounted biologgers to provide the necessary information to train supervised learning models on accelerometer data. Wijers et al. (2018) used audio recordings from acoustic sensors to label time-synced accelerometry data and train an RF model to identify five behavioural states in free-ranging African lions, and a similar approach was adopted by Studd et al. (2021) to identify predation and feeding events by Canada lynx (*Lynx canadensis*).

Where labelled accelerometer or other biologging data suitable for model training are unavailable, unsupervised classification approaches can provide a valuable alternative. These methods do not require the researcher to define a set number of behavioural states *a priori*, instead relying on inherent structure in the data to identify classes that explain as much variation in the data set as possible. Such approaches included unsupervised HMM (Leos-Barajas et al. 2017) and k-means clustering (van Moorter et al. 2010). The latter method has been successfully employed using a combination of accelerometer data and movement characteristics (step lengths and turn angles) from GPS locations to identify several behavioural states (e.g. feeding, resting, and multiple movement modes) in leopards (*Panthera pardus*, Wilmers et al. 2017) and African lions (Suraci et al. 2019c).

In addition to reducing time allocation to fitness-enhancing behaviours, actively avoiding risk from humans can itself be an energetically expensive activity, potentially forcing carnivores to move through more rugged (and thus energetically demanding) terrain or engage in inefficient movements to minimize the risk of encountering people. By correlating oxygen consumption by animals in laboratory settings with overall dynamic body acceleration (Qasem et al. 2012) measured from collar-mounted accelerometers, researchers have estimated energetic expenditure of free-living carnivores—including pumas (*Puma concolor*; Williams et al. 2014, Dunford et al. 2022), wolves (Bryce et al. 2022), and brown bears (*Ursus arctos*; Finnegan et al. 2023)—as they move across the landscape. For pumas in a human-dominated landscape in California, these methods have revealed substantial energetic costs of being pushed into more rugged terrain (Dunford et al. 2022) and of avoiding humans at finer scales (e.g. when in close proximity to developed areas), the latter leading to high cumulative movement costs and smaller home range sizes for males in heavily modified environments (Nickel et al. 2021). To investigate the NCEs of behaviour change and energetic costs, survival analyses (Cox 1972) can distinguish the contribution of direct (e.g. retaliatory killing, vehicle collisions) and indirect (e.g. starvation from lost feeding opportunities or energetic costs) anthropogenic mortality sources to carnivore survival (Bastianelli et al. 2021).

Analysis of occurrence data

While the temporal and spatial resolution of occurrence data (e.g. camera trap detections or direct observations) are typically much lower than those for biologging data, occurrence data sets have the advantage of potentially capturing space use, temporal activity patterns, and behaviour of a wide range of species, and can be used to monitor a larger proportion of individuals in a given carnivore population than can feasibly be tagged in many biologging studies. Time-stamped occurrence data allow researchers to address a range of questions, from relatively subtle impacts of human activity on spatial and/or temporal activity patterns to potential changes in density of carnivore populations across gradients of human disturbance. Camera trap data are perhaps the most common occurrence data type for apex carnivores given their relative ease of collection and non-invasive nature, and many of the methods described in this section are particularly tractable with camera data, though data from other passive sensors or environmental samples (e.g. hair snares, acoustic monitors, fecal DNA) often lend themselves to similar analytical approaches.

Occupancy modelling (MacKenzie et al. 2002, Burton et al. 2015) is perhaps the most common approach for quantifying the risk effects of humans on carnivore spatial distribution and habitat use via occurrence data. Occupancy models rely on repeated sampling at survey locations to derive detection histories (series of detections or non-detections during each sampling event) for one or more species of interest. These data can then be used to model an unobserved occupancy process, describing whether a focal species occurs at a given site, while accounting for imperfect detection, i.e. non-detection of a species when it was actually present (MacKenzie et al. 2004, Royle and Dorazio 2008). Importantly, both the occupancy and detection processes can be modelled as a function of environmental or other covariates, allowing ample opportunity to test whether anthropogenic impacts on the landscape (e.g. development, agriculture, roads), or the presence of humans themselves (e.g. recreationists captured on camera) affect the probability that an apex carnivore species uses a given location and/or the probability of detecting that species if present (Schuette et al. 2013, Nickel et al. 2020, Granados et al. 2023). Multi-species occupancy models can quantify interspecific interactions by modelling the occupancy probability of one species as conditional on the occupancy of another (Devarajan et al. 2020). Such models can be used to test hypotheses regarding whether carnivore interactions with their competitors or prey (i.e. co-occurrence probability) change across gradients of human disturbance (Wang et al. 2015, Lamichhane et al. 2023), although the degree to which co-occurrence captures true 'interactions' has been questioned (Blanchet et al. 2020).

Human disturbance can also shift carnivore temporal activity patterns (Gaynor et al. 2018), and analyses of camera trap data have played an important role in describing these temporal shifts. In a commonly used method originally described by Ridout and Linkie (2009), the timestamps of carnivore detections on camera can be used to fit kernel density curves describing periods of relatively low and high diel activity patterns. By comparing two such curves using bootstrapped circular statistics (Ridout and Linkie 2009, Linkie and Ridout 2011), researchers have estimated the degree of temporal overlap between humans and carnivores (e.g. Haswell et al. 2020), a potential proxy for the risk of human–carnivore conflict, as well as the shift in diel activity patterns by a given carnivore species in areas of low vs. high human disturbance and how this may affect their temporal overlap with their prey (e.g. Mills and Harris 2020). A potential shortcoming of this overlap method for measuring the impacts of humans on temporal activity patterns is that it requires human disturbance to be discretized into two or more categories (e.g. natural vs. developed areas) so that separate kernel density curves can be fit to animal detections taken from locations in each category and compared using circular statistics. Newly developed continuous-time occupancy models (Kellner et al. 2022) require no such assumptions. This approach uses time-stamped detection data to model a continuous 'detection intensity' process, which can be allowed to vary throughout the 24-hour day to capture changes in temporal activity (and therefore detection probability) and allows for the quantification of changes in diel activity patterns as a function of a continuously varying human disturbances (e.g. housing density; Parsons et al. 2022). Continuous time occupancy models have the added advantage of allowing researchers to capture both spatial and temporal impacts of humans in the same analysis by modelling both the (binary) occupancy process and continuous detection intensity as a function of human covariates. Other relevant approaches for simultaneously capturing spatial and temporal avoidance of humans by wildlife include time-to-event models and avoidance–attraction ratios (Niedballa et al. 2019, Parson et al. 2016), both of which rely on calculating the time between the detection of a human and the subsequent detection of the focal species at a camera site.

As noted earlier, when risk effects of humans are sufficiently strong or persistent, they could lead to NCEs on carnivore population size or density (Smith et al. 2024). Several methods have been developed to estimate population density for carnivores and other species from occurrence data, including spatial capture-recapture (SCR; Efford 2004). SCR extends traditional mark-recapture techniques by explicitly modelling the target species' use of space and allows density to be estimated continuously across the study area as a function of environmental covariates, including human disturbance. Mammalian carnivores (particularly felids and ursids) have been the most common subjects of SCR modelling (Tourani 2022) and, given the requirement of these models that animals are individually identifiable, carnivore SCR models have frequently been fit using data from carnivore hair snares, fecal DNA, or other genetic samples (e.g. Proffitt et al. 2015, Roffler et al. 2019). For situations in which individual identifiability is not possible, as is the case in many camera-trapping studies, modifications of the SCR approach have been developed (e.g. unmarked SCR), though with potential compromises in model precision (Augustine et al. 2019). Integrated species distribution models (iSDMs) based on an underlying Poisson point process (Isaac et al. 2020) provide another promising avenue for estimating continuous changes in (relative) population density across a study area as a function of human disturbance. These models provide the opportunity to model multiple occurrence data types (e.g. presence only, presence-absence, count data; Fletcher et al. 2019, Landau et al. 2022) as manifestations of the same underlying density process, and thus allow researchers to leverage data from large online, participatory repositories such as GBIF, iNaturalist, or Wildlife Insights in conjunction with data from structured surveys.

A brief note on quantifying human impacts

Many of the approaches just described for analysing biologging and occurrence data (e.g. HSA, occupancy models, iSDMs) draw on environmental covariates, including estimates of human activity and/or landscape modification, to model aspects of carnivore space use, distribution, or population density. The accuracy and precision of estimates of human impacts on carnivores are therefore only as good as the anthropogenic covariate data themselves. Fortunately, the availability and quality of anthropogenic covariates is continuously improving, in many cases allowing researchers to match the spatial and temporal scale of the covariate to that of the processes under study (e.g. changes in carnivore habitat use over metres or days). A range of remotely sensed or continuously collected data sets on 'human mobility' are available, some of which occur at very fine spatial and/or temporal scales (Ellis-Soto et al. 2023). High-resolution data sets, including daily estimates of nighttime light pollution from NASA's VIIRS satellite (Ditmer et al. 2021) and community-contributed movement paths of recreating humans from the Strava platform (Corradini et al. 2021), have already been incorporated into analyses of human impacts on carnivore behaviour and space use (Bolas et al. 2025). Other data sets provide annual maps of human land use (e.g. MapBiomas, Souza et al. 2020) or estimates of the intensity of human modification at multiple time points (e.g. Theobald et al. 2020) produced at relatively high resolution (e.g. as fine as 30 m) and over national or global scales. When matched to the dates of carnivore biologging or detection data, these data sets can be used to ensure that land use covariates appropriately describe the landscape as experienced by the individuals under study. Computing power and data set management can be major obstacles to using large-scale, high-resolution data sets when examining risk effects on carnivores. However, cloud platforms such as Google Earth Engine (Gorelick et al. 2017), which hosts several dozen petabytes of spatial data sets and facilitates very large spatial computations on cloud servers, have substantially reduced the barriers to working with such data sets. The combination of advances in our ability to collect and model data on carnivores and substantial increases in the availability and quality of covariate data on human disturbances provides enormous opportunity for increasing our understanding of human risk effects on carnivore populations.

Application of contemporary methods to address risk effects of humans on carnivores in the Anthropocene

Development and urbanization

Many apex carnivores are highly sensitive to development and habitat transformation. Avoidance of the built environment can be observed through investigations of occupancy, habitat selection, movement behaviour, and activity patterns. For example, in wolves and brown bears, HMMs were used to separate movement tracks into behavioural states, which were integrated into SSFs to measure behaviour-specific avoidance of development (Whittington et al. 2022). Similarly, habitat selection (RSF) of field-investigated sites representing different behaviours (i.e. kill sites, communication sites, and dens) showed that pumas are more sensitive to development when engaging in social and reproductive activities (Wilmers et al. 2013). Additional nuance has been found when comparing data resolution, site, and demographic classes of carnivores in their response to disturbance, in addition to testing for nonlinear effects (Serieys et al. 2023). A case study of pumas in coastal California has implemented analyses of habitat selection (Wilmers et al. 2013, Smith et al. 2019, Yovovich et al. 2020, Nisi et al. 2022), home range size (Nickel et al. 2021), activity (Wang et al. 2017), movement behaviour and energetics (Wang et al. 2017, Suraci et al. 2019a, b, Nickel et al. 2021, Nisi et al. 2021), occupancy (Nickel et al. 2020), and foraging behaviour (Smith et al. 2015, 2016) to develop a comprehensive assessment of the risk effects of development on pumas using camera traps, ABRs, GPS collars, and accelerometers (Figure 8.2).

Recreation

The shared use of trails by humans and wildlife has facilitated the collection of numerous camera trap studies on carnivore spatial and temporal responses to recreation. Camera trap studies have revealed that apex carnivore activity (Nickel et al. 2020, Sytsma et al. 2022, Procko et al. 2023, Burton et al. 2024), occupancy and habitat use (Erb et al. 2012, Procko et al. 2022), intensity of use (Suraci et al. 2021), and overlap with prey (Muhly et al. 2011) are affected by nonlethal recreation on trails. Biologging data also informs understandings of carnivore responses to recreation. GPS locations have been leveraged to illustrate that movement speeds of brown bears and wolves increase in areas of high trail density (Whittington et al. 2022). A HSA for wolverines (*Gulo gulo*; measured using RSFs) found that these carnivores avoid both motorized (e.g. snowmobile) and non-motorized (e.g. skiing) winter recreation (Heinemeyer et al. 2019). Similarly, a wolf SSF indicated that wolves avoid roads in Yellowstone more during times of high recreation (Anton et al. 2020). Increased resolution of carnivore movement data, as well as availability of crowd-sourced recreation databases (e.g. Strava), will continue to improve inferences regarding the scale-dependence of risk effects of human activities on carnivores (e.g. Thorsen et al. 2022b, Bolas et al. 2025).

Human–carnivore conflict and working landscapes

Carnivore behaviour in working landscapes is driven by a combination of lethal risk from retaliatory killings, disturbance by nonlethal human activity, attraction to resource subsidies, and behaviour modulation due to habitat modification (e.g. reduction in vegetative cover). If we focus on the diversity of methods employed to examine a single species, the African lion, risk responses to livestock production and associated disturbances have been investigated using GPS locations to quantify temporal patterns of space use (Oriol-Cotterill et al. 2015), accelerometers and SSFs to measure behaviour-specific avoidance of humans (Suraci et al. 2019c), PathSFs (Burak et al. 2023) and occupancy analyses (Everatt et al. 2015) to assess landscape-level patterns of space use, and home range and daily distance travelled analyses to examine effects of humans on space use within a protected area (Lesilau et al. 2021). Innovative molecular techniques can also be used to examine the behaviour of apex carnivores in working landscapes; for example, DNA metabarcoding of scats revealed that leopard diets are dominated by livestock in Pakistan (Shehzad et al. 2015).

Anthropogenic cues

Auditory, olfactory, and visual cues associated with human activity are widespread drivers of NCEs in wildlife (Chapter 7, this volume). Although responses of elusive and wide-ranging apex carnivores to human activity are more often measured by examining movement and space use behaviour from GPS locations, recent methodological advances have allowed for the direct testing of the effect of anthropogenic cues on carnivore behaviour. Much of this work has tested

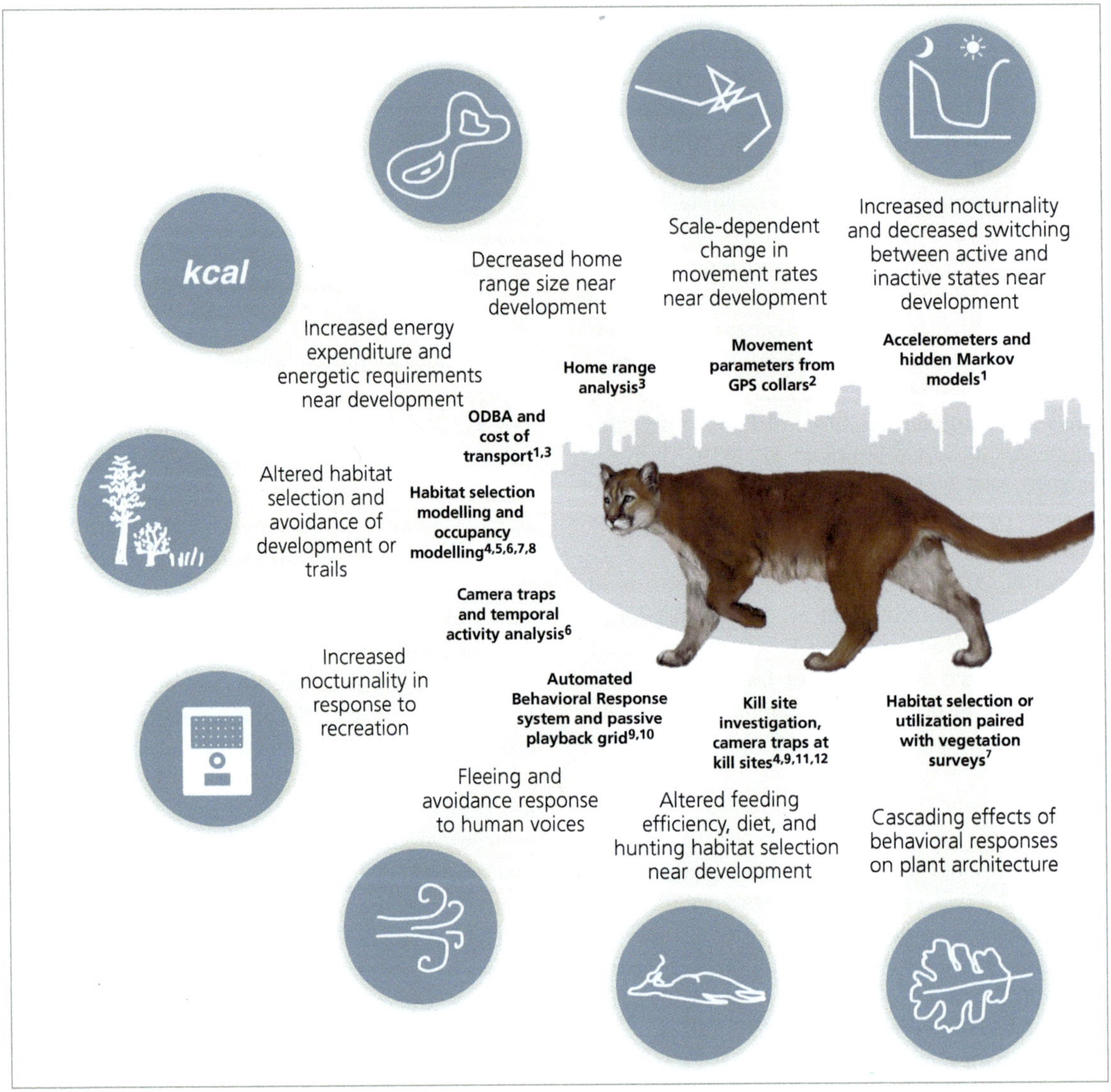

Figure 8.2 Case Study: risk effects in pumas of the Santa Cruz Mountains, C.A. Anthropogenic risk effects have been studied in pumas in the Santa Cruz Mountains at the behavioural, physiological, individual fitness, and population levels. Methods are shown in the inside ring and the associated risk effect findings are show in the outside ring. Puma illustration by Amy Koehler. [1]Wang et al. 2017, [2]Nisi et al. 2021, [3]Nickel et al. 2021, [4]Wilmers et al. 2013, [5]Smith et al. 2019, [6]Nickel et al. 2020. [7]Yovovich et al. 2021, [8]Nisi et al. 2022, [9]Smith et al. 2017, [10]Suraci et al. 2019b, [11]Smith et al. 2015, [12]Smith et al. 2016.

responses to auditory cues. A passive playback array combined with GPS locations of collared pumas revealed that pumas avoid areas where human voices, but not other sounds, were broadcasted (Suraci et al. 2019b; Figure 8.2). In the same population, an implementation of ABRs indicated that at kill sites, pumas flee, take longer to return, and reduce overall feeding time in response to human voices, but not barking dogs or a control sound of chirping frogs (Smith et al. 2017, Suraci et al. 2019a). While tests of visual and olfactory cues are rare, a study on flight initiation distance (FID) of wolves to approaching researchers found that FIDs were largest in 'no wind' conditions, indicating that auditory cues, rather than olfactory or visual cues, mediated wolf responsiveness to humans due to differences in detectability (Karlsson et al. 2007). Conversely, FIDs of Eurasian lynx (*Lynx lynx*) appear to be driven primarily by visual cues, as tolerance distance is strongly correlated with visibility (Sunde et al. 1998).

Other nonlethal effects

In addition to the impacts of perceived risk on carnivore behaviour, humans have a range of other nonlethal effects on carnivores mediated by rapid anthropogenic changes in ecological communities and the environment. The novel data streams and analytical approaches described here are similarly shedding light on these other non-lethal effects. For instance, resource depletion is a leading cause of apex carnivore declines (Ripple et al. 2014), leading to reduced nutritional condition and reproductive success at the individual level and lower carrying capacity at the population level. Camera traps have been used to investigate relative densities of carnivores and their prey in human-dominated landscapes, highlighting the indirect link between anthropogenic prey depletion, carnivore occupancy, and access to specific prey types (Xiao et al. 2018, Mills and Harris 2020). Conversely, resource subsidies are often discussed in the context of increasing access to high-calorie foods (Newsome et al. 2014), but despite this potential benefit, the exploitation of subsidies can exert substantial costs. For example, consumption of low-quality anthropogenic foods by coyotes (*Canis latrans*), assessed by stable isotope analysis of guard hairs, was associated with increased mange prevalence (Murray et al. 2015), and GPS and stable isotope analyses showed that black bears (*Ursus americananus*) using developed areas for foraging had increased body fat but reduced cub and adult survival from vehicle collisions and other anthropogenic sources of mortality (Johnson et al. 2020). The behavioural and physiological impacts of anthropogenic climate change on apex carnivore populations are only beginning to be explored, but impacts on carnivore space use and seasonal activity patterns are starting to emerge. For example, a combination of GPS and accelerometer data on head tilt rates were used to determine den entry and exit dates in brown bears, revealing that warmer temperature was the strongest predictor of denning duration and emergence timing (Johnson et al. 2017), setting up the potential for phenological mismatch between denning and resource availability. Recent increases in the frequency and magnitude of wildfires have provided additional insights into the impact of climate change on carnivores. Habitat selection (SSF), activity (HMM from accelerometer data), daily distance travelled (CTMM), and home range overlap analyses showed that pumas were exposed to greater anthropogenic risk when avoiding burned zones (Blakey et al. 2022). Long-term monitoring of carnivore habitat associations and occupancy post-fire can also be achieved with camera trap arrays (Lewis et al. 2022, Calhoun et al. 2023). Novel technological tools, including customization and recovery of biologging tags (e.g. SensorDrop, Rafiq et al. 2023a), are likely to improve estimates of carnivore behavioural and physiological responses to rapid environmental changes, including temperature anomalies, drought, fire, and other climatic events.

Looking forward

Quantifying anthropogenic risk effects in apex carnivores requires careful selection of human, carnivore, and environmental metrics. Human disturbances can include both lethal and non-lethal activities that cause phenotypic change in carnivores. The behavioural or physiological phenotypes examined should be those likely to be affected by human activities (see Chapter 7, this volume). Measuring consequences of risk-induced trait responses in carnivores requires the study of fitness or density metrics of the carnivore population (NCE; see Chapter 9, this volume) or of a third species with which the carnivore interacts (TMIE; see Chapter 11, this volume). Although it is common for researchers to select single metrics for analyses of risk effects (Moll et al. 2017), such approaches are likely to miss important pathways of risk effects in carnivores. Combining novel methods in biologging, non-invasive sampling, and remote sensing will provide rich opportunities in the study of risk effects across scales. This integrative work is necessary to inform conservation and management plans that seek to improve population or community outcomes by mitigating human impacts on carnivore behaviour. Particularly as synergistic effects of human activity increasingly threaten carnivore populations (see Chapter 15, this volume), leveraging diverse methodologies to disentangle the mechanisms of human-induced risk effects will both advance understandings of risk effects in nature (Peacor et al. 2020) and their application to conservation (Smith et al. 2024).

To achieve improved inference of carnivore risk effects, we suggest the following directions for method development and application:

1. Integrate experiments with the study of naturally occurring patterns to link mechanism to process
2. Design experiments to test a range of magnitudes of human disturbance to calibrate estimates of impacts that can be applied at the landscape level
3. Draw on large biologging data sets to examine spatiotemporal risk effects across contexts (habitat, anthropogenic disturbance, carnivore species, diversity of carnivore and prey communities)
4. Investigate continuity across resolutions and scales of environmental and biologging data
5. Explore individual variation in risk responses and integrate individual variation into modelling approaches
6. Employ multiple metrics of anthropogenic disturbance, carnivore phenotypic response, and population or fitness costs

References

Abrahms, B., N. R. Jordan, K. A. Golabek, J. W. McNutt, A. M. Wilson, and J. S. Brashares. 2016. Lessons from integrating behaviour and resource selection: activity-specific responses of African wild dogs to roads. *Animal Conservation* 19:247–255.

Augustine, B. C., J. A. Royle, S. M. Murphy, R. B. Chandler, J. J. Cox, and M. J. Kelly. 2019. Spatial capture–recapture for categorically marked populations with an application to genetic capture–recapture. *Ecosphere* 10:e02627.

Alston, J. M., et al. 2022. Mitigating pseudoreplication and bias in resource selection functions with autocorrelation-informed weighting. *Methods in Ecology and Evolution* 14:643–654.

Andersen, G. E., H. W. McGregor, C. N. Johnson, and M. E. Jones. 2020. Activity and social interactions in a wide-ranging specialist scavenger, the Tasmanian devil (*Sarcophilus harrisii*), revealed by animal-borne video collars. *PLoS One* 15:e0230216.

Anton, C. B., D. W. Smith, J. P. Suraci, D. R. Stahler, T. P. Duane, and C. C. Wilmers. 2020. Gray wolf habitat use in response to visitor activity along roadways in Yellowstone National Park. *Ecosphere* 11:e03164.

Ashenafi, Z. T., T. Coulson, C. Sillero-Zubiri, and N. Leader-Williams. 2005. Behaviour and ecology of the Ethiopian wolf (*Canis simensis*) in a human-dominated landscape outside protected areas. *Animal Conservation* 8:113–121.

Avgar, T., S. R. Lele, J. L. Keim, and M. S. Boyce. 2017. Relative selection strength: quantifying effect size in habitat- and step-selection inference. *Ecology and Evolution* 7:5322–5330.

Avgar, T., J. R. Potts, M. A. Lewis, and M. S. Boyce. 2016. Integrated step selection analysis: bridging the gap between resource selection and animal movement. *Methods in Ecology and Evolution* 7:619–630.

Bastianelli, M. L., et al. 2021. Survival and cause-specific mortality of European wildcat (*Felis silvestris*) across Europe. Biological Conservation 261:109239.

Beltran, R., et al. 2025. Maximizing biological insights from instruments attached to animals. *Trends in Ecology and Evolution* 40(1):37–46.

Benson, J. F., K. J. Mills, K. M. Loveless, and B. R. Patterson. 2013. Genetic and environmental influences on pup mortality risk for wolves and coyotes within a Canis hybrid zone. *Biological Conservation* 166:133–141.

Bidder, O.R., et al. 2020. Monitoring canid scent marking in space and time using a biologging and machine learning approach. *Scientific Reports* 10:588.

Blakey, R. V., J. A. Sikich, D. T. Blumstein, and S. P. D. Riley. 2022. Mountain lions avoid burned areas and increase risky behavior after wildfire in a fragmented urban landscape. *Current Biology* 32:4762–4768.e5.

Blanchet, F. G., K. Cazelles, and D. Gravel. 2020. Co-occurrence is not evidence of ecological interactions. *Ecology Letters* 23(7):1050–1063.

Blanchet, L., et al. 2019. Movement and heart rate in the Scandinavian brown bear (*Ursus arctos*). *Anim. Biotelemetry* 7:18.

Bolas, E. C., et al. 2025. Human recreation influences activity of a large carnivore in an urban landscape. *Biological Conservation* 301:110812.

Börger, L. and J. Fryxell. 2012. Quantifying individual differences in dispersal using net squared displacement. In *Dispersal Ecology and Evolution*, edited by J. Clobert, M. Baguette, T. G. Benton, and J. M. Bullock. Oxford University Press, Oxford.

Boyce, M. S. and L. L. McDonald. 1999. Relating populations to habitats using resource selection functions. *Trends in Ecology & Evolution* 14(7):268–272.

Brockman, C. J., W. B., Collins, J. M. Welker, D. E. Spalinger, and B. W. Dale. 2017. Determining kill rates of ungulate calves by brown bears using neck-mounted cameras. *Wildlife Society Bulletin* 41:88–97.

Brown, D. D., R. Kays, M. Wikelski, R. Wilson, and A. P. Klimley. 2013. Observing the unwatchable through acceleration logging of animal behavior. *Animal Biotelemetry* 1:20.

Bryce, C. M., C. E. Dunford, A. M. Pagano, Y. Wang, B. L. Borg, S. M. Arthur, and T. M. Williams. 2022. Environmental correlates of activity and energetics in a wide-ranging social carnivore. *Animal Biotelemetry* 10:1.

Burak, M. K., et al. 2023. Spatiotemporal patterns of lion (*Panthera leo*) space use in a human–wildlife system. *Ecological Solutions and Evidence* 4:e12276.

Burton, A. C., et al. 2024. Mammal responses to global changes in human activity vary by trophic group and landscape. *Nature Ecology Evolution* 8(5): 924–935.

Burton, A. C., et al. 2015. Wildlife camera trapping: a review and recommendations for linking surveys to ecological processes. *Journal of Applied Ecology* 52: 675–685.

Bushaw, J. D., K. M. Ringelman, and F. C. Rohwer. 2019. Applications of unmanned aerial vehicles to survey mesocarnivores. *Drones* 3:28.

Byrne, M., C. J. McCoy, and J. Hinton. 2014. Using dynamic Brownian bridge movement modelling to measure temporal patterns of habitat selection. *Journal of Animal Ecology* 83(5):1234–1243.

Calabrese, J. M., C. H. Fleming, M. J. Noonan, and X. Dong. 2021. ctmmweb: a graphical user interface for autocorrelation-informed home range estimation. *Wildlife Society Bulletin* 45:162–169.

Calhoun, K. L., B. R. Goldstein, K. M. Gaynor, A. McInturff, L. Solorio, and J. S. Brashares. 2023. Mammalian resistance to megafire in western U.S. woodland savannas. *Ecosphere* 14:e4613.

Carroll, S. L., G. M. Schmidt, J. S. Waller, and T. A. Graves. 2024. Evaluating density-weighted connectivity of black bears (*Ursus americanus*) in Glacier National Park with spatial capture–recapture models. *Movement Ecology* 12:8.

Chrétien, L.-P., J. Théau, and P. Ménard. 2015. Wildlife multispecies remote sensing using visible and thermal infrared imagery acquired from an unmanned aerial vehicle (UAV). *ISPRS—International Archives of Photogrammetry, Remote Sensing, and Spatial Information Sciences* XL-1/W4:241–248.

Clapp, J. G., et al. 2022. Multi-model application informs prey composition of mountain lions Puma concolor. *Wildlife Biology* 2022(5):e01035.

Clapp, J. G., J. D. Holbrook, and D. J. Thompson. 2021. GPSeqClus: an R package for sequential clustering of animal location data for model building, model application, and field site investigations. *Methods in Ecology and Evolution* 12(5):787–793.

Corradini, A., et al. 2021. Effects of cumulated outdoor activity on wildlife habitat use. *Biological Conservation* 253:108818.

Corcoran, E., M. Winsen, A. Sudholz, and G. Hamilton. 2021. Automated detection of wildlife using drones: Synthesis, opportunities and constraints. *Methods in Ecology and Evolution* 12:1103–1114.

Courbin, N., D. Fortin, C. Dussault, V. Fargeot, and R. Courtois. 2013. Multi-trophic resource selection function enlightens the behavioural game between wolves and their prey. *Journal of Animal Ecology* 82:1062–1071.

Cox, D. R. 1972. Regression models and life-tables. *Journal of the Royal Statistical Society: Series B (Methodological)* 34(2):187–202.

Craighead, J. J., F. C. Craighead, J. R. Varney, and C. E. Cote. 1971. Satellite monitoring of black bear. *BioScience* 21:1206–1212.

Craiu, R. V. and T. Duchesne. 2018. A scalable and efficient covariate selection criterion for mixed effects regression models with unknown random effects structure. *Computational Statistics & Data Analysis* 117:154–161.

Craiu, R. V., T. Duchesne, D. Fortin, and S. Baillargeon. 2011. Conditional logistic regression with longitudinal follow-up and individual-level random coefficients: a stable and efficient two-step estimation method. *Journal of Computational and Graphical Statistics* 20:767–784.

Crawshaw, G. J., K. J. Mills, C. Mosley, and B. R. Patterson. 2007. Field implantation of intraperitoneal radiotransmitters in eastern wolf (*Canis lycaon*) pups using inhalation anesthesia with sevoflurane. *Journal of Wildlife Diseases* 43:711–718.

Creel, S., D. Christianson, and P. Schuette. 2013a. Glucocorticoid stress responses of lions in relationship to group composition, human land use, and proximity to people. *Conservation Physiology* 1:cot021.

Creel, S., et al. 2018. Changes in African large carnivore diets over the past half-century reveal the loss of large prey. *Journal of Applied Ecology* 55:2908–2916.

Creel, S., J. A. Winnie, and D. Christianson. 2013b. Underestimating the frequency, strength and cost of antipredator responses with data from GPS collars: an example with wolves and elk. *Ecology & Evolution* 3:5189–5200.

Cross, P. C., et al. 2013. Female elk contacts are neither frequency nor density dependent. *Ecology* 94:2076–2086.

Devarajan, K., T. L. Morelli, and S. Tenan. 2020. Multi-species occupancy models: review, roadmap, and recommendations. *Ecography* 43:1612–1624.

DeYoung, R. W. and R. L. Honeycutt. 2005. The molecular toolbox: genetic techniques in wildlife ecology and management. *The Journal of Wildlife Management* 69:1362–1384.

Dickie, M., S. R. McNay, G. D. Sutherland, M. Cody, and T. Avgar. 2020. Corridors or risk? Movement along, and use of, linear features varies predictably among large mammal predator and prey species. *Journal of Animal Ecology* 89:623–634.

Ditmer, M. A., et al. 2021. Artificial nightlight alters the predator–prey dynamics of an apex carnivore. *Ecography* 44:149–161.

Ditmer, M. A., et al. 2019. Bears habituate to the repeated exposure of a novel stimulus, unmanned aircraft systems. *Conservation Physiology* 7:coy067.

Efford, M. 2004. Density estimation in live-trapping studies. *Oikos* 106(3):598–610.

Elbroch, L. M., N. Lagos, J. Cárdenas, D. Goic, R. Moraga, and O. Ohrens. 2023. Tourism and human computers offer new tools to monitor Patagonia's top carnivore. *Science of the Total Environment* 877:162916.

Elbroch, L. M., B. Lowrey, and H. U. Wittmer. 2018. The importance of fieldwork over predictive modeling in quantifying predation events of carnivores marked with GPS technology. *Journal of Mammalogy* 99:223–232.

Ellis-Soto, D., et al. 2023. A vision for incorporating human mobility in the study of human–wildlife interactions. *Nature Ecology & Evolution* 7:1362–1372.

Engebretsen, K. N., D. DeBloois, and J. K. Young. 2023. Use of radio-linked VHF technology to monitor neonate carnivores. *Wildlife Society Bulletin* 47:1–10

Erb, P. L., W. J. McShea, and R. P. Guralnick. 2012. Anthropogenic influences on macro-level mammal occupancy in the Appalachian trail corridor. *PLoS One* 7:e42574.

Evans, M. N., S. Guerrero-Sanchez, M. S. A. Bakar, P. Kille, and B. Goossens. 2016. First known satellite collaring of a viverrid species: preliminary performance and implications of GPS tracking Malay civets (*Viverra tangalunga*). *Ecology Research* 31:475–481.

Everatt, K. T., L. Andresen, and M. J. Somers. 2015. The influence of prey, pastoralism and poaching on the hierarchical use of habitat by an apex predator. *South African Journal of Wildlife Research* 45:187–196.

Farhadinia, M. S., et al. 2020. Vertical relief facilitates spatial segregation of a high density large carnivore population. *Oikos* 129(3):346–355.

Fieberg, J. and L. Börger. 2012. Could you please phrase 'home range' as a question? *Journal of Mammalogy* 93:890–902.

Fieberg, J. and C. O. Kochanny. 2005. Quantifying home-range overlap: the importance of the utilization distribution. *Journal of Wildlife Management* 69:1346–1359.

Finnegan, S. P., A. M. Pagano, N. J. Svoboda, S. L. Schooler, and J. L. Belant. 2023. Energy landscapes of Kodiak brown bears: a comparison of accelerometer and global positioning system-derived estimates. *Animal Biotelemetry* 11:7.

Fleming, C. H., W. F. Fagan, T. Mueller, K. A. Olson, P. Leimgruber, and J. M. Calabrese. 2015. Rigorous home range estimation with movement data: a new autocorrelated kernel density estimator. *Ecology* 96: 1182–1188.

Fletcher, R. J., T. J. Hefley, E. P. Robertson, B. Zuckerberg, R. A. McCleery, and R. M. Dorazio. 2019. A practical guide for combining data to model species distributions. *Ecology* 100:e02710.

Fortin, D., H. L. Beyer, M. S. Boyce, D. W. Smith, T. Duchesne, and J. S. Mao. 2005. Wolves influence elk movements: behavior shapes a trophic cascade in Yellowstone National Park. *Ecology* 86:1320–1330.

Gaynor, K. M., C. E. Hojnowski, N. H. Carter, and J. S. Brashares. 2018. The influence of human disturbance on wildlife nocturnality. *Science* 360(6394):1232–1235.

Giese, L., et al. 2021. Using machine learning for remote behaviour classification—verifying acceleration data to infer feeding events in free-ranging cheetahs. *Sensors* 21:5426.

Gorelick, N., M. Hancher, M. Dixon, S. Ilyushchenko, D. Thau, and R. Moore. 2017. Google Earth Engine: planetary-scale geospatial analysis for everyone. *Remote Sensing of Environment* 202:18–27.

Granados, A., et al. 2023. Mammalian predator and prey responses to recreation and land use across multiple scales provide limited support for the human shield hypothesis. *Ecology and Evolution* 13:e10464.

Gurarie, E., R. D. Andrews, and K. L. Laidre. 2009. A novel method for identifying behavioural changes in animal movement data. *Ecology Letters* 12:395–408.

Gurarie, E., et al. 2017. A framework for modelling range shifts and migrations: asking when, whither, whether and will it return. *Journal of Animal Ecology* 86(4):943–959.

Haswell, P. M., J. Kusak, K. A. Jones, and M. W. Hayward. 2020. Fear of the dark? A mesopredator mitigates large carnivore risk through nocturnality, but humans moderate the interaction. *Behavioral Ecology and Sociobiology* 74:62.

Heinemeyer, K., J. Squires, M. Hebblewhite, J. J. O'Keefe, J. D. Holbrook, and J. Copeland. 2019. Wolverines in winter: indirect habitat loss and functional responses to backcountry recreation. *Ecosphere* 10:e02611.

Horne, J. S., E. O. Garton, S. M. Krone, and J. S. Lewis. 2007. Analyzing animal movements using Brownian bridges. *Ecology* 88:2354–2363.

Isaac, N. J. B., et al. 2020. Data integration for large-scale models of species distributions. *Trends in Ecology & Evolution* 35:56–67.

Johnson, H. E., D. L. Lewis, and S. W. Breck. 2020. Individual and population fitness consequences associated with large carnivore use of residential development. *Ecosphere* 11:e03098.

Johnson, H. E., et al. 2017. Increases in residential and energy development are associated with reductions in recruitment for a large ungulate. *Global Change Biology* 23:578–591.

Karlsson, J., M. Eriksson, and O. Liberg. 2007. At what distance do wolves move away from an approaching human? *Canadian Journal of Zoology* 85:1193–1197.

Kays, R., M. C. Crofoot, W. Jetz, and M. Wikelski. 2015. Terrestrial animal movement as an eye on life and planet. *Science* 348:aaa2478.

Kellner, K. F., A. W. Parsons, R. Kays, J. J. Millspaugh, and C. T. Rota. 2022. A two-species occupancy model with a continuous-time detection process reveals spatial and temporal interactions. *Journal of Agricultural, Biological and Environmental Statistics* 27:321–338.

Klappstein, N. J., L. Thomas, and T. Michelot. 2023. Flexible hidden Markov models for behaviour-dependent habitat selection. *Movement Ecology* 11:30.

Ladds, M. A., A. P. Thompson, J.-P. Kadar, D. J Slip, D. P Hocking, and R. G Harcourt. 2017. Super machine learning: improving accuracy and reducing variance of behaviour classification from accelerometry. *Animal Biotelemetry* 5:8.

Lamichhane, S., et al. 2023. Non-exploitative human disturbance provides shelter for prey from predator. *Ecology and Evolution* 13:e10200.

Landau, V. A., B. R. Noon, D. M. Theobald, N. T. Hobbs, and C. K. Nielsen. 2022. Integrating presence-only and occupancy data to model habitat use for the northernmost population of jaguars. *Ecological Applications* 32:e2619.

Laske, T. G., et al. 2018. Development and utilization of implantable cardiac monitors in free-ranging American black and Eurasian brown bears: system evolution and lessons learned. *Animal Biotelemetry* 6:13.

Leos-Barajas, V., et al. 2017. Analysis of animal accelerometer data using hidden Markov models. *Methods in Ecology and Evolution* 8:161–173.

Lesilau, F., S. Verschueren, M. van't Zelfde, K. C. J. M. Musters, G. R. de Snoo, and H. H. de Iongh. 2021. Spatial ecology of lions in a small, semi-fenced park surrounded by dense human populations: the case study of Nairobi National Park, Kenya. *Mammalia* 85:198–207.

Lewis, J. S., L. LeSueur, J. Oakleaf, and E. S. Rubin. 2022. Mixed-severity wildfire shapes habitat use of large herbivores and carnivores. *Forest Ecology and Management* 506:119933.

Linkie, M. and M. S. Ridout. 2011. Assessing tiger–prey interactions in Sumatran rainforests. *Journal of Zoology* 284:224–229.

Linklater, W. L. and J. V. Gedir. 2011. Distress unites animal conservation and welfare towards synthesis and collaboration: distress, conservation and welfare. *Animal Conservation* 14:25–27.

Linklater, W. L., E. A. MacDonald, J. R. B. Flamand, and N. M. Czekala. 2010. Declining and low fecal corticoids are associated with distress, not acclimation to stress, during the translocation of African rhinoceros. *Animal Conservation* 13:104–111.

Long, R. A., T. M. Donovan, P. Mackay, W. J. Zielinski, and J. S. Busas. 2007. Effectiveness of scat detection dogs for detecting forest carnivores. *Journal of Wildlife Management* 71:2007–2017.

Lyons, A. J., W. C. Turner, and W. M. Getz. 2013. Home range plus: a space-time characterization of movement over real landscapes. *Movement Ecology* 1:2.

MacKenzie, D. I., J. D. Nichols, G. B. Lachman, S. Droege, J. A. Royle, and C. A. Langtimm. 2002. Estimating site occupancy rates when detection probabilities are less than one. *Ecology* 83:2248–2255.

MacKenzie, D. I., L. L. Bailey, and J. D. Nichols. 2004. Investigating species co-occurrence patterns when species are detected imperfectly. *Journal of Animal Ecology* 73:546–555.

Manly, B. F. J., L. L. McDonald, and D. L. Thomas. 1993. *Resource Selection by Animals: Statistical Design and Analysis for Field Studies*. Chapman & Hall, London, England.

McDonald, R. S., J. D. Roth, and W. G. Anderson. 2018. Prey cortisol affects the usefulness of fecal glucocorticoid metabolite concentration as an indicator of stress in a carnivore. *Canadian Journal of Zoölogy* 96: 367–371.

McInturff, A., J. R. B. Miller, K. M. Gaynor, and J. S. Brashares. 2020. Patterns of coyote predation on sheep in California: a socio-ecological approach to mapping risk of livestock–predator conflict. *Conservation Science and Practice* 3(3):e175.

Mech, L. D. 1980. Making the most of radio tracking—a summary of wolf studies in Northeastern Minnesota. In *A Handbook on Biotelemetry and Radio Tracking*, edited by C. J. Amlaner and D. W. Macdonald. Pergamon, Oxford.

Merkel, B., J. Aars, K. L. Laidre, and J. W. Fox. 2023. Light-level geolocation as a tool to monitor polar bear (*Ursus maritimus*) denning ecology: a case study. *Animal Biotelemetry* 11:11.

Metz, M. C., et al. 2020. Accounting for imperfect detection in observational studies: modeling wolf sightability in Yellowstone National Park. *Ecosphere* 11:e03152.

Middleton, A. D., et al. 2013. Linking anti-predator behaviour to prey demography reveals limited risk effects of an actively hunting large carnivore. *Ecology Letters* 16:1023–1030.

Mills, K. L. and N. C. Harris. 2020. Humans disrupt access to prey for large African carnivores. *Elife* 9:e60690.

Moll, R. J., et al. 2017. The many faces of fear: a synthesis of the methodological variation in characterizing predation risk. *Journal of Animal Ecology* 86:749–765.

Morelle, K., N. Bunnefeld, P. Lejeune, and S. A. Oswald. 2017. From animal tracks to fine-scale movement modes: a straightforward approach for identifying multiple spatial movement patterns. *Methods in Ecology and Evolution* 8:1488–1498.

Mueller, M. A., D. Drake, and M. L. Allen. 2019. Using citizen science to inform urban canid management. *Landscape and Urban Planning* 189:362–371.

Muff, S., J. Signer, and J. Fieberg. 2020. Accounting for individual-specific variation in habitat-selection studies: efficient estimation of mixed-effects models using Bayesian or frequentist computation. *Journal of Animal Ecology* 89:80–92.

Muhly, T. B., C. Semeniuk, A. Massolo, L. Hickman, and M. Musiani. 2011. Human activity helps prey win the predator-prey space race. *PloS One* 6:e17050.

Muhly, T. B., et al. 2019. Functional response of wolves to human development across boreal North America. *Ecology and Evolution* 9:10801–10815.

Mumma, M. A., et al. 2017. Examining spatial patterns of selection and use for an altered predator guild. *Oecologia* 185:725–735.

Murray, M., M. A. Edwards, B. Abercrombie, and C. C. St. Clair. 2015. Poor health is associated with use of anthropogenic resources in an urban carnivore. *Proceedings of the Royal Society of London: Series B, Biological Sciences* 282:20150009.

Naganuma, T., et al. 2021. Animal-borne video systems provide insight into the reproductive behavior of the Asian black bear. *Ecology and Evolution* 11:9182–9190.

Nathan, R., O. Spiegel, S. Fortmann-Roe, R. Harel, M. Wikelski, and W. M. Getz. 2012. Using tri-axial acceleration data to identify behavioral modes of free-ranging animals: general concepts and tools illustrated for griffon vultures. *Journal of Experimental Biology* 215:986–996.

Newsome, T. M., et al. 2014. The ecological effects of providing resource subsidies to predators: resource subsidies and predators. *Global Ecology and Biogeography* 24:1–11.

Nickel, B. A., J. P. Suraci, M. L. Allen, and C. C. Wilmers. 2020. Human presence and human footprint have non-equivalent effects on wildlife spatiotemporal habitat use. *Biological Conservation* 241:108383.

Nickel, B. A., J. P. Suraci, A. C. Nisi, and C. C. Wilmers. 2021. Energetics and fear of humans constrain the spatial ecology of pumas. *Proceedings of the National Acadamedy of Sciences* 118:e2004592118.

Niedballa, J., A. Wilting, R. Sollmann, H. Hofer, and A. Courtiol. 2019. Assessing analytical methods for detecting spatiotemporal interactions between species from camera trapping data. *Remote Sensing in Ecology and Conservation* 5(3):272–285.

Nisi, A. C., J. F. Benson, and C. C. Wilmers. 2022. Puma responses to unreliable human cues suggest an ecological trap in a fragmented landscape. *Oikos* 2022(5):e09051.

Nisi, A. C., et al. 2021. Temporal scale of habitat selection for large carnivores: balancing energetics, risk and finding prey. *Journal of Animal Ecology* 91(1):182–195.

Noonan, M. J., et al. 2021. Estimating encounter location distributions from animal tracking data. *Methods in Ecology and Evolution* 12:1158–1173.

Noonan, M. J., et al. 2019. A comprehensive analysis of autocorrelation and bias in home range estimation. *Ecological Monographs* 89:e01344.

Oriol-Cotterill, A., D. W. Macdonald, M. Valeix, S. Ekwanga, and L. G. Frank. 2015. Spatiotemporal patterns of lion space use in a human-dominated landscape. *Animal Behaviour* 101:27–39.

Pagano, A. M., et al. 2018. High-energy, high-fat lifestyle challenges an Arctic apex predator, the polar bear. *Science* 359:568–572.

Pagano, A. M., et al. 2017. Using tri-axial accelerometers to identify wild polar bear behaviors. *Endangered Species Research* 32:19–33.

Palmer, M. S., C. Wang, J. Plucinski, and R. M. Pringle. 2022. BoomBox: an automated behavioural response (ABR) camera trap module for wildlife playback experiments. *Methods in Ecology and Evolution* 13: 611–618.

Parsons, A. W., et al. 2016. The ecological impact of humans and dogs on wildlife in protected areas in eastern North America. *Biological Conservation* 203:75–88.

Parsons, A. W., K. F. Kellner, C. T. Rota, S. G. Schuttler, J. J. Millspaugh, and R. W. Kays. 2022. The effect of urbanization on spatiotemporal interactions between gray foxes and coyotes. *Ecosphere* 13:e3993.

Paun, I., D. Husmeier, J. G. C. Hopcraft, M. M. Masolele, and C. J. Torney. 2022. Inferring spatially varying animal movement characteristics using a hierarchical continuous-time velocity model. *Ecology Letters* 25: 2726–2738.

Peacor, S. D., B. T. Barton, D. L. Kimbro, A. Sih, and M. J. Sheriff. 2020. A framework and standardized terminology to facilitate the study of predation-risk effects. *Ecology* 101(12):e03152.

Petroelje, T. R., J. L. Belant, D. E. Beyer, and N. J. Svoboda. 2020. Identification of carnivore kill sites is improved by verified accelerometer data. *Animal Biotelemetry* 8:18.

Platonov, N. G., I. N. Mordvintsev, and V. V. Rozhnov. 2013. The possibility of using high resolution satellite images for detection of marine mammals. *Biology Bulletin* 40:197–205.

Pollock, T. I., D. O. Hunter, D. P. Hocking, and A. R. Evans. 2022. Eye in the sky: observing wild dingo hunting behaviour using drones. *Wildlife Research* 50:212–223.

Povlsen, P., D. Bruhn, C. Pertoldi, and S. Pagh. 2023. A novel scouring method to monitor nocturnal mammals using uncrewed aerial vehicles and thermal cameras—a comparison to line transect spotlight counts. *Drones* 7:661.

Prange, S., S. D. Gehrt, and S. Hauver. 2011. Frequency and duration of contacts between free-ranging raccoons: uncovering a hidden social system. *Journal of Mammalogy* 92:1331–1342.

Procko, M., R. Naidoo, V. LeMay, and A. C. Burton. 2022. Human impacts on mammals in and around a protected area before, during, and after COVID-19 lockdowns. *Conservation Science and Practice* 4:e12743.

Procko, M., R. Naidoo, V. LeMay, and A. C. Burton. 2023. Human presence and infrastructure impact wildlife nocturnality differently across an assemblage of mammalian species. *PLoS One* 18:e0286131.

Prokopenko, C. M., M. S. Boyce, and T. Avgar. 2017. Characterizing wildlife behavioural responses to roads using integrated step selection analysis. *Journal of Applied Ecology* 54:470–479.

Proffitt, K. M., et al. 2015. Integrating resource selection into spatial capture-recapture models for large carnivores. *Ecosphere* 6:art239.

Prugh, L. R., et al. 2019. Designing studies of predation risk for improved inference in carnivore-ungulate systems. *Biological Conservation* 232:194–207.

Qasem, L., et al. 2012. Tri-axial dynamic acceleration as a proxy for animal energy expenditure; should we be summing values or calculating the vector? *PLoS One* 7:e31187.

Rafiq, K., R. G. Appleby, A. Davies, and B. Abrahms. 2023a. SensorDrop: a system to remotely detach individual sensors from wildlife tracking collars. *Ecology and Evolution* 13:e10220.

Rafiq, K., N. R. Jordan, K. Golabek, J. W. McNutt, A. Wilson, and B. Abrahms. 2023b. Increasing ambient temperatures trigger shifts in activity patterns and temporal partitioning in a large carnivore guild. *Proceedings of the Royal Society B: Biological Sciences* 290:20231938.

Resheff, Y. S., S. Rotics, R. Harel, O. Spiegel, and R. Nathan. 2014. AcceleRater: a web application for supervised learning of behavioral modes from acceleration measurements. *Movement Ecology* 2:27.

Ridout, M. S., and M. Linkie. 2009. Estimating overlap of daily activity patterns from camera trap data. *Journal of Agricultural, Biological, and Environmental Statistics* 14:322–337.

Ripple, W. J., et al. 2014. Status and ecological effects of the world's largest carnivores. *Science* 343:1241484.

Roffler, G. H., J. N. Waite, K. L. Pilgrim, K. E. Zarn, and M. K. Schwartz. 2019. Estimating abundance of a cryptic social carnivore using spatially explicit capture–recapture. *Wildlife Society Bulletin* 43:31–41.

Royle, J. A. and R. M. Dorazio. 2008. *Hierarchical Modeling and Inference in Ecology: The Analysis of Data from Populations, Metapopulations and Communities*. Elsevier, Cambridge, M. A.

Rutishauser, M., et al. 2011. CARNIVORE: a disruption-tolerant system for studying wildlife. *EURASIP Journal on Wireless Communications and Networking* 2011:40817.

Rutz, C., et al. 2015. Calibrating animal-borne proximity loggers. *Methods in Ecology and Evolution* 6:656–667.

Sarkar, M. S., R. Niyogi, R. L. Masih, P. Hazra, L. Maiorano, and R. John. 2021. Long-distance dispersal and home range establishment by a female sub-adult tiger (*Panthera tigris*) in the Panna landscape, central India. *European Journal of Wildlife Research* 67:54.

Schlägel, U. E., et al. 2019. Estimating interactions between individuals from concurrent animal movements. *Methods in Ecology and Evolution* 10:1234–1245.

Schuette, P., A. P. Wagner, M. E. Wagner, and S. Creel. 2013. Occupancy patterns and niche partitioning within a diverse carnivore community exposed to anthropogenic pressures. *Biological Conservation* 158: 301–312.

Serieys, L. E. K., et al. 2023. Anthropogenic activities and age class mediate carnivore habitat selection in a human-dominated landscape. *iScience* 26:107050.

Shehzad, W., et al. 2015. Forest without prey: livestock sustain a leopard *Panthera pardus* population in Pakistan. *Oryx* 49:248–253.

Shehzad, W., et al. 2012. Carnivore diet analysis based on next-generation sequencing: application to the leopard cat (*Prionailurus bengalensis*) in Pakistan. *Molecular Ecology* 21:1951–1965.

Signer, J., J. Fieberg, and T. Avgar. 2017. Estimating utilization distributions from fitted step-selection functions. *Ecosphere* 8:e01771.

Silva, I., et al. 2022. Autocorrelation-informed home range estimation: a review and practical guide. *Methods in Ecology and Evolution* 13:534–544.

Smith, J. A., et al. 2020a. Where and when to hunt? Decomposing predation success of an ambush carnivore. *Ecology* 101:e03172.

Smith, J.A., T. P. Duane, and C. C. Wilmers. 2019. Moving through the matrix: promoting permeability for large carnivores in a human-dominated landscape. *Landscape Urban Plan* 183:50–58.

Smith J. A., et al. 2024. Population and community consequences of perceived risk from humans in wildlife. *Ecology Letters* 27(6):e14456.

Smith, J. A., et al. 2017. Fear of the human 'super predator' reduces feeding time in large carnivores. *Proceedings of the Royal Society of London. Series B, Biological Sciences* 284:20170433.

Smith, J. A., et al. 2020b. Zooming in on mechanistic predator–prey ecology: integrating camera traps with experimental methods to reveal the drivers of ecological interactions. *Journal of Animal Ecology* 89:1997–2012.

Smith, J. A., A. C. Thomas, T. Levi, Y. Wang, and C. C. Wilmers. 2018. Human activity reduces niche partitioning among three widespread mesocarnivores. *Oikos* 127:890–901.

Smith, J. A., Y. Wang, and C. C. Wilmers. 2015. Top carnivores increase their kill rates on prey as a response to human-induced fear. *Proceedings of the Royal Society B: Biological Sciences*, 282:20142711.

Smith, J. A., Y. Wang, and C. C. Wilmers. 2016. Spatial characteristics of residential development shift large carnivore prey habits. *Journal of Wildlife Management* 80:1040–1048.

Souza, C. M., et al. 2020. Reconstructing three decades of land use and land cover changes in Brazilian biomes with Landsat archive and Earth Engine. *Remote Sensing* 12:2735.

Srivathsa, A., et al. 2020. Opportunities for prioritizing and expanding conservation enterprise in India using a guild of carnivores as flagships. *Environmental Research Letters* 15:064009.

Steyaert, S. M. J. G., et al. 2016. Human shields mediate sexual conflict in a top predator. *Proceedings of the Royal Society B. Biological Sciences* 283:20160906.

Street, G. M., et al. 2021. Solving the sample size problem for resource selection functions. *Methods in Ecology and Evolution* 12:2421–2431.

Studd, E. K., et al. 2021. The purr-fect catch: using accelerometers and audio recorders to document kill rates and hunting behaviour of a small prey specialist. *Methods in Ecology and Evolution* 12:1277–1287.

Sunde, P., S. Ø. Stener, and T. Kvam. 1998. Tolerance to humans of resting lynxes Lynx lynx in a hunted population. *Wildlife Biology* 4:177–183.

Suraci, J. P., et al. 2017. A new automated behavioural response system to integrate playback experiments into camera trap studies. *Methods in Ecology and Evolution* 8:957–964.

Suraci, J. P., J. A. Smith, M. Clinchy, L. Y. Zanette, and C. C. Wilmers. 2019a. Humans, but not their dogs, displace pumas from their kills: an experimental approach. *Scientific Reports* 9:12214.

Suraci, J. P., M. Clinchy, L. Y. Zanette, and C. C. Wilmers. 2019b. Fear of humans as apex predators has landscape-scale impacts from mountain lions to mice. *Ecology Letters* 22:1578–1586.

Suraci, J. P., L. G. Frank, A. Oriol-Cotterill, S. Ekwanga, T. M. Williams, and C. C. Wilmers. 2019c. Behavior-specific habitat selection by African lions may promote their persistence in a human-dominated landscape. *Ecology* 100:e02644.

Suraci, J. P., et al. 2021. Disturbance type and species life history predict mammal responses to humans. *Global Change Biology* 27:3718–3731.

Suraci, J. P., L. M. Lacey, P. T. Freeman, A. Stratton, C. Kupar, K. Sager-Fradkin, D. Bergman, B. Ackerman, K. A. Phillips, S. Murphie, C. Sullivan, and L. M. Elbroch. 2025. Puma habitat preferences when moving and feeding predict the potential for human–carnivore conflict in shared landscapes. *Ecological Applications* 35:e70101.

Sytsma, M. L. T., T. Lewis, B. Gardner, and L. R. Prugh. 2022. Low levels of outdoor recreation alter wildlife behaviour. *People and Nature* 4:1547–1559.

Tezuka, S., et al. 2023. Comparing information derived on food habits of a terrestrial carnivore between animal-borne video systems and fecal analyses methods. *Journal of Mammalogy* 104:184–193.

Theobald, D. M., C. Kennedy, B. Chen, J. Oakleaf, S. Baruch-Mordo, and J. Kiesecker. 2020. Earth transformed: detailed mapping of global human modification from 1990 to 2017. *Earth System Science Data* 12:1953–1972.

Thorsen, N. H., J. E. Hansen, O.-G. Støen, J. Kindberg, A. Zedrosser, and S. C. Frank. 2022a. Movement and habitat selection of a large carnivore in response to human infrastructure differs by life stage. *Movement Ecology* 10:52.

Thorsen, N. H., R. Bischof, J. Mattisson, T. R. Hofmeester, J. D. C. Linnell, and J. Odden. 2022b. Smartphone app reveals that lynx avoid human recreationists on local scale, but not home range scale. *Scientific Reports* 12:4787.

Tilberg, M. and P. M. Dixon. 2022. Statistical inference for the utilization distribution overlap index (UDOI). *Methods in Ecology and Evolution* 13:1082–1092.

Tourani, M. 2022. A review of spatial capture–recapture: Ecological insights, limitations, and prospects. *Ecology and Evolution* 12:e8468.

Van Meter, P. E., J. A. French, S. M. Dloniak, H. E. Watts, J. M. Kolowski, and K. E. Holekamp. 2009. Fecal glucocorticoids reflect socio-ecological and anthropogenic stressors in the lives of wild spotted hyenas. *Hormones and Behavior* 55:329–337.

Van Moorter, B., D. R. Visscher, C. L. Jerde, J. L. Frair, and E. H. Merrill. 2010. Identifying movement states from location data using cluster analysis. *Journal of Wildlife Management* 74:588–594.

Van Scoyoc, A., J. Smith, K. Gaynor, K. Barker, and J. Brashares. 2023. The influence of human activity on predator-prey interactions. *Journal of Animal Ecology* 92(6):1124–1134.

Vardi, R., O. Berger-Tal, and U. Roll. 2021. iNaturalist insights illuminate COVID-19 effects on large mammals in urban centers. *Biological Conservation* 254:108953–108953.

Wang, Y., M. L. Allen, and C. C. Wilmers. 2015. Mesopredator spatial and temporal responses to large predators and human development in the Santa Cruz Mountains of California. *Biological Conservation* 190:23–33.

Wang, Y., J. A. Smith, and C. C. Wilmers. 2017. Residential development alters behavior, movement, and energetics in an apex predator, the puma. *Plos One* 12: e0184687.

Watanabe, Y. Y. and Y. P. Papastamatiou. 2023. Biologging and biotelemetry: tools for understanding the lives and environments of marine animals. *Annual Review of Animal Biosciences* 11:247–267.

Whittington, J., M. Hebblewhite, R. W. Baron, A. T. Ford, and J. Paczkowski. 2022. Towns and trails drive carnivore movement behaviour, resource selection, and connectivity. *Movement Ecology* 10:17.

Wiig, Ø., et al. 2017. Performance and retention of lightweight satellite radio tags applied to the ears of polar bears (*Ursus maritimus*). *Animal Biotelemetry* 5:9.

Wijers, M., A. Loveridge, D. W. Macdonald, and A. Markham. 2021. CARACAL: a versatile passive acoustic monitoring tool for wildlife research and conservation. *Bioacoustics* 30:41–57.

Wijers, M., et al. 2018. Listening to lions: animal-borne acoustic sensors improve bio-logger calibration and behaviour classification performance. *Frontiers in Ecology and Evolution* 6:171.

Wilkinson, C. E., J. S. Brashares, A. C. Bett, and M. Kelly. 2021. Examining drivers of divergence in recorded and perceived human-carnivore conflict hotspots by integrating participatory and ecological data. *Frontiers in Conservation Science* 2:681769.

Williams, T. M., et al. 2014. Instantaneous energetics of puma kills reveal advantage of felid sneak attacks. *Science* 346:81–85.

Wilmers, C. C., L. A. Isbell, J. P. Suraci, and T. M. Williams. 2017. Energetics-informed behavioral states reveal the drive to kill in African leopards. *Ecosphere* 8(6):e01850.

Wilmers, C. C., B. Nickel, C. M. Bryce, J. A. Smith, R. E. Wheat, and V. Yovovich. 2015. The golden age of bio-logging: how animal-borne sensors are advancing the frontiers of ecology. *Ecology* 96:1741–1753.

Wilmers, C. C., et al. 2013. Scale dependent behavioral responses to human development by a large predator, the puma. *PLoS One* 8(4):e60590.

Wilson, M. W., A. D. Ridlon, K. M. Gaynor, S. D. Gaines, A. C. Stier, and B. S. Halpern. 2020. Ecological impacts of human-induced animal behaviour change. *Ecology Letters* 23:1522–1536.

Winner, K., et al. 2018. Statistical inference for home range overlap. *Methods in Ecology and Evolution* 9:1679–1691.

Xiao, W., et al. 2018. Relationships between humans and ungulate prey shape Amur tiger occurrence in a core protected area along the Sino-Russian border. *Ecology and Evolution* 8:11677–11693.

Yovovich, V., M. L. Allen, L. T. Macaulay, and C. C. Wilmers. 2020. Using spatial characteristics of apex carnivore communication and reproductive behaviors to predict responses to future human development. *Biodiversity and Conservation* 29:2589–2603.

Zanette, L. Y., et al. 2023. Fear of the human 'super predator' pervades the South African savanna. *Current Biology* 33:4689–4696.e4.

CHAPTER 9

Risk Effects Impact Carnivore Population Persistence

Jonathan N. Pauli, Shotaro Shiratsuru, Matthew M. Smith, and Mauriel Rodriguez Curras

Introduction

Humans and pre-human hominids have caused the extinction or extirpation of large carnivores (species of Mammalia: Carnivora >44 kg) for >200,000 years (Smith et al. 2018, Chapter 2, this volume). While wild carnivores (i.e. excluding pets) compose <0.0001% of the total biomass on Earth (Bar-On et al. 2018, Greenspoon et al. 2023), these species can have disproportionate effects on ecosystems functioning and biodiversity (Estes et al. 2011, Ripple et al. 2014); thus, the persistence of carnivores is of global conservation importance (Wolf and Ripple 2018). Largely due to conservation initiatives and public interest, some carnivore populations are rebounding in the twentieth century, with notable and sometimes surprising increases within human-altered landscapes (Chapron et al. 2014). Accordingly, there is growing interest in understanding the lethal (i.e. from hunting, poaching, or other human take) and risk impacts (i.e. risk-induced trait responses; Peacor et al. 2020) that humans have on large carnivore populations and their persistence.

A primary goal in evolutionary ecology and wildlife management is to predict the future of populations and to ensure the viability of these populations. Population growth rate (λ)—calculated from demographic vital rates—is a standard metric to predict a population's future size and viability (Morris and Doak 2002). These two parameters are not only central to assessing persistence but are also the fitness components for a population or group of organisms (Caswell 1989, Charnov 1990). In assessing the risk effects of humans on large carnivore persistence, it is most useful to explore how the fitness and growth potential of a population (i.e. survival and reproduction) are affected. Due to the slow life histories of most carnivores, proportionally the most influential vital rates on population growth and persistence are typically survival, especially of adults. Thus, much attention in carnivore recovery and management focuses on forces shaping survival. Nevertheless, fecundity is a necessary component of growth and fitness and is often a vital rate that exhibits much more natural variation (Mills et al. 2001, Norris and McCulloch 2003). Consequently, there can be more opportunities for population response and change for factors that influence fecundity. In assessing the risk effects of humans on carnivore fitness and persistence, it is useful to explore the linkage between behaviourally and physiologically mediated effects on both survival and reproduction (often called nonconsumptive effects; Peacor et al. 2020, see Figure 8.1, Chapter 8, this volume). Although several approaches have been used to address the perception of risk on a landscape (e.g. the landscape of fear, giving up densities, actual-perceived risk; Laundré et al. 2010, Brown et al. 1999, Rodriguez Curras et al. 2021), few have linked how the perception of risk impacts survival and reproduction, especially for large carnivores.

Jonathan N. Pauli et al., *Risk Effects Impact Carnivore Population Persistence*. In: *Apex Predators in the Anthropocene*. Edited by: Robert A. Montgomery and Remington J. Moll, Oxford University Press. © Oxford University Press (2026). DOI: 10.1093/9780191996665.003.0009

Humans are the most lethal species in evolutionary history (Darimont et al. 2015), with distinct top-down, lethal effects on animal populations, especially large carnivores (Estes et al. 2011, Ripple et al. 2014). The role of humans, in many instances, is considered analogous to predation risk (Frid and Dill 2002); the disturbance stimuli exerted by humans are expected to induce similar behavioural and physiological responses in animals (Frid and Dill 2002, Dantzer et al. 2014; Figure 9.1). The risk-disturbance hypothesis (Frid and Dill 2002) has been supported by empirical evidence: humans shifting species habitat use to lower-quality areas (Gill et al. 1996, Nellemann et al. 2000), increasing the level of vigilance (Duchesne et al. 2000, Frid 2003), triggering physiological stress responses (Dantzer et al. 2014) and inducing important effects on reproduction and survival (Pauli and Buskirk 2007). Therefore, behavioural responses of animals to human disturbance appear to follow similar pathways and can result in similar risk effects. Although risk effects can be equal in magnitude to lethal effects (Preisser and Bolnick 2008), the fitness consequences from behaviourally mediated effects of human activities to large carnivores remain understudied (Montgomery et al. 2020). Nevertheless, the established framework for studying risk effects of predators can bring insight into understanding the same effects of humans on carnivore population persistence.

Herein, we review current literature within two main categories of risk effects: those induced by carnivores and those induced by humans on fitness (e.g. survival and recruitment) and population persistence. Given the dearth of literature that has linked risk effects on carnivore persistence, we first draw upon concepts (in a non-exhaustive literature review) for nonlethal interactions between carnivores and their prey. Second, given that humans and large carnivores have competed for a shared prey base for much of our evolutionary history (Milošević 2020), we also explore what is known from non-lethal interactions between carnivores with effects on persistence, which are characteristically structured by competition (Palomares and Caro 1999, Donadio and Buskirk 2006). Finally, we review the demographic consequences of risk effects of humans on our prey and review all available literature on human to carnivore risk effects on carnivore fitness and persistence. We conclude this chapter by offering a framework—through the lens of error management theory—to conceptualize expectations of nonlethal effects of humans on large carnivore fitness and population persistence.

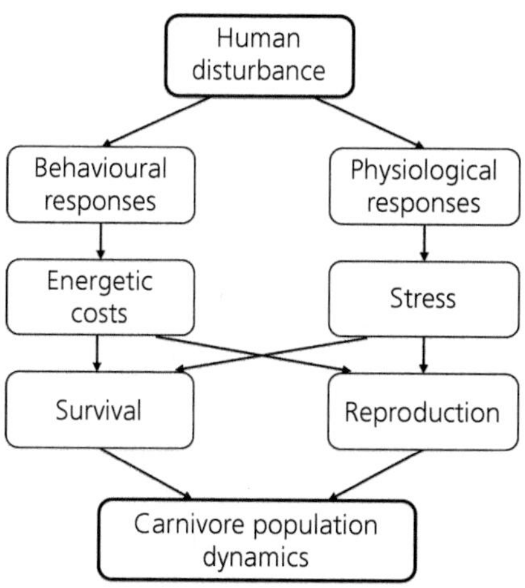

Figure 9.1 The mechanisms by which human risk effects have negative fitness and demographic consequences for carnivore populations. See Figure 8.1 in Chapter 8 of this volume for a complementary schematic that links some of these factors to the technologies and methods used to quantify them in large carnivore populations.

How risk can influence populations

Organisms alter their behaviours, especially in relation to foraging, in response to spatial and temporal variations in perceived predation risk. For example, species may be forced to forage in less favourable patches when their preferred foraging locations coincide with risk (Hernández and Laundré 2005, Fortin et al. 2005). Some species may not forego foraging in the presence of risk, but instead increase their vigilance (Laundré et al. 2001, Dröge et al. 2017a). In either case, there is a trade-off between energy intake and the likelihood of mortality; simply put, behavioural responses should come at energetic costs (Creel 2018). Such costs can reduce the nutritional state of individuals (Sinclair and Arcese 1995), and in some cases impact survival and reproduction (Christianson and Creel 2010).

Prey exposed to acute risk, like when encountering or being pursued by a predator, may also exhibit a physiological (neuroendocrine) response (Crespi et al. 2013). Specifically, activation of the hypothalamic pituitary adrenal (HPA) axis will trigger a complex set of positive and negative physiological feedback between the hypothalamus, pituitary gland, and adrenal gland to prepare an individual to fight or flight in short-term high-risk situations (Hawlena and Schmitz 2010). Indeed, persistent exposure to risk and chronically elevated physiological stress can also negatively affect reproduction and survival (Hawlena and Schmitz 2010, Crespi et al. 2013).

Ultimately, these fitness costs incurred from behavioural and physiological responses to risk can have important implications on population dynamics (Creel 2018) and might be comparable, or even exceed, lethal effects (Preisser et al. 2005). The magnitude of risk effects, however, are context dependent (Wirsing et al. 2021), and their relative importance for population dynamics is under debate (Peers et al. 2018).

Carnivore-induced risk effects on populations

Carnivore–prey

Numerous studies have examined antipredator responses (Lima and Dill 1990) and the resulting fitness and demographic consequences in predator–prey interactions (Lima 1998, Preisser et al. 2005), revealing that risk effects of carnivores on prey demography—including reduced survival and reproduction—are primarily driven by the costs of behavioural and physiological antipredator responses (Creel 2018). However, studies that directly tested the effect of predation risk on demographic parameters of prey in mammalian carnivore–prey interactions are nevertheless scarce.

Impacts of predation risk on reproductive parameters (e.g. fecundity, litter size, body size of offspring, and offspring survival) of mammalian prey have been reported in some carnivore-prey interactions. For example, snowshoe hares (*Lepus americanus*) showed higher levels of cortisol, deteriorated body condition, and fewer offspring per female when Canada lynx (*Lynx canadensis*) were abundant, though demographic consequences of negative fitness were mitigated in a predator exclosure experiment (Boonstra et al. 1998). Similarly, Sheriff et al. (2009) found that the level of faecal cortisol metabolite of female hares in the wild was negatively correlated with the number, size, and mass of offspring, and that females chronically exposed to a simulated predator (domestic dog *Canis lupus familiaris*) showed higher levels of faecal cortisol, higher abortion rates, and reduced mass of offspring. Higher predator occurrence increased faecal glucocorticoids and hair cortisol in a population of mountain goats (*Oreamnos americanus*), which reduced the number of reproductive females (Dulude-de Broin et al. 2020). Increased ovulation and lactation in female white-tailed deer (*Odocoileus virginianus*) were reported following population decline in coyote (*Canis latrans*), implying that chronic predation risk of coyote negatively impacts fecundity and recruitment of deer (Cherry et al. 2016). Beyond stress-mediated effects, predation risk also negatively impacts the reproduction of prey through food-mediated effects by triggering behavioural responses of prey that result in lower food intake or less efficient energy intake (Creel 2018). For example, the presence of wolves (*Canis lupus*) induced lower energy intake and a higher degree of starvation in elk (*Cervus elaphus*), which ultimately led to lower pregnancy rates (Christianson and Creel 2010, Creel et al. 2011).

The risk effects on prey fecundity have been experimentally (Korpimaki et al. 1994, Sheriff et al. 2009) and observationally (Boonstra et al. 1998, Cherry et al. 2016, Dulude-de Broin et al. 2020) demonstrated in a variety of species; however, compared to the negative effects of predation risk on prey reproduction, the survival costs of predation risk have not been thoroughly evaluated in carnivore–prey interactions (Peers et al. 2018). Indeed, most evidence for these effects comes from invertebrate populations (Schmitz et al. 1997, Schmitz 2008). Researchers have examined the relationship between the level of physiological stress induced by predation risk and prey survival, but conclusive evidence for a negative effect of physiological responses on long-term prey survival is lacking (Crespi et al. 2013). Some studies have reported negative effects of chronic predation risk on prey survival, though the mechanistic pathways are not clearly understood. For example, chronic exposure to a domestic dog resulted in a higher mortality rate of pregnant snowshoe hares and a lower survival rate of their offspring from birth to weaning (MacLeod et al. 2018). European roe deer (*Capreolus capreolus*) inhabiting areas with higher resource selection function values of Eurasian lynx (*Lynx lynx*) exhibited higher mortality rates unrelated to lynx predation (Gehr et al. 2018).

In addition to the scarcity of studies directly examining the impact of predation risk on demographic parameters of prey, empirical evidence for risk effects on prey population persistence is lacking for mammalian carnivore-prey interactions (Sheriff et al. 2020). However, research has shown that the magnitude of antipredator responses and their fitness costs can be mediated by various biotic and abiotic factors (Wirsing et al. 2021). One of the most important factors that influence the magnitude of antipredator responses is predator hunting mode (Preisser et al. 2007, Schmitz 2008). Specifically, sedentary predators such as ambush predators provide more persistent and predictable cues, and thereby trigger stronger behavioural antipredator responses that result in greater risk effects (Preisser et al. 2007, Schmitz 2007, 2008). In addition, functional traits of predator and prey, energetic state of prey, resource availability, climate and landscape conditions are all expected to drive the magnitude of antipredator responses and ultimately the context dependency of risk effects (Brown and Kotler 2004, Wirsing et al. 2021). Understanding such context dependency will bring insight into the risk effects of humans on carnivores, considering that different types of human disturbance likely provide different cues, which in turn shape the magnitude and fitness costs of carnivore responses.

Carnivore–carnivore

The lower resource availability at higher trophic levels (Elton 1927) tends to aggregate carnivores around a limited and shared resource. Interspecific competition is especially apparent between carnivores because their evolved morphological, physiological, and behavioural traits to become highly effective killers (Gittleman 1989) are transferable to interactions with other guild members. Accordingly, competitive interactions often manifest as interference, whereby the dominant guild members—species that are ~2–5 times larger than their competitors—display aggression or even kill subordinate species (Palomares and Caro 1999, Donadio and Buskirk 2006, Prugh and Sivy 2020). Interspecific killing is an important cause of death among subordinate carnivores, accounting for up to 54% mortality in some cases (Lindström et al. 1995) but can be highly variable (Mean = 11%, Range = 0–54%; see Prugh and Sivy 2020 for a recent review and meta-analysis). To avoid interspecific killing, carnivores exhibit a number of risk avoidance strategies, reacting to direct cues of predation risk or avoiding the habitat associations of their competitors (Leo et al. 2015, Rodriguez Curras et al. 2021). Carnivores navigate these competitive interactions by behaviourally mitigating their space use (Fedriani et al. 2000, Berger and Gese 2007), diel activity (Hayward and Slotow 2009, Di Bitetti et al. 2010), and resource use (Crooks and Van Vuren 1995, Karanth and Sunquist 1995). To minimize the risk of fatal interactions, subordinate carnivores will often proactively avoid the larger guild members in space (Fedriani et al. 2000), especially where resources are limited (Robinson et al. 2014) and where species exhibit constrained diel activity (Balme et al. 2017). For example, in the Tanzania's Serengeti National Park, the persistence of cheetahs (*Acinonyx jubatus*) was impacted by interspecific competition with lions (*Panthera leo*) and hyenas (*Hyanea* spp. and *Crocuta* spp.; Durant et al. 2004). Due to their smaller body size, cheetahs lose ~15% of their kills to larger carnivores, most often to hyenas (Hunter et al. 2007). Furthermore, >70% of cheetah cubs can be depredated by lions and hyenas, limiting recruitment (Laurenson 1994). In response, cheetahs avoided areas occupied by lions, and to a lesser extent, hyenas—likely due to their less predictable habitat domain (Durant 2000a, Broekhuis et al. 2013). However, as the lion numbers tripled over a 30-year period in Serengeti National Park, Tanzania, the cheetah population remained relatively stable (Swanson et al. 2014), suggesting that behavioural mediation is sufficient to cope with the risk of mortality. Indeed, the most experienced and successful female cheetahs are best at managing their relationships with lions and hyenas to survive, reproduce, and critically, recruit new individuals into the population (Durant 2000b).

Although the impacts of risk effects, especially behavioural changes in subordinate species, are often central to studies on competitive interactions between carnivores (Wang et al. 2015, Dröge et al. 2017b, Rodriguez Curras et al. 2021), quantifying their impacts on survival and reproduction is difficult because they are confounded by trophic facilitation (Prugh and Sivy 2020, Ruprecht et al. 2021). Large carnivores can provision >1000 kg of carrion per individual per year (Prugh and Sivy 2020) to meso- and small carnivores, however, this trophic facilitation often brings subordinate species in proximity with dominant guild members. Accordingly, subordinate carnivores often navigate potentially deadly encounters with dominant carnivores (Berger et al. 2008) while also being attracted to high-risk–high-reward resources (i.e. carrion; Sivy et al. 2017); these two opposing forces (i.e. suppression and facilitation) are hypothesized to create a 'fatal attraction' for subordinate carnivores (Sivy et al. 2017, Prugh and Sivy 2020). Although risk avoidance in carnivores is an important driver of carnivore community structure (Ritchie and Johnson 2009), scavenging (Selva and Fortuna 2007, Pereira et al. 2014) and kleptoparasitism (Cusack et al. 2017) have evolved within Carnivora as stable evolutionary strategies. For example, hyenas (*Crocuta crocuta*) in areas with lower lion densities do not typically exhibit changes in litter size or juvenile survival, but experience higher rates of food intake by successfully usurping lion kills, which can result in higher lifetime reproductive success due to an earlier weaning age and a shorter inter-birth interval (Watts and Holekamp 2008). However, whether attenuated lethal or risk effects due to lower lion densities have an effect on survival or reproduction remains unclear (Preisser et al. 2005). The positive interactions associated with resource facilitation complicate our ability to quantify risk effects, though the highly conserved nature of these competitive interactions suggests that behaviourally mediating predation risk is an effective means of avoiding interspecific killing while maximizing access to provisioned resources. In northwestern Oregon, coyotes were frequently observed scavenging from puma (*Puma concolor*) kills—so much so that their diets contained similar quantities of ungulate—though they behaviourally minimize potential encounters (Ruprecht et al. 2021). Nonetheless, pumas killed nearly one-quarter of the coyote population annually, complicating the net benefits gained from scavenging. These multifaceted (suppression and facilitation) and complicated (scavenging and kleptoparasitism) interactions make it difficult to untangle the impacts of risk effects on fitness and population persistence because the benefits from scavenging can be especially impactful to fecundity.

Much of the emphasis of risk effects in carnivore–carnivore systems has been on changes in the behaviour of the subordinate species (Dröge et al. 2017, Rodriguez Curras et al. 2021); although it is often implied, there is limited evidence of physiological impacts on subordinate carnivores. One of the strongest impacts may be for subordinate species to switch prey to avoid interspecific aggression (Sivy et al. 2018) or to avoid their interspecific competitors in space or in the times they are active (Leo et al. 2015, Dröge et al. 2017a, b, Rodriguez Curras et al. 2021). These dynamic risk effects are what we expect in human-large carnivore interactions *a priori* because the impacts of humans are likened to large carnivore impacts (Frid and Dill 2002), allowing us to make predictions from carnivore–carnivore interactions.

Human-induced risk effects on populations

Human–prey

Humans are ubiquitous in ecological communities (Greenspoon et al. 2023) and can exert strong lethal and risk effects on large carnivore populations (Frid and Dill 2002). Antipredator responses to humans are often considered to be analogous to those of non-human predators, inducing similar behavioural and physiological responses. For humans, though, these cues are not only limited to our presence or our habitat preferences, but can also involve novel structures, development, and technology that we are associated with (Frid and Dill 2002). Several reviews have explored human-induced antipredator responses in animal populations (e.g. Coetzee and Chown 2016, Wilson et al. 2020, Lasky and Bombaci 2023), which highlight notable changes in animal activity (Gilbert et al. 2023), foraging strategies (Smith et al. 2018), and habitat use (Tucker et al. 2018). Altered behaviours from human actions—especially hunting—have been shown to induce nutritional changes too: Féret et al. (2003) and Roy and Woolf (2001) found that game birds subjected to intensive hunting increased the time spent flying and decreased the time spent foraging, with negative consequences for body condition. Hjeljord and Histøl (1999) found that the body mass of moose (*Alces alces*) in Norway was correlated negatively with hunting intensity. Less work, however, has connected the antipredator responses to humans to the demographic effects or population persistence (see Smith et al. 2021 for a review).

When hunting, humans can induce important behaviourally mediated nonconsumptive effects on their prey. In a before-and-after control-impact experiment, Pauli and Buskirk (2007) found that highly social prairie dogs (*Cynomys ludovicianus*) subjected to hunting altered their behaviour—foraging less and seeking underground refuges more—with subsequent negative effects on nutritional condition. Reduced nutritional condition led to a collapse in reproductive output among the treatment colonies. The presence of humans engaging in non-lethal recreational activities (e.g. hiking) can also induce antipredator responses that lead to nonconsumptive effects. Human disturbance from hikers has been linked to reduced foraging rates among an endangered corvid, which negatively impacted juvenile survival, potentially jeopardizing the viability of a population (Kerbiriou et al. 2009). Similarly, recreational use has been identified as a factor reducing the number of migratory shorebirds at a post-breeding stop-over area; further, the authors found no evidence of eventual habituation to human recreation (Martin et al. 2015). Finally, the risk effects of whale watching activities have been assessed for various species of cetaceans (Christiansen et al. 2016) with mixed consequences. For example, the foraging rates of bottlenose dolphins (*Tursiops aduncus*) have been shown to be negatively impacted by whale watching, with observed decreases in body condition (Christiansen et al. 2010). However, although the behavioural disruptions caused by whale watching boats substantially impacted the feeding patterns of minke whales (*Balaenoptera acutorostrata*), the cumulative exposure of individuals to boats was relatively low, resulting in an effect on foetal growth no different from natural variability (Christiansen and Lusseu 2015). The nonconsumptive effects resulting from proactive responses via human-modification of habitat have also been observed in ungulate reproduction responding to infrastructure development (Nellemann et al. 2003), nest success in birds due to vehicle disturbance (Spaul and Heath 2016), and amphibian abundance due to human hiking (Rodríguez-Prieto and Fernández-Juricic 2005).

There are, however, notable exceptions where human-induced risk effects did not affect population-level processes: pronghorn (*Antilocapra americana*) showed no fitness consequences from energy development (Beckmann and Seilder 2009), and human disturbance did not influence the activity budgets or the population size of sifakas (*Propithecus verreauxi*; Chen-Kraus et al. 2022). Similarly, past experiments have shown that chronic (human-associated) noise disturbance will lead to birds flushing their nests, but such behavioural responses did not translate to changes in reproductive success (Delaney et al. 2011, Meillere et al. 2015). Finally, while human activities have also been linked to physiological responses like increased heart rates (Ellenberg et al. 2006) or stress levels (Pauli and Buskirk 2007), such physiological responses to human disturbance are by no means universal (e.g. Morgan et al. 2012, Northrup et al. 2021) and their linkage to fitness or demographic declines are variable (Villanueva et al. 2012). Overall, then, human-induced risk effects are context dependent and do not show consistent patterns across human disturbance types (Smith et al. 2021). Furthermore, because many human cues of risk are novel in evolutionary timescales (Williams and Jackson 2007), these stimuli are unpredictable for most species (Guiden et al. 2019).

Human–carnivore

Like the other interactions we have described, large carnivores must balance the risk of human encounters with critical activities such as hunting and feeding. Indeed, large carnivores change their spatiotemporal movement, activity, and foraging to avoid humans, and such changes can compromise individual fitness and influence populations (Nickel et al. 2020, Moll et al. 2021, Smith et al. 2021). The behavioural avoidance of humans to reduce the risk of encounters can alter their movement (Suraci et al. 2019, Tucker et al. 2018), habitat selection (Abrahms et al. 2016, Heinmeyer et al. 2019, Nisi et al. 2022, Oriol-Cotterill et al. 2015, Wilmers et al. 2013), and feeding behaviour (Smith et al. 2017). However, complete avoidance can be prohibitively costly, and large carnivores may instead rely on temporal partitioning to minimize the risk of encounters (Kronfeld-Schor and Dayan 2003). Several studies have shown large carnivores avoid human encounters by partitioning time, often resulting in an increase in nocturnality across multiple taxa (Ordiz et al. 2012, Rasmussen and Macdonald 2012, Wang et al. 2015, Wheat and Wilmers 2016, Gaynor et al. 2018). Changes in spatiotemporal movement, activity, and foraging to avoid humans have been found across large carnivore taxa, including pumas (*Puma concolor*; Smith et al. 2017, Blecha et al. 2018, Wang et al. 2017), African lions (Valeix et al. 2012, Loveridge et al. 2017), bears (*Ursus* spp. Ordiz et al. 2011, Hertel et al. 2017), tigers (*Panthera tigris*; Carter et al. 2012), Eurasian lynx (Gehr et al. 2017), spotted hyenas (Boydston et al. 2003), wolves (Rogala et al. 2011), and wolverines (*Gulo gulo*; Heinmeyer et al. 2019).

Carnivore response to humans not only varies across species but also varies between individuals. For example, adult male brown bears (*Ursus arctos*) in British Columbia, Canada, exhibited strong avoidance of human recreation that allowed additional feeding opportunities of high-quality resources (e.g. protein from fish) for females with cubs. This increase in protein consumption was positively correlated with female mass and litter size, which could have influenced population growth, although this was not explicitly shown (Nevin and Gilbert 2005). In contrast, increasing tourism in southwest Kenya reduced cheetah cub recruitment (Broekhuis 2018). Among pumas in southern California, USA, females showed reduced prey consumption time, and in response, an increased kill rates compared to males in areas with higher human density (Smith et al. 2015). Similarly, female wolverines showed stronger avoidance of motorized and non-motorized winter recreation areas (Heinmeyer et al. 2019). However, the individual fitness and population trajectory consequences of risk effects of human presence have largely gone unexplored, with few examples present in the current literature.

Linking population-level responses of large carnivores to non-lethal human activity has been frequently called for (Smith et al. 2015, Hertel et al. 2017) but rarely carried out, with a few notable exceptions (Sheriff et al. 2018). For example, French et al. (2011) compared reproductive rates of California sea lions (*Zalophus californianus*) across a gradient of human activity by measuring the ratio of pups to females and the frequency of human activities within 50 m of colonies. They found reduced reproductive rates across multiple years as the frequency of human exposure increased and predicted a decline in population growth (λ) as disturbance increased (French et al. 2011). In contrast, an increase in human presence may not always precipitate a pronounced population decline among large carnivores. For example, Boydston et al. (2003) calculated several demographic measures of spotted hyenas from two time periods that featured an increase in human activity. Despite the increase in livestock grazing, spotted hyena density did not decline, likely due to unchanged adult female survival and fecundity (Boydston et al. 2003), although population growth rate was not calculated. The authors attributed this result to the predictability and intensity of this form of human presence (Dheer et al. 2022). In addition, human presence in the form of tourism and researchers had not contributed to declining populations of the southern elephant seal (*Mirounga leonina*), with no effect on population growth, mortality, or fecundity (Wilkinson and Bester 1988, McMahon et al. 2005). However, permanent human presence in the form of housing density along shorelines was correlated to increased perinatal death for the ringed seal (*Pusa hispida*), and long-term persistence was hypothesized to be impacted but was not explicitly quantified (Liukkonen et al. 2017).

Large carnivores respond differently to various types of human disturbance, and might employ a combination of simultaneous spatial avoidance, temporal partitioning, or switching prey. For example, across a wildland-urban interface in southern California, USA, pumas spatially avoided areas with a high human footprint but temporally partitioned areas of human presence temporally rather than spatially (Nickel et al. 2020). In this system without human-caused lethal effects (i.e. no hunting), pumas avoided human infrastructure, and reduced their activity and feeding time in areas of high human presence (Nickel et al. 2021, Smith et al. 2015, 2017, Wilmers et al. 2013). The strong behavioural responses led to energetic costs driven by an increased kill rate in pumas due to kill-site abandonment following human auditory stimuli (talking; Smith et al. 2017). In this population, the growth rate (λ) was predicted to be <1 in 42.1% of the study area due to higher female mortality in exurban areas (25 buildings/km^2); however, fecundity could not be evaluated due to insufficient reproductive data across levels of housing density (Nisi et al. 2023). Chapter 8 in this volume describes additional details of these human-triggered behavioural responses in California pumas. In another wildland-urban interface in Colorado, USA, both male and female pumas exhibited lower survival rates but consumed more synanthropic wildlife and domestic species with increasing housing density (Moss et al. 2016).

Like resource facilitation that large carnivores provide to meso- and small carnivores, humans provide resource subsidies to large carnivores in the form of domestic prey (Moss et al. 2016), refuse and crops (e.g. black bears *Ursus americanus*; Kirby et al. 2019), and waste sites (Petroelje et al. 2019). Indeed, food subsidies contribute to over a billion tonnes of food waste per year to ecosystems globally (Kaza et al. 2018, Oro et al. 2013). The fitness consequences of behavioural responses to humans, then, are complicated by these resources that humans provide and the fact that consumption of human food subsidies varies among large carnivores. For example, Johnson et al. (2020) predicted population growth rate for black bears in southwestern Colorado from field estimates of adult female survival, cub survival, and fecundity across levels of human development. Survival of females and cubs declined with increasing human development while fecundity increased, but the population was predicted to decline overall with increasing human development (Johnson et al. 2020). Similarly, female black bears in the Lake Tahoe Basin, USA, had higher mortality rates, but higher fecundity rates in a more urban landscape (likely due to food subsidies from human refuse) compared to an adjacent wildland area (Beckmann and Lackey 2008). Furthermore, black bears in Colorado consumed more human foods and reduced hibernation length, with the reduction in hibernation associated with greater cellular ageing (Kirby et al. 2019). Thus, the benefits to large carnivores from human food subsidies requires balancing the reward of more foraging opportunities against the cost of more frequent encounters that may impact survival and reproduction differently and must be carefully considered in future studies of risk effects of humans.

Synthesis of human–carnivore risk effects on populations

Research evaluating the population-level effects (e.g. population growth and persistence) of the risk of human presence on large carnivores remains limited. Many studies have alluded to negative population consequences, but few (if any) have directly quantified the change in population growth and persistence. In part, the need for vital rates of species with slow life histories, large spatial requirements, and low population densities often require long-term data that is logistically challenging to collect. This difficulty can result in studies with small sample sizes, a focus on a singular demographic aspect, and broad categorizations of human presence. Moreover, the form, scale, and intensity of non-lethal human presence can vary substantially both spatially and temporally (Moll et al. 2021). Regardless, humans can induce risk effects on large carnivores through the same individual fitness pathways (e.g. behavioural and physiological) as in other predator–prey systems. However, as in carnivore–carnivore interactions, humans can also alter the foraging of carnivores through food subsidies, which may in turn confound risk effects on population trajectories. We hypothesize that the intensity and predictability of the non-lethal human activity are critical axes to evaluate the response of large carnivores. Consequently, the ultimate effect on population trajectories is likely to be context dependent. High intensity and continuous human presence (e.g. urbanization; Nisi et al. 2023) will likely elicit a stronger negative response at the population level compared to low intensity and temporary activity (e.g. pastoralism; Dheer et al. 2022)—while some of these habitats may be effectively inaccessible for some species, mitigated by their ecological traits and life-history strategies. For example, although human diel activity patterns are relatively similar across human-inhabited areas (Gaynor et al. 2018), the response that large carnivores exhibit to such activity varies across species. Indeed, large carnivores vary substantially in their behaviours, sociality, and diets. Species traits and evolutionary histories likely inform the potential tolerance of human activity and how a species may respond to changes in habitat availability, the temporal partitioning required to mediate interactions with humans, and ability to exploit new resources. As the intensity of non-lethal human activity increases and becomes more spatially expansive and continuous in nature it will be important to consider species traits that may constrain population trajectories or allow for greater population growth within human-dominated landscapes.

To explore how the intensity of human development and presence may influence persistence of large carnivores as well as the traits associated with occurrence within human presence, we compiled occurrence and species trait data—combining data from GBIF, the Amniote database (Myhrvold et al. 2015), and the COMBINE database (Soria et al. 2021)—for large carnivores globally and overlaid these with the Human Footprint Index (HFI; Venter et al. 2016). We detected a broad human impact at the species level, quantified by the IUCN classification and the percentage of range loss experienced by large carnivores (Figure 9.2, a and b), where species that experience a higher level of human intensity tend to be classified as 'Least Concern' by the IUCN and have experienced a lower proportional loss to their overall distribution. Those species also tended to be nocturnal and possess broad/generalized diets (Figure 9.2, c and d). These relationships make intuitive sense but highlight (1) the impacts to large carnivores and (2) the behavioural traits that are most effective in human landscapes. Beyond these species-level observations, we suspect that changes to these ecological traits would also be present at the population level even for the same species but at different locations. For example, the diets of large carnivores are more varied in human landscapes, where they consume refuse (Moss et al. 2016) or human-associated pets and livestock (Yirga et al. 2012). Furthermore, species also tend to shift towards nocturnality in the presence of humans (Griffiths and Schaik 1993, Gaynor et al. 2018). Nevertheless, we generally predict that the species that are less demographically impacted by humans tend to be those with high behavioural and life-history plasticity, which allows large carnivores to alter their behaviours or life-history strategies in unpredictable human landscapes and thereby mitigates the risk effects of humans (Guiden et al. 2019).

It is also important to consider the detailed mechanism by which large carnivores perceive and behaviourally respond to novel cues of risk, in addition to behavioural and life-history plasticity of the species (see Chapter 7, this volume). From a fitness perspective, animals should optimally trade off the benefits and costs of increasing their responsiveness to novel risk cues (Sih 2013). Moreover, whether the responses of large carnivores to perceived novel cues of risk are adaptive or maladaptive depends on their past environments or evolutionary history (Sih 2013). Therefore, species traits and evolutionary history likely inform the potential tolerance of human activity and how a species can respond to changes in habitat availability and the temporal partitioning required to mediate interactions with humans. Ecological traits may confer a high tolerance of human activity when increased false-positive responses result in a low cost to survival or reproduction, while there is no change in the consequences of false-negative errors to non-lethal human presence. As non-lethal human activities become spatially expansive and increasingly continuous in nature, it will be increasingly important to consider species traits that may confer a high cost to population growth and the positive errors that lead to population declines.

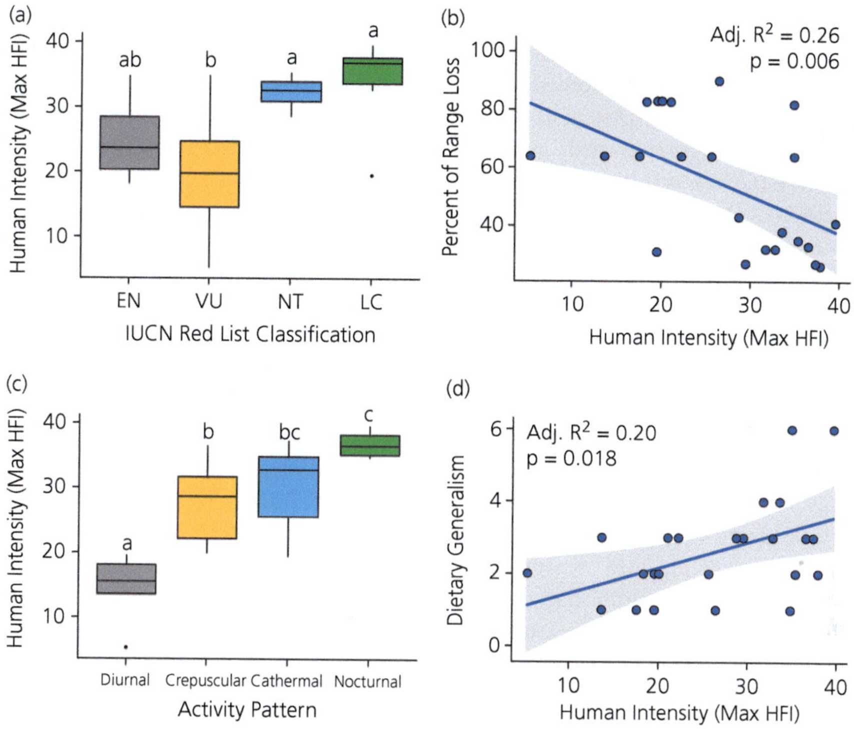

Figure 9.2 Human intensity (quantified as the maximum level of HFI that a large carnivore experiences throughout its range) compared across (1) two variables we used a proxy for persistence, (a) the IUCN Red List Classifications and (b) the percent of range loss experienced by the species (data from Ripple et al. 2014, supplemented with IUCN data) and (2) two variables we used as a proxy for behavioural response of carnivores, (c) the activity pattern, and (d) diet generalism. Letters above boxplots represent significant differences between groups resulting from an ANOVA at the $p \leq 0.05$ level.

Human–carnivore risk effects on populations: Future directions and error management

A gradient from highly specialized (e.g. a predator–prey arms race; Abrams 1986) to broadly generalizable (e.g. increasing body size of prey species; Baker et al. 2022) adaptations has evolved for species to cope with predation risk. Antipredator responses to novel predators can be effective if the resulting errors associated between the past and present dangers are linked (Abrams 1994)—that is, the predators share some common traits such as hunting mode (e.g. Preisser et al. 2007) or habitat domain (e.g. Schmitz et al. 2017). While non-lethal human risk can be analogous to risk effects of predation (Frid and Dill 2002), humans pose novel threats to species (Sih et al. 2011), which are often mismatched with the threats they have experienced throughout their evolutionary history (Williams and Jackson 2007, Guiden et al. 2019). This mismatch is compounded by the decoupling between contemporary human stimuli and the disproportionate strength of the actual risk we pose (Trimmer et al. 2017). Thus, the true risk associated with humans is context dependent (Moll et al. 2021). Indeed, the risk experienced by an individual may not be useful for future encounters and individuals may erroneously use past encounters to inform present threats. Such events can compound and lead to higher errors when species assess human risk. The confounded nature of these inflated errors results in species (and individuals) disproportionately assigning risk to humans when no actual risk exists and simultaneously—and paradoxically—assuming humans are safe when they pose a great risk. For example, bear hunting in Wisconsin is limited to a brief 4–5-day window in the autumn; however, hunters are allowed to create feeding stations year-round (Kirby et al. 2017). This habituation to a risk-free reward, coupled with non-exploitative (i.e. non-hunting) recreationists throughout much of the year, can lead to a temporal mismatch between false-positive and false-negative errors. Accordingly, the cues that large carnivores use to assess risk from their natural predators and competitors frequently fail as indicators of human risk (Sih et al. 2011, Ciuti et al. 2012), and can result in population declines (Smith et al. 2021). Large carnivores have high capacities for novel technical problem-solving (Benson-Amram et al. 2016), which may explain their ability to exploit (Moss et al. 2016, Rodriguez Curras et al. 2022)—and often persist (Kirby et al. 2016, Smith et al. 2016)—in human landscapes. Additionally, some forms of human risk are predictable. For example, humans display relatively constrained daily activity times (Gaynor et al. 2018), space use (Nickel et al. 2020), sounds (Ware et al. 2015), and scents (Pannunzi and Nowotny 2019). This predictability presents opportunities for large carnivore populations to adapt successful strategies that minimize costly errors (Johnson et al. 2013). The adaptive capacity of large carnivores is perhaps best exemplified by lion behaviour in response to safari excursions in Kruger National Park, South Africa, where humans are (rightly) regarded as a minimal threat (Petracca et al. 2019). Understanding how large carnivores cope with novel human risk, will be critical for designing interventions to mitigate unintended consequences (Smith et al. 2021). Accordingly, the best strategy humans can employ may be to increase the predictability of our behaviours, thereby allowing large carnivores (and other species) to learn and adequately respond to our actions.

Despite numerous contributions that have advanced our understanding of the impacts of risk effects on wildlife populations, the actual quantification of these effects on populations persistence remains relatively unexplored (Sheriff et al. 2020, Peacor et al. 2022). In particular, the application of existing conceptual frameworks to predict the effects of human-induced risk on large carnivore populations To that end, we propose to utilize a well-established framework in other scientific disciplines (i.e. economics and psychology) to help address the gaps in our knowledge of risk effects of humans on large carnivore population persistence; error management theory (*sensu* Johnson et al. 2013, EMT Figure 9.1).

As we reviewed, individuals face trade-offs between avoiding predators and other fitness-enhancing activities (Lima and Dill 1990). Although assessing predation risk poses a similar problem to foraging—traditionally studied as optimization problems (Macarthur and Pianka 1966, Sih and Christensen 2001)—acquiring or updating information on predation risk can be lethal (Brown 1988, Abrams 1994). This is because the actual risk of predation can never be precisely predicted (Bouskila and Blumstein 1992, Brown et al. 1999) and the cost of incorrectly assessing predation risk is exceedingly high (Bouskila and Blumstein 1992). Central to error management theory, the false-negative (not perceiving risk, when there is risk present; FN) and false-positive (perceiving risk, when there is no risk; FP) error rates in perceiving risked are inherently linked and the perception of risk should always exceed the actual level of risk that individuals encounter (Abrams 1994), resulting in a bias towards false-positive errors (i.e. determining there is risk, when there is no actual risk present). In other words, species, broadly, and large carnivores, specifically, have evolved for millennia to minimize their FN error rates, which increases the probability of FPs. Accordingly, given that the fitness costs of these types of error differ (likely such that $FN \gg FP$) and the background probability of nonlethal human risk is largely unpredictable, species persistence should ultimately be dictated by the relative importance of FN and FP (i.e. FN/FP—what we call 'tolerance'; Figure 9.3). Furthermore, because our interest lies in understanding the nonlethal effects of humans, a key variable is understanding the FP rate (Figure 9.4). As FP increases, there is a higher incurred cost in the form of missed opportunities (foraging or otherwise), classically quantified by giving up densities (Brown et al. 1999) translated into landscapes of fear (Laundré et al. 2010, Gaynor et al. 2019). This heightened perception of risk serves to maximize fitness via skewing risk assessment away from more costly false-negative

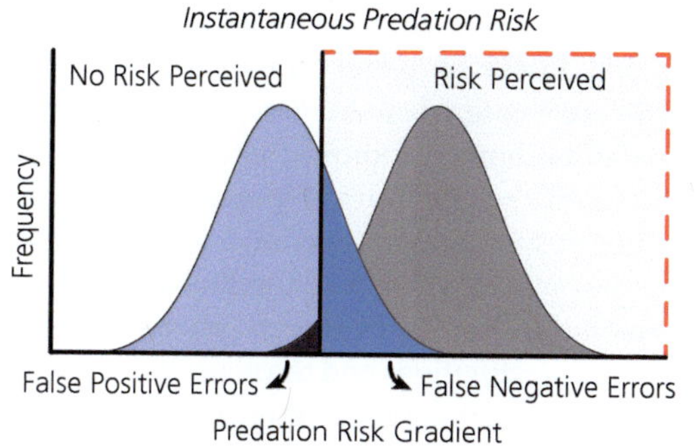

Figure 9.3 Frequency distributions of the two possible perceptions of risk (no risk perceived [blue] or risk perceived [grey]) are plotted, showing an arbitrary value of instantaneous predation risk (black diagonal line) along a gradient. To the right of the observed level of risk (grey square) a species should perceive risk (risk is perceived and risk is present; a true positive; TP), whereas to the left, there is no risk so the appropriate response would be to not perceive there is risk (a true negative; TN). However, risk perception is often misaligned, as influenced by the likelihood of an error in risk assessment, that is, the false positive, FP (determining there is risk, when there is no risk [black portion of the 'Risk Perceived' frequency distribution]), or a false-negative error, FN (failing to determine risk, when there is risk present [dark blue portion of the 'No Risk Perceived' frequency distribution]).

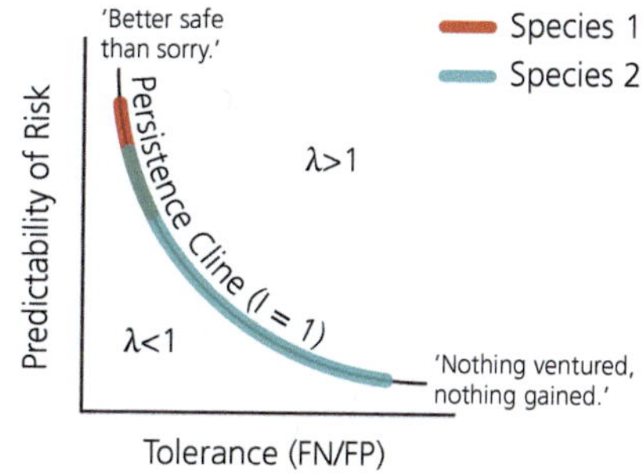

Figure 9.4 Predictions from the 'threshold of risk' hypothesis showing the tolerance of a species (FN/FP; x-axis) and the predictability of risk (y-axis) in a given landscape provide a persistence cline for individuals of a population. This 'threshold of risk' may prove especially useful when thinking of the plethora of risk sources in human landscapes that can be highly unpredictable because they have not been experienced across the evolutionary history of most species. The persistence isocline (black line) exists such that species that exhibit lower values of tolerance (high FP rates) can only persist in highly predictable environments, whereas species that are more tolerant can persist in highly unpredictable environments. Alternative strategies of coping with risk showing a narrow threshold of risk (Species 1; red) and a broad threshold of risk (Species 2; green)—species with a broad threshold of risk are better able to cope with unpredictable environments and, therefore, are more likely to persist in human landscapes. The clines represent a continuum from 'Better safe than sorry' to 'Nothing ventured, nothing gained'(*sensu* Johnson et al. 2013).

errors rather than serving to optimize expected payoffs *per se* (e.g. reproduction or nutrition).

Both empirical (Brown 1988) and theoretical (Bouskila and Blumstein 1992) work has shown that assessing predation risk is especially biased towards false-positive errors, and that overestimating risk is marginally costly to individual fitness over time. We propose the species 'threshold of risk' hypothesis, within the confines of error management theory. Under this framework, we hypothesize that more tolerant species or individuals—those with high tolerance quantified as higher values of observed FN/FP error rates—are best able to cope in unpredictable environments and thereby have higher persistence in human landscapes. On the other hand, species or individuals that have a generally low threshold of risk may not cope with unpredictability broadly, especially those associated with human landscapes. Because FN and FP error rates are linked and, over evolutionary timescales, species minimize their FN rates, the cost is to increase their FP rates (i.e. $FN/FP \ll 1$; the central tenet of EMT; Johnson et al. 2013). The extremely high level of depredation of humans on large carnivores should theoretically lead to an opportunity for selection for accurate assessments of risk (Bouskila and Blumstein 1992), but technological innovations have removed any stimulus from the act of killing (Montgomery et al. 2020). Alternatively, low predation levels—such as those that some large carnivores experience in human landscapes—should result in higher levels of inaccurate risk assessment, benefiting individuals that have a higher tolerance for false-negative errors. Accordingly, it is most likely that variability in tolerance, rather than a higher FN or FP *per se*, has the greatest impact on persistence in unpredictable (e.g. human) landscapes. Under the threshold of risk hypothesis, we predict that the species- and individual-level variability in tolerance (FN/FP) is associated with ecological traits such as constrained diel activity patterns (Gaynor et al. 2018) and dietary specialization (Jesmer et al. 2020; which we identified at the species level in Figure 9.2). Furthermore, we suspect that social structure (Crabtree and Sheldon 1999), hunting mode (Preisser et al. 2007), and habitat domain (Schmitz et al. 2017) will be important mediators of tolerance, though these relationships may be more complicated or indirectly related (Figure 9.4).

In applying this framework, researchers must quantify the fitness of large carnivores in human landscapes, and, as noted earlier, long-term data will be most useful in this endeavour. Using our framework, targeted research can untangle the false-positive and false-negative error rates in response to specific stimuli (e.g. when do large carnivores avoid human stimuli that [i] does not pose a risk to them or [ii] poses a risk to them, but they do not interpret it as a risk). Specifically, much work has been done on quantifying the FP rates of species to perceived levels of risk (Gaynor et al. 2019, Smith et al. 2022), however, very little work quantifies the FN rates that species experience because understanding when mistakes in risk perception led to death are inherently difficult to quantify; it is easier to backtrack the decisions that lead to success. Identifying these false-positive and false-negative errors will be an effective means for managing large carnivores, or—conversely—to better understand the types of cues we should use to mitigate human–carnivore conflict (i.e. stimuli that large carnivores can cue in on to avoid human structures, agricultural areas, and human foods).

We identified that there is uncertainty in the non-lethal effects that humans induce on large carnivores, though we found general trends (Figure 9.2) and a potential path forward (the threshold of risk hypothesis; Figure 9.3). First, extinction is an evolutionary process shaped by error management—large carnivores will continue to co-occur with humans and, likely, populations will exhibit key traits to cope with the uncertainty of risk associated with humans, as they have for millennia. Second, the importance of persistence will emerge on relatively short time scales as human population grows and large carnivore populations continue to integrate into human-dominated landscapes. Accordingly, we must act now in our efforts to mitigate the extinction risk by attempting to build predictable sources of risk into human landscapes. Finally, these effects will be inherently context dependent and geographically variable. The projected reductions and losses of large carnivores (in particular) will accelerate where human populations grow or where the human footprint intensifies. By contrast, regions where human populations decline or where resource extraction and agriculture are sourced from elsewhere (i.e. the global west) will continue to see some carnivoran recoveries. These will involve restorations and recolonizations, particularly of species that do not threaten human interests or species that are able to tolerate human risk. Using the framework, we predict that species traits will mitigate a species' (and individual) tolerance of humans in error-prone landscapes.

References

Abrahms, B., N. R. Jordan, K. A. Golabek, J. W. McNutt, A. M. Wilson, and J. S. Brashares. 2016. Lessons from integrating behaviour and resource selection: activity-specific responses of African wild dogs to roads. *Animal Conservation* 19:247–255.

Abrams, P. A. 1986. Adaptive responses of predators to prey and prey to predators: the failure of the arms-race analogy. *Evolution* 40:1229–1247.

Abrams, P. A. 1994. The fallacies of 'ratio-dependent' predation. *Ecology* 75:1842–1850.

Baker, H. K., C. E. F. Bruggeman, and J. B. Shurin. 2022. Population niche width is driven by within-individual niche expansion and individual specialization in introduced brook trout in mountain lakes. *Oecologia* 200(1): 1–10.

Balme, G. A., R. T. Pitman, H. S. Robinson, J. R. B. Miller, P. J. Funston, and L. T. B. Hunter. 2017. Leopard distribution and abundance is unaffected by interference competition with lions. *Behavioral Ecology* 28:1348–1358.

Bar-On, Y. M., R. Phillips, and R. Milo. 2018. The biomass distribution on Earth. *Proceedings of the National Academy of Sciences of the United States of America* 115:6506–6511.

Beckmann, J. P., and C. Lackey. 2008. Carnivores, urban landscapes, and longitudinal studies: a case history of black bears. *Human-Wildlife Interactions* 2:168–174.

Beckmann, J. P. and R. G. Seidler. 2009. *Wildlife and Energy Development—Pronghorn of the Upper Green River Basin—Year 4 Summary*. Wildlife Conservation Society, Bronx, N. Y.

Benson-Amram, S., B. Dantzer, G. Stricker, E. M. Swanson, and K. E. Holekamp. 2016. Brain size predicts problem-solving ability in mammalian carnivores. *Proceedings of the National Academy of Sciences of the United States of America* 113:2532–2537.

Berger, K. M. and E. M. Gese. 2007. Does interference competition with wolves limit the distribution and abundance of coyotes? *Journal of Animal Ecology* 76: 1075–1085.

Berger, K. M., E. M. Gese, and J. Berger. 2008. Indirect effects and traditional trophic cascades: a test involving wolves, coyotes, and pronghorn. *Ecology* 89:818–828.

Blecha, K. A., R. B. Boone, and M. W. Alldredge. 2018. Hunger mediates apex predator's risk avoidance response in wildland-urban interface. *The Journal of Animal Ecology* 87:609–622.

Boonstra, R., D. Hik, G. R. Singleton, and A. Tinnikov. 1998. The impact of predator-induced stress on the snowshoe hare cycle. *Ecological Monographs* 68: 371–394.

Bouskila, A. and D. T. Blumstein. 1992. Rules of thumb for predation hazard assessment: predictions from a dynamic model. *The American Naturalist* 139: 161–176.

Boydston, E. E., K. M. Kapheim, H. E. Watts, M. Szykman, and K. E. Holekamp. 2003. Altered behaviour in spotted hyenas associated with increased human activity. *Animal Conservation* 6:207–219.

Broekhuis, F. 2018. Natural and anthropogenic drivers of cub recruitment in a large carnivore. *Ecology and Evolution* 8:6748–6755.

Broekhuis, F., G. Cozzi, M. Valeix, J. W. McNutt, and D. W. Macdonald. 2013. Risk avoidance in sympatric large carnivores: reactive or predictive? *Journal of Animal Ecology* 82:1098–1105.

Brown, J. S. 1988. Patch use as an indicator of habitat preference, predation risk, and competition. *Behavioral Ecology and Sociobiology* 22:37–47.

Brown, J. S. 1999. Vigilance, patch use and habitat selection: foraging under predation risk. *Evolutionary Ecology Research* 1:49–71.

Brown, J. S. and B. P. Kotler. 2004. Hazardous duty pay and the foraging cost of predation. *Ecology Letters* 7: 999–1014.

Carter, N. H., B. K. Shrestha, J. B. Karki, N. M. B. Pradhan, and J. Liu. 2012. Coexistence between wildlife and humans at fine spatial scales. *Proceedings of the National Academy of Sciences of the United States of America* 109:15360–15365.

Caswell, H. 1989. Analysis of life table response experiments I. Decomposition of effects on population growth rate. *Ecological Modelling* 46:221–237.

Chapron, G., et al. 2014. Recovery of large carnivores in Europe's modern human-dominated landscapes. *Science* 346:1517–1519.

Charnov, E. L. 1990. On evolution of age of maturity and the adult lifespan. *Journal of Evolutionary Biology* 3: 139–144.

Chen-Kraus, C., N. A. Raharinoro, M. A. Randrianirinarisoa, D. J. Anderson, R. R. Lawler, D. P. Watts, and A. F. Richard. 2022. Human-lemur coexistence in a multiple-use landscape. *Frontiers in Ecology and Evolution* 10:779861.

Cherry, M. J., K. E. Morgan, B. T. Rutledge, L. M. Conner, and R. J. Warren. 2016. Can coyote predation risk induce reproduction suppression in white-tailed deer? *Ecosphere* 7:e01481.

Christiansen, F. and D. Lusseau. 2015. Linking behavior to vital rates to measure the effects of non-lethal disturbance on wildlife. *Conservation Letters* 8: 424–431.

Christiansen, F., D. Lusseau, E. Stensland, and P. Berggren. 2010. Effects of tourist boats on the behaviour of Indo-Pacific bottlenose dolphins off the south coast of Zanzibar. *Endangered Species Research* 11:91–99.

Christianson, D. A., and S. Creel. 2010. A nutritionally mediated risk effect of wolves on elk. *Ecology* 91: 1184–1191.

Ciuti, S., J. M. Northrup, T. B. Muhly, S. Simi, M. Musiani, J. A. Pitt, and M. S. Boyce. 2012. Effects of humans on behaviour of wildlife exceed those of natural predators in a landscape of fear. *PLoS One* 7:e50611.

Coetzee, B. W. T. and S. L. Chown. 2016. A meta-analysis of human disturbance impacts on antarctic wildlife. *Biological Reviews* 91:578–596.

Crabtree, R. L., and J. W. Sheldon. 1999. Coyotes and canid coexistence. In *Carnivores in Ecosystems: the Yellowstone Experience*, edited by T. W. Clark, A. P. Curlee, S. C. Minta, and P. Kareiva. Yale University Press, New Haven, C. T.

Creel, S. 2018. The control of risk hypothesis: reactive vs. proactive antipredator responses and stress-mediated vs. food-mediated costs of response. *Ecology Letters* 21:947–956.

Creel, S., D. A. Christianson, and J. A. Winnie Jr. 2011. A survey of the effects of wolf predation risk on pregnancy rates and calf recruitment in elk. *Ecological Applications* 21:2847–2853.

Crespi, E. J., T. D. Williams, T. S. Jessop, and B. Delehanty. 2013. Life history and the ecology of stress: how do glucocorticoid hormones influence life-history variation in animals? *Functional Ecology* 27:93–106.

Christiansen, F., et al. 2016. Food provisioning increases the risk of injury in a long-lived marine top predator. *Royal Society Open Science* 3(12):160560.

Crooks, K. R. and D. Van Vuren. 1995. Resource utilization by two insular endemic mammalian carnivores, the island fox and island spotted skunk. *Oecologia* 104: 301–307.

Cusack, J. J., et al. 2017. Revealing kleptoparasitic and predatory tendencies in an African mammal community using camera traps: a comparison of spatiotemporal approaches. *Oikos* 126:812–822.

Dantzer, B., Q. E. Fletcher, R. Boonstra, and M. J. Sheriff. 2014. Measures of physiological stress: a transparent or opaque window into the status, management and conservation of species? *Conservation Physiology* 2:cou023.

Darimont, C. T., C. H. Fox, H. M. Bryan, and T. E. Reimchen. 2015. The unique ecology of human predators. *Science* 349:858–860.

Delaney, D. K., L. L. Pater, L. D. Carlile, E. W. Spadgenske, T. A. Beaty, and R. H. Melton. 2011. Response of red-cockaded woodpeckers to military training operations. *Wildlife Monographs* 177:1–38.

Dheer, A., et al. 2022. Diurnal pastoralism does not reduce juvenile recruitment nor elevate allostatic load in spotted hyenas. *Journal of Animal Ecology* 91:2289–2300.

Di Bitetti, M. S., C. D. De Angelo, Y. E. Di Blanco, and A. Paviolo. 2010. Niche partitioning and species coexistence in a neotropical felid assemblage. *Acta Oecologica* 36:403–412.

Donadio, E. and S. W. Buskirk. 2006. Diet, morphology, and interspecific killing in carnivora. *The American Naturalist* 167:524–536.

Dröge, E., S. Creel, M. S. Becker, and J. M'soka. 2017a. Risky times and risky places interact to affect prey behaviour. *Nature Ecology & Evolution* 1:1123–1128.

Dröge, E., S. Creel, M. S. Becker, and J. M'soka. 2017b. Spatial and temporal avoidance of risk within a large carnivore guild. *Ecology and Evolution* 7:189–199.

Duchesne, M., S. D. Côté, and C. Barrette. 2000. Responses of woodland caribou to winter ecotourism in the Charlevoix Biosphere Reserve, Canada. *Biological Conservation* 96:311–317.

Dulude-de Broin, F., S. Hamel, G. F. Mastromonaco, and S. D. Côté. 2020. Predation risk and mountain goat reproduction: evidence for stress-induced breeding suppression in a wild ungulate. *Functional Ecology* 34: 1003–1014.

Durant, S. M. 2000a. Living with the enemy: avoidance of hyenas and lions by cheetahs in the Serengeti. *Behavioral Ecology* 11:624–632.

Durant, S. M. 2000b. Predator avoidance, breeding experience and reproductive success in endangered cheetahs, *Acinonyx jubatus*. *Animal Behaviour* 60: 121–130.

Durant, S. M., M. Kelly, and T. M. Caro. 2004. Factors affecting life and death in serengeti cheetahs: environment, age, and sociality. *Behavioral Ecology* 15:11–22.

Ellenberg, U., T. Mattern, P. J. Seddon, and G. L. Jorquera. 2006. Physiological and reproductive consequences of human disturbance in Humboldt penguins: the need for species-specific visitor management. *Biological Conservation* 133:95–106.

Elton, C. S. 1927. *Animal Ecology*. Macmillan, New York.

Estes, J. A., et al. 2011. Trophic downgrading of planet Earth. *Science* 333:301–306

Fedriani, J. M., T. K. Fuller, R. M. Sauvajot, and E. C. York. 2000. Competition and intraguild predation among three sympatric carnivores. *Oecologia* 125:258–270.

Féret, M., G. Gauthier, A. Béchet, J.-F. Giroux, and K. A. Hobson. 2003. Effect of a spring hunt on nutrient storage by greater snow geese in Southern Quebec. *The Journal of Wildlife Management* 67:796–807.

Fortin, D., H. L. Beyer, M. S. Boyce, D. W. Smith, T. Duchesne, and J. S. Mao. 2005. Wolves influence elk movements: behavior shapes a trophic cascade in Yellowstone National Park. *Ecology* 86:1320–1330.

French, S. S., M. González-Suárez, J. K. Young, S. Durham, and L. R. Gerber. 2011. Human disturbance influences reproductive success and growth rate in California Sea Lions (*Zalophus californianus*). *PLoS One* 6:e17686.

Frid, A. 2003. Dall's sheep responses to overflights by helicopter and fixed-wing aircraft. *Biological Conservation* 110:387–399.

Frid, A., and L. Dill. 2002. Human-caused disturbance stimuli as a form of predation risk. *Conservation Ecology* 6:11.

Gaynor, K. M., C. E. Hojnowski, N. H. Carter, and J. S. Brashares. 2018. The influence of human disturbance on wildlife nocturnality. *Science* 360:1232–1235.

Gaynor, K. M., J. S. Brown, A. D. Middleton, M. E. Power, and J. S. Brashares. 2019. Landscapes of fear: spatial patterns of risk perception and response. *Trends in Ecology & Evolution* 34:355–368.

Gehr, B., E. J. Hofer, A. Ryser, E. Vimercati, K. Vogt, and L. F. Keller. 2018. Evidence for nonconsumptive effects from a large predator in an ungulate prey? *Behavioral Ecology* 29:724–735.

Gehr, B., E. J. Hofer, S. Muff, A. Ryser, E. Vimercati, K. Vogt, and L. F. Keller. 2017. A landscape of coexistence for a large predator in a human dominated landscape. *Oikos* 126:1389–1399.

Gilbert, N. A., et al 2023. Daily activity timing in the Anthropocene. *Trends in Ecology & Evolution* 38:324–336.

Gill, J. A., W. J. Sutherland, and A. R. Watkinson. 1996. A method to quantify the effects of human disturbance on animal populations. *Journal of Applied Ecology* 33: 786–792.

Gittleman, J. L., ed. 1989. *Carnivore Behavior, Ecology, and Evolution*. Springer, Boston, M. A.

Greenspoon, L., et al. 2023. The global biomass of wild mammals. *Proceedings of the National Academy of Sciences of the United States of America* 120:e2204892120.

Griffiths, M. and C. P. Schaik. 1993. The impact of human traffic on the abundance and activity periods of Sumatran rain forest wildlife. *Conservation Biology* 7: 623–626.

Guiden, P. W., S. L. Bartel, N. W. Byer, A. A. Shipley, and J. L. Orrock. 2019. Predator–prey interactions in the Anthropocene: reconciling multiple aspects of novelty. *Trends in Ecology & Evolution* 34:616–627.

Hawlena, D. and O. J. Schmitz. 2010. Physiological stress as a fundamental mechanism linking predation to ecosystem functioning. *The American Naturalist* 176: 537–556.

Hayward, M. W. and R. Slotow. 2009. Temporal partitioning of activity in large African carnivores: tests of multiple hypotheses. *South African Journal of Wildlife Research* 39:109–125.

Heinemeyer, K., J. Squires, M. Hebblewhite, J. J. O'Keefe, J. D. Holbrook, and J. Copeland. 2019. Wolverines in winter: indirect habitat loss and functional responses to backcountry recreation. *Ecosphere* 10:e02611.

Hernández, L. and J. W. Laundré. 2005. Foraging in the 'landscape of fear' and its implications for habitat use and diet quality of elk *Cervus elaphus* and bison *Bison bison*. *Wildlife Biology* 11:215–220.

Hertel, A. G., J. E. Swenson, and R. Bischof. 2017. A case for considering individual variation in diel activity patterns. *Behavioral Ecology* 28:1524–1531.

Hjeljord, O. and T. Histøl. 1999. Range-body mass interactions of a northern ungulate—a test of hypothesis. *Oecologia* 119:326–339.

Hunter, J. S., S. M. Durant, and T. M. Caro. 2007. Patterns of scavenger arrival at cheetah kills in Serengeti National Park Tanzania. *African Journal of Ecology* 45:275–281.

Jesmer, B. R., M. J. Kauffman, M. A. Murphy, and J. R. Goheen. 2020. A test of the niche variation hypothesis in a ruminant herbivore. *Journal of Animal Ecology* 89:2825–2839.

Johnson, D. D. P., D. T. Blumstein, J. H. Fowler, and M. G. Haselton. 2013. The evolution of error: error management, cognitive constraints, and adaptive decision-making biases. *Trends in Ecology & Evolution* 28: 474–481.

Johnson, H. E., D. L. Lewis, and S. W. Breck. 2020. Individual and population fitness consequences associated with large carnivore use of residential development. *Ecosphere* 11:e03098.

Karanth, K. U. and M. E. Sunquist. 1995. Prey selection by tiger, leopard and dhole in tropical forests. *Journal of Animal Ecology* 64:439–450.

Kaza, S., L. Yao, P. Bhada-Tata, and F. Van Woerden. 2018. *What a Waste 2.0: A Global Snapshot of Solid Waste Management to 2050*. World Bank, Washington, D. C.

Kerbiriou, C., I. Le Viol, A. Robert, E. Porcher, F. Gourmelon, and R. Julliard. 2009. Tourism in protected areas can threaten wild populations: from individual response to population viability of the chough *Pyrrhocorax pyrrhocorax*. *Journal of Applied Ecology* 46: 657–665.

Kirby, R., M. W. Alldredge, and J. N. Pauli. 2016. The diet of black bears tracks the human footprint across a rapidly developing landscape. *Biological Conservation* 200:51–59.

Kirby, R., D. M. Macfarland, and J. N. Pauli. 2017. Consumption of intentional food subsidies by a hunted carnivore. *Journal of Wildlife Management* 81(7): 1161–1169.

Kirby, R., H. E. Johnson, M. W. Alldredge, and J. N. Pauli. 2019. The cascading effects of human food on hibernation and cellular aging in free-ranging black bears. *Scientific Reports* 9:2197.

Korpimaki, E., K. Norrdahl, and J. Valkama. 1994. Reproductive investment under fluctuating predation risk: microtine rodents and small mustelids. *Evolutionary Ecology* 8:357–368.

Kronfeld-Schor, N. and T. Dayan. 2003. Partitioning of time as an ecological resource. *Annual Review of Ecology, Evolution, and Systematics* 34:153–181.

Lasky, M. and S. Bombaci. 2023. Human-induced fear in wildlife: a review. *Journal for Nature Conservation* 74:126448.

Laundré, J. W., L. Hernández, and K. B. Altendorf. 2001. Wolves, elk, and bison: reestablishing the 'landscape of fear' in Yellowstone National Park, U.S.A. *Canadian Journal of Zoology* 79:1401–1409.

Laundré, J. W., L. Hernández, and W. Ripple. 2010. The landscape of fear: ecological implications of being afraid. *The Open Ecology Journal* 3:1–7.

Laurenson, M. K. 1994. High juvenile mortality in cheetahs (*Acinonyx jubatus*) and its consequences for maternal care. *Journal of Zoology* 234:387–408.

Leo, V., R. P. Reading, and M. Letnic. 2015. Interference competition: odours of an apex predator and conspecifics influence resource acquisition by red foxes. *Oecologia* 179:1033–1040.

Lima, S. L. 1998. Nonlethal effects in the ecology of predator-prey interactions: what are the ecological effects of anti-predator decision-making? *BioScience* 48:25–34.

Lima, S. L. and L. M. Dill. 1990. Behavioral decisions made under the risk of predation: a review and prospectus. *Canadian Journal of Zoology* 68:619–640.

Lindström, E. R., S. M. Brainerd, J. O. Helldin, and K. Overskaug. 1995. Pine marten–red fox interactions: a case of intraguild predation? *Annales Zoologici Fennici* 32:123–130.

Liukkonen, L., et al. 2017. Long-term effects of land use on perinatal mortality in the endangered Saimaa ringed seal population. *Endangered Species Research* 34:283–291.

Loveridge, A. J., M. Valeix, N. B. Elliot, and D. W. Macdonald. 2017. The landscape of anthropogenic mortality: how African lions respond to spatial variation in risk. *Journal of Applied Ecology* 54:815–825.

MacArthur, R. H. and E. R. Pianka. 1966. On optimal use of a patchy environment. *The American Naturalist* 100:603–609.

MacLeod, K. J., C. J. Krebs, R. Boonstra, and M. J. Sheriff. 2018. Fear and lethality in snowshoe hares: the deadly effects of non-consumptive predation risk. *Oikos* 127:375–380.

Martín, B., S. Delgado, A de la Cruz, S. Tirado, and M. Ferrer. 2015. Effects of human presence on the long-term trends of migrant and resident shorebirds: evidence of local population declines. *Animal Conservation* 18:73–81.

Mcmahon, C. R., M. N. Bester, H. R. Burton, M. A. Hindell, and C. J. A. Bradshaw. 2005. Population status, trends and a re-examination of the hypotheses explaining the recent declines of the southern elephant seal *Mirounga leonina*. *Mammal Review* 35:82–100.

Meillère, A., F. Brischoux, C. Parenteau, and F. Angelier. 2015. Influence of urbanization on body size, condition, and physiology in an urban exploiter: a multi-component approach. *PLoS ONE* 10:e0135685.

Mills, L. S., D. F. Doak, and M. J. Wisdom. 2001. Elasticity analysis for conservation decision making: reply to Ehrlén et al. *Conservation Biology* 15:281–283.

Milošević, S. 2020. *Competition Between Humans and Large Carnivores*. British Archaeological Reports, Oxford.

Moll, R. J., A. K. Killion, M. W. Hayward, and R. A. Montgomery. 2021. A framework for the Eltonian niche of humans. *BioScience* 71:928–941.

Montgomery, R. A., D. W. Macdonald, and M. W. Hayward. 2020. The inducible defences of large mammals to human lethality. *Functional Ecology* 34:2426–2441.

Morgan, G. M., T. E. Wilcoxen, M. A. Rensel, and S. J. Schoech. 2012. Are roads and traffic sources of physiological stress for the Florida scrub-jay?. *Wildlife Research* 39(4):301–310.

Morris, W. F. and D. F. Doak. 2002. *Quantitative Conservation: Theory and Practice of Population Viability Analysis*. Sinauer Associates, Sunderland.

Moss, W. E., M. W. Alldredge, and J. N. Pauli. 2016. Quantifying risk and resource use for a large carnivore in an expanding urban–wildland interface. *Journal of Applied Ecology* 53:371–378.

Myhrvold, N., E. Baldridge, B. Chan, D. Sivam, D. Freeman, and S. K. M. Ernest. 2015. An amniote life-history database to perform comparative analyses with birds, mammals, and reptiles. *Ecology* 96:3109–000.

Nellemann, C., I. Vistnes, P. Jordhøy, O. Strand, and A. Newton. 2003. Progressive impact of piecemeal infrastructure development on wild reindeer. *Biological Conservation* 113:307–317.

Nellemann, C., P. Jordhøy, O.-G. Støen, and O. Strand. 2000. Cumulative impacts of tourist resorts on wild reindeer (*Rangifer tarandus tarandus*) during winter. *Arctic* 53:9–17.

Nevin, O. T. and B. K. Gilbert. 2005. Measuring the cost of risk avoidance in brown bears: further evidence of positive impacts of ecotourism. *Biological Conservation* 123:453–460.

Nickel, B. A., J. P. Suraci, A. C. Nisi, and C. C. Wilmers. 2021. Energetics and fear of humans constrain the spatial ecology of pumas. *Proceedings of the National Academy of Sciences of the United States of America* 118:e2004592118.

Nickel, B. A., J. P. Suraci, M. L. Allen, and C. C. Wilmers. 2020. Human presence and human footprint have non-equivalent effects on wildlife spatiotemporal habitat use. *Biological Conservation* 241:108383.

Nisi, A. C., J. F. Benson, R. King, and C. C. Wilmers. 2023. Habitat fragmentation reduces survival and drives source–sink dynamics for a large carnivore. *Ecological Applications* 33:e2822.

Nisi, A. C., et al. 2022. Temporal scale of habitat selection for large carnivores: balancing energetics, risk and finding prey. *Journal of Animal Ecology* 91: 182–195.

Norris, K. and N. McCulloch. 2003. Demographic models and the management of endangered species: a case study of the critically endangered Seychelles magpie robin. *Journal of Applied Ecology* 40:890–899.

Northrup, J. M., C. R. Anderson Jr., B. D. Gerber, and G. Wittemyer. 2021. Behavioral and demographic responses of mule deer to energy development on winter range. *Wildlife Monographs* 208:1–37.

Ordiz, A., O.-G. Støen, M. Delibes, and J. E. Swenson. 2011. Predators or prey? Spatio-temporal discrimination of human-derived risk by brown bears. *Oecologia* 166:59–67.

Ordiz, A., O.-G. Støen, S. Sæbø, J. Kindberg, M. Delibes, and J. E. Swenson. 2012. Do bears know they are being hunted? *Biological Conservation* 152:21–28.

Oriol-Cotterill, A., M. Valeix, L. G. Frank, C. Riginos, and D. W. Macdonald. 2015. Landscapes of coexistence for terrestrial carnivores: the ecological consequences of being downgraded from ultimate to penultimate predator by humans. *Oikos* 124:1263–1273.

Oro, D., M. Genovart, G. Tavecchia, M. S. Fowler, and A. Martínez-Abraín. 2013. Ecological and evolutionary implications of food subsidies from humans. *Ecology Letters* 16:1501–1514.

Palomares, F. and T. M. Caro. 1999. Interspecific killing among mammalian carnivores. *The American Naturalist* 153:492–508.

Pannunzi, M. and T. Nowotny. 2019. Odor stimuli: not just chemical identity. *Frontiers in Physiology* 10:1428.

Pauli, J. N. and S. W. Buskirk. 2007. Risk-Disturbance overrides density dependence in a hunted colonial rodent, the black-tailed prairie dog *Cynomys ludovicianus*. *Journal of Applied Ecology* 44:1219–1230.

Peacor, S. D., Barton, B. T., Kimbro, D. L., Sih, A., and Sheriff, M. J. 2020. A framework and standardized terminology to facilitate the study of predation-risk effects. *Ecology* 101:e03152.

Peacor, S. D., et al. 2022. A skewed literature: few studies evaluate the contribution of predation-risk effects to natural field patterns. *Ecology Letters* 25:2048–2061.

Peers, M. J. L., et al. 2018. Quantifying fear effects on prey demography in nature. *Ecology* 99:1716–1723.

Pereira, L. M., N. Owen-Smith, and M. Moleón. 2014. Facultative predation and scavenging by mammalian carnivores: seasonal, regional and intra-guild comparisons. *Mammal Review* 44:44–55.

Petracca, L. S., et al. 2019. The effectiveness of hazing African lions as a conflict mitigation tool: implications for carnivore management. *Ecosphere* 10(12):e02967.

Petroelje, T. R., J. L. Belant, D. E. Beyer, and N. J. Svoboda. 2019. Subsidies from anthropogenic resources alter diet, activity, and ranging behavior of an apex predator (*Canis lupus*). *Scientific Reports* 9:13438.

Preisser, E. L. and D. I. Bolnick. 2008. The many faces of fear: comparing the pathways and impacts of nonconsumptive predator effects on prey populations. *PLoS One* 3:e2465.

Preisser, E. L., D. I. Bolnick, and M. F. Benard. 2005. Scared to death? The effects of intimidation and consumption in predator–prey interactions. *Ecology* 86:501–509.

Preisser, E. L., J. L. Orrock, and O. J. Schmitz. 2007. Predator hunting mode and habitat domain alter nonconsumptive effects in predator–prey interactions. *Ecology* 88:2744–2751.

Prugh, L. R. and K. J. Sivy. 2020. Enemies with benefits: integrating positive and negative interactions among terrestrial carnivores. *Ecology Letters* 23:902–918.

Rasmussen, G. S. A. and D. W. Macdonald. 2012. Masking of the zeitgeber: African wild dogs mitigate persecution by balancing time. *Journal of Zoology* 286:232–242.

Ripple, W. J., et al. 2014. Status and ecological effects of the world's largest carnivores. *Science* 343:1241484.

Ritchie, E. G. and C. N. Johnson. 2009. Predator interactions, mesopredator release and biodiversity conservation. *Ecology Letters* 12:982–998.

Robinson, Q. H., D. Bustos, and G. W. Roemer. 2014. The application of occupancy modeling to evaluate intraguild predation in a model carnivore system. *Ecology* 95:3112–3123.

Rodriguez Curras, M., E. Donadío, A. D. Middleton, and J. N. Pauli. 2021. Perceived risk structures the space use of competing carnivores. *Behavioral Ecology* 32:1380–1390.

Rodriguez Curras, M., E. Donadío, A. D. Middleton, and J. N. Pauli. 2022. Carnivore niche partitioning in a human landscape. *The American Naturalist* 199:496–509.

Rodríguez-Prieto, I. and E. Fernández-Juricic. 2005. Effects of direct human disturbance on the endemic Iberian frog *Rana iberica* at individual and population levels. *Biological Conservation* 123:1–9.

Rogala, J., M. Hebblewhite, J. Whittington, C. White, J. Coleshill, and M. Musiani. 2011. Human activity differentially redistributes large mammals in the Canadian Rockies national parks. *Ecology and Society* 16:16.

Roy, C. and A. Woolf. 2001. Effects of hunting and hunting-hour extension on mourning dove foraging and physiology. *The Journal of Wildlife Management* 65:808–815.

Ruprecht, J., et al. 2021. Variable strategies to solve risk–reward tradeoffs in carnivore communities. *Proceedings of the National Academy of Sciences of the United States of America* 118:e2101614118.

Schmitz, O. J. 2007. Predator diversity and trophic interactions. *Ecology* 88:2415–2426.

Schmitz, O. J. 2008. Effects of predator hunting mode on grassland ecosystem function. *Science* 319:952–954.

Schmitz, O. J., A. P. Beckerman, and K. M. O'Brien. 1997. Behaviorally mediated trophic cascades: effects of predation risk on food web interactions. *Ecology* 78: 1388–1399.

Schmitz, O. J., J. R. B. Miller, A. M. Trainor, and B. Abrahms. 2017. Toward a community ecology of landscapes: predicting multiple predator–prey interactions across geographic space. *Ecology* 98:2281–2292.

Selva, N. and M. A. Fortuna. 2007. The nested structure of a scavenger community. *Proceedings of the Royal Society B: Biological Sciences* 274:1101–1108.

Sheriff, M. J., B. Dantzer, O. P. Love, and J. L. Orrock. 2018. Error management theory and the adaptive significance of transgenerational maternal-stress effects on offspring phenotype. *Ecology and Evolution* 8:6473–6482.

Sheriff, M. J., C. J. Krebs, and R. Boonstra. 2009. The sensitive hare: sublethal effects of predator stress on reproduction in snowshoe hares. *Journal of Animal Ecology* 78:1249–1258.

Sheriff, M. J., S. D. Peacor, D. Hawlena, and M. Thaker. 2020. Non-consumptive predator effects on prey population size: a dearth of evidence. *Journal of Animal Ecology* 89:1302–1316.

Sih, A. 2013. Understanding variation in behavioural responses to human-induced rapid environmental change: a conceptual overview. *Animal Behaviour* 85:1077–1088.

Sih, A. and B. Christensen. 2001. Optimal diet theory: when does it work, and when and why does it fail? *Animal Behaviour* 61:379–390.

Sih, A., M. C. O. Ferrari, and D. J. Harris. 2011. Evolution and behavioural responses to human-induced rapid environmental change. *Evolutionary Applications* 4:367–387.

Sinclair, A. R. E., and P. Arcese. 1995. Population consequences of predation-sensitive foraging: the Serengeti wildebeest. *Ecology* 76:882–891.

Sivy, K. J., C. B. Pozzanghera, J. B. Grace, and L. R. Prugh. 2017. Fatal attraction? Intraguild facilitation and suppression among predators. *The American Naturalist* 190:663–679.

Sivy, K. J., C. B. Pozzanghera, K. E. Colson, M. A. Mumma, and L. R. Prugh. 2018. Apex predators and the facilitation of resource partitioning among mesopredators. *Oikos* 127:607–621.

Smith, F. A., R. E. Elliott Smith, S. K. Lyons, and J. L. Payne. 2018. Body size downgrading of mammals over the late quaternary. *Science* 360:310–313.

Smith, J. A., et al. 2017. Fear of the human 'super predator' reduces feeding time in large carnivores. *Proceedings of the Royal Society B: Biological Sciences* 284:20170433.

Smith, J. A., K. M. Gaynor, and J. P. Suraci. 2021. Mismatch between risk and response may amplify lethal and non-lethal effects of humans on wild animal populations. *Frontiers in Ecology and Evolution* 9:604973.

Smith, J. A., Y. Wang, and C. C. Wilmers. 2015. Top carnivores increase their kill rates on prey as a response to human-induced fear. *Proceedings of the Royal Society B: Biological Sciences* 282:20142711.

Smith, J. A., Y. Wang, and C. C. Wilmers. 2016. Spatial characteristics of residential development shift large carnivore prey habits. *The Journal of Wildlife Management* 80:1040–1048.

Soria, C. D., M. Pacifici, M. Di Marco, S. M. Stephen, and C. Rondinini. 2021. Combine: a coalesced mammal database of intrinsic and extrinsic traits. *Ecology* 102:e03344.

Spaul, R. J. and J. A. Heath. 2016. Nonmotorized recreation and motorized recreation in shrub-steppe habitats affects behavior and reproduction of golden eagles (*Aquila chrysaetos*). *Ecology and Evolution* 6:8037–8049.

Suraci, J. P., M. Clinchy, L. Y. Zanette, and C. C. Wilmers. 2019. Fear of humans as apex predators has landscape-scale impacts from mountain lions to mice. *Ecology Letters* 22:1578–1586.

Swanson, A., et al. 2014. Cheetahs and wild dogs show contrasting patterns of suppression by lions. *Journal of Animal Ecology* 83:1418–1427.

Trimmer, P. C., S. M. Ehlman, and A. Sih. 2017. Predicting behavioural responses to novel organisms: state-dependent detection theory. *Proceedings of the Royal Society B: Biological Sciences* 284:20162108.

Tucker, M. A., et al. 2018. Moving in the Anthropocene: global reductions in terrestrial mammalian movements. *Science* 359:466–469.

Valeix, M., G. Hemson, A. J. Loveridge, G. Mills, and D. W. Macdonald. 2012. Behavioural adjustments of a large carnivore to access secondary prey in a human-dominated landscape. *Journal of Applied Ecology* 49: 73–81.

Venter, O., et al. 2016. Sixteen years of change in the global terrestrial human footprint and implications for biodiversity conservation. *Nature Communications* 7:1–11.

Villanueva, C., B. G. Walker, and M. Bertellotti. 2012. A matter of history: effects of tourism on physiology, behaviour and breeding parameters in magellanic penguins (*Spheniscus magellanicus*) at two colonies in Argentina. *Journal of Ornithology* 153: 219–228.

Wang, Y., J. A. Smith, and C. C. Wilmers. 2017. Residential development alters behavior, movement, and energetics in an apex predator, the puma. *PLoS One* 12:e0184687.

Wang, Y., M. L. Allen, and C. C. Wilmers. 2015. Mesopredator spatial and temporal responses to large predators and human development in the Santa Cruz Mountains of California. *Biological Conservation* 190:23–33.

Ware, H. E., C. J. W. McClure, J. D. Carlisle, and J. R. Barber. 2015. A phantom road experiment reveals traffic noise is an invisible source of habitat degradation. *Proceedings of the National Academy of Sciences of the United States of America* 112:12105–12109.

Watts, H. E. and K. E. Holekamp. 2008. Interspecific competition influences reproduction in spotted hyenas. *Journal of Zoology* 276:402–410.

Wheat, R. E. and C. C. Wilmers. 2016. Habituation reverses fear-based ecological effects in brown bears (*Ursus arctos*). *Ecosphere* 7:e01408.

Wilkinson, I. and M. Bester. 1988. Is onshore human activity a factor in the decline of the southern elephant seal? *South African Journal of Antarctic Research* 18:14–17.

Williams, J. W. and S. T. Jackson. 2007. Novel climates, no-analog communities, and ecological surprises. *Frontiers in Ecology and the Environment* 5:475–482.

Wilmers, C. C., et al. 2013. Scale dependent behavioral responses to human development by a large predator, the puma. *PLoS One* 8:e60590.

Wilson, M. W., A. D. Ridlon, K. M. Gaynor, S. D. Gaines, A. C. Stier, and B. S. Halpern. 2020. Ecological impacts of human-induced animal behaviour change. *Ecology Letters* 23:1522–1536.

Wirsing, A. J., M. R. Heithaus, J. S. Brown, B. P. Kotler, and O. J. Schmitz. 2021. The context dependence of non-consumptive predator effects. *Ecology Letters* 24: 113–129.

Wolf, C. and W. J. Ripple. 2018. Rewilding the world's large carnivores. *Royal Society Open Science* 5:172235.

Yirga, G., H. H. De Iongh, H. Leirs, K. Gebrihiwot, J. Deckers, and H. Bauer. 2012. Adaptability of large carnivores to changing anthropogenic food sources: diet change of spotted hyena (*Crocuta crocuta*) during Christian fasting period in Northern Ethiopia. *Journal of Animal Ecology* 81:1052–1055.

SECTION FOUR

Implications of Human-Carnivore Interactions on Ecosystem Functioning

CHAPTER 10

Cascading Effects of Humans on Trophic Dynamics

Laura R. Prugh, Samantha E.S. Kreling, Taylor R. Ganz, Calum X. Cunningham, Yasmine Hentati, Benjamin K. Sullender, Catherine M. Breen, and Michael Procko

Introduction

In the age of the Anthropocene, human impacts represent a dominant force capable of altering ecosystem dynamics worldwide. A mere 3% of the terrestrial land surface is estimated to harbour intact faunal communities (Plumptre et al. 2021). While human impacts are pervasive across trophic levels (Figure 10.1; Dirzo et al. 2014, Gray et al. 2018), persecution of apex predators has been particularly intense throughout human history, and large carnivores have thus experienced disproportionately large range contractions and population declines (Wolf and Ripple 2017). Large carnivore densities are reduced by people through a combination of prey depletion (Wolf and Ripple 2016), human-induced mortality, and subsequently strong behavioural avoidance of human-dominated areas (Woodroffe 2000, Berger 2007, Darimont et al. 2015). Recent efforts to restore large carnivore populations have led to range expansions of wolves (*Canis lupus*), cougars (*Puma concolor*), and brown bears (*Ursus arctos*) in North America and Europe, thereby reversing the declining population trends of prior decades (Chapron et al. 2014, Wilmers et al. 2025). As large carnivores recover, their functional roles in these vastly altered landscapes remain uncertain (Kuijper et al. 2016). Human-caused mortality can reduce large carnivore density even when legal protections are in place (Benson et al. 2023), and in parks where carnivores are fully protected from harvest and vehicle collisions are rare, populations can still decline due to poaching and source-sink dynamics with adjacent unprotected areas (Woodroffe and Ginsberg 1998, Novaro et al. 2005, Durant et al. 2017, Rija et al. 2020).

Although human impacts on large carnivore density are overwhelmingly negative, they can be positive when lethal effects are minimal and anthropogenic food subsidies are substantial, which can occur in highly urbanized areas (Newsome et al. 2015). For example, the density of leopards (*Panthera pardis*) was estimated to be five times higher in an urban park in Mumbai than in a nearby protected area, and this difference was attributed to higher abundance of the leopard's primary prey (e.g. feral dogs, *Canis familiaris*) in the city (Surve et al. 2022). Similarly, abundant populations of black bears (*Ursus americanus*) can occur in exurban areas where anthropogenic subsidies are readily available (Merkle et al. 2013). Because these cases of human influences leading to higher densities of large carnivores are relatively rare, this chapter will focus primarily on the consequences of reduced large carnivore densities that more typically result from anthropogenic effects.

Human-caused reductions in apex predator densities can have strong cascading effects on lower trophic levels (Estes et al. 2011). For example, the absence of top-down control can lead to overabundant ungulate populations that degrade habitat through overbrowsing, with ecosystem-wide

Laura R. Prugh et al., *Cascading Effects of Humans on Trophic Dynamics*. In: *Apex Predators in the Anthropocene*. Edited by: Robert A. Montgomery and Remington J. Moll, Oxford University Press. © Oxford University Press (2026). DOI: 10.1093/9780191996665.003.0010 impacts such as reduced bird diversity (Allombert et al. 2005). Cascading effects of high ungulate densities can also include negative societal impacts, such as agricultural losses (Cote et al. 2004) and increased vehicle collisions (Gilbert et al. 2017, Raynor et al. 2021). Released mesopredator populations can similarly lead to undesirable knock-on effects, such as declines in small mammal and bird populations (Crooks and Soulé 1999, Prugh et al. 2009). Predicting the cascading effects of large carnivores on ecosystem dynamics when humans are considered can be difficult because humans have strong direct and indirect effects on all trophic levels, their impacts on carnivores can be a mix of positive and negative, and effects cascade to lower trophic levels via multiple pathways (Figure 10.1).

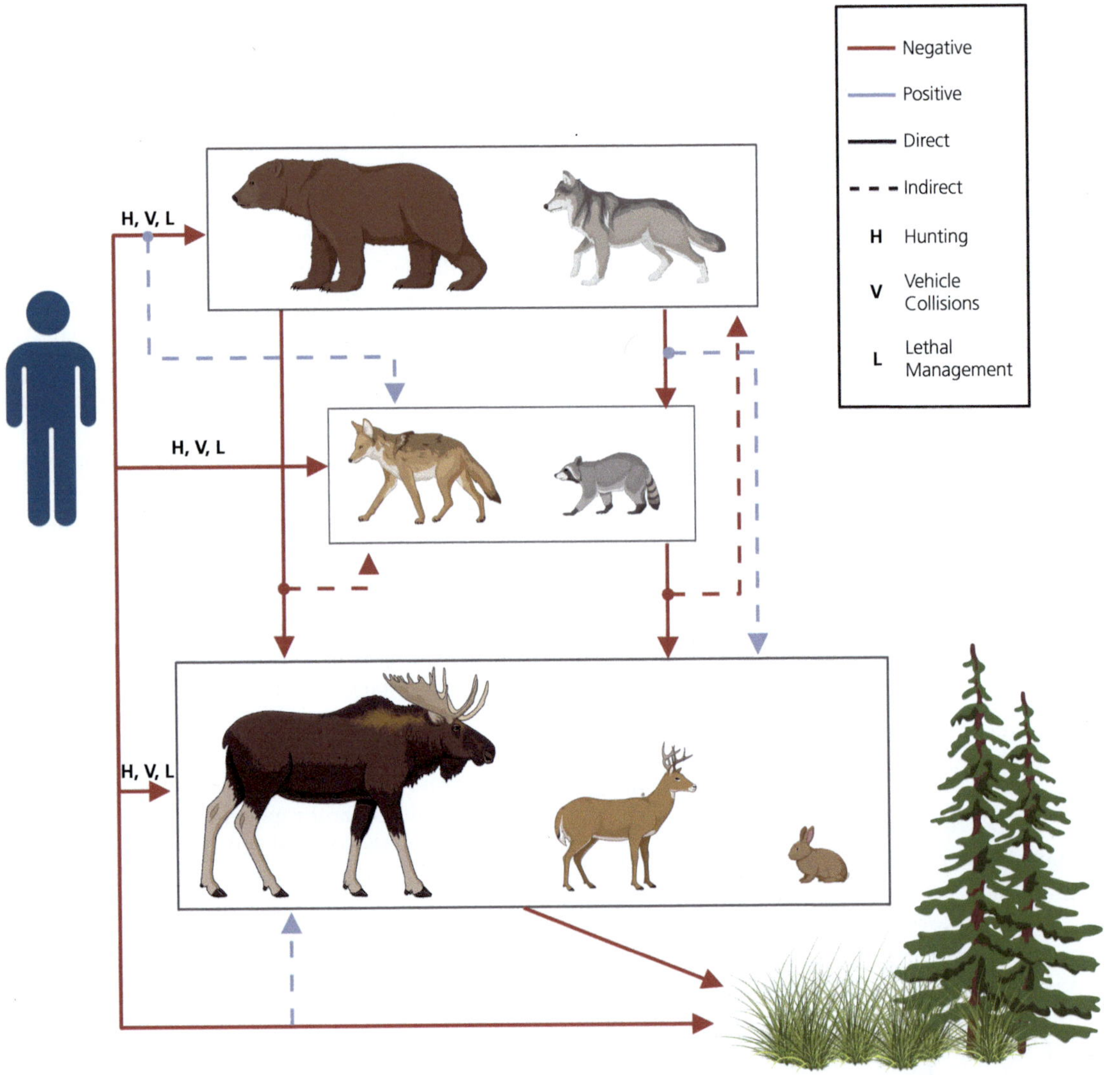

Figure 10.1 Pervasive influence of humans on trophic networks. Humans reduce densities of apex predators (bear, wolf), mesopredators (coyote, raccoon) and herbivores (moose, deer, rabbit) through harvest (H), vehicle collisions (V), and lethal management (L). Humans also directly affect vegetation through activities such as logging, farming, and development. These anthropogenic effects mediate the strength of direct (solid lines) and indirect (dashed lines) positive (blue) and negative (red) interactions among species within each trophic level. This diagram has been simplified by grouping species within trophic levels together for visual clarity.

Source: Created with BioRender.com

In this chapter, we propose that the top-down effects of humans differ from top-down effects of other predators, because the pervasive direct effects of humans on lower trophic levels can alter cascading effects of large carnivores in fundamental ways. We begin by distinguishing classic trophic cascade chains from triangular trophic interactions that are more common in food webs, such as intraguild predation and apparent competition. Next, we highlight the interconnectedness of top-down and bottom-up human influences that can mediate cascading effects. We then examine how a critically important process that is often left out of food web investigations—scavenging—is also affected by humans. Finally, we explore how climate change can alter the cascading effects of large carnivores.

Trophic triangles

Traditional trophic cascades have typically been conceptualized as linear in nature. For example, apex predators influence the abundance and behaviour of herbivores, which in turn influence the abundance and composition of plant communities. In an ecosystem with these three trophic levels, apex predators exert top-down control on herbivore populations, releasing plant communities from the top-down control of herbivores and fostering a 'green world' where primary producers are able to thrive (Figure 10.2a; Hairston et al. 1960). The loss of top predators, whether by habitat fragmentation, disease, lethal removal, or other means, creates a ripple effect where the lack of predator pressure releases the herbivore populations, which in turn consume higher amounts of vegetation (Ripple et al. 2016). However, these linear trophic chains do not account for the complexity inherent in real-world ecological food webs. In reality, food webs are characterized by omnivory and many triangular interactions that involve a mixture of competition and predation (Polis and Strong 1996, Arim and Marquet 2004, Thompson et al. 2007). For example, intraguild predation is a pervasive triangular interaction in which a top predator consumes an intermediate predator and both predators consume shared prey (Polis and Holt 1992). Apparent competition is a similarly pervasive triangular interaction in which a shared predator consumes two prey species, thereby leading to an indirect negative interaction between the prey species (Holt and Lawton 1994, DeCesare et al. 2010). These 'trophic triangles' complicate predictions regarding cascading effects of large carnivores. In this section, we show how the inclusion of humans in trophic food webs affects the strength and directionality of top-down forcing from apex predators. These effects have the potential to introduce new trophic triangles between humans and other predators, and to alter the dynamics within existing triangles.

In the context of a traditional linear trophic cascade, introducing human pressure on apex predators should create a 'brown world' where plant communities experience increased herbivory due to the reduction of predators and subsequent increase in herbivores (Figure 10.2b). Perhaps the most well-studied example of the consequences of a trophic cascade disrupted by humans is that of Yellowstone National Park, W. Y., USA, where the extirpation of grey wolves via lethal removal contributed to the expansion of the Rocky Mountain elk (*Cervus elaphus*) population between the late nineteenth and late twentieth centuries (Vucetich et al. 2005). In the elk population's northern range, sensitive riparian vegetation and growing tree stands could not support the ongoing intensity of elk browse. Consequently, recruitment of riparian woody species such as aspen and cottonwood (*Populus* spp.) stagnated throughout the century (Beschta and Ripple 2020). Humans reintroduced wolves in 1995 following a federal mandate to restore the species to the American West, such that elk populations were once again subject to predation pressure and riparian vegetation could potentially recover, shifting the ecosystem from a 'brown world' to a 'green world' (Beschta and Ripple 2016). However, some argue that this example is exaggerated by sampling biases and the tendency to attribute vegetation changes solely to reintroduced wolves when other species such as beavers, trout, and bears may have played key roles (Hobbs et al. 2024). For example, the introduction of invasive lake trout (*Salvelinus namaycush*) by humans drove populations of native cutthroat trout (*Oncorhynchus clarkii bouvieri*) to low levels by the

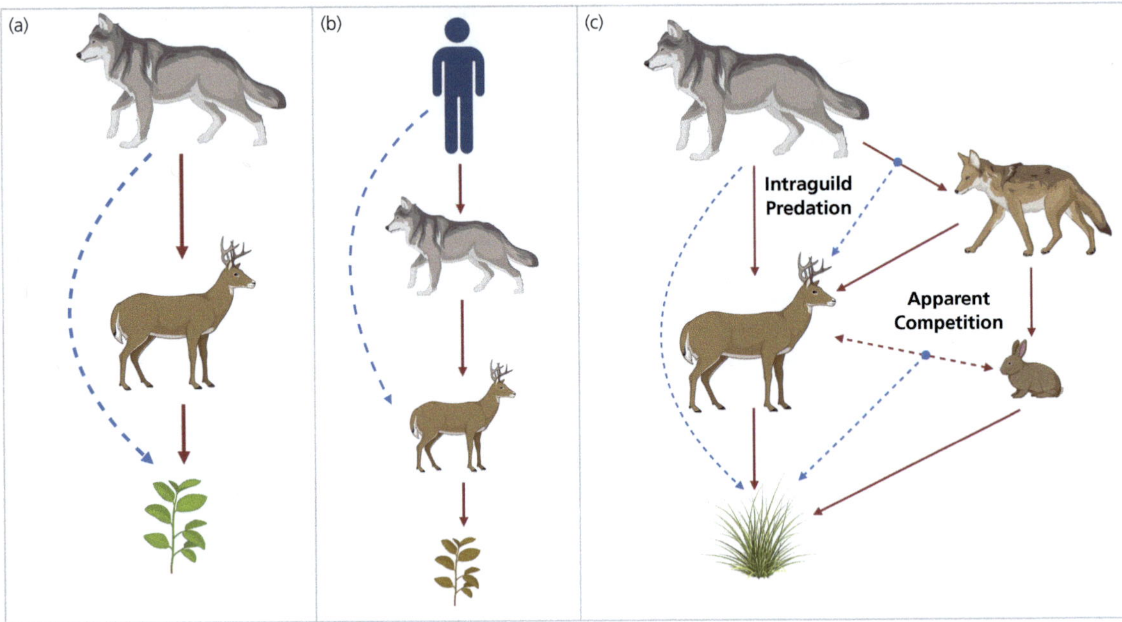

Figure 10.2 Trophic triangles can reduce the strength of linear cascades. (a) In the absence of humans or triangular interactions, apex predators suppress herbivores (red arrow), indirectly benefiting plants (dashed blue arrow) and creating a 'green' world. (b) Humans suppress apex predators, indirectly benefiting herbivores and creating a 'brown' world. (c) Inclusion of trophic triangles, such as intraguild predation involving mesopredators and apparent competition with other herbivores can weaken the strength of linear cascades.

Source: Created with BioRender.com

late 1980s, which reduced the availability of this key food source for grizzly bears and likely caused bears to increase predation rates on elk calves (Middleton et al. 2013). This example illustrates not only the unexpected ways that human actions (e.g. trout introduction) can alter the top-down effects of large carnivores, but also the importance of accounting for triangular trophic interactions (e.g. trout-bear-elk) to understand how anthropogenic effects cascade through food webs.

Trophic triangles such as intraguild predation and apparent competition should generally reduce the strength of linear cascades, although the specific effects will depend on the strengths of each interaction pathway, and opposing effects may cancel each other out. Consider, for example, a hypothetical linear cascade involving wolves, deer, and willows (Figure 10.2a). If these were the only three species in the ecosystem, all interaction pathways would be strong, leading to a positive indirect effect of wolves on willows. Adding humans to the top of this linear cascade would shift the indirect benefit from humans to deer, but the interactions should all remain strong (Figure 10.2b). However, adding coyotes (*Canis latrans*) to this food web introduces a trophic triangle, whereby wolves indirectly benefit deer by suppressing coyote populations that also prey on deer (Figure 10.2c). If this indirect benefit is strong enough, the direct negative effect of wolves on deer via the linear cascade could even be cancelled out. Indeed, simulations based on empirical data of wolves, coyotes, and their shared prey (Dall sheep, *Ovis dalli dalli*) in Alaska showed that wolf control programmes intended to increase ungulate populations could have the reverse effect if wolf suppression of coyotes was strong (Prugh and Arthur 2015). Similarly, adding another prey species consumed by coyotes, such as rabbits (*Sylvilagus sp.*), would introduce an apparent competition triangle to the food web that could further counteract the linear cascade (Figure 10.2c).

Food webs have typically been conceptualized and constructed without humans, perhaps in part because it is difficult to construct a visually coherent diagram that includes the many interactions humans have with species in every trophic level. For example, we omitted direct paths from humans to deer and from humans to plants in Figure 10.2 for visual clarity, but those pathways would exist in a realistic system, introducing additional trophic triangles that would alter the net strength of species interactions. Empirical estimates of interaction strengths in trophic networks that include humans are rare, but a study in Transylvania, Romania did so and reported that top-down pressure on apex predators from humans (via lethal management and hunting) and indirect human suppression of predators (via domestic dogs) had a stronger effect on ungulate populations than apex predators did (Dorresteijn et al. 2015).

In some cases (e.g. within protected areas), human presence can disproportionately impact apex predators, providing a shield for prey and mesopredators to evade top-down pressures (Chapter 11). Outside of protected areas, however, wildlife are subject to mortality via hunting and trapping, vehicle collisions, and conflict removals. Accordingly, prey or subordinate predators that shift their activity towards humans to avoid large carnivores may unknowingly place themselves in an ecological trap where rates of human-caused mortality are greater than predation rates in natural systems. For example, bobcats (*Lynx rufus*) and coyotes in northern Washington took refuge in areas with a greater human footprint to avoid wolves and cougars, demonstrating a stronger perceived fear of apex predators than of humans (Prugh et al. 2023). However, humans killed four times more bobcats and three times more coyotes than large carnivores did (Prugh et al. 2023). This case illustrates how the behavioural response of subordinate species to apex predators can amplify lethal anthropogenic effects on lower trophic levels. Furthermore, the trophic triangle between humans, apex predators, and mesopredators may counteract the trophic triangle between apex predators, mesopredators, and shared prey (Figure 10.2c). Indeed, no evidence of an indirect benefit of wolves or cougars on white-tailed deer (*Odocoileus virginianus*) via mesopredator suppression was found in this northern Washington system (Ganz et al. 2024).

Interconnection between top-down and bottom-up human impacts

Anthropogenic land use changes, such as habitat loss, degradation, and fragmentation contribute to the decline of large carnivores while simultaneously affecting other species throughout food webs (Newbold et al. 2020). Both human population density and species diversity are positively correlated with primary productivity at regional-to-global scales, so anthropogenic impacts have especially strong potential to affect wildlife communities via changes in resource availability (Luck 2007). Dual forcing of resource availability (i.e. bottom-up factors) and predation (i.e. top-down factors) interact to limit prey populations, though the relative strength of each is system specific (Ballard et al. 2001, Forrester and Wittmer 2013), and bottom-up factors can modulate the strength of trophic cascades (Muhly et al. 2013). Therefore, accounting for the impacts of anthropogenic habitat change on lower trophic levels is necessary to fully understand the consequences of carnivore decline on ecosystem structure and functioning (Dorresteijn et al. 2015). For instance, high primary productivity and mild climates reduced the effects of predation by grey wolves and Eurasian lynx (*Lynx lynx*) on roe deer (*Capreolus capreolus*) across Europe (Melis et al. 2009). Furthermore, experimental removal of predators only weakly increases prey abundance in regions where forage is a key limiting factor (Clark and Hebblewhite 2021). In this section, we explore the interconnectedness between top-down and bottom-up human impacts, focusing on the: (1) outcomes when humans increase primary productivity; (2) effects when humans decrease primary productivity; (3) implications of accompanying changes to landscape structure; and (4) consequences of these dual forcings in more complex food webs.

To understand these dynamics, we once again consider a simple three-level food chain including plants, herbivorous prey, and a single species of predator as in a classic trophic cascade (Figure 10.3). If humans decrease predator density, then top-down limits on prey density should weaken, strengthening bottom-up effects (Muhly et al. 2013). As a result, human landscape modifications that

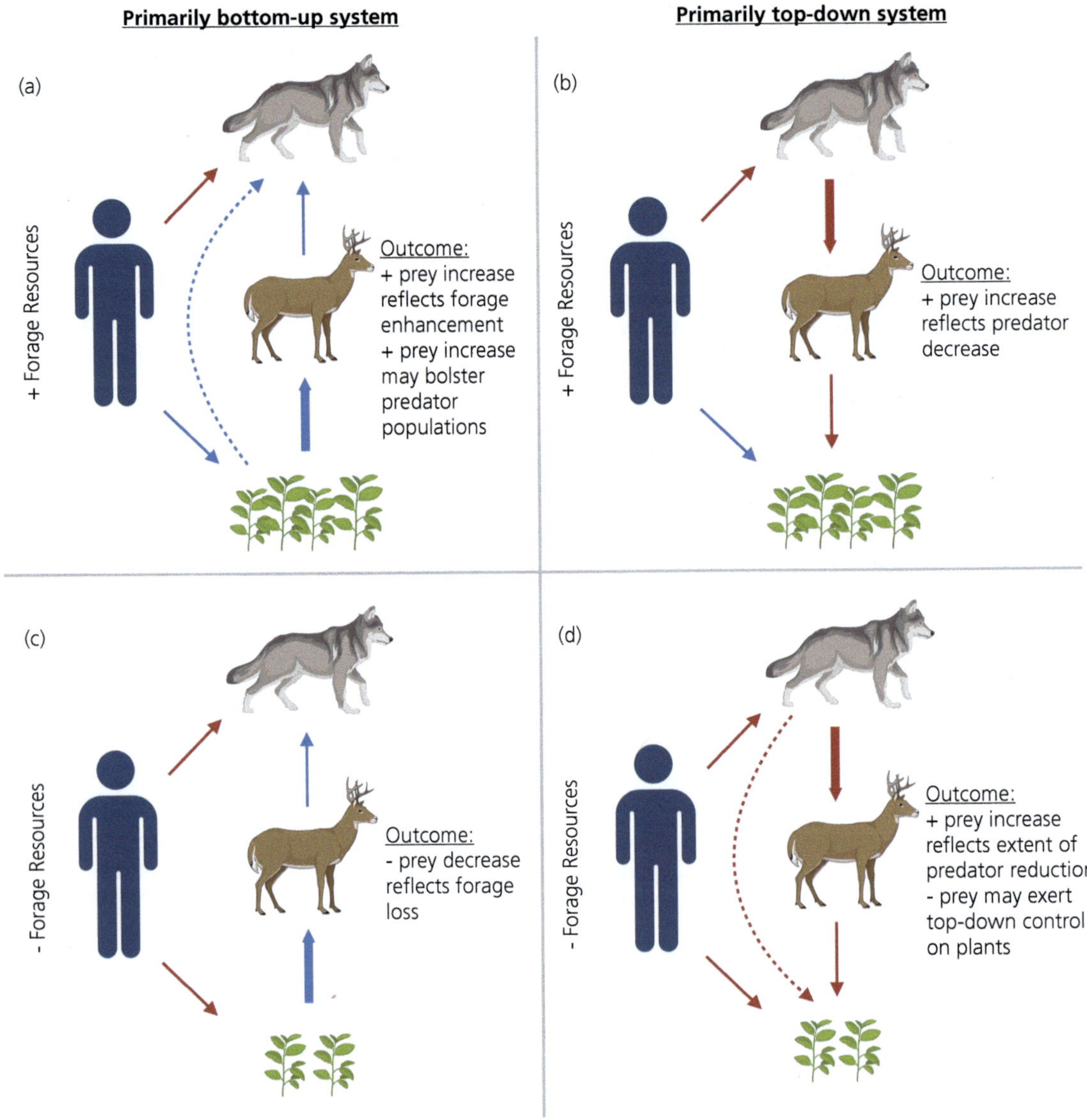

Figure 10.3 Human alterations of the forage landscape influence trophic relationships and can alter the implications of reduced predator densities. Solid lines show direct effects, dashed lines show indirect effects, positive effects are blue and negative are red. (a) In bottom-up driven systems where humans increase forage resources, predators may be better able to withstand persecution due to enhanced prey population growth. (b) In top-down driven systems where humans increase forage resources, prey populations may be especially likely to become overabundant when predator densities are reduced. (c) When human activities reduce forage resources in bottom-up systems, prey populations should decline, potentially amplifying the negative effects of humans on predators. (d) When humans reduce forage resources and predator densities in top-down systems, prey populations may still increase in response to relaxed predation, further reducing forage biomass.

Source: Created with BioRender.com

increase forage should increase herbivore abundance (Figure 10.3a). For example, timber harvest has increased forage availability for deer by creating early seral habitat (Hull et al. 2020, Hayes et al. 2022), leading to booming white-tailed deer populations in Alberta, Canada (Latham et al. 2011). McShea (2012) argued that deer populations across much of North America exceed historic densities owing to agricultural food subsidies coupled with the near-extirpation of apex predators. Notably, increases in prey abundance following forage enhancement can flow up the food chain and increase predator resilience to human persecution. In western Canada, high deer and moose (*Alces alces*) abundance in logged landscapes has increased wolf density, despite efforts to reduce wolf abundance via predator removal (Latham et al. 2011). Woodland caribou (*Rangifer tarandus caribou*) are threatened with extinction in these altered landscapes via a classic case of apparent competition, whereby elevated deer and moose populations support wolves and cause unsustainably high predation rates on caribou (Lamb et al. 2024). Likewise, Tasmanian devils (*Sarcophilus harrisii*) occur at their highest densities in human-modified landscapes because of the high prey availability therein (Cunningham et al. 2021). However, if anthropogenic activities increase food resources but the prey population remains primarily top-down controlled, prey response should generally reflect the reduction in predator density rather than changing forage dynamics (Figure 10.3b).

Humans can also reduce forage availability to prey directly, by converting suitable habitat to anthropogenic use or otherwise degrading habitat, and indirectly, by causing species to avoid areas of suitable forage (i.e. behaviourally mediated forage-loss; Dwinnell et al. 2019). In these cases, the net outcome for prey should depend on the relative strength of the effects of reduced forage and decreased predator density. Net effects would be deleterious to prey if the loss of forage outweighed a reduction in predator density, which would occur for bottom-up limited prey populations (Figure 10.3c). For instance, plains zebra (*Equus quagga*) in northern Kenya were five times less abundant when cattle ranching was the primary land use and predators were suppressed, relative to a period when land use shifted to conservation and ecotourism (Georgiadis et al. 2007). Georgiadis et al. (2007) determined that competition between zebras and cattle for forage had stronger negative effects on zebras than predation did when predator numbers were higher. In contrast, prey could benefit if the effects of predator declines exceed forage reduction, which would occur in primarily top-down limited systems (Figure 10.3d). In such cases, theoretical models suggest that herbivores could subsequently exert strong top-down control on plants (Ryall and Fahrig 2006).

Changes to habitat structure are likely to accompany, if not drive, changes in primary productivity, which should mediate interactions between predators and prey contingent on the traits of the interacting species (Janssen et al. 2007, Wirsing et al. 2021). If anthropogenic landscape changes increase the hunting efficiency of predators even as their densities are reduced, predation rates may not change. Australia offers an example of how habitat structure can affect predation rates. There, the increase in high-intensity fires following European arrival has reduced hiding cover for prey, resulting in high rates of feral cat (*Felis catus*) predation and threatening endemic small mammal populations with extinction (Leahy et al. 2016, Legge et al. 2023). In contrast, a reduction in hunting habitat or an increase in prey refugia coupled with a loss of predators could amplify the effects of predator loss on prey. Additionally, the spatial distribution of human impacts may influence predator–prey outcomes, as patchy food subsidies can create ecological traps by congregating prey where they may be more easily exploited by predators (Prevedello et al. 2013).

The dynamics discussed here provide a useful framework for predicting outcomes of interactions with simple three-level food chains (plants, herbivorous prey, and a single predator species) modified by humans, yet accounting for complexity beyond predator–prey dyads is necessary to predict the extent of trophic cascades in many anthropogenic systems (Montgomery et al. 2019). In food webs that include mesopredators, for instance, anthropogenic activities that affect productivity play important roles in mediating the strength of mesopredator release by altering prey availability (Litvaitis and Villafuerte 1996). In Sweden, the near-extirpation of lynx and wolves released red foxes (*Vulpes vulpes*) from suppression in high productivity regions converted to agriculture where prey became highly abundant (Elmhagen and Rushton 2007). However, mesopredator release was minimal in low productivity boreal and alpine regions where prey availability was lower (Elmhagen and Rushton 2007). These findings indicate that agricultural expansion coupled with predator persecution shifted the ecosystem from top-down to bottom-up control such that the strength of mesopredator release was inversely related to ecosystem productivity (Elmhagen and Rushton 2007, Elmhagen et al. 2010).

Human subsidies more broadly (e.g. refuse, intentional feeding, and domestic animals) can also alter food web dynamics among predators, with the strength of effects likely varying across urbanization gradients. As in the case of roadkill, human subsidies may increase intraguild predation by bringing predators into the same space for a concentrated food source (Manlick and Pauli 2020, Prugh and Sivy 2020). Mesopredator populations are often primarily regulated by intraguild predation from apex predators in more 'natural' systems such as national parks (Sivy et al. 2017, Jachowski et al. 2020), whereas urban mesopredator populations that lack apex predators generally have higher densities, especially in the presence of anthropogenic subsidies (Fedriani et al. 2001, Graser et al. 2012). The addition of anthropogenic subsidies may have a strong bottom-up effect on mesopredators in heavily modified areas and therefore relieve the effects of intraguild predation on and between mesopredator populations. Because human influences manifest across all trophic levels, accounting for the multiple pathways of these effects is necessary to fully understand the implications of apex predator decline. Investigations into the effects of anthropogenic predator decline should explicitly consider how related habitat modifications may mediate the effects of changes in predator density.

Scavenging dynamics and services

Interactions among carnivores involve a complex interplay between competition, intra-guild killing, and carrion provisioning (Ritchie and Johnson 2009, Allen et al. 2014, Sivy et al. 2017, Prugh and Sivy 2020). Because almost all carnivore species scavenge to some degree, carcasses provide essential resources but are also hotspots of dangerous interactions. Human-induced changes in predator density can thus affect ecosystems through changes in both the resource and risk landscapes.

Anthropogenic suppression of large carnivores causes shifts in the sources of carrion. Large carnivores provision ecosystems with substantial carrion subsidies, estimated on average at 1351 kg annually per large carnivore (Prugh and Sivy 2020). Thus, suppression of large carnivores results in fewer carrion inputs from predator kills, but proportionally more carrion inputs from natural mortality, roadkill, human hunting, and livestock (Newsome et al. 2015). These sources of carrion have different characteristics than predator-derived carrion. Natural mortality and roadkill, for example, enters the carrion pool intact, thus providing substantially more food for scavengers than is typical of predator kills, but this food may be difficult for some scavengers to access beneath the unopened hide. In contrast to the year-round nature of predator kills, carrion inputs from natural mortality, roadkill and human hunting are often aggregated in space and/or time (Wilmers et al. 2003b, Fielding et al. 2022). Following the reintroduction of wolves to Yellowstone National Park, for instance, carrion inputs shifted from a pulse of winter kill to a steady trickle, increasing the temporal consistency of carrion availability for scavengers (Wilmers et al. 2003a, Wilmers et al. 2003b, Walker et al. 2018). Similarly, mass mortality of kangaroos in Australia often occurs during droughts (Robertson 1986), and this pulse of carrion is likely further pronounced in areas where kangaroo densities have been vastly inflated by human suppression of dingoes (Letnic et al. 2012, Rees et al. 2019). The altered distribution and abundance of carcasses can have pervasive cascading effects by shifting nutrient pools and altering predator–prey interactions (Wilson and Wolkovich 2011, Johnson-Bice et al. 2023). For example, carcass sites in the Białowieża Forest, Poland, had increased encounter rates between red foxes and brown hares (*Lepus europaeus*), and the abundance of hares and red squirrels (*Sciurus vulgaris*) were lower at these sites (Cortes-Avizanda et al. 2009).

In addition to shifts in resources, changes in large carnivore densities also alter the risk landscape. Using wolf control in Alaska as a natural experiment, Sivy et al. (2017) found that small predators were attracted to wolf kills at a local scale, but were seemingly suppressed by wolves at a landscape scale. In Oregon, Ruprecht et al. (2021) showed that carrion provisioning by cougars had a dominant influence on coyote movements and resource acquisition, but cougars were also a large source of mortality, killing an estimated 23% of coyotes annually (Ruprecht et al. 2021). Even considering such high mortality of mesopredators, it is difficult to determine whether scavenging on large predator kills is a fitness-maximizing strategy because the optimal choice is a complex function of risk versus reward (evaluated with imperfect knowledge), in conjunction with the energetic state of the scavenger (Ruprecht et al. 2021). In locations where humans have reduced large carnivore abundance, scavenging should frequently be the optimal choice because carrion will often carry high rewards relative to risk (Figure 10.4). While perhaps not as risky as scavenging from large carnivore kills, scavenging from roadkill can increase the chance that the scavenger will become roadkill itself (Grilo et al. 2015, Planillo et al. 2018), and vehicular collisions can be an important source of mortality for carnivores (Benson et al. 2023, Moore et al. 2023). Humans may also intentionally poison carcasses to reduce predator populations (Ntemiri et al. 2018, Pauli et al. 2018).

Anthropogenic reductions to carnivore densities have the capacity to reduce the ecosystem service provided by carnivores in the form of carcass removal. Scavenger communities are less diverse in areas severely altered by humans, likely reducing their efficiency (Sebastián-González et al. 2019). The effect of carnivores on carcass removal rates will likely depend on their dietary preferences and population densities. While most carnivore species scavenge to some degree, there is wide interspecific variation in their propensities to hunt versus scavenge (Moleón et al. 2014, Pereira et al. 2014). Canids, for instance, are reported visiting carcasses threefold more often than felids, and their diets contain 15-fold more carrion than felids (Prugh and Sivy 2020). Species such as bears, Tasmanian devils, and hyenas likewise engage in relatively more scavenging than typical predators (Zager and Beecham 2006, Yarnell et al. 2013, Allen et al. 2015, Andersen

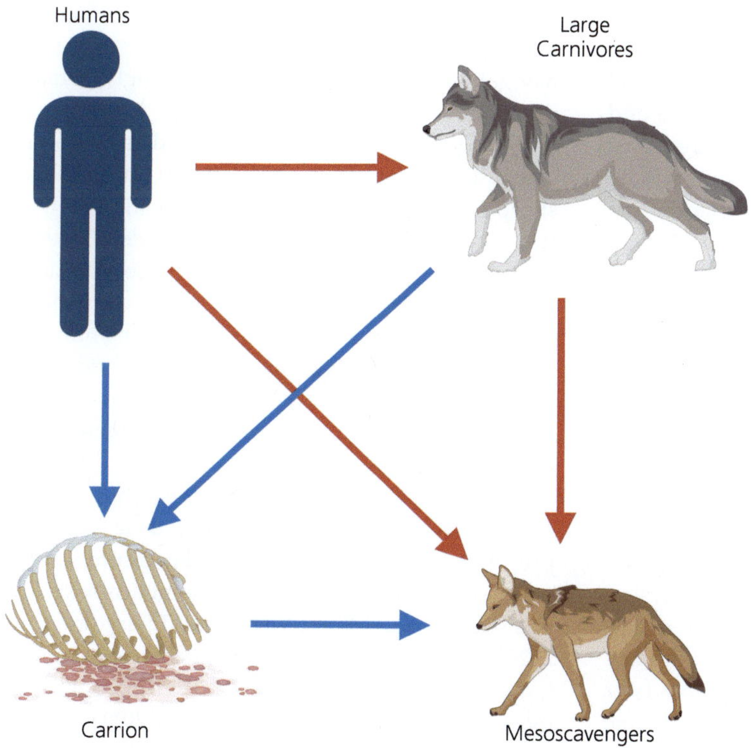

Figure 10.4 Humans influence scavenging dynamics through several main pathways (blue = positive, red = negative). First, by reducing large carnivore densities, humans reduce carrion inputs from large carnivores, but also reduce the risk of smaller scavengers encountering larger predators at carcasses. Second, humans provision ecosystems with different sources of carrion, including roadkill and remains from hunting, which likely benefit a subset of the scavenger community. The direct lethal effect of humans on mesoscavengers further complicates this web of interactions.

Source: Created with BioRender.com

et al. 2020) and can reach relatively high population densities (Mattson et al. 2005, Cunningham et al. 2021). Thus, they likely play important roles in efficiently removing carrion from the environment, and in turn, reducing reservoirs of disease (Sonawane et al. 2021). Tasmanian devil population declines, for example, led to a 3.1-fold increase in the persistence times of carcasses, in turn allowing smaller carnivores more scavenging opportunities (Cunningham et al. 2018, Fielding et al. 2022). The mesoscavenger release hypothesis posits that these findings hold more generally, with global declines of apex scavengers hypothesized to increase carcass persistence times and release mesoscavengers from resource limitation (O'Bryan et al. 2019).

In sum, human-induced changes in carnivore density influence both the risk and resource landscapes, leading to changes in interspecific interactions at carcasses that scale up to population-level effects, and leading to changes in scavenging dynamics and efficiency.

Climate change and cascading effects

The impacts of climate change on trophic dynamics can stem from two broad categories: directional climate change and extreme events (Coumou and Rahmstorf 2012, Harris et al. 2018). Directional change in the long-term patterns of temperature or precipitation eventually alters the distribution of species, ecosystems, and ecotones (Brito-Morales et al. 2018, Smith and Goetz 2021). In addition to sustained shifts, both observed and forecasted climate change trends show increased frequency and magnitude of extreme events, such as heatwaves, heavy precipitation, droughts, tropical storms, and combinations of these events (McPhillips et al. 2018, Walsh et al. 2020). Biodiversity can increase the resistance of communities to climate extremes (Isbell et al. 2015), and thus the loss of apex predators could reduce ecosystem resistance and resilience to climate change. Depending on local physical processes, directional changes and extreme events may manifest in opposite directions, and both categories of climate change impacts shape predator access to resources and the cascading effects of large carnivores.

Directional climate change has density-related effects that depend largely on local context, including climate change velocity and biome composition (Loarie et al. 2009, Brito-Morales et al. 2018). As abiotic constraints on primary productivity intensify (e.g. increased aridity and desertification; Huang et al. 2016) or are reduced (e.g. Arctic greening; Dial et al. 2022), distributions of key prey species will shift and change in extent. Although forage expansion and increased primary productivity can reduce bottom-up limitations on some herbivore populations, such as moose in Alaska (Tape et al. 2016, Zhou et al. 2020), whether these climate-driven boosts to prey population numbers result in increased carnivore densities have not been examined. A study of wolf denning phenology and success in northwestern North America found that the timing of denning did not change despite a 2-week advance in the start-of-season during the 18-year period (Mahoney et al. 2020). Although this finding raises the potential for a trophic mismatch with the timing of peak prey availability, wolf reproductive success did not decline during the study, indicating wolves may be resilient to shifting seasonality. Carnivores may not be as resilient to the shifting ranges of their prey, which can require navigation through novel and increasingly human-dominated habitats. For example, Aryal et al. (2014) documented a climate-induced reduction in vegetation productivity in Nepal that brought blue sheep (*Pseduois nayaur*) to lower elevations where they raided crops, and snow leopards (*Panthera uncia*) followed suit, which increased livestock depredations. A recent synthesis found compelling evidence that climate change is amplifying human–wildlife conflict worldwide due to increased resource scarcity leading to more frequent encounters, especially with carnivores (Abrahms et al. 2023).

Directional climate change may also shift the balance between apex predators and their ungulate prey by affecting hunting success. For example, seasonal snow cover presents an energetically costly landscape for wildlife to navigate (Ellington et al. 2020, Martin et al. 2020). Climate change is reducing the extent and duration of deep snow in most regions (Liston and Hiemstra 2011), which may reduce movement costs for large animals.

However, snow is also an integral driver of predation success for many snow-adapted carnivores (Peers et al. 2020). Foot-loading of carnivores is typically lower than that of ungulates, giving carnivores a movement advantage in deeper, softer snow (Sullender et al. 2023). Thus, a future climate with shallower, denser snow may generally favour ungulates and reduce carnivore hunting success. Additional anthropogenic pressure, such as lethal control or incidental take, may destabilize populations already facing significant pressure as a result of the changing climate (Wilson and Durner 2020). For example, long-term observations of wolves in Scandinavia indicate their recovery is hampered by a combination of poaching and low hunting success due to reduced snow cover (Sand et al. 2006, Ordiz et al. 2015).

In contrast to directional climate change, extreme weather events tend towards more density-independent effects such as mass mortality of wildlife (Fey et al. 2015). These events can affect multiple trophic levels simultaneously, such as the ecosystem-wide mortality caused by the megafires of 2020 in Australia (Legge et al. 2023). An intense drought in California from 2012 to 2015 likewise reduced abundances of species in all trophic levels, but carnivores exhibited the most severe declines (Prugh et al. 2018). Mass mortality of apex predators can lead to reduced predation rates and longer carrion persistence, releasing prey and scavenger populations from potential top-down regulation (Barton et al. 2022). Conversely, density-independent die-offs at lower trophic levels may provide a short-term resource pulse for scavenging carnivores, but will reduce their carrying capacity in the longer-term (Hansen et al. 2013). Ungulate die-offs are more likely to result from extreme climate events when their densities are near carrying capacity, which is most likely to occur when predators are absent or rare. For example, a predator control programme aiming to increase Dall sheep numbers in the Wrangell Mountains of Alaska was initially successful, but it ultimately resulted in a sheep population crash following a winter with deep, icy snow (Mitchell et al. 2015). In contrast, sheep did not decline in an adjacent site where predator numbers were not reduced. Reindeer (*Rangifer tarandus platyrhynchus*) on Svaalbard in the Norwegian high Arctic live in a predator-free environment, and their responses to rain-on-snow events are stronger when densities are high (Aanes et al. 2003, Pedersen et al. 2021). Thus, human-caused reductions of large carnivore populations can exacerbate the destabilizing effects of extreme climate events on wildlife communities.

Conclusion

Humans are often omitted from trophic networks, and including them reveals a seemingly intractable tangle of interactions due to the pervasive and multifaceted impacts that affect species within every trophic level. However, some general conclusions can be drawn from considering these complexities. First, when apex predator densities are reduced by humans, linear cascading effects should be weakened both by triangular interactions and by the direct effects of humans on lower trophic levels. Managing for diversity throughout food webs should thus increase the resistance of ecosystems to anthropogenic pressure. Second, these cascading effects should be strongest in productive systems, such as human-altered landscapes that provide more food for herbivores or mesopredators. Maintaining apex predators may thus be especially beneficial in such systems, because these mid-level species have strong potential to become overabundant when apex predators are removed. Third, carrion biomass should increase in abundance while becoming more patchily distributed in space and time where apex predator densities are reduced, which can reduce the efficiency of scavenging services and alter intraguild and predator–prey interactions. Fourth, climate change is amplifying human–carnivore conflicts and may also reduce the hunting success of carnivores in snowscapes. Both of these impacts should exacerbate carnivore declines, thereby increasing the strength of cascading effects. Holistic consideration of these multiple pathways, while daunting, will improve our understanding of anthropogenic impacts on trophic dynamics. Conservation and management strategies that account for these multiple pathways should be most effective in achieving positive outcomes for carnivores in a changing world.

References

Aanes, R., B.-E. Sæther, E. J. Solberg, S. Aanes, O. Strand, and N. A. Øritsland. 2003. Synchrony in Svalbard reindeer population dynamics. *Canadian Journal of Zoology* 81:103–110.

Abrahms, B., et al. 2023. Climate change as a global amplifier of human–wildlife conflict. *Nature Climate Change* 13:224–234.

Allen, M. L., L. M. Elbroch, C. C. Wilmers, and H. U. Wittmer. 2014. Trophic facilitation or limitation? Comparative effects of pumas and black bears on the scavenger community. *PLoS One* 9:e102257.

Allen, M. L., L. M. Elbroch, C. C. Wilmers, and H. U. Wittmer. 2015. The comparative effects of large carnivores on the acquisition of carrion by scavengers. *American Naturalist* 185:822–833.

Allombert, S., A. J. Gaston, and J. L. Martin. 2005. A natural experiment on the impact of overabundant deer on songbird populations. *Biological Conservation* 126:1–13.

Andersen, G. E., H. W. McGregor, C. N. Johnson, and M. E. Jones. 2020. Activity and social interactions in a wide-ranging specialist scavenger, the Tasmanian devil (Sarcophilus harrisii), revealed by animal-borne video collars. *PLoS One* 15:e0230216.

Arim, M. and P. A. Marquet. 2004. Intraguild predation: a widespread interaction related to species biology. *Ecology Letters* 7:557–564.

Aryal, A., D. Brunton, and D. Raubenheimer. 2014. Impact of climate change on human-wildlife-ecosystem interactions in the Trans-Himalaya region of Nepal. *Theoretical and Applied Climatology* 115:517–529.

Ballard, W. B., D. Lutz, T. W. Keegan, L. H. Carpenter, and J. C. deVos, Jr. 2001. Deer-predator relationships: a review of recent North American studies with emphasis on mule and black-tailed deer. *Wildlife Society Bulletin* 29:99–115.

Barton, P. S., A. Reboldi, S. Bonat, P. Mateo-Tomas, and T. M. Newsome. 2022. Climate-driven animal mass mortality events: is there a role for scavengers? *Environmental Conservation* 50(1): 1–6.

Benson, J. F., et al. 2023. The ecology of human-caused mortality for a protected large carnivore. *Proceedings of the National Academy of Sciences* 120:e2220030120.

Berger, J. 2007. Fear, human shields and the redistribution of prey and predators in protected areas. *Biology Letters* 3:620–623.

Beschta, R. L. and W. J. Ripple. 2016. Riparian vegetation recovery in Yellowstone: the first two decades after wolf reintroduction. *Biological Conservation* 198:93–103.

Beschta, R. L. and W. J. Ripple. 2020. Large carnivore extirpation linked to loss of overstory aspen in Yellowstone. *Food Webs* 22:e00140.

Brito-Morales, I., et al. 2018. Climate velocity can inform conservation in a warming world. *Trends in Ecology & Evolution* 33:441–457.

Chapron, G., et al. 2014. Recovery of large carnivores in Europe's modern human-dominated landscapes. *Science* 346:1517–1519.

Clark, T. J., and M. Hebblewhite. 2021. Predator control may not increase ungulate populations in the future: a formal meta-analysis. *Journal of Applied Ecology* 58: 812–824.

Cortes-Avizanda, A., N. Selva, M. Carrete, and J. A. Donazar. 2009. Effects of carrion resources on herbivore spatial distribution are mediated by facultative scavengers. *Basic and Applied Ecology* 10:265–272.

Cote, S. D., T. P. Rooney, J. P. Tremblay, C. Dussault, and D. M. Waller. 2004. Ecological impacts of deer overabundance. *Annual Review of Ecology Evolution and Systematics* 35:113–147.

Coumou, D. and S. Rahmstorf. 2012. A decade of weather extremes. *Nature Climate Change* 2:491–496.

Crooks, K. R. and M. E. Soulé. 1999. Mesopredator release and avifaunal extinctions in a fragmented system. *Nature* 400:563–566.

Cunningham, C. X., et al. 2021. Quantifying 25 years of disease-caused declines in Tasmanian devil populations: host density drives spatial pathogen spread. *Ecology Letters* 24:958–969.

Cunningham, C. X., C. N. Johnson, L. A. Barmuta, T. Hollings, E. J. Woehler, and M. E. Jones. 2018. Top carnivore decline has cascading effects on scavengers and carrion persistence. *Proceedings of the Royal Society B-Biological Sciences* 285:20181582.

Darimont, C. T., C. H. Fox, H. M. Bryan, and T. E. Reimchen. 2015. The unique ecology of human predators. *Science* 349:858–860.

DeCesare, N. J., M. Hebblewhite, H. S. Robinson, and M. Musiani. 2010. Endangered, apparently: the role of apparent competition in endangered species conservation. *Animal Conservation* 13:353–362.

Dial, R. J., C. T. Maher, R. E. Hewitt, and P. F. Sullivan. 2022. Sufficient conditions for rapid range expansion of a boreal conifer. *Nature* 608:546–551.

Dirzo, R., H. S. Young, M. Galetti, G. Ceballos, N. J. B. Isaac, and B. Collen. 2014. Defaunation in the Anthropocene. *Science* 345:401–406.

Dorresteijn, I., et al. 2015. Incorporating anthropogenic effects into trophic ecology: predator–prey interactions in a human-dominated landscape. *Proceedings of the Royal Society B: Biological Sciences* 282:20151602.

Durant, S. M., et al. 2017. The global decline of cheetah *Acinonyx jubatus* and what it means for conservation. *Proceedings of the National Academy of Sciences* 114: 528–533.

Dwinnell, S. P. H., H. Sawyer, J. E. Randall, J. L. Beck, J. S. Forbey, G. L. Fralick, and K. L. Monteith. 2019. Where to forage when afraid: does perceived risk impair use of the foodscape? Ecological Applications 29:e01972. https://doi.org/10.1002/eap.1972.

Ellington, E. H., E. M. Muntz, and S. D. Gehrt. 2020. Seasonal and daily shifts in behavior and resource selection: how a carnivore navigates costly landscapes. *Oecologia* 194:87–100.

Elmhagen, B., G. Ludwig, S. P. Rushton, P. Helle, and H. Linden. 2010. Top predators, mesopredators and their prey: interference ecosystems along bioclimatic productivity gradients. *Journal of Animal Ecology* 79:785–794.

Elmhagen, B. and S. P. Rushton. 2007. Trophic control of mesopredators in terrestrial ecosystems: top-down or bottom-up? *Ecology Letters* 10:197–206.

Estes, J. A., et al. 2011. Trophic downgrading of planet Earth. *Science* 333:301–306.

Fedriani, J. M., T. K. Fuller, and R. M. Sauvajot. 2001. Does availability of anthropogenic food enhance densities of omnivorous mammals? An example with coyotes in southern California. *Ecography* 24:325–331.

Fey, S. B., et al. 2015. Recent shifts in the occurrence, cause, and magnitude of animal mass mortality events. *Proceedings of the National Academy of Sciences of the United States of America* 112:1083–1088.

Fielding, M. W., et al. 2022. Dominant carnivore loss benefits native avian and invasive mammalian scavengers. *Proceedings of the Royal Society B: Biological Sciences* 289:20220521.

Forrester, T. D. and H. U. Wittmer. 2013. A review of the population dynamics of mule deer and black-tailed deer Odocoileus hemionus in North America. *Mammal Review* 43:292–308.

Ganz, T., et al. 2024. White-tailed deer population dynamics in a multi-predator landscape shaped by humans. *Ecological Applications* 34(5):e3003.

Georgiadis, N. J., F. Ihwagi, J. G. N. Olwero, and S. S. Romañach. 2007. Savanna herbivore dynamics in a livestock-dominated landscape. II: ecological, conservation, and management implications of predator restoration. *Biological Conservation* 137:473–483.

Gilbert, S. L., et al. 2017. Socioeconomic benefits of large carnivore recolonization through reduced wildlife-vehicle collisions. *Conservation Letters* 10:431–439.

Graser, W. H., S. D. Gehrt, L. L. Hungerford, and C. Anchor. 2012. Variation in demographic patterns and population structure of raccoons across an urban landscape. *Journal of Wildlife Management* 76:976–986.

Gray, T. N. E., et al. 2018. The wildlife snaring crisis: an insidious and pervasive threat to biodiversity in Southeast Asia. *Biodiversity and Conservation* 27: 1031–1037.

Grilo, C., D. J. Smith, and N. Klar. 2015. Carnivores: struggling for survival in roaded landscapes. In *Handbook of Road Ecology*, edited by R. van der Ree, D. J. Smith, and C. Grilo. John Wiley & Sons, New York, N. Y.

Hairston, N. G., F. E. Smith, and L. B. Slobodkin. 1960. Community structure, population control and competition. *American Naturalist* 94:421–425.

Hansen, B. B., et al. 2013. Climate events synchronize the dynamics of a resident vertebrate community in the high Arctic. *Science* 339:313–315.

Harris, R. M. B., et al. 2018. Biological responses to the press and pulse of climate trends and extreme events. *Nature Climate Change* 8:579–587.

Hayes, T. A., N. J. DeCesare, C. J. Peterson, C. J. Bishop, and M. S. Mitchell. 2022. Trade-offs in forest disturbance management for plant communities and ungulates. *Forest Ecology and Management* 506:119972.

Hobbs, N. T., D. B. Johnston, K. N. Marshall, E. C. Wolf, and D. J. Cooper. 2024. Does restoring apex predators to food webs restore ecosystems? Large carnivores in Yellowstone as a model system. *Ecological Monographs* 94(2):e1598.

Holt, R. D. and J. H. Lawton. 1994. The ecological consequences of shared natural enemies. *Annual Review of Ecology and Systematics* 25:495–520.

Huang, J. P., H. P. Yu, X. D. Guan, G. Y. Wang, and R. X. Guo. 2016. Accelerated dryland expansion under climate change. *Nature Climate Change* 6:166–171.

Hull, I. T., L. A. Shipley, S. L. Berry, C. Loggers, and T. R. Johnson. 2020. Effects of fuel reduction timber harvests on forage resources for deer in northeastern Washington. *Forest Ecology and Management* 458:117757.

Isbell, F., et al. 2015. Biodiversity increases the resistance of ecosystem productivity to climate extremes. *Nature* 526:574–U263.

Jachowski, D. S., A. Butler, R. Y. Y. Eng, L. Gigliotti, S. Harris, and A. Williams. 2020. Identifying mesopredator release in multi-predator systems: a review of evidence from North America. *Mammal Review* 50:367–381.

Janssen, A., M. W. Sabelis, S. Magalhães, M. Montserrat, and T. van der Hammen. 2007. Habitat structure affects intraguild predation. *Ecology* 88:2713–2719.

Johnson-Bice, S. M., T. D. Gable, J. D. Roth, and J. K. Bump. 2023. Patchy indirect effects of predation: predators contribute to landscape heterogeneity and ecosystem function via localized pathways. *Oikos* 2023(10):e10065.

Kuijper, D. P. J., et al. 2016. Paws without claws? Ecological effects of large carnivores in anthropogenic landscapes. *Proceedings of the Royal Society B: Biological Sciences* 283:20161625.

Lamb, C. T., et al. 2024. Effectiveness of population-based recovery actions for threatened southern mountain caribou. *Ecological Applications* 34(4):e2965.

Latham, A. D. M., M. C. Latham, N. A. Mccutchen, and S. Boutin. 2011. Invading white-tailed deer change wolf–caribou dynamics in northeastern Alberta. *The Journal of Wildlife Management* 75:204–212.

Leahy, L., S. M. Legge, K. Tuft, H. W. McGregor, L. A. Barmuta, M. E. Jones, and C. N. Johnson. 2016. Amplified predation after fire suppresses rodent populations in Australia's tropical savannas. *Wildlife Research* 42: 705–716.

Legge, S., L. Rumpff, S. T. Garnett, and J. C. Z. Woinarski. 2023. Loss of terrestrial biodiversity in Australia: magnitude, causation, and response. *Science* 381:622–631.

Letnic, M., E. G. Ritchie, and C. R. Dickman. 2012. Top predators as biodiversity regulators: the dingo *Canis lupus* dingo as a case study. *Biological Reviews* 87: 390–413.

Liston, G. E. and C. A. Hiemstra. 2011. The changing cryosphere: pan-Arctic snow trends (1979–2009). *Journal of Climate* 24:5691–5712.

Litvaitis, J. A. and R. Villafuerte. 1996. Intraguild predation, mesopredator release, and prey stability. *Conservation Biology* 10:676–677.

Loarie, S. R., P. B. Duffy, H. Hamilton, G. P. Asner, C. B. Field, and D. D. Ackerly. 2009. The velocity of climate change. *Nature* 462:1052–U1111.

Luck, G. W. 2007. The relationships between net primary productivity, human population density and species conservation. *Journal of Biogeography* 34:201–212.

Mahoney, P. J., et al. 2020. Denning phenology and reproductive success of wolves in response to climate signals. *Environmental Research Letters* 15(12):125001.

Manlick, P. J. and J. N. Pauli. 2020. Human disturbance increases trophic niche overlap in terrestrial carnivore communities. *Proceedings of the National Academy of Sciences* 117:26842–26848.

Martin, M. E., K. M. Moriarty, and J. N. Pauli. 2020. Forest structure and snow depth alter the movement patterns and subsequent expenditures of a forest carnivore, the Pacific marten. *Oikos* 129:356–366.

Mattson, D. J., S. Herrero, and T. Merrill. 2005. Are black bears a factor in the restoration of North American grizzly bear populations? *Ursus* 16:11–30.

McPhillips, L. E., et al. 2018. Defining extreme events: a cross-disciplinary review. *Earths Future* 6:441–455.

McShea, W. J. 2012. Ecology and management of white-tailed deer in a changing world. *Annals of the New York Academy of Sciences* 1249:45–56.

Melis, C., et al. 2009. Predation has a greater impact in less productive environments: variation in roe deer, *Capreolus capreolus*, population density across Europe. *Global Ecology and Biogeography* 18:724–734.

Merkle, J. A., H. S. Robinson, P. R. Krausman, and P. Alaback. 2013. Food availability and foraging near human developments by black bears. *Journal of Mammalogy* 94:378–385.

Middleton, A. D., et al. 2013. Grizzly bear predation links the loss of native trout to the demography of migratory elk in Yellowstone. *Proceedings of the Royal Society B: Biological Sciences* 280:20130870.

Wilmers, C. C., T. Levi, L. R. Prugh, J. Ruprecht, and D. R. Stahler. 2025. The ecological impacts of large-carnivore recovery in North America. *Annual Review of Ecology, Evolution, and Systematics* 56:337–363.

Mitchell, C. D., R. Chaney, K. Aho, J. G. Kie, and R. T. Bowyer. 2015. Population density of Dall's sheep in Alaska: effects of predator harvest? *Mammal Research* 60:21–28.

Moleón, M., J. A. Sánchez-Zapata, N. Selva, J. A. Donázar, and N. Owen-Smith. 2014. Inter-specific interactions linking predation and scavenging in terrestrial vertebrate assemblages. *Biological Reviews* 89: 1042–1054.

Montgomery, R. A., R. J. Moll, E. Say-Sallaz, M. Valeix, and L. R. Prugh. 2019. A tendency to simplify complex systems. *Biological Conservation* 233:1–11.

Moore, L. J., et al. 2023. Demographic effects of road mortality on mammalian populations: a systematic review. *Biological Reviews* 98:1033–1050.

Muhly, T. B., M. Hebblewhite, D. Paton, J. A. Pitt, M. S. Boyce, and M. Musiani. 2013. Humans strengthen bottom-up effects and weaken trophic cascades in a terrestrial food web. *PLoS One* 8:e64311.

Newbold, T., et al. 2020. Global effects of land use on biodiversity differ among functional groups. *Functional Ecology* 34:684–693.

Newsome, T. M., et al. 2015. The ecological effects of providing resource subsidies to predators. *Global Ecology and Biogeography* 24:1–11.

Novaro, A. J., M. C. Funes, and R. S. Walker. 2005. An empirical test of source-sink dynamics induced by hunting. *Journal of Applied Ecology* 42:910–920.

Ntemiri, K., et al. 2018. Animal mortality and illegal poison bait use in Greece. *Environmental Monitoring and Assessment* 190:488.

O'Bryan, C. J., M. H. Holden, and J. E. M. Watson. 2019. The mesoscavenger release hypothesis and implications for ecosystem and human well-being. *Ecology Letters* 22:1340–1348.

Ordiz, A., et al. 2015. Wolves, people, and brown bears influence the expansion of the recolonizing wolf population in Scandinavia. *Ecosphere* 6(12): 1–14.

Pauli, J. N., E. Donadio, and S. A. Lambertucci. 2018. The corrupted carnivore: how humans are rearranging the return of the carnivore-scavenger relationship. *Ecology* 99:2122–2124.

Pedersen, A. O., L. T. Beumer, R. Aanes, and B. B. Hansen. 2021. Sea or summit? Wild reindeer spatial responses to changing high-arctic winters. *Ecosphere* 12(12):e03883.

Peers, M. J., et al. 2020. Climate change increases predation risk for a keystone species of the boreal forest. *Nature Climate Change* 10(12):1149–1153.

Pereira, L. M., N. Owen-Smith, and M. Moleón. 2014. Facultative predation and scavenging by mammalian carnivores: seasonal, regional and intra-guild comparisons. *Mammal Review* 44:44–55.

Planillo, A., C. Mata, A. Manica, and J. E. Malo. 2018. Carnivore abundance near motorways related to prey and roadkills. *The Journal of Wildlife Management* 82:319–327.

Plumptre, A. J., et al. 2021. Where might we find ecologically intact communities? *Frontiers in Forests and Global Change* 4:626635.

Polis, G. A. and R. D. Holt. 1992. Intraguild predation—the dynamics of complex trophic interactions. *Trends in Ecology & Evolution* 7:151–154.

Polis, G. A. and D. R. Strong. 1996. Food web complexity and community dynamics. *American Naturalist* 147: 813–846.

Prevedello, J. A., C. R. Dickman, M. V. Vieira, and E. M. Vieira. 2013. Population responses of small mammals to food supply and predators: a global meta-analysis. *Journal of Animal Ecology* 82:927–936.

Prugh, L. R. and S. M. Arthur. 2015. Optimal predator management for mountain sheep conservation depends on the strength of mesopredator release. *Oikos* 124: 1241–1250.

Prugh, L. R., et al. 2023. Fear of large carnivores amplifies human-caused mortality for mesopredators. *Science* 380:754–758.

Prugh, L. R., et al. 2018. Ecological winners and losers of extreme drought in California. *Nature Climate Change* 8:819–824.

Prugh, L. R. and K. J. Sivy. 2020. Enemies with benefits: integrating positive and negative interactions among terrestrial carnivores. *Ecology Letters* 23:902–918.

Prugh, L. R., et al. 2009. The rise of the mesopredator. *BioScience* 59:779–791.

Raynor, J. L., C. A. Grainger, and D. P. Parker. 2021. Wolves make roadways safer, generating large economic returns to predator conservation. *Proceedings of the National Academy of Sciences of the United States of America* 118:e2023251118.

Rees, J. D., R. T. Kingsford, and M. Letnic. 2019. Changes in desert avifauna associated with the functional extinction of a terrestrial top predator. *Ecography* 42:67–76.

Rija, A. A., R. Critchlow, C. D. Thomas, and C. M. Beale. 2020. Global extent and drivers of mammal population declines in protected areas under illegal hunting pressure. *PLoS One* 15:e0227163.

Ripple, W. J., et al. 2016. What is a trophic cascade? *Trends in Ecology & Evolution* 31:842–849.

Ritchie, E. G. and C. N. Johnson. 2009. Predator interactions, mesopredator release and biodiversity conservation. *Ecology Letters* 12:982–998.

Robertson, G. 1986. The mortality of kangaroos in drought. *Australian Wildlife Research* 13:349–354.

Ruprecht, J., et al. 2021. Variable strategies to solve risk-reward tradeoffs in carnivore communities. *Proceedings of the National Academy of Sciences* 118:e2101614118.

Ryall, K. L. and L. Fahrig. 2006. Response of predators to loss and fragmentation of prey habitat: a review of theory. *Ecology* 87:1086–1093.

Sand, H., C. Wikenros, P. Wabakken, and O. Liberg. 2006. Effects of hunting group size, snow depth and age on the success of wolves hunting moose. *Animal Behaviour* 72(4):781–789.

Sebastián-González, E., et al. 2019. Scavenging in the Anthropocene: human impact drives vertebrate scavenger species richness at a global scale. *Global Change Biology* 25:3005–3017.

Sivy, K. J., C. B. Pozzanghera, J. B. Grace, and L. R. Prugh. 2017. Fatal attraction? Intraguild facilitation and suppression among predators. *The American Naturalist* 190:663–679.

Smith, A. J., and E. M. Goetz. 2021. Climate change drives increased directional movement of landscape ecotones. *Landscape Ecology* 36:3105–3116.

Sonawane, C., G. Yirga, and N. H. Carter. 2021. Public health and economic benefits of spotted hyenas Crocuta crocuta in a peri-urban system. *Journal of Applied Ecology* 58(12):2892–2902.

Sullender, B. K., C. X. Cunningham, J. D. Lundquist, and L. R. Prugh. 2023. Defining the danger zone: critical snow properties for predator–prey interactions. *Oikos* 2023(10):e09925.

Surve, N. S., S. Sathyakumar, K. Sankar, D. Jathanna, V. Gupta, and V. Athreya. 2022. Leopards in the city: the tale of Sanjay Gandhi National Park and Tungareshwar Wildlife Sanctuary, two protected areas in and adjacent to Mumbai, India. *Frontiers in Conservation Science* 3:787031.

Tape, K. D., D. D. Gustine, R. W. Ruess, L. G. Adams, and J. A. Clark. 2016. Range expansion of moose in Arctic Alaska linked to warming and increased shrub habitat. *PLoS One* 11:e0152636.

Thompson, R. M., M. Hemberg, B. M. Starzomski, and J. B. Shurin. 2007. Trophic levels and trophic tangles: the prevalence of omnivory in real food webs. *Ecology* 88:612–617.

Vucetich, J. A., D. W. Smith, and D. R. Stahler. 2005. Influence of harvest, climate and wolf predation on Yellowstone elk, 1961–2004. *Oikos* 111:259–270.

Walker, L. E., et al. 2018. Population responses of common ravens to reintroduced gray wolves. *Ecology and Evolution* 8:11158–11168.

Walsh, J. E., et al. 2020. Extreme weather and climate events in northern areas: a review. *Earth-Science Reviews* 209:103324.

Wilmers, C. C., R. L. Crabtree, D. W. Smith, K. M. Murphy, and W. M. Getz. 2003a. Trophic facilitation by introduced top predators: grey wolf subsidies to scavengers in Yellowstone National Park. *Journal of Animal Ecology* 72:909–916.

Wilmers, C. C., D. R. Stahler, R. L. Crabtree, D. W. Smith, and W. M. Getz. 2003b. Resource dispersion and consumer dominance: scavenging at wolf- and hunter-killed carcasses in Greater Yellowstone, USA. *Ecology Letters* 6:996–1003.

Wilson, E. E. and E. M. Wolkovich. 2011. Scavenging: how carnivores and carrion structure communities. *Trends in Ecology & Evolution* 26:129–135.

Wilson, R. R. and G. M. Durner. 2020. Seismic survey design and effects on maternal polar bear dens. *Journal of Wildlife Management* 84(2):201–212.

Wirsing, A. J., M. R. Heithaus, J. S. Brown, B. P. Kotler, and O. J. Schmitz. 2021. The context dependence of non-consumptive predator effects. *Ecology Letters* 24: 113–129.

Wolf, C. and W. J. Ripple. 2016. Prey depletion as a threat to the world's large carnivores. *Royal Society Open Science* 3:160252.

Wolf, C. and W. J. Ripple. 2017. Range contractions of the world's large carnivores. *Royal Society Open Science* 4:170052–170052.

Woodroffe, R. 2000. Predators and people: using human densities to interpret declines of large carnivores. *Animal Conservation* 3:165–173.

Woodroffe, R. and J. R. Ginsberg. 1998. Edge effects and the extinction of populations inside protected areas. *Science* 280:2126–2128.

Yarnell, R. W., et al. 2013. The influence of large predators on the feeding ecology of two African mesocarnivores: the black-backed jackal and the brown hyaena. *South African Journal of Wildlife Research* 43: 155–166.

Zager, P. and J. Beecham. 2006. The role of American black bears and brown bears as predators on ungulates in North America. *Ursus* 17:95–108.

Zhou, J., K. D. Tape, L. Prugh, G. Kofinas, G. Carroll, and K. Kielland. 2020. Enhanced shrub growth in the Arctic increases habitat connectivity for browsing herbivores. *Global Change Biology* 26:3809–3820.

CHAPTER 11

The Trait-Mediated Indirect Effects of Human Interaction

Kaggie D. Orrick, Mary K. Burak, and Oswald J. Schmitz

Introduction

> Among wild quadrupeds the hind appears to be pre-eminently intelligent; for example, in its habit of bringing forth its young on the sides of public roads, where the fear of man forbids the approach of wild animals.
>
> **Aristotle: *History of Animals Book IX***

The quoted insight from ancient philosophy and science, where prey use public roads to protect their young from predators, finds a remarkable parallel in the **human shield hypothesis** in contemporary conservation science, as famously observed in the case of moose (*Alces alces*) (Berger 2007). The human shield hypothesis occurs when human displacement of predators indirectly benefits prey species by reducing the perception of predation risk either spatially or temporally; effectively, decoupling predators from prey (Muhly et al. 2011). This common thread between ancient and current science underscores the intelligence of animals as they engage in an evolutionary ecological game known as the 'landscape of fear'. In this adaptive game, differing contexts of predator occurrence and hunting behaviour elicit different levels of anti-predator responses by prey that, in turn, can shape the demography, landscape distribution, and abundance of predators and prey (Brown et al. 1999, Schmitz 2005, Creel and Christianson 2008, Laundré et al. 2014). Originally suggested to describe predator–prey dynamics, the scientific community now recognizes humans as an additional player in this adaptive game, resulting in apex predators having to develop their own landscape of fear in response to human activity and potential anthropogenic mortality risk. Examples such as Aristotle's observations and Berger (2007) highlight how human actions, coupled with the built environment, can modify the landscape of fear and, thereby, affect predator–prey dynamics.

Humans, as such, are inherently intertwined members of an ecological community, participating in a complex three-way interaction module consisting of humans, predators, and prey (Berger 2007, Miller and Schmitz 2019). These interactions can have density-mediated or trait-mediated effects. **Density-mediated effects** occur when the impacts of a species in an ecosystem are modified by changes in the abundance of an intermediary species with which it directly interacts (see Chapter 10, this volume). **Trait-mediated effects** occur when the impacts of one species in an ecosystem are modified by changes in the phenotype (behaviour, morphology, physiology) of a second, intermediary species with which it interacts. The important distinction is that a density-mediated effect emerges when the local density of an intermediary species is lowered by direct *mortality* from predation. Within this context, a trait-mediated effect emerges when the local population density of an intermediary species changes via *spatial or temporal redistribution* across a landscape in response to a perception of direct *risk* of predation. When the secondary species with an altered phenotype then affects a third species, this becomes an indirect interaction, also known as a **trait-mediated indirect interaction**, or behavioural

Kaggie D. Orrick, Mary K. Burak, and Oswald J. Schmitz, *The Trait-Mediated Indirect Effects of Human Interaction*. In: *Apex Predators in the Anthropocene*. Edited by: Robert A. Montgomery and Remington J. Moll, Oxford University Press.
DOI: 10.1093/9780191996665.003.0011

indirect interaction (Abrams 1984, 1995, Miller and Kerfoot 1987). This top-down effect, which describes the indirect interactions between species within the same food web, is aptly called a trophic cascade (Ripple et al. 2016). Apex predators have top-down effects that have reverberating impacts throughout their ecological community (Ritchie and Johnson 2009, Ripple et al. 2014, Prugh and Sivy 2020).

Trait-mediated indirect effects caused by human agency and the built environment can be pervasive across landscapes, and the trophic cascade can be insidious. Thus, it is important that conservation practitioners are cognizant that trait-mediated indirect effects could unintentionally (Berger 2007) or intentionally (Miller and Schmitz 2019) emerge when conserving or restoring landscapes dominated by humans. In the case study by Berger (2007) regarding humans, brown bears (*Ursos arctos)*, and moose, the built environment (roads) led to fewer predators occurring near such landscape features (a negative direct effect of humans on predators), which in turn creates a predation refuge for those prey species (a positive indirect effect of humans on prey; Figure 11.1C. But the same type of built environment (e.g. infrastructure such as roads and linear features) can generate opposite outcomes. For instance, the construction of roads and associated clear-cutting opens access into habitat interiors thereby promoting ease of predator movement and increasing encounters with prey, such as with wolves (*Canis lupus*) and woodland caribou (*Rangifer tarandus*; Whittington et al. 2011; Leclerc et al. 2012; McKay et al. 2021; Figure 11.1C). It underlines the point that trait-mediated effects of anthropogenic disturbance (e.g. human presence, activity and land use, and infrastructure) can be highly context-dependent (Kuijper et al. 2016, Ordiz et al. 2021).

In this chapter, we review and summarize the indirect effects that humans have on ecological communities by exploring trait-mediated impacts on large carnivores. We first discuss the foundational literature on predator-driven trait-mediated indirect effects in ecosystems. We then outline the innumerable ways in which humans initiate indirect effects on community composition and functioning by causing trait changes in co-occurring predators. We review the current literature to provide examples and general takeaways from these studies. We then discuss how this work is currently understudied, and there are several additional, potential indirect, cascade effects on community composition that humans may instigate through predator trait-mediated effects. We conclude by highlighting the ways in which humans can act as novel apex predators within an ecosystem, recognizing that the trait-mediated effects humans have on predators have repercussions across ecosystem functioning and composition that are unique and context-specific (Lewis et al. 2017). Throughout, we highlight how humans can act as novel 'apex' predators, thereby impacting non-human apex predators and modifying their effects on prey. We attribute this context-dependency to the flexibility in trait expression as organisms adaptively respond to different environmental contexts (e.g. changes in resource quality or consumer pressure).

Background

The concept of trait-mediated effects emerged from early explorations of the link between animal behaviour and community ecology (Abrams 1984, Sih 1980), which introduced the idea that the mere perception of risk (or non-consumptive effect) of predation could determine the community ecological consequences of predator–prey interactions as much or more so than via classic predator consumptive effects. The idea that animal species ought to respond to perceived predation risk emerges from a general principle in evolutionary ecology that to maximize fitness, all prey species effectively must reconcile a trade-off between resource consumption and avoiding becoming resources of their predators. Numerous empirical syntheses (Lima and Dill 1990, Werner and Peacor 2003, Preisser et al. 2005, Peckarsky et al. 2008) have affirmed the ubiquity and importance of risk effects in predator–prey interactions, leading to the widely accepted idea that predators create a 'landscape of fear' for prey, in which non-consumptive risk effects can play an important hand in determining prey traits (demography, spatial and temporal movement behaviour, stress physiology) with attendant effects on the spatial distribution and abundance of animals across a landscape (Brown et al. 1999, Creel and Christianson 2008, Laundré et al. 2014).

The landscape of fear concept further highlights that there is considerable spatial and temporal variation in prey perception of predation risk depending on landscape features and habitat structure (Gaynor et al. 2018, 2019); in particular for large vertebrate predator effects on prey (Laundré et al. 2014, Schmitz et al. 2017, Bleicher 2017).

The landscape of fear concept is now being advanced as a key consideration in wildlife conservation and management to understand how human modifications to landscapes alter predator and prey spatial distribution and abundance (Atkins et al. 2017, Allen et al. 2019, Miller and Schmitz 2019). This stems from the rising number of studies showing that predators and prey may 'adapt' their behaviour (or other traits) to accommodate perceived risks from human-caused changes to landscapes due to, for example, human activity cycles, land use transformation, habitat alteration, and built infrastructure (Fardell et al. 2021, Wooster et al. 2022, Thapa et al. 2022, Moleón and Sánchez-Zapata 2023).

Human-induced predator trait-mediated effects on ecological communities

In this section, we present documented evidence of humans exerting trait-mediated influences on apex predators, leading to subsequent indirect repercussions on prey populations, landscapes, and ecosystems, more broadly. These trait-mediated indirect effects have the potential to propagate through ecosystems via various pathways. While the recognition of human-induced trait-mediated effects is becoming more prevalent in the scientific literature, the ecological ramifications within the trophic pyramid remain inadequately explored and documented (Gaynor et al. 2018, Hoeks et al. 2020,

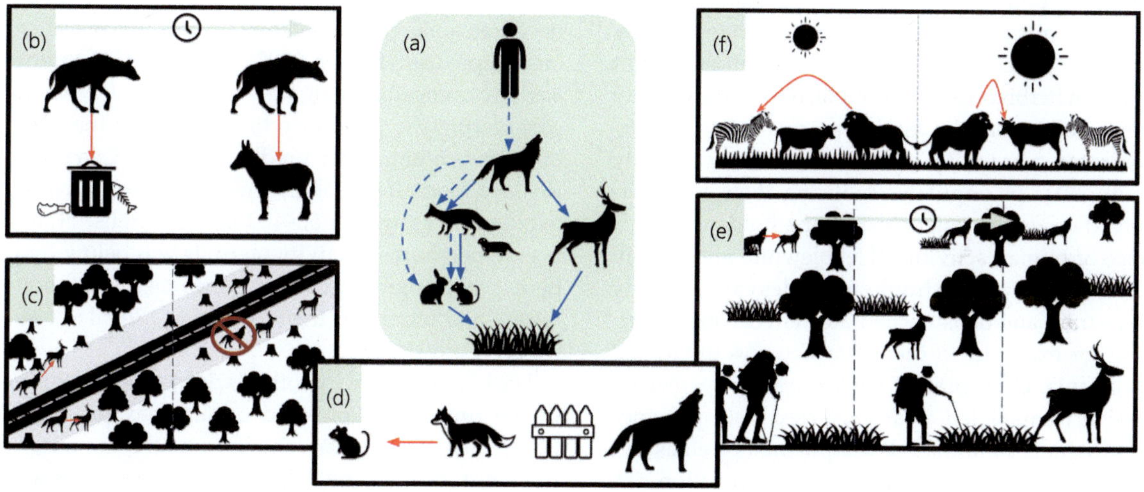

Figure 11.1 (a) Humans can indirectly affect predators, which have direct and cascading direct and indirect effects on the meso-predator's prey and vegetation in its environment. For example: (b) Human presence can increase the availability of food resources and consequently shift the diets of apex predators. In one study, hyenas temporally transitioned from scavenging meat-filled trash to preying upon available donkeys during various periods of human fasting (Yirga 2012); (c) Clear cutting of vegetation and heightened human activity along roads can either decrease predator presence, creating a shield effect for prey species, or it can increase hunting success for predators (Berger 2007, Whittington et al. 2011, Leclerc et al. 2012, McKay et al. 2021); (d) Predator-proof fences has led to meso-predator increases within fenced areas, having cascading effects on their prey (Doherty et al. 2015); (e) Intensity of human activity can alter the presence of predators and prey in an environment. For example, wolves have been found to avoid human hiking trails at all times, whereas prey take refuge from predators near trails at times when human activity is low (Rogala et al. 2011); and (f) Increased surface temperatures caused by climate change have contributed to energy inefficiencies among predators, leading to a behavioural shift in feeding choices. For example, lions have been found to switch prey preferences to livestock, which are easier to catch and thus less energy intensive (Schiess-Meier et al. 2007).

Source: Icons sourced from the following creators at Noun Project: Fence Icon, Aeliz; Rodent, Delwar Hossain; Hiker (both), Kangrif; Grass (1), Free Fair & Healthy, Grass (2), Melinda Courey; Deer, VectorBakery; Cow (1), Bakunetsu Kaito; Cow (2), parkjisun; Lion, Aleks; Sun, Lilik Sofiyanti; Fish bone, Gilda Martini; Butten chicken leg, Podgornaia Elena; Rabbit, David Khai; Tree, Milinda Courey; Hyena, Matthew Davis; Wolf, IronSV; Fox, Hafid Firman Syarif; Weasel, Victoruler; Donkey, Giulia; Zebra, T. Kiefer Robertson; Trash, Arif Fajar Yulianto

Wilson et al. 2020, Ordiz et al. 2021). There is empirical support indicating that human activities have, in certain cases, disrupted the traditional dynamics of predator–prey interactions by modifying or constraining a predator's capacity to induce behavioural-mediated effects in prey (Ordiz et al. 2013). While key principles can be deduced across studies, it is imperative to acknowledge that each ecological system has distinct contextual intricacies necessitating careful scrutiny.

Human shield hypothesis

As humans impact predators through a **landscape of fear (see Chapters 7–9, this volume)**, this can create the **human shield hypothesis**—also called a **landscape of refugia**—for other prey. It has been suggested that the separation of predator and prey species contributes to prey 'winning' the predator–prey 'space race' (Sih 2005). The *spatial* human shield hypothesis occurs when predators avoid humans, providing refugia near human activity for other species (Table 11.1). These refugia might result in herbivores, for example, remaining equally dispersed across the landscape regardless of concentrated areas with high human activity (Muhly et al. 2011) or ungulates reducing their own perceived predation risk by exploiting predator avoidance of human activities (Hebblewhite and Merrill 2007). Notably, these trophic cascades can indirectly have trait- and density-mediated effects on producers. For example, in the Santa Cruz Mountains of California, USA, pumas' *(Puma concolor)* perceived risk of humans led to an increased deer presence in proximity to human development as well as significantly increased foraging at low-predation risk sites (Yovovich et al. 2021). In the Caribbean, mongoose (*Herpestes javanicus*) avoidance of humans subsequently decreased their activity in nesting habitats for hawksbill sea turtles (*Eretmochelys imbricata*), leading to increased nest survival (Leighton et al. 2010). Linear infrastructure such as fences, roads, and pipelines disrupt wildlife movement and migration patterns. This is particularly true of large apex predators whose body size and long-distance movement rates increase their likelihood of, first, encountering such infrastructure and, second, being impeded by it. Linear infrastructure, particularly fences, can change species distributions, affecting both predator access to resources as well as prey distribution (McInturff et al. 2020, Wilkinson et al. 2023). For example, whereas human-modified landscapes have been found to have a higher presence of invasive meso-predators (e.g. feral cats, foxes), livestock fencing alternatively restricted meso-predator movement, which indirectly benefited the abundance of smaller meso-predators and prey (Doherty et al. 2015; Figure 11.1D). Additionally, Cherry et al. (2016) demonstrated how predator fear of linear infrastructure, such as roads, can induce behaviourally mediated trophic cascades in savanna ecosystems.

Sometimes, behaviourally mediated cascades only affect individuals in a population during certain life history stages. For example, in Yellowstone National Park, USA, calving female moose were found to give birth nearer to paved roads, which was interpreted as a behavioural decision to avoid traffic-averse brown bears, while mothers in bear-free areas or non-birthing calves did not alter landscape use (Berger 2007). Prey species might also dynamically navigate such trade-offs when considering the proximity to humans for the benefits of avoiding other predators. In the Canadian Rocky Mountains, wolves and elk (*Cervus elaphus*) avoided areas within 50–400 m from trails. However, when human activity on trails was below two people per hour, elk used this area as a predation refuge and increased their attraction (Rogala et al. 2011; Figure 11.1E).

The human shield hypothesis does not only apply to the prey being pursued by an apex predator. Linked to the human shield hypothesis, meso-predators can also be affected by changes in apex predator activity or spatial use, thereby resulting in **the meso-predator release hypothesis** (also known as the **meso-carnivore release hypothesis**). It originally hypothesized that density-mediated changes of apex predators—often focused upon apex predator removal or population declines—would alter the distribution, behaviour, or abundance of subordinate meso-predator, such as foxes (*Vulpes vulpes*) and coyotes (*Canis latrans*). In turn, this could have cascading effects on lower trophic levels (Soule et al. 1988, Ritchie et al. 2012, Moll et al. 2021). However, the meso-predator release hypothesis can also occur

Table 11.1 Trait-mediated indirect effects transmitted from humans to predators that influence ecological communities

CASCADE EXAMPLE	DESCRIPTION	CITATION
HUMAN SHIELD	Humans decreased habitat use by wolves, with cascading effects across the trophic pyramid.	(Hebblewhite et al. 2005)
	Moose prefer closer proximity to roads during calving season.	(Berger 2007)
	Human activity displaces predators but not prey.	(Muhly et al. 2011)
	Wolves avoided trails with human activity, while elk avoided trails with high human activity but did not avoid trails with low human activity.	(Rogala et al. 2011)
	Urban areas act as a refuge for foxes by displacing coyotes.	(Moll et al. 2018)
	Higher human activity decreased puma presence. This increased mule deer densities which decreased recruitment of riparian cottonwood trees, increased bank erosion and decreased abundance of terrestrial and aquatic species.	(Ripple and Beschta 2006)
	Deer increased activity levels near human developments to avoid puma presence. In turn, pumas had increased consumption in areas close to humans than comparable sites farther away.	(Yovovich et al. 2021)
	Elk selected for areas near human activity to decrease predation risk from wolves. Dependent on whether they were migrant elk or resident elk.	(Hebblewhite and Merrill 2007)
	Temporal avoidance of humans may facilitate human-wildlife coexistence, as well, but it also alters community interactions, beyond disrupting diel patterns of activity of affected species.	(Gaynor et al. 2018)
MESO-PREDATOR RELEASE	Dominant/generalist meso-predators encroach on niche space of subordinate meso-predators in areas of high human activity; resource use may be related to human-induced activity shifts.	(Smith et al. 2018)
	Literature review on carnivore niche partitioning; human disturbance can have contrasting effects.	(Sévêque et al. 2020)
	Human activity decreased wolf activity during the day. Meso-predators increased diurnal diel cycles.	(Frey et al. 2020)
	Recolonized wolves changed the temporal activity of coyote and bobcats.	(Shores et al. 2019)
ANTHROPOGENIC RESOURCES	Lions seasonally moved closer or further to livestock.	(Oriol-Cotterill et al. 2015)
	Brown bears decreased use of native forage (ants, forbs, berries) when there was access to sheep carcasses.	(Dahle et al. 1998)
	Spotted hyenas increased predation on donkeys when waste foods unavailable.	(Yirga et al. 2012)
	Human-provided food had a direct impact on predators and indirect impact on co-occurring species.	(Newsome et al. 2015)
WASTEFUL CONSUMPTION	Pumas made more frequent kills of deer in urbanized landscape due to fear of humans.	(Smith et al. 2015, 2017)

because of trait-mediated changes in apex predator behaviour. In Israel, apex predators were negatively affected by proximity to a scenic road heavily used by tourists, including high traffic volumes, leading to an increase in the meso-predator population (Shamoon et al. 2018). In the Canadian Rocky Mountains, human activity decreased apex predator activity during the day, leading to behavioural changes in meso-predator diel cycles (Frey et al. 2020). The human shield hypothesis has also been seen among small mammals in the Santa Cruz Mountains. An experimental playback study, for instance, demonstrated that human recordings had suppressive effects on pumas and medium-sized carnivores and increased small mammal abundance (Suraci et al. 2019).

Humans also impact apex predators' daily activity and can develop a *temporal* human shield hypothesis with trophic cascade effects (Gaynor et al. 2018). In the Bale Mountains National Park, Ethiopia, spotted hyena (*Crocuta crocuta*) occurred less frequently in areas closer to human settlements, leading to mountain nyala using the settlements as both a spatial refuge (relocating to human settlements closer than average) and a temporal refuge (coming closer to settlements during hyena hunting hours at night) (Atickem et al. 2014). Temporal human shields, including the trophic cascade effects, are less well studied than the effect of spatial human shields.

Feeding patterns and hunting success

Humans exhibit other trait-mediated effects on apex predators which can trigger cascade mechanisms. Stress emanating from the perception of risk of humans can decrease puma time spent feeding at a kill, lowing the energetic return per kill and requiring predators to kill more individuals (Smith et al. 2015, 2017). Pumas in the Santa Cruz Mountains of California were found to kill more frequently and spend less time at kills in residential areas (Smith et al. 2015) due to a risk-induced response from humans (Smith et al. 2017). Importantly, sometimes surplus killing by predators (a negative trait-mediated effect of human presence) is wrongly mistaken for optimal foraging theory (Stephens and Krebs 1986). Herein, predators may kill multiple prey if the prey are of poor quality and/or if predators have easy access to better prey (Lincoln and Quinn 2019).

Paradox of enrichment

Human-provided food, also referred to as human subsidies, can have a direct impact on predators and an indirect impact on co-occurring species (Newsome et al. 2015). Indirect effects can include changes in food sources, predation, and competition (Newsome et al. 2015). Changes in food sources include increases in depredation on livestock when native prey are not available (Valeix et al. 2012, Yirga et al. 2012) or decrease in native forage when livestock carcasses are available. The unintended creation of feral animals, such as donkeys in Death Valley, can also have an impact on predator diet and hunting strategies (Lundgren et al. 2022). Human-caused effects can be caused by cultural events. For example, in Ethiopia, spotted hyenas have demonstrated behavioural plasticity in response to human fasting periods, times of the year when there is a stark decline in human demand for meat, by scavenging urban waste or hunting donkeys (Yirga et al. 2012; Figure 11.1B).

Climate change

Although human-induced land use change has long been considered the most pressing threat to terrestrial predator survival, climate change is now rapidly surpassing this as one of the most widespread human-driven trait-mediated effects on predators. Rising annual mean temperatures can affect predator activity budgets (Bryce et al. 2022) and space use (Aryal et al. 2014), leading to trade-offs between quality and accessible resources. As a result, climate change could influence the occurrence and intensity of human-carnivore conflicts (Aryal et al. 2014, Abrahms 2021). In Botswana, climate change has reduced vegetation and prey availability, making it more difficult for lions (*Panthera leo*) to hunt and thus increasing the number of lion attacks on livestock (Schiess-Meier et al. 2007; Figure 11.1F). In response to lower levels of sea ice, polar bears *(Ursus maritimus)* have increased the amount of hunting that they do on land, resulting in higher human–wildlife conflict (Atwood et al. 2016). Ultimately, while there have been few studies that have addressed the potential effect of climate change on trait-mediated effects on large carnivores and its impact on community composition and functioning, the literature demonstrating climate change trait-mediated effects on animals has been reviewed elsewhere

(Knowlton and Graham 2010, Harmon and Barton 2013, Beever et al. 2017).

Landscape- and ecosystem-level considerations for trait-mediated indirect effects

Direct, interspecific interactions such as competition and predation impact ecological communities (Bolnick and Preisser 2005). However, trait-mediated effects of apex predators can generate equivalent, or even larger, effects on ecosystem functioning and composition. As highlighted throughout this book, there are innumerable ways in which carnivores affect other species within their ecosystem. In the previous section, we have demonstrated how human activities can induce trait changes in co-occurring predators that, in turn, have indirect effects on community composition and ecosystem function. However, a disproportionate amount of research has examined trait-based effects of humans on predators without subsequently examining potential cascading effects on community composition and ecosystem function (Lasky and Bombaci 2023).

There are also several examples in which human-induced trait-mediated effects on carnivores can also impact community composition and ecosystem functioning. These effects manifest in alterations to prey physiology (Zanette et al. 2014), modifications in prey foraging behaviour (Trussell et al. 2006), and shifts in prey demography (Preisser et al. 2007). Additionally, trait-mediated effects can extend to less-explored aspects of ecosystem functioning, such as the role of predators in seed dispersal, exemplified by bears (Ford et al. 2014) and pumas (Sarasola et al. 2016, Ordiz et al. 2021). Human disturbance can also alter movement patterns of numerous predators that have been linked to nutrient transfer in separate natural systems studies, including wolves (Ashenafi et al. 2005, Hebblewhite et al. 2005), pumas (Barry et al. 2019), and bears (Hilderbrand et al. 1999). In short, predators have several ways in which they impact the rest of the ecosystem and landscape. While the studies of the trait-mediated impacts of humans on predators have yet to identify these specific linkages, we can safely assume that these relationships exist and have yet to be discovered. Thus, we acknowledge this as an important area of future research inquiry.

In conclusion, trait-mediated effects in apex predators induced by humans could play a crucial role in shaping community dynamics and ecosystem functioning. Additionally, much of our knowledge about this has involved research within protected areas, where species that spatiotemporally overlap with humans do not face lethal effects. However, these behavioural responses can have different outcomes based on the spectrum of anthropogenic mortality risk (e.g. unprotected areas that do/not allow hunting). Relevant research should, first, identify the trait-mediated effect in play and, second, identify the landscape- or ecosystem-level process in question. Such research, ideally, requires spatial and temporal variation in the trait-mediated change to determine any correlation with the process being measured as well.

Towards a general conception of human-induced trait-mediated effects across landscape

Given the body of work that underscores the multifarious impacts of humans on predators, and how predators, in turn, impact community functioning and composition, it is difficult to predict how future human-induced trait-mediated indirect effects on wildlife could impact landscapes and ecosystems. Predictive models can help hypothesize cascade effects on trait-mediated shifts in predators (Becker and Hall 2014, Gil and Hein 2017). Notably, there is untapped potential in exploring pathways related to foraging and movement, which can lead to a multitude of consequences, such as nutrient cycling, primary production, and habitat modification (Wilson et al. 2020).

Humans hold some characteristics of an apex predator in particular ways for particular systems (Lennox et al. 2022), however they also provide unique effects that modify and impact the rest of the natural communities. Importantly, there is still much to be desired in understanding the full breadth and depth of how human trait-mediated effects on carnivores contribute to ecosystem trophic cascades. Without intentional studies using controlled experiments to study these trophic cascades, we cannot conclusively determine human impacts on ecosystem functioning and composition.

What is obvious is that, broadly, as a scientific community, it is unclear the breadth and depth that trait-mediated indirect effects are transmitted from humans to predators that influence ecological communities (Lennox et al. 2022). However, there is clear documentation of the role apex predators play in ecosystem function and composition (e.g. Miller et al. 2001; Davies et al. 2016; Atkins et al. 2019; Jiménez et al. 2019), as well as a clear demonstration that there are trait-mediated effects of humans on apex carnivores (**Chapters 7–9,** this volume). Further work needs to be done on these potentially cascading effects and system-wide trophic consequences (**Chapter 12, this volume**). There is also potential for some of these impacts to manifest or impact human–wildlife coexistence. The context of a landscape is critical, and historical, place-based histories and context should be considered (Gadsden et al. 2023, Orrick et al. 2024).

Importantly, it is critical to recognize the dynamic nature of human-induced risk in predators. Compared to other species, keystone species may have more detectable changes on an ecosystem (Wilson et al. 2020). Habituation to human presence can cause species to lose their risk of humans and shift their behaviour with subsequent impacts on trophic system function (Magle et al. 2014, Alldredge et al. 2019, Lasky and Bombaci 2023). Notably, there is a significant lack of research that has, as yet, explored the effect of gradients of human activity. Indeed, humans are also dynamic players in this framework. We also reiterate that human activities, such as land use changes, linear infrastructure, noise, pollution, and water alterations, can have far-reaching consequences for apex predators and the entire ecosystem. Understanding these impacts is essential for effective conservation and management strategies to maintain the balance and integrity of ecosystems. Future research should continue to explore these interactions, particularly in various land-use settings and human-dominated landscapes, to inform sustainable conservation practices.

Conclusion

Classical conservation posits that apex predators need vast areas to roam to pursue their biological nature and fulfil their functional roles in ecosystems. However, apex predators have been demonstrated to co-occur with humans, studies are increasingly demonstrating that apex predators do not necessarily require vast open spaces. But conservation practitioners and scientists must understand how humans can change an apex predator's autonomous ability to fulfil their functional roles in ecosystems (Riggio et al. 2018). Behaviour plasticity can act as a coping mechanism to human-induced change. Animal behaviour change might not behave linearly to human activity and remains context-dependent. Apex predators that represent keystone species—such as wolves—may cause faster changes in ecosystem function and composition than other species. As a coping mechanism, some behaviour changes might support viable populations to remain despite increased human disturbance. Wildlife and conservation managers may be able to use what we know about behavioural shifts to benefit human–wildlife coexistence and preserve ecosystem function (Miller and Schmitz 2019, Wilkinson et al. 2020, Wilson et al. 2020, Gaynor et al. 2021, p. 202).

References

Abrahms, B. 2021. Human-wildlife conflict under climate change. *Science* 373:484–485.

Abrams, P. A. 1984. Foraging time optimization and interactions in food webs. *The American Naturalist* 124:80–96.

Abrams, P. A. 1995. Implications of dynamically variable traits for identifying, classifying, and measuring direct and indirect effects in ecological communities. *The American Naturalist* 146:112–134.

Alldredge, M. W., F. E. Buderman, and K. A. Blecha. 2019. Human–cougar interactions in the wildland–urban interface of Colorado's front range. *Ecology and Evolution* 9:10415–10431.

Allen, B. L., et al. 2019. Animal welfare considerations for using large carnivores and guardian dogs as vertebrate biocontrol tools against other animals. *Biological Conservation* 232:258–270.

Aryal, A., D. Brunton, and D. Raubenheimer. 2014. Impact of climate change on human-wildlife-ecosystem interactions in the Trans-Himalaya region of Nepal. *Theoretical and Applied Climatology* 115:517–529.

Ashenafi, Z. T., T. Coulson, C. Sillero-Zubiri, and N. Leader-Williams. 2005. Behaviour and ecology of the Ethiopian wolf (*Canis simensis*) in a human-dominated landscape outside protected areas. *Animal Conservation* 8:113–121.

Atickem, A., L. E. Loe, and N. Chr. Stenseth. 2014. Individual heterogeneity in use of human shields by mountain nyala. *Ethology* 120:715–725.

Atkins, A., S. M. Redpath, R. M. Little, and A. Amar. 2017. Experimentally manipulating the landscape of fear to manage problem animals: experimentally manipulating the landscape of fear. *The Journal of Wildlife Management* 81:610–616.

Atkins, J. L., et al. 2019. Cascading impacts of large-carnivore extirpation in an African ecosystem. *Science* 364:173–177.

Atwood, T. C., et al. 2016. Rapid environmental change drives increased land use by an arctic marine predator. *PLOS ONE* 11:e0155932.

Barry, J. M., et al. 2019. Pumas as ecosystem engineers: ungulate carcasses support beetle assemblages in the Greater Yellowstone ecosystem. *Oecologia* 189: 577–586.

Becker, D. J., and R. J. Hall. 2014. Too much of a good thing: resource provisioning alters infectious disease dynamics in wildlife. *Biology Letters* 10:20140309.

Beever, E. A., et al. 2017. Behavioral flexibility as a mechanism for coping with climate change. *Frontiers in Ecology and the Environment* 15:299–308.

Berger, J. 2007. Fear, human shields and the redistribution of prey and predators in protected areas. *Biology Letters* 3:620–623.

Bleicher, S. S. 2017. The landscape of fear conceptual framework: definition and review of current applications and misuses. *Peer J* 5:e3772.

Bolnick, D. I., and E. L. Preisser. 2005. Resource competition modifies the strength of trait-mediated predator-prey interactions: a meta-analysis. *Ecology* 86:2771–2779.

Brown, J. S., J. W. Laundré, and M. Gurung. 1999. The ecology of fear: optimal foraging, game theory, and trophic interactions. *Journal of Mammalogy* 80:385–399.

Bryce, C. M., et al. 2022. Environmental correlates of activity and energetics in a wide-ranging social carnivore. *Animal Biotelemetry* 10:1.

Cherry, M. J., R. J. Warren, and L. Mike Conner. 2016. Fear, fire, and behaviorally mediated trophic cascades in a frequently burned savanna. *Forest Ecology and Management* 368:133–139.

Creel, S., and D. Christianson. 2008. Relationships between direct predation and risk effects. *Trends in Ecology & Evolution* 23:194–201.

Dahle, B., O. J. Sørensen, E. H. Wedul, J. E. Swenson, and F. Sandegren. 1998. The diet of brown bears *Ursus arctos* in central Scandinavia: effect of access to free-ranging domestic sheep *Ovis aries*. *Wildlife Biology* 4:147–158.

Davies, A. B., C. J. Tambling, G. I. H. Kerley, and G. P. Asner. 2016. Limited spatial response to direct predation risk by African herbivores following predator reintroduction. *Ecology and Evolution* 6:5728–5748.

Doherty, T. S., C. R. Dickman, D. G. Nimmo, and E. G. Ritchie. 2015. Multiple threats, or multiplying the threats? Interactions between invasive predators and other ecological disturbances. *Biological Conservation* 190:60–68.

Fardell, L. L., C. E. M. Nano, C. R. Pavey, and C. R. Dickman. 2021. Small prey animal habitat use in landscapes of fear: effects of predator presence and human activity along an urban disturbance gradient. *Frontiers in Ecology and Evolution* 9. DOI:10.3389/fevo.2021.750094.

Ford, A. T., et al. 2014. Large carnivores make savanna tree communities less thorny. *Science* 346(6207):346–349.

Frey, S., J. P. Volpe, N. A. Heim, J. Paczkowski, and J. T. Fisher. 2020. Move to nocturnality not a universal trend in carnivore species on disturbed landscapes. *Oikos* 129:1128–1140.

Gadsden, G. I., N. Golden, and N. C. Harris. 2023. Place-based bias in environmental scholarship derived from social–ecological landscapes of fear. *BioScience* 73: 23–35.

Gaynor, K. M., J. S. Brown, A. D. Middleton, M. E. Power, and J. S. Brashares. 2019. Landscapes of fear: spatial patterns of risk perception and response. *Trends in Ecology & Evolution* 34:355–368.

Gaynor, K. M., et al. 2021. An applied ecology of fear framework: linking theory to conservation practice. *Animal Conservation* 24:308–321.

Gaynor, K. M., C. E. Hojnowski, N. H. Carter, and J. S. Brashares. 2018. The influence of human disturbance on wildlife nocturnality. *Science* 360:1232–1235.

Gil, M. A., and A. M. Hein. 2017. Social interactions among grazing reef fish drive material flux in a coral reef ecosystem. *Proceedings of the National Academy of Sciences* 114:4703–4708.

Harmon, J. P., and B. T. Barton. 2013. On their best behavior: how animal behavior can help determine the combined effects of species interactions and climate change. *Annals of the New York Academy of Sciences* 1297:139–147.

Hebblewhite, M., and E. H. Merrill. 2007. Multiscale wolf predation risk for elk: does migration reduce risk?. *Oecologia* 152:377–387.

Hebblewhite, M., et al. 2005. Human activity mediates a trophic cascade caused by wolves. *Ecology* 86:2135–2144.

Hilderbrand, G. V., T. A. Hanley, C. T. Robbins, and C. C. Schwartz. 1999. Role of Brown Bears (*Ursus arctos*) in the flow of marine nitrogen into a terrestrial ecosystem. *Oecologia* 121:546–550.

Hoeks, S., M. A. J. Huijbregts, M. Busana, M. B. J. Harfoot, J.-C. Svenning, and L. Santini. 2020. Mechanistic insights into the role of large carnivores for ecosystem structure and functioning. *Ecography* 43: 1752–1763.

Jiménez, J., et al. 2019. Restoring apex predators can reduce mesopredator abundances. *Biological Conservation* 238:108234.

Knowlton, J. L., and C. H. Graham. 2010. Using behavioral landscape ecology to predict species responses to land-use and climate change. *Biological Conservation* 143:1342–1354.

Kuijper, D. P. J., et al. 2016. Paws without claws? Ecological effects of large carnivores in anthropogenic landscapes. *Proceedings of the Royal Society B: Biological Sciences* 283:20161625.

Lasky, M., and S. Bombaci. 2023. Human-induced fear in wildlife: a review. *Journal for Nature Conservation* 74:126448.

Laundré, J. W., et al. 2014. The landscape of fear: the missing link to understand top-down and bottom-up controls of prey abundance?. *Ecology* 95:1141–1152.

Leclerc, M., C. Dussault, and M.-H. St-Laurent. 2012. Multiscale assessment of the impacts of roads and cutovers on calving site selection in woodland caribou. *Forest Ecology and Management* 286:59–65.

Leighton, P. A., J. A. Horrocks, and D. L. Kramer. 2010. Conservation and the scarecrow effect: can human activity benefit threatened species by displacing predators?. *Biological Conservation* 143:2156–2163.

Lennox, R. J., et al. 2022. The roles of humans and apex predators in sustaining ecosystem structure and function: contrast, complementarity and coexistence. *People and Nature* 4:1071–1082.

Lewis, P.-M., G. L. Burns, and D. Jones. 2017. Response and responsibility: humans as apex predators and ethical actors in a changing societal environment. *Food Webs* 12:49–55.

Lima, S. L., and L. M. Dill. 1990. Behavioral decisions made under the risk of predation: a review and prospectus. *Canadian Journal of Zoology* 68:619–640.

Lincoln, A. E., and T. P. Quinn. 2019. Optimal foraging or surplus killing: selective consumption and discarding of salmon by Brown Bears. *Behavioral Ecology* 30:202–212.

Lundgren, E. J., et al. 2022. A novel trophic cascade between cougars and feral donkeys shapes desert wetlands. *Journal of Animal Ecology* 91:2348–2357.

Magle, S. B., L. S. Simoni, E. W. Lehrer, and J. S. Brown. 2014. Urban predator–prey association: coyote and deer distributions in the Chicago metropolitan area. *Urban Ecosystems* 17:875–891.

McInturff, A., W. Xu, C. E. Wilkinson, N. Dejid, and J. S. Brashares. 2020. Fence ecology: frameworks for understanding the ecological effects of fences. *BioScience* 70:971–985.

McKay, T. L., K. E. Pigeon, T. A. Larsen, and L. A. Finnegan. 2021. Close encounters of the fatal kind: landscape features associated with central mountain caribou mortalities. *Ecology and Evolution* 11: 2234–2248.

Miller, B., B. Dugelby, and D. Foreman. 2001. The importance of large carnivores to healthy ecosystems. *Endangered Species Update* 18:202–210.

Miller, J. R. B., and O. J. Schmitz. 2019. Landscape of fear and human-predator coexistence: applying spatial predator-prey interaction theory to understand and reduce carnivore-livestock conflict. *Biological Conservation* 236:464–473.

Miller, T. E., and W. Kerfoot. 1987. *Redefining Indirect Effects. Predation: Direct and Indirect Impacts on Aquatic Communities*. University Press of New England, Hanover, New Hampshire, USA: 33–37.

Moleón, M., and J. A. Sánchez-Zapata. 2023. Extending the dynamic landscape of fear in a human-dominated world. *Trends in Ecology & Evolution* 38:215–216.

Moll, R. J., et al. 2018. Humans and urban development mediate the sympatry of competing carnivores. *Urban Ecosystems* 21:765–778.

Moll, R. J., et al. 2021. An apex carnivore's life history mediates a predator cascade. *Oecologia* 196:223–234.

Muhly, T. B., C. Semeniuk, A. Massolo, L. Hickman, and M. Musiani. 2011. Human activity helps prey win the predator-prey space race. *PLoS ONE* 6:e17050.

Newsome, T. M., et al. 2015. The ecological effects of providing resource subsidies to predators. *Global Ecology and Biogeography* 24:1–11.

Ordiz, A., M. Aronsson, J. Persson, O.-G. Støen, J. E. Swenson, and J. Kindberg. 2021. Effects of human disturbance on terrestrial apex predators. *Diversity* 13:68.

Ordiz, A., R. Bischof, and J. E. Swenson. 2013. Saving large carnivores, but losing the apex predator?. *Biological Conservation* 168:128–133.

Oriol-Cotterill, A., D. W. Macdonald, M. Valeix, S. Ekwanga, and L. G. Frank. 2015. Spatiotemporal patterns of lion space use in a human-dominated landscape. *Animal Behaviour* 101:27–39.

Orrick, K., M. Dove, and O. J. Schmitz. 2024. Human–nature relationships: an introduction to social–ecological practice theory for human–wildlife interactions. *Ambio* 53:201–211.

Peckarsky, B. L., et al. 2008. Revisiting the classics: considering nonconsumptive effects in textbook examples of predator-prey interactions. *Ecology* 89:2416–2425.

Preisser, E. L., D. I. Bolnick, and M. F. Benard. 2005. Scared to death? The effects of intimidation and consumption in predator-prey interactions. *Ecology* 86:501–509.

Preisser, E. L., J. L. Orrock, and O. J. Schmitz. 2007. Predator hunting mode and habitat domain alter nonconsumptive effects in predator-prey interactions. *Ecology* 88:2744–2751.

Prugh, L. R., and K. J. Sivy. 2020. Enemies with benefits: integrating positive and negative interactions among terrestrial carnivores. *Ecology Letters* 23:902–918.

Riggio, J., H. Kija, E. Masenga, F. Mbwilo, F. Van de Perre, and T. Caro. 2018. Sensitivity of Africa's larger mammals to humans. *Journal for Nature Conservation* 43:136–145.

Ripple, W. J., and R. L. Beschta. 2006. Linking a cougar decline, trophic cascade, and catastrophic regime shift in Zion National Park. *Biological Conservation* 133: 397–408.

Ripple, W. J., et al. 2014. Status and ecological effects of the world's largest carnivores. *Science* 343:1241484.

Ripple, W. J., et al. 2016. What is a trophic cascade?. *Trends in Ecology & Evolution* 31:842–849.

Ritchie, E. G., B. Elmhagen, A. S. Glen, M. Letnic, G. Ludwig, and R. A. McDonald. 2012. Ecosystem restoration with teeth: what role for predators?. *Trends in Ecology & Evolution* 27:265–271.

Ritchie, E. G., and C. N. Johnson. 2009. Predator interactions, mesopredator release and biodiversity conservation. *Ecology Letters* 12:982–998.

Rogala, J. K., M. Hebblewhite, J. Whittington, C. A. White, J. Coleshill, and M. Musiani. 2011. Human activity differentially redistributes large mammals in the Canadian Rockies National Parks. *Ecology and Society* 16:art16.

Sarasola, J. H., J. I. Zanón-Martínez, A. S. Costán, and W. J. Ripple. 2016. Hypercarnivorous apex predator could provide ecosystem services by dispersing seeds. *Scientific Reports* 6:19647.

Schiess-Meier, M., S. Ramsauer, T. Gabanapelo, and B. König. 2007. Livestock predation—insights from problem animal control registers in Botswana. *The Journal of Wildlife Management* 71:1267–1274.

Schmitz, O. J. 2005. Behavior of predators and prey and links with population-level processes. In *Ecology of Predator-Prey Interactions*, edited by P. Barbosa and I. Castellanos. Oxford University Press, Oxford, U. K.

Schmitz, O. J., J. R. B. Miller, A. M. Trainor, and B. Abrahms. 2017. Toward a community ecology of landscapes: predicting multiple predator-prey interactions across geographic space. *Ecology* 98:2281–2292.

Sévêque, A., L. K. Gentle, J. V. López-Bao, R. W. Yarnell, and A. Uzal. 2020. Human disturbance has contrasting effects on niche partitioning within carnivore communities. *Biological Reviews* 95:1689–1705.

Shamoon, H., S. Cain, U. Shanas, A. Bar-Massada, Y. Malihi, and I. Shapira. 2018. Spatio-temporal activity patterns of mammals in an agro-ecological mosaic with seasonal recreation activities. *European Journal of Wildlife Research* 64:35.

Shores, C. R., J. A. Dellinger, E. S. Newkirk, S. M. Kachel, and A. J. Wirsing. 2019. Mesopredators change temporal activity in response to a recolonizing apex predator. *Behavioral Ecology* 30:1324–1335.

Sih, A. 1980. Optimal behavior: can foragers balance two conflicting demands?. *Science* 210(4473):1041–1043.

Sih, A. 2005. Predator-prey space use as an emergent outcome of a behavioral response race. *Ecology of Predator-Prey Interactions* 256:78.

Smith, F. A., R. E. E. Smith, S. K. Lyons, and J. L. Payne. 2018. Body size downgrading of mammals over the late quaternary. *Science* 360:310–313.

Smith, J. A., et al. 2017. Fear of the human 'super predator' reduces feeding time in large carnivores. *Proceedings of the Royal Society B: Biological Sciences* 284:20170433.

Smith, J. A., Y. Wang, and C. C. Wilmers. 2015. Top carnivores increase their kill rates on prey as a response to human-induced fear. *Proceedings of the Royal Society B: Biological Sciences* 282:20142711.

Soule, M. E., D. T. Bolger, A. C. Alberts, J. Wrights, M. Sorice, and S. Hill. 1988. Reconstructed dynamics of rapid extinctions of chaparral-requiring birds in urban habitat islands. *Conservation Biology* 2:75–92.

Stephens, D. W., and J. R. Krebs. 1986. *Foraging Theory*. Princeton University Press, Princeton, N. J.

Suraci, J. P., M. Clinchy, L. Y. Zanette, and C. C. Wilmers. 2019. Fear of humans as apex predators has landscape-scale impacts from mountain lions to mice. *Ecology Letters* 22:1578–1586.

Thapa, S. K., J. F. de Jong, A. R. Hof, N. Subedi, Y. Liefting, and H. H. T. Prins. 2022, 22 September. *Integration of the Landscape of Fear Concept in Terrain Management—An Experimental Study on Deer in Bardia National Park, Nepal*. Research Square.

Trussell, G. C., P. J. Ewanchuk, and C. M. Matassa. 2006. Habitat effects on the relative importance of trait- and density-mediated indirect interactions. *Ecology Letters* 9:1245–1252.

Valeix, M., G. Hemson, A. J. Loveridge, G. Mills, and D. W. Macdonald. 2012. Behavioural adjustments of a large carnivore to access secondary prey in a human-dominated landscape: wild prey, livestock and lion ecology. *Journal of Applied Ecology* 49:73–81.

Werner, E. E., and S. D. Peacor. 2003. A review of trait-mediated indirect interactions in ecological communities. *Ecology* 84:1083–1100.

Whittington, J., et al. 2011. Caribou encounters with wolves increase near roads and trails: a time-to-event approach. *Journal of Applied Ecology* 48:1535–1542.

Wilkinson, C. E., P. F. Jones, and A. F. Jakes. 2023. Editorial: disentangling the complexity of fence effects on wildlife and ecosystems. *Frontiers in Conservation Science* 4. DOI:10.3389/fcosc.2023.1147486.

Wilkinson, C. E., et al. 2020. An ecological framework for contextualizing carnivore–livestock conflict. *Conservation Biology* 34:854–867.

Wilson, M. W., A. D. Ridlon, K. M. Gaynor, S. D. Gaines, A. C. Stier, and B. S. Halpern. 2020. Ecological impacts of human-induced animal behaviour change. *Ecology Letters* 23:1522–1536.

Wooster, E. I. F., et al. 2022. Predator protection dampens the landscape of fear. *Oikos* 2022:e09059.

Yirga, G., H. H. De Iongh, H. Leirs, K. Gebrihiwot, J. Deckers, and H. Bauer. 2012. Adaptability of large carnivores to changing anthropogenic food sources: diet change of Spotted Hyena (*Crocuta crocuta*) during Christian fasting period in northern Ethiopia: adaptability of large carnivores to changing anthropogenic food sources. *Journal of Animal Ecology* 81:1052–1055.

Yovovich, V., M. Thomsen, and C. C. Wilmers. 2021. Pumas' fear of humans precipitates changes in plant architecture. *Ecosphere* 12:e03309.

Zanette, L. Y., M. Clinchy, and J. P. Suraci. 2014. Diagnosing predation risk effects on demography: can measuring physiology provide the means?. *Oecologia* 176: 637–651.

CHAPTER 12

The Consequences of Trophic Downgrading, and the Promise of Trophic Upgrading

Kaitlyn M. Gaynor and Nicholas S. Wright

This chapter is dedicated to Jim Estes (1945–2025), whose pioneering research on trophic cascades and the consequences of apex predator loss transformed our understanding of ecosystems and their conservation.

Introduction

Apex consumers play a key role in ecosystems through the top-down regulation of food webs, controlling the biomass of other species through the combined effects of predation and predation risk (Terborgh and Estes 2013). Apex consumer loss can therefore have profound ecological consequences. In 2011, Estes and colleagues introduced the concept of 'trophic downgrading', which they defined as 'removing large apex consumers from nature' (Estes et al. 2011). Their paper highlighted the pathways through which apex consumer loss transforms ecosystems, and has since shaped the way that ecologists understand ecosystem functioning in the Anthropocene and informed approaches to conservation and restoration.

The loss of apex consumers, which include large carnivores, is a key feature of the human-driven sixth mass extinction (Cowie et al. 2022). Large carnivores are disproportionately affected by human disturbance, as their lower densities make populations more vulnerable, and their dietary requirements and large home ranges bring them into contact and conflict with people (Ripple et al. 2014, Ingeman et al. 2022). As explored in Section 2 of this book, there are myriad drivers of large carnivore declines and extirpation, including direct persecution, utilization, habitat loss, and depletion of prey (Ripple et al. 2014). Given that large carnivores often play important roles as keystone species, their loss is of particular concern for the conservation of entire ecosystems (Worm and Paine 2016).

Since the publication of the foundational trophic downgrading paper in 2011, there has been growing evidence of the effects of apex consumer removal on ecosystem functioning. There has also been significant debate within the scientific community about the extent to which downgrading fundamentally transforms ecosystems, the contexts in which downgrading is most impactful, and the ability to reverse the effects of downgrading through carnivore reintroductions (Allen et al. 2017). The outsized conservation implications of these debates warrant a critical evaluation of the evidence and an appreciation for the complexity and context-dependency of trophic downgrading and its effects.

In this chapter, we review the consequences of trophic downgrading for biotic, abiotic, and socioecological processes, and examine how the broader ecological context mediates these effects of apex consumer loss, with a focus on large carnivores. We also explore the potential to 'upgrade' systems through the restoration of large carnivores,

Kaitlyn M. Gaynor and Nicholas S. Wright, *The Consequences of Trophic Downgrading, and the Promise of Trophic Upgrading*.
In: *Apex Predators in the Anthropocene*. Edited by: Robert A. Montgomery and Remington J. Moll, Oxford University Press.
 DOI: 10.1093/9780191996665.003.0012

Box 12.1 Sea otters in the North Atlantic

The iconic relationship between sea otters (*Enhydra lutris*), sea urchins (*Strongylocentrotus spp.*), and kelp forests has been central to ecologists' understanding of trophic downgrading. After years in the Aleutian Islands, James Estes recognized that islands where sea otters were extirpated for the fur trade were often the same islands where sea urchins overgrazed kelp forests, creating desolate urchin barrens (Estes and Palmisano 1974). Subsequent research revealed the role of sea otters as a keystone species in coastal ecosystems: by suppressing urchin populations, otters protect kelp forests and the biodiverse communities they harbour.

Although this quintessential trophic cascade appears simple, researchers have continued to uncover its complexity. The hunting-induced collapse of sea otters likely contributed to the extinction of one of the great marine megafauna species (Estes et al. 2016). Massive post-downgrading declines in kelp may have spelt the end for Stellar's sea cows (*Hydrodamalis gigas*), which grazed on kelp, in the eighteenth century. Later, in the 1980s, sea otters again declined precipitously, driven by an increase in predation by orcas (*Orcinus orca*) after the population collapse of their preferred pinniped prey (Estes et al. 1998). Similar to the effects of hunting, orca suppression of sea otters led to rising urchin populations and dramatic reductions in kelp density (Estes et al. 1998). Recent work has linked the sea otter cascade to another apex consumer: after depressing populations of their black-tailed deer (*Odocoileus hemionus*) prey, grey wolves (*Canis lupus*) in Southeast Alaska turned to sea otters (Roffler et al. 2023). This marine subsidy supported high densities of wolves, and in an example of apparent competition, wolves almost hunted the deer to local extinction. This predation of otters by both orcas and wolves reveals the powerful effects of adding a fourth link to a trophic cascade, and highlights trophic links between oceanic, coastal, and terrestrial systems.

Over the last 30 years, sea otters have moved into salt marsh and seagrass ecosystems (Silliman et al. 2018). In California, sea otters benefit seagrass through a four-level cascade in which otters suppress crabs, releasing populations of grazing invertebrates that clean algae off of seagrass and facilitate its growth (Hughes et al. 2013). Similar studies in more biodiverse ecosystems in British Columbia (Hessing-Lewis et al. 2018) and Alaska (Raymond et al. 2021) found weaker effects of sea otters on seagrass biomass, hinting at the importance of context-dependency in determining the effects of trophic downgrading.

and the limitations to such approaches. Throughout, we draw on case studies to highlight the nuances of trophic downgrading and its effects, and the challenges of studying complex and changing ecosystems.

The ecological consequences of trophic downgrading

Trophic downgrading often leads directly to changes in the densities and behaviours of other species that interact with the apex consumer (see Chapters 10, 11, this volume), including herbivores and meso-predators (Prugh et al. 2009, Ritchie and Johnson 2009, Suraci et al. 2016). Indirect interactions known as trophic cascades link the fates of species across trophic levels (Ripple et al. 2016), and the effects of predator loss can therefore ripple throughout food webs. Consequently, 'downgraded' ecosystems may look very different from those with large carnivores, and even shift into alternative stable states (Estes et al. 2011). We draw on evidence from models, experiments, and observational studies to explore how trophic cascades altered by predator loss or declines alter biotic, abiotic, and socioecological processes (see Figure 12.1). While we focus here on studies that have documented downgrading effects, they represent a non-exhaustive selection of examples, rather than general rules. At the end of the section, we discuss how the broader ecological context shapes the downstream effects (or lack thereof) of trophic downgrading.

Biotic processes

Herbivory

Where predators exert top-down control on food webs, predation changes the density, community composition, and traits of herbivores, with consequences for plant communities. Trophic downgrading can alter density-mediated trophic

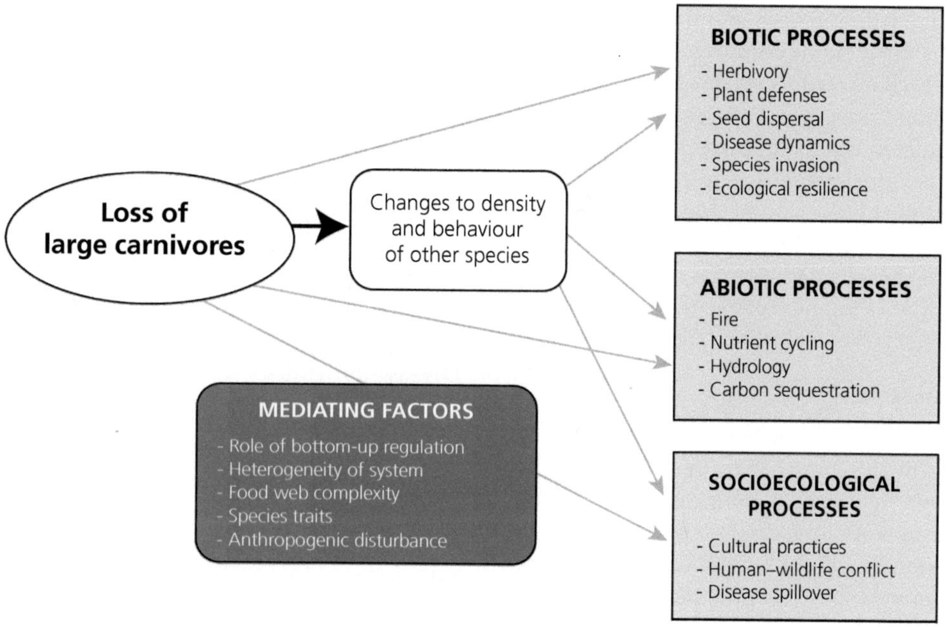

Figure 12.1 Trophic downgrading can alter a myriad of biotic, abiotic, and socio-ecological processes, due to the loss of large carnivores and subsequent changes to the density and behaviour of other species. The extent to which large carnivore loss shapes trophic cascades depends on factors related to the abiotic and biotic characteristics of the system.

cascades: herbivore populations increase after release from predation, in turn leading to a decrease in plant biomass (Power 1990). In food chains with four links, downgrading instead releases secondary consumers, leading to herbivore declines and therefore an increase in plant biomass (Power 1990). Where apex carnivore loss leads to increases in large mammalian herbivore density, a strong negative effect is expected on plant growth, survival, and cover, and a shift in the competitive balance from trees towards grass (Pringle et al. 2023). Given that carnivores can mediate the coexistence of a diverse herbivore guild by suppressing dominant competitors, their loss can drive declines in herbivore diversity (Henke and Bryant 1999) with corresponding changes to plant communities. Meso-predator release, in which the removal of large carnivores increases the densities of smaller carnivores due to relaxed competition and intraguild predation (Prugh et al. 2009), also facilitates trophic cascades that ultimately change herbivory and shape vegetation communities. In Australia, dingo (*Canis lupus dingo*) removal led to an increase in meso-predator densities, which in turn reduced small mammal populations, driving the encroachment of woody shrubs into grasslands (Gordon et al. 2017). In addition to these changes in herbivore densities, trophic downgrading may lead to behaviourally mediated trophic cascades, in which the behavioural responses of herbivores to changes in predation risk (Pace et al. 1999) shape vegetation composition and structure (Beckerman et al. 1997, Burkholder et al. 2013).

Following the extirpation of a dominant large carnivore, changes to herbivory can even drive vegetation communities into novel states. The loss of wolves in North America, for example, has led to intense sapling herbivory by white-tailed deer (*Odocoileus virginianus*) and created a demographic bottleneck on tree recruitment that suppresses palatable species, limits forest regeneration, and drives regime shifts (Bradshaw and Waller 2016). Livestock grazing in systems with intense predator control also highlights the role that herbivory can play in driving vegetation states in the absence of top-down regulation, as seen in shifts from semiarid grassland to shrub-dominated desert in the American Southwest (Grover and Musick 1990). Even after cattle ranching stopped, these ecosystems remained shrub-dominated due to positive feedback cycles, caused by the unpalatability of desert shrubs to herbivores and changes to water and nutrient fluxes (Mueller et al. 2007). Landscape-scale ecological regime shifts caused by trophic downgrading can be sustained by bioclimatic feedbacks and experience hysteresis effects that make them difficult to restore.

Plant defences

By modifying patterns of herbivory, the removal of large carnivores can change the physical structure of plants that exhibit morphological defences in response to herbivory, through both plastic phenotypic responses and through selection. In African savannas, impala browse preferentially in areas with lower predation risk from African wild dogs (*Lycaon pictus*). Trees in these highly browsed areas developed more defensive thorns than trees growing in areas with higher risk, highlighting the role of carnivores in shaping plant defences (Ford et al. 2014). In Venezuela, the construction of a dam created predator-free islands which experienced a subsequent explosion in prey populations (Terborgh et al. 2001). Released from top-down regulation, the hyperabundant herbivores dramatically reduced tree seedling recruitment and diversity, and shifted the community composition of these downgraded islands towards trees with strong mechanical and chemical defences.

Seed dispersal

Trophic downgrading can affect plant communities by changing patterns of frugivory and seed dispersal. When large carnivores are themselves omnivorous and consume fruit, such as bears (Harrer and Levi 2018) or maned wolves (*Chrysocyon brachyurus*; Motta Junior and Martins 2002), their removal reduces seed dispersal (Draper et al. 2022). By increasing large herbivore density, downgrading can increase long-distance dispersal of large-seeded plants (Pringle et al. 2023). Meso-predator release can also increase frugivory, given that many meso-predators consume fruit. In Spain, for example, red foxes (*Vulpes vulpes*) and stone martens (*Martes foina*) increased their frugivory in the absence of Iberian lynx (*Lynx pardinus*; Burgos et al. 2022). However, meso-predator release may instead limit seed dispersal as meso-predators prey on important seed dispersers like rodents and birds. These effects can be behaviourally-mediated: the experimental placement of coyote (*Canis latrans*) scat at artificial seed depots decreased seed predation by fearful rodents, but attracted arthropods that preyed on only the smaller seeds (Bartel and Orrock 2021).

Disease dynamics

By releasing populations of herbivores and meso-predators that serve as disease vectors, trophic downgrading can increase the incidence of disease. Furthermore, carnivores often selectively prey on sick individuals because they are easier to catch (Hoy et al. 2022, Tanner et al. 2019), so the removal of predation pressure can lead to disease outbreaks and reduce overall health in prey populations (Packer et al. 2003, Brandell et al. 2022). Additionally, many large carnivores, like grizzly bears (*Ursus arctos horribilis*) and hyenas (*Crocuta crocuta*), are facultative scavengers and consume carcasses. By reducing the amount of rotting meat on the landscape, carnivores help control the spread of pathogens (O'Bryan et al. 2018), and thus downgrading generally increases disease risk. However, if released meso-predator populations increase their scavenging in the absence of apex carnivores, there may be a neutral or even no net effect of downgrading on disease transmission from carcasses, or possibly even a reduction (O'Bryan et al. 2019).

Species invasion

Trophic downgrading can make ecosystems more vulnerable to invasion, with cascading consequences for native species. Extensive browsing from abundant herbivores in downgraded ecosystems, such as white-tailed deer in the northeastern United States, can facilitate invasions by well-defended or unpalatable plant species that outcompete the over-browsed native plants (Russell et al. 2017). In the United Kingdom, the extirpation of pine martens (*Martes martes*) contributed to grey squirrel (*Sciurus carolinensis*) invasion, to the detriment of native red squirrels (*Sciurus vulgaris*; Bamber et al. 2020), while in Tasmania, disease-driven declines in Tasmanian devils (*Sarcophilus harrisii*) enabled the invasion of feral cats (Hollings et al. 2016). Large carnivore reintroductions have therefore been proposed as a management technique to control invasive species, through predation or interference competition.

Meso-predators may expand into new areas in the absence of apex carnivores. In Fennoscandia, the extirpation of lynx (*Lynx lynx*) and wolves, in conjunction with niche expansion due to climate change, contributed to the northward range expansion of red foxes, which then outcompeted and suppressed the breeding of Arctic foxes (*Vulpes lagopus*; Ritchie et al. 2012). Like many islands, the Azores archipelago lacks endemic apex carnivores, and this system has experienced invasion by cats, ferrets (*Mustela putorius furo*), and least weasels (*Mustela nivalis*), leading to declines in native sea birds (Lamelas-López et al. 2020). On the other hand, downgraded ecosystems experiencing meso-predator release may be more resistant to invasion from rodents. In the case of the Azores, for example, the invasive meso-predators suppress the populations of invasive rats, another threat to nesting seabirds (Lamelas-López and Santos 2021).

Stability and resilience

Predators can stabilize ecosystems by dampening population fluctuations of lower trophic levels (Rezende et al. 2021). By shifting ecosystems from top-down to bottom-up control, trophic downgrading may destabilize ecosystems and compromise their ability to respond to perturbations in the abiotic environment (Britten et al. 2014). In downgraded ecosystems, bottom-up processes like soil nutrients and rainfall regulate populations, leading to greater fluctuations of producer, herbivore, and meso-predator populations, while top-down regulated systems may be more stable, given that carnivores are long-lived and respond more slowly to interannual fluctuations. As extreme disturbances increase in frequency and intensity under climate change, it is important to consider how trophic downgrading can weaken ecological stability and resilience.

Abiotic processes

Fire

By inducing changes in herbivory that alter plant biomass and composition, terrestrial trophic downgrading has context-specific effects on fire dynamics. The removal of carnivores is generally predicted to decrease fire frequency and intensity, given that herbivory can reduce fuel loads (Van Langevelde et al. 2003, Rouet-Leduc et al. 2021). More complex effects emerge when herbivory shifts vegetation composition, as in central Alaska, where moose (*Alces alces*) preferentially browse broadleaf trees, accelerating succession from deciduous forest to more flammable spruce forest (Feng et al. 2012). Because wolves suppress moose densities, they can slow the transition to coniferous forest, decreasing landscape-scale flammability. Behaviourally-mediated trophic cascades can also potentially alter fire regimes: in a longleaf pine (*Pinus palustris*) savanna in the southeastern United States, experimental coyote exclusion led to changes in deer habitat selection and reduced browsing on oak saplings, which ultimately increased the recruitment of oak trees that inhibit fires (Cherry et al. 2016). By occupying the predator niche that was once filled by red wolves (*Canis rufus*) in this ecosystem, coyote predation risk helps maintain a frequent burn regime that supports diverse fire-adapted savanna species.

Nutrient cycling

Trophic downgrading can alter nutrient cycling throughout food webs, and alter the transfer of nutrients between ecosystems, including marine and terrestrial systems. Endangered Galápagos sea lions (*Zalophus wollebaecki*) hunt fish in the Atlantic's productive Humboldt Current, and transfer nutrients to the arid, low-productivity Galápagos islands (Fariña et al. 2003). Continued declines of sea lions could therefore be catastrophic for endemic coastal flora that relies on these subsidies (Páez-Rosas et al. 2020). In terrestrial ecosystems, large carnivore loss can shape spatial patterns of carcass deposition, and of herbivore concentrations and thus dung deposition, typically resulting in the homogenization of soil nutrients across landscapes (Monk and Schmitz 2022). Changes in prey density after downgrading can affect soil nutrients, as in the case of Australia's Strzelecki Desert. After dingo extirpation, red kangaroo (*Macropus rufus*) populations were released, and increased herbivory caused declines in leaf litter deposition, which in turn decreased soil nitrogen, carbon, and phosphorus concentrations (Morris and Letnic 2017).

In four-link marine food chains, the loss of carnivores can shift systems to eutrophic states. In the Baltic Sea, hunting and pollution led to seal declines, which, along with nutrient-loading from agricultural fertilizer, drove the system to a eutrophic, hypoxic state, characterized by algal blooms. The loss of seals may have further contributed to these blooms through a trophic cascade: zooplanktivorous fish were released from seal predation, possibly leading to declines in zooplankton that would otherwise consume phytoplankton. Abiotic oceanographic nutrient cycling mechanisms have seemingly stabilized this regime shift to eutrophic conditions (Österblom et al. 2007). In other cases, however, restoring carnivores can reduce surpluses of anthropogenic nutrients and restore ecosystems to pre-disturbance states, as seen via a trophic cascade involving sea otters, crabs, invertebrate grazers, epiphytes, and seagrass in California (Hughes et al. 2013, 2016).

Hydrology

By altering patterns of herbivory, trophic downgrading can alter freshwater habitats. In Utah, cougar (*Puma concolor*) declines led to an increase in the density of mule deer, which increased browsing pressure on riparian cottonwoods. The subsequent reduction in tree recruitment led to increased streambank erosion and wider, shallower channels, which in turn caused declines in aquatic biodiversity (Ripple and Beschta 2006). In the California desert, cougars were also linked to a behaviourally mediated trophic cascade reducing feral donkey (*Equus asinus*) activity at wetlands, mitigating the degradation of desert water sources (Lundgren et al. 2022). Trophic downgrading may also promote the creation of wetlands by releasing ecosystem engineers. In northern Minnesota, wolves prey heavily on dispersing beavers (*Castor canadensis*), affecting the establishment of beaver dams (Gable et al. 2020). Wolf removal may therefore drive landscape-scale changes to the distribution of wetlands in these boreal forests. Given that these effects of downgrading can interact with changes to water regimes in the face of climate change and human water use, it is important to understand the context-specific role of trophic cascades in hydrological systems.

Carbon sequestration

Trophic downgrading has important interactions with climate change, given its effects on carbon sequestration and storage. The loss of sea otters, and subsequent explosions in urchin populations, greatly reduced kelp biomass across the western coast of North America (Box 12.1). Recovering sea otters promote the sequestration of a tremendous amount of carbon, valued at \$200–\$400 million (Wilmers et al. 2012). Preventing trophic downgrading in vulnerable coastal ecosystems that play an outsized role in carbon sequestration is an effective and low-cost climate mitigation strategy.

In terrestrial ecosystems, on the other hand, there is mixed evidence for the effects of apex carnivores and trophic downgrading on carbon fluxes. In Isle Royale National Park, the selective herbivory of moose on hardwoods releases conifers, which return less nitrogen to the soil through leaf litter and decrease ecosystem productivity. By reducing moose density, wolves can therefore increase carbon sequestration in boreal forests (Wilmers and Schmitz 2016). In contrast, elk (*Cervus canadensis*) grazing stimulates plant productivity in the grasslands of Yellowstone, and the reintroduction of wolves and the declining elk herds appear to have decreased the rate of carbon sequestration (Wilmers and Schmitz 2016). Broadly, large carnivores affect the balance between carbon storage by old, mature forests and carbon sequestration by fast-growing young forests in complex ways that manifest over decades.

Socioecological processes

Large carnivores are often intentionally removed from ecosystems in the interest of human livelihoods and safety. In downgraded ecosystems, communities often directly benefit from reductions in livestock predation and threats to human lives. However, trophic downgrading can also have cascading socioecological consequences that negatively affect people. Carnivores play an important cultural role in many human societies (see Chapters 13, 14, this volume), and their loss can have profound impacts on people who value them for their meat, fur, or other products, or for their spiritual or cultural significance. Large carnivores provide other 'predation services' to society, including direct services like hunting and tourism, and indirect services that arise from the role that large carnivores play in ecosystems (O'Bryan et al. 2018, Gilbert et al. 2021). By disrupting biotic and abiotic processes, trophic downgrading can remove some of these benefits.

Human–wildlife conflict

In the absence of large carnivores, the growth in populations of herbivores or meso-predators increases the incidence of human–wildlife conflicts involving these species (see also Chapter 5 and 13 in this volume). The loss of mountain lions in the eastern United States led to explosions in deer abundance and high rates of deer-vehicle collisions. The reintroduction of mountain lions could reduce the number of collisions by an estimated 22%, preventing 21,400 human injuries and over $2 billion within 30 years (Gilbert et al. 2017).

Higher densities of herbivores or meso-predators in downgraded ecosystems may increase foraging on anthropogenic food sources, including crops, forestry seedlings, residential landscaping, or livestock (Gilbert et al. 2021). In Ghana, the loss of lions (*Panthera leo*) and leopards (*Panthera pardus*) led to increases in baboon (*Papio anubis*) populations (Brashares et al. 2010). An increase in crop foraging by baboons led to disease transmission between baboons and people, compromised livelihoods and food security, and contributed to reduction in education as children were kept from school to guard crops. In the eastern Himalayas, cascading risk responses reduce crop foraging by herbivores (Thinley et al. 2018). Tigers (*Panthera tigris*) avoid people in agricultural areas; leopards and dholes (*Cuon alpinus*) avoid tigers (and thus spend more time near farms); herbivores avoid leopards and dholes (and thus avoid farms). It is possible that the removal of apex carnivores from this system could disrupt these dynamics and contribute to an increase in human–wildlife conflict.

Disease transmission

As explored earlier, large carnivores can reduce the incidence of disease in their prey populations, and trophic downgrading can therefore increase the transmission of zoonotic diseases. In Mumbai, India, urban leopards prey on stray dogs, which transmit rabies to humans. By suppressing dogs, leopards are predicted to prevent around 90 human deaths from rabies each year (Braczkowski et al. 2018). In the Tigray region of Ethiopia, spotted hyenas consume dead livestock and even human corpses, potentially mitigating disease risk (Yirga et al. 2015).

Given that rodents are an important reservoir for many pathogens, predator removal can affect zoonotic disease transmission by initiating trophic cascades that alter rodent density (e.g. via meso-predator release that suppresses rodents; Ostfeld and Holt 2004). The extirpation of grey wolves from the eastern United States may be contributing to the dramatic rise of Lyme disease, via a complex cascade (Box 12.2). Wolf eradication helped coyotes expand their range, contributing to the widespread decline of the red fox (Berger and Gese 2007, Newsome and Ripple 2015). Foxes are major predators of rodent species like deer mice (*Peromyscus* spp.), which host ticks that transmit Lyme disease. In New York, coyote abundance, fox rarity, and Lyme disease prevalence are closely correlated, illustrating the effects of downgrading on disease (Levi et al. 2012). Large carnivores also shape the prevalence of rodent-borne disease by suppressing herbivore densities. In Kenya, rodent densities doubled when herbivores were experimentally excluded, leading to a doubling of fleas infected with the pathogenic bacteria *Bartonella* (Young et al. 2015).

Context-dependency of effects

While our understanding of the breadth and magnitude of the ecological impacts of apex predator declines has grown since Estes and colleagues (2011) published their 'trophic downgrading' paper, ecologists have also come to recognize the context-dependency of these effects. Even if the loss of apex

Box 12.2 Grey wolves in Yellowstone National Park

The loss and reintroduction of wolves in Yellowstone National Park has become a classic and contested example of the effects of trophic downgrading. Following the extirpation of wolves, fearless elk browsed heavily on exposed riparian saplings (Ripple and Beschta 2004). Declines in streamside tree cover led to erosion, warmer water, and fewer beavers, driving landscape-scale degradation. Wolves were reintroduced in 1995, and the effects of downgrading appeared to reverse: elk changed their browsing in response to risk, allowing riparian trees to regenerate, which altered stream flow (Ripple and Beschta 2003). Further research indicated that elk also suppressed aspen (*Populus tremuloides*) growth in upland sites, where they preferred to browse given lower wolf predation risk (Ripple and Beschta 2007). Wolf extirpation may have also affected disease in aspen by leading to more bark stripping by elk, which facilitates fungal infection (Beschta and Ripple 2020). This groundbreaking research suggested that a terrestrial apex carnivore could create a 'landscape of fear' that altered herbivore foraging, initiating a behaviourally-mediated trophic cascade that rippled throughout the ecosystem. Since the documentation of the Yellowstone cascade, subsequent research has documented wolf-driven trophic cascades across North America (Hebblewhite et al. 2005, Beschta and Ripple 2009, Flagel et al. 2016).

Alongside growing evidence for wolf-driven trophic cascades, the role of wolves in Yellowstone has been complicated by further research that has provided additional explanations for observed ecological changes. The increase in riparian cottonwoods may be related to snowmelt floods that supported seedling recruitment, which coincidentally occurred around the same time as wolf reintroduction (Rose and Cooper 2017). Furthermore, an experiment by Kauffman et al. (2010) suggested that the declining recruitment of aspen during the period of wolf absence was more likely caused by increased elk numbers than behavioural shifts, and presented evidence that aspen recruitment in Yellowstone has not actually increased following the reintroduction of wolves. There may also be a limited physiological cost of wolf predation risk on elk: despite great variation in the number of wolf encounters experienced by individuals, predation risk was not correlated with body fat or pregnancy rates, providing additional evidence that the lethal effects of wolves might be more important than risk effects (Middleton et al. 2013). A recent study pointed to a potential mechanism that can account for a lack of wolf predation risk on elk physiology and vegetation: elk can avoid wolves not just in space, but in time (Kohl et al. 2018). By utilizing risky open habitats while crepuscular wolves are resting, and retreating to forests by day, elk can navigate the landscape of fear without strong ecological effects. Ultimately, it appears that wolves play an important ecological role, but their effects are mediated by climate, hydrology, and other species in this complex community.

carnivores has measurable impacts on prey species, for example, trophic cascades can attenuate at the plant-herbivore interface, and trophic downgrading may not influence plant biomass, demography, composition, or defences (Shurin et al. 2002). The loss of a top trophic level therefore does not always lead to wholesale changes in ecosystem functioning, and the cascading consequences of predator loss vary with characteristics of the ecosystem (Haswell et al. 2017).

The impacts of trophic downgrading should be stronger in simpler food webs or more homogenous systems, (i.e., aquatic systems, relative to terrestrial systems; Strong 1992, Halaj and Wise 2001). Complex food webs, or those with higher predator diversity, should be more resilient to the loss of a single carnivore species (Schmitz et al. 2000, Finke and Denno 2004). Ecosystem complexity has been suggested as an explanation for weaker carnivore-ungulate trophic cascades in multi-predator, multi-prey African savannas, as compared to North American systems (Comley et al. 2020). While predation may shape patterns of herbivory for one prey species, other herbivores may not display the same patterns of risk-sensitive foraging and thus consume plants that would otherwise be released (Ford et al. 2015).

Trophic downgrading should have larger impacts on top-down regulated ecosystems, as compared to bottom-up regulated ecosystems. In systems that are resource-limited, even if only seasonally, the removal of large carnivores may not have measurable changes on their herbivore prey, or on plant biomass (Maron and Pearson 2011). Using a mechanistic global ecosystem model, Hoeks et al. (2020) simulated the removal of large carnivores from ecosystems and found that the impacts of trophic downgrading on ecosystems increased with primary productivity in accordance with the exploitation ecosystem hypothesis, which predicts greater responses of herbivores to large carnivore loss in productive ecosystems. Other factors like fire and competition may play a larger bottom-up role in seedling recruitment than the top-down effects of herbivory (Donaldson et al. 2022). The role of top-down and bottom-up control, and thus the effects of trophic downgrading, can vary temporally in the same ecosystem, due to fluctuations in abiotic conditions (Meserve et al. 2003, Piovia-Scott et al. 2017). The effect of primary productivity on trophic downgrading does not always conform to predictions from theory, given the complexity of ecosystems and organisms (Borer et al. 2005).

Life history and other traits can explain species-specific responses to predator extirpation or reintroduction. In Spain, the recovery of the Iberian lynx had greater effect on the stone marten, whose wide dietary breadth allowed them to shift their spatial distribution to avoid lynx, while the genet (*Genetta genetta*) remained inside lynx territories where their preferred prey was present (Burgos et al. 2023). The hunting modes and life history traits associated with the carnivore species themselves also matter. Sit-and-wait predators (such as mountain lions) generate stronger risk effects than cursorial predators (such as wolves) (Preisser et al. 2007, Schmitz 2008).

The broader community composition further mediates the ecological effects of downgrading. For example, systems where unpalatable plants replace those that have been overgrazed by abundant herbivores may be more likely to switch into a novel state (Borer et al. 2005). Trophic cascades are stronger in aquatic systems because producers have a higher production-to-biomass ratio, are more palatable, and are smaller relative to herbivores (Shurin and Seabloom 2005). In terrestrial systems, the presence of megaherbivores dampens cascades, as large-bodied species are not as vulnerable to predation. Experimental manipulation of predation risk in an African savanna altered mesoherbivore space use and foraging patterns but did not alter nutrient redistribution, given that megaherbivores continued to defecate in high-risk areas (le Roux et al. 2018). Migratory herbivores also display less risk-sensitive foraging and widely outnumber predators, and thus disrupt trophic cascades (Donaldson et al. 2022).

Challenges and opportunities for 'trophic upgrading'

Restoring apex carnivores provides opportunities for 'trophic upgrading', in which downgraded systems are restored as closely as possible to the pre-downgraded state (Ritchie et al. 2012, Stier et al. 2016, Svenning et al. 2016) (Box 12.3). This 'upgrading' falls under the umbrella of 'trophic rewilding', a restoration approach with the goal of promoting self-sustaining ecosystem functioning, typically through the reintroduction of megafauna (Fernández et al. 2017). The efficacy and ethics of trophic rewilding have sparked debate, particularly with regard to proposals to reintroduce novel assemblages (Lorimer et al. 2015). Here, we focus specifically on the potential to restore ecosystem functioning through the reintroduction of extirpated apex carnivores to downgraded systems (see example in Figure 12.2).

Restoring large carnivores

In some cases, apex carnivores return naturally via dispersal from nearby areas, if the disturbances that led to their extirpation are reduced or the ecosystem is otherwise restored (Silliman et al. 2018). This natural recolonization has been occurring in Europe, with lynx, brown bear (*Ursus arctos*), grey wolves, and wolverines (*Gulo gulo*) returning to areas of their native range from which they had been extirpated (Chapron et al. 2014). Predators may even move into what appear to be novel habitats to which they were in fact native, such as sea otters expanding into estuarine marshes, wolves into coastal habitats, and harp seals (*Pagophilus groenlandicus*) into subtropical climates (Silliman et al. 2018). This passive approach to trophic upgrading relies on the proximity to nearby source populations with dispersing individuals, and landscape connectivity that facilitates movement. Restoring

Box 12.3 Dingoes in Australia

Dingoes, the only extant large mammalian carnivore in Australia, were introduced 4000 years ago and have been viewed as a pest since European colonization, targeted by eradication campaigns and excluded from southeastern Australia's rangelands by a 5600 km 'dingo fence' (Prowse et al. 2015). Dingoes are now absent from much of their former range, and their loss has been linked to two distinct and extensively researched trophic cascades.

Dingo declines led to increases in the densities of the dominant herbivore, kangaroos, and subsequent decreases in grass biomass. This link between dingoes and kangaroo density is stronger in more productive regions, while in arid regions, kangaroo populations are bottom-up regulated (Letnic and Crowther 2013). This trophic cascade appears to be weakened when dingoes switch to reptile prey as kangaroos decline (Choquenot and Forsyth 2013), though prey switching may prevent dingo populations from crashing and stabilize top-down effects on kangaroos over the long term. Top-down control of herbivores by dingoes also benefits cattle by reducing competition for pasture, and dingoes lead to higher and more stable profits for ranchers, despite their extermination in the name of protecting livestock (Prowse et al. 2015).

The second trophic cascade is shaped by dingo's effect on invasive meso-predators and has proved harder to untangle, with over 130 peer-reviewed studies on the subject. Two invasive meso-predator predators—feral cats and red foxes—have driven the extinction of many endemic Australian marsupial species, particularly in arid regions (Johnson and Isaac 2009). By preying on and competing with cats and foxes, dingoes may drive a trophic cascade benefiting threatened small mammal populations, which would justify dingo protection as a biodiversity conservation strategy. Over the last several decades, many studies have found that dingoes increase small mammal abundance (Letnic and Koch 2010). There is evidence that dingoes suppress cat populations and push them to hunt at non-optimal times (Brook et al. 2012), which increases the abundance and foraging efficiency of dusky hopping mice (*Notomys fuscus*; Gordon et al. 2015). However, other studies have not found links between dingoes, meso-predators, and small mammals (Fancourt et al. 2019, Castle et al. 2021, Kreplins et al. 2021). One recent meta-analysis found that dingo removal typically increased fox but not cat abundance, and fox declines facilitated the population growth of small mammals (Hunter et al. 2018). In contrast, another meta-analysis found that scientific evidence for meso-predator release is equivocal and based largely on correlative studies, with the few causal studies finding no evidence of dingo suppression of meso-predators (Castle et al. 2023). The case of the dingo highlights the complex effects of apex predators on ecosystems through their interactions both with herbivores and meso-predators, and the challenges for ecologists in deciphering those effects.

connectivity and corridors is therefore a promising avenue for facilitating large carnivore restoration (Bhatt et al. 2023).

In many cases, carnivore restoration requires direct management intervention, including translocations and intentional reintroductions. Large carnivore restorations have been increasingly common globally, with some success (Thomas et al. 2023; but see Marshall et al. 2016). However, where downgrading drove carnivores extinct, reintroducing the same species becomes impossible. Moreover, given variation in modern human tolerance of large carnivores, it might only be possible to restore such species at densities much lower than those occurring historically. If the goal is to restore ecological functioning, the functional role of a predator is more important than its identity or origin (Ritchie et al. 2012), although there is debate about the ethics and effectiveness of replacing lost species with functional equivalents (Rubenstein and Rubenstein 2016).

Given species extinctions, ambiguous historical records, and other long-term changes to ecosystems, it is often difficult to identify relevant ecological baselines for restoration. While we focus here on recent losses of apex predators, human-caused trophic downgrading has been occurring since the Pleistocene extinctions (Malhi et al. 2016, Smith et al. 2022, Chapter 2, this volume). Although there are challenges in disentangling the effects of megacarnivore declines with those of climatic change and human impacts, prehistoric evidence suggests that this megafauna loss profoundly reshaped ecosystems worldwide (Malhi et al. 2016). It is therefore important to acknowledge that humans are often restoring ecosystems to a shifted baseline that had already undergone Pleistocene downgrading (Corlett 2013).

Figure 12.2 The loss of large carnivores can alter ecosystems via trophic cascades. It may be possible to reverse the effects of trophic downgrading through the reintroduction of large carnivores. However, in some cases, these "upgraded" ecosystems may not resemble the original, intact ecosystems due to hysteresis and to the ongoing impacts of human disturbance. In Yellowstone, the riparian ecosystem in the early 1900s existed in a beaver-willow state (A), in which wolf predation suppressed elk numbers and allowed tall willow trees to flourish. These trees provided material for beaver dams, and in turn, the dams created wetlands that allowed willow seeds to germinate. When wolves were extirpated in the 1920s, elk populations increased and they overbrowsed young willows, preventing tall trees from growing. Without tall trees the beavers abandoned these rivers and the wetlands they created vanished, forcing the system into an elk-grassland alternate stable state (B). When wolves were reintroduced in the 1990s, elk populations declined, but willow recruitment remained low, in the absence of flooding from beaver dams, and the few willows that survived were still overbrowsed. Without tall willows, beavers have not returned and the system appears to have entered a novel alternate stable state (C). Wolf extirpation thus disrupted a positive-feedback cycle between ecosystem engineer and vegetation that could not be restored by wolf reintroduction (*Hobbs et al. 2024; Fig. 12.2*). Illustration by Jacelyn Shu.

Ecological potential for trophic upgrading

Even when large carnivores successfully re-establish populations, the degree to which restoration is capable of truly 'resetting' a downgraded ecosystem to its natural functioning remains a subject of debate (Kuijper et al. 2016). As discussed earlier in the context of trophic downgrading, the existence and nature of trophic cascades will vary widely across systems, and the same principles apply to upgrading. Furthermore, trophic downgrading and trophic upgrading are not necessarily reciprocal processes. A meta-analysis found that reintroduction of apex predators did not consistently affect other trophic levels, including meso-predators, omnivores, herbivores, and primary producers (Alston et al. 2019).

Ecosystems may fail to recover to their pre-downgraded states following large carnivore reintroductions. Systems might have low ecological resistance or resilience, and therefore be unable to absorb or bounce back from shocks or disturbances such as apex consumer loss, without fundamentally changing states (Holling 1973). As Estes et al. (2011) noted, 'downgraded' ecosystems without apex consumers may represent alternative stable states that cannot be easily reversed. Upgrading will not reverse the effects of downgrading if ecosystems have undergone regime shifts and crossed tipping points, experiencing hysteresis (the existence of a different state under the same environmental conditions) maintained by feedback cycles (Scheffer et al. 2001, Folke et al. 2004). There may also be time lags and transient dynamics between the reintroduction of apex predators and the reestablishment of stable ecosystem dynamics (Alston et al. 2019).

Schweiger et al. (2019) emphasize the importance of ecological memory, which is related to the path-dependency that leads to hysteresis, in determining the success of restoration. Ecological memory consists of an ecosystem's accumulated abiotic and biotic material, which shapes the present landscape and ecological networks, and of information legacies from past dynamics. These legacies can include the behavioural or morphological traits of resident species like prey or meso-predators which influence their response to reintroduced carnivores. For example, naïve prey or meso-predators often fail to respond to risk posed by reintroduced carnivores, with consequences for the lethal and trait-altering effects of predation (Suraci et al. 2016). The importance of ecological memory for trophic upgrading should depend on the time since the species was lost, the speed of biotic and abiotic turnover, species interaction strength, and the capacity of surrounding source ecosystems to compensate (Schweiger et al. 2019).

The extent to which a reintroduced carnivore species is capable of restoring ecological dynamics is related to the traits and ecological characteristics of that species. Generally, species that were responsible for the greatest ecological transformation when lost will have the potential for greatest transformation when reintroduced. The 'impact potential' of a species will be higher for species with greater population densities and home range sizes (and thus greater space use), higher per capita lifetime capacity for ecosystem engineering, and low niche overlap with resident species (resulting in low functional redundancy; Schweiger et al. 2019). Even if the densities of apex carnivores recover, the behaviour of translocated individuals (e.g. social behaviour) may differ from baseline conditions in ways that prevent the species from stepping into the same ecological niche (Ritchie et al. 2012).

Trophic upgrading in the Anthropocene

It is important to ensure that the threats that originally led to apex carnivore loss are mitigated, or reintroduced populations will again decline. Even so, there will always be challenges to restoring apex consumers to ecosystems that are experiencing high degrees of anthropogenic disturbance (Kuijper et al. 2016). Efforts to upgrade ecosystems often involve the reintroduction of carnivores to landscapes that look very different from those they previously inhabited (Chapron et al. 2014). In highly modified landscapes like those in much of Europe, Kuijper et al. (2016) argue that carnivore reintroductions have limited ecological impacts, due to anthropogenic changes in both top-down and bottom-up ecological processes.

Where humans exert top-down effects on food webs, large carnivores play the role of the penultimate rather than apex predator (Oriol-Cotterill et al. 2015). Humans impose both lethal and risk effects on both predators and prey, reshaping the nature of predator-prey interactions. The lethal removal of carnivores to mitigate human–wildlife conflict can keep predator densities low enough that their ecological role is limited (Ordiz et al. 2013, Oriol-Cotterill et al. 2015). Kuijper et al. (2016) suggest that there is therefore greater potential for carnivore reintroductions in human-modified landscapes to initiate behaviourally-mediated trophic cascades, relative to density-mediated cascades, given that behavioural responses of prey can arise even when predator densities are low (Laundré et al. 2001). However, behavioural responses of prey to highly predictable patterns of human hunting may be stronger than those in response to carnivores, in cases where prey are forced to trade off risk, thus weakening behavioural and physiological risk effects of predators (Proffitt et al. 2009). In some cases, however, anthropogenic impacts actually amplify the effects of predators on ecosystems via risk effects on carnivores; for example, mountain lions abandoned kills more near human development and increased their kill rates to compensate (Smith et al. 2015).

By changing the playing field of predator-prey interactions, human activity can further reshape trophic cascades. Anthropogenic land conversion may homogenize habitat or otherwise alter physical habitat structure in ways that make it easier for predators to hunt (Dickie et al. 2020), for prey to escape, or for prey to adjust their space use to mitigate risk (Sahlén et al. 2015). Subsequent changes in the magnitude of lethal and risk effects alter trophic cascades. Anthropogenic food subsidies also complicate the ecological consequences of carnivore reintroductions (Oro et al. 2013), as resource subsidies to prey (e.g. agriculture, supplementary feeding) can weaken density-mediated effects of apex carnivores on prey populations (Muhly et al. 2013). While humans can provide subsidies to predators in the form of livestock or refuse, human attitudes towards carnivores often ultimately limit carnivore population densities, regardless of resource availability (Kuijper et al. 2016).

Socioeconomic considerations of upgrading

Even if it were possible to reach desired ecological restoration targets through apex carnivore reintroductions, the financial and logistical hurdles

Box 12.4 Large carnivores in Gorongosa National Park

Mozambique's Gorongosa National Park historically supported diverse and abundant large mammals. During Mozambique's civil conflict (1977–1992), hunting decimated Gorongosa's large mammal populations, extirpating most large carnivore species. Where there had once been 500 lions, there were six (Yeats 2010), and African wild dogs, leopards, and hyenas were completely extirpated. While large herbivores had dramatically declined during the conflict, their populations rebounded quickly in the absence of predation and competition (Stalmans et al. 2019). By 2018, 95% of Gorongosa's pre-war large herbivore biomass had been restored (Stalmans et al. 2019). However, the herbivore community composition had changed dramatically: once dominant megaherbivores, like elephants (*Loxodonta africana*), hippos (*Hippopotamus amphibius*), zebras (*Equus quagga*), and buffalo (*Syncerus caffer*) were rare, while mesoherbivore antelope species, like waterbuck (*Kobus ellipsiprymnus*), bushbuck (*Tragelaphus sylvaticus*), and impala (*Aepyceros melampus*), underwent rapid population growth (Stalmans et al. 2019). This asymmetric recovery may be due in part to the extirpation of large carnivores that hunt mesoherbivores.

In the absence of predation risk, some antelope species expanded into new habitats that they previously avoided. The typically forest-dwelling bushbuck expanded into the treeless floodplain grassland, where they suppressed a common floodplain plant species (Atkins et al. 2019). Bushbuck responded to experimental predator cues by moving back into forest habitats and reducing their consumption of floodplain plants (Atkins et al. 2019). This experiment suggests that changes to herbivore behaviour, habitat use, and plant suppression in response to trophic downgrading may be reversed by the restoration of a landscape of fear.

Recognizing that the lack of predators had shifted Gorongosa into a novel, antelope-dominated, downgraded state, the park has made trophic upgrading a restoration priority. With limited dispersal corridors connecting Gorongosa to other carnivore habitats, most predator species were not able to return naturally to the park (Bouley et al. 2018). Over the last five years, the Gorongosa Restoration Project has reintroduced African wild dogs, leopards, and hyenas, and reduced the threat of snaring through increased ranger patrols. African wild dogs, in particular, have undergone rapid population growth (Bouley et al. 2018) and there are now close to 200 individuals in the park. It remains to be seen whether predation pressure from this dynamically changing large carnivore assemblage will shift the balance between mesoherbivores and megaherbivores back to the pre-war baseline. The trophic upgrading of Gorongosa offers a window into the ecological impact of predator recovery following extreme anthropogenic disturbance.

involved are non-trivial (Rohr et al. 2018). It is typically expensive to reintroduce a single animal, and restoring a population can have prohibitive costs (Weise et al. 2014). Some have argued for the ethical obligation to restore lost species regardless of cost, and regardless of a net benefit to humans or biodiversity (Lewis et al. 2017). However, trophic interactions can introduce trade-offs with carnivore reintroductions and other conservation objectives, such as the maintenance of vulnerable prey populations (Kojola et al. 2009, Rominger 2018).

There are important ethical debates about introducing large carnivores that negatively affect people. Large carnivores often pose a threat to human lives or livelihoods, and the interests, values, and attitudes of different groups of people must therefore be weighed (Bauer et al. 2009). In the case of sea otter recovery, for example, declining shellfish fisheries bear the costs of trophic upgrading, while finfish fisheries and ecotourism benefit (Gregr et al. 2020). The burden of living alongside large carnivores is disproportionately experienced by poor, rural communities in the Global South (Braczkowski et al. 2023). Carnivore reintroduction can further exacerbate these inequalities, posing important environmental justice considerations (McInturff et al. 2021) and emphasizing the need for centrally considering human wellbeing, communities, and cultural heritage when considering reintroductions (Montgomery et al. 2020).

Conclusions

Since Estes et al. brought attention to 'trophic downgrading', there has been a growing body of work on the ecological consequences of large carnivore loss. However, evidence remains limited, with much of our understanding coming from several highly-studied systems (Allen et al. 2017). For many carnivore species and systems, ecologists still lack sufficient understanding of species interactions and trophic linkages to identify or predict the effects of apex carnivore removal (Ripple et al. 2014). In other cases, our evidence may not be as valuable as we think, given that studies are often designed to find evidence of trophic downgrading effects, with insufficient exploration of alternative hypotheses for observed phenomena (Allen et al. 2017). Sometimes, follow-up studies on initial reports of strong restoration effects have suggested that some such early findings have been overstated (Hobbs et al. 2024).

It can be challenging to document the ecological consequences of trophic downgrading, given the complexity of ecological interaction pathways and anthropogenic disturbances (Ford and Goheen 2015, Montgomery et al. 2019). The ecological consequences of trophic downgrading are revealed only after apex consumers have been extirpated, and there may be significant time lags in indirect effects on lower trophic levels (Piovia-Scott et al. 2017). The ability to fully quantify ecological change in these cases is often limited by a lack of pre-extirpation data and the often unavoidable lack of system-level replication. Insights into the potential consequences of trophic downgrading are often inferred from the experimental exclusion of predators and comparisons of areas with and without predators, which all have limitations in their ability to fully simulate or predict the long-term consequences of trophic downgrading (Bell et al. 2003). However, even with long-term, consistent monitoring efforts across trophic levels before and after downgrading, it can be difficult to infer causation from observational, correlative studies (Allen et al. 2017).

While there are many cases of downgrading dramatically altering biotic and abiotic processes, there are also many where predator loss does not measurably affect the populations or behaviour of herbivores or meso-predators. Even where predator loss does affect these trophic levels immediately below, trophic downgrading may not have a cascading impact on biotic or abiotic processes, perhaps because other predators compensate for their loss or because bottom-up processes are more important. The context-dependent consequences of downgrading are evident, given that the impacts of losing the same carnivore species are dramatically different between ecosystems, climates, seasons, and species assemblages.

The case studies highlighted throughout the chapter reveal the immense complexity of attempting to untangle the effects of trophic downgrading. The effect of wolf predation risk on limiting elk herbivory and thus riparian forests, long considered the archetypal example of a North American behaviourally-mediated trophic cascade, has proven hard to confirm despite two decades of research. The direct effect of dingoes on invasive meso-predators and indirect effect on imperilled small mammals, a matter of urgent conservation importance in Australia and the subject of an extensive body of research, remains elusive and appears to vary between meso-predators and aridity levels. Understanding the effects of trophic upgrading in Mozambique's Gorongosa National Park (Box 12.4) is especially challenging given that it is an extremely diverse megafaunal assemblage with multiple predator species being reintroduced simultaneously, and populations of other species still recovering from the civil conflict, amidst seasonal and interannual disturbance. The sea otter-urchin-kelp food chain has been highlighted as a clear-cut example of a trophic cascade that can drive an ecological regime shift, likely in part because it is an aquatic system, with a palatable dominant producer, and a relatively small number of different species involved. Yet, even this seemingly simple dynamic continues to hold big surprises, including recent evidence of the role of wolves in this trophic cascade.

These challenges reveal the need for long-term ecological studies of trophic downgrading that incorporate data on biotic and abiotic processes, and use experimental methods to establish casualty where possible. It is only through these time and energy-intensive methods that committed researchers have been able to shed light on the complexity of trophic downgrading. Documenting and predicting the impacts of both downgrading and upgrading across systems will advance our fundamental understanding of ecology while informing management in the midst of ongoing apex predator loss and reintroduction.

References

Allen, B. L., et al. 2017. Can we save large carnivores without losing large carnivore science?. *Food Webs* 12(12): 64–75.

Alston, J. M., et al. 2019. Reciprocity in restoration ecology: when might large carnivore reintroduction restore ecosystems?. *Biological Conservation* 234(June): 82–89.

Atkins, J. L., et al. 2019. Cascading impacts of large-carnivore extirpation in an African ecosystem. *Science* 364(6436):173–177.

Bamber, J. A., C. M. Shuttleworth, M. W. Hayward, and D. J. Everest. 2020. Reinstating trophic cascades as an applied conservation tool to protect forest ecosystems from invasive grey squirrels (*Sciurus Carolinensis*). *Food Webs* 25 (December):e00164.

Bartel, S. L. and J. L. Orrock. 2021. An omnivorous mesopredator modifies predation of omnivore-dispersed seeds. *Ecosphere* 12(2):e03369.

Bauer, N., A. Wallner, and M. Hunziker. 2009. The change of European landscapes: human-nature relationships, public attitudes towards rewilding, and the implications for landscape management in Switzerland. *Journal of Environmental Management, Environmental and Landscape Change* 90(9):2910–2920.

Beckerman, A. P., M. Uriarte, and O. J. Schmitz. 1997. Experimental evidence for a behavior-mediated trophic cascade in a terrestrial food chain. *Proceedings of the National Academy of Sciences of the United States of America* 94(20):10735–10738.

Bell, T., W. E. Neill, and D. Schluter. 2003. The effect of temporal scale on the outcome of trophic cascade experiments. *Oecologia* 134(4):578–586.

Berger, K. M. and E. M. Gese. 2007. Does interference competition with wolves limit the distribution and abundance of coyotes?. *Journal of Animal Ecology* 76(6): 1075–1085.

Beschta, R. L. and W. J. Ripple. 2009. Large predators and trophic cascades in terrestrial ecosystems of the western United States. *Biological Conservation* 142(11):2401–2414.

Beschta, R. L. and W. J. Ripple. 2020. Large carnivore extirpation linked to loss of overstory aspen in Yellowstone. *Food Webs* 22(March):e00140.

Bhatt, T. R., J. G. Castley, H. S. Baral, and A. Chauvenet. 2023. A synthesis of priorities, patterns, and gaps in large carnivore corridor research. *Frontiers in Conservation Science* 4:1094443.

Borer, E. T., et al. 2005. What determines the strength of a trophic cascade?. *Ecology* 86(2):528–537.

Bouley, P., M. Poulos, R. Branco, and N. H. Carter. 2018. Post-war recovery of the African lion in response to large-scale ecosystem restoration. *Biological Conservation* 227(November):233–242.

Braczkowski, A. R., et al. 2023. The unequal burden of human-wildlife conflict. *Communications Biology* 6(1):1–9.

Braczkowski, A. R, C. J. O'Bryan, M. J. Stringer, J. E. M. Watson, H. P. Possingham, and H. L. Beyer. 2018. Leopards provide public health benefits in Mumbai, India. *Frontiers in Ecology and the Environment* 16(3): 176–182.

Bradshaw, L. and D. M. Waller. 2016. Impacts of white-tailed deer on regional patterns of forest tree recruitment. *Forest Ecology and Management* 375(September): 1–11.

Brandell, E. E., et al. 2022. Examination of the interaction between age-specific predation and chronic disease in the greater Yellowstone ecosystem. *Journal of Animal Ecology* 91(7):1373–1384.

Brashares, J., L. Prugh, C. J. Stoner, and C. Epps. 2010. Ecological and conservation implications of mesopredator release. In *Trophic Cascades: Predators, Prey, and the Changing Dynamics of Nature*, edited by J. Terborgh and J. A. Estes. Oxford University Press, Oxford.

Britten, G. L., M. Dowd, C. Minto, F. Ferretti, F. Boero, and H. K. Lotze. 2014. Predator decline leads to decreased stability in a coastal fish community. *Ecology Letters* 17(12):1518–1525.

Brook, L. A., C. N. Johnson, and E. G. Ritchie. 2012. Effects of predator control on behaviour of an apex predator and indirect consequences for mesopredator suppression. *Journal of Applied Ecology* 49(6):1278–1286.

Burgos, T., et al. 2022. Predation risk can modify the foraging behaviour of frugivorous carnivores: implications of rewilding apex predators for plant–animal mutualisms. *Journal of Animal Ecology* 91(5):1024–1035.

Burgos, T., et al. 2023. Top-down and bottom-up effects modulate species co-existence in a context of top predator restoration. *Scientific Reports* 13(1):4170.

Burkholder, D. A., M. R. Heithaus, J. W. Fourqurean, A. Wirsing, and L. M. Dill. 2013. Patterns of top-down control in a seagrass ecosystem: could a roving apex predator induce a behaviour-mediated trophic cascade? *Journal of Animal Ecology* 82(6):1192–1202.

Castle, G., M. S. Kennedy, and B. L. Allen. 2023. Stuck in the mud: persistent failure of 'the science' to provide reliable information on the ecological roles of Australian dingoes. *Biological Conservation* 285(September):110234.

Castle, G., D. Smith, L. R. Allen, and B. L. Allen. 2021. Terrestrial mesopredators did not increase after top-predator removal in a large-scale experimental test of mesopredator release theory. *Scientific Reports* 11(1):18205.

Chapron, G., et al. 2014. Recovery of large carnivores in Europe's modern human-dominated landscapes. *Science* 346(6216):1517–1519.

Cherry, M. J., R. J. Warren, and L. M. Conner. 2016. Fear, fire, and behaviorally mediated trophic cascades in a frequently burned savanna. *Forest Ecology and Management* 368(May):133–139.

Choquenot, D. and D. M. Forsyth. 2013. Exploitation ecosystems and trophic cascades in non-equilibrium systems: pasture—red kangaroo—dingo interactions in arid Australia. *Oikos* 122(9):1292–1306.

Comley, J., C. Joubert, N. Mgqatsa, and D. Parker. 2020. Lions do not change rivers: complex African savannas preclude top-down forcing by large carnivores. *Journal for Nature Conservation* 56(May):125844.

Corlett, R. T. 2013. The shifted baseline: prehistoric defaunation in the tropics and its consequences for biodiversity conservation. *Biological Conservation* 163 (July):13–21.

Cowie, R. H., P. Bouchet, and B. Fontaine. 2022. The sixth mass extinction: fact, fiction or speculation? *Biological Reviews* 97(2):640–663.

Dickie, M., S. R. McNay, G. D. Sutherland, M. Cody, and T. Avgar. 2020. Corridors or risk? Movement along, and use of, linear features varies predictably among large mammal predator and prey species. *Journal of Animal Ecology* 89(2):623–634.

Donaldson, J. E., R. Holdo, J. Sarakikya, and T. M. Anderson. 2022. Fire, grazers, and browsers interact with grass competition to determine tree establishment in an African savanna. *Ecology* 103(8):e3715.

Draper, J. P., J. K. Young, E. W. Schupp, N. G. Beckman, and T. B. Atwood. 2022. Frugivory and seed dispersal by carnivorans. *Frontiers in Ecology and Evolution* 10:864864.

Estes, J. A., M. T. Tinker, T. M. Williams, and D. F. Doak. 1998. Killer whale predation on sea otters linking oceanic and nearshore ecosystems. *Science* 282(5388):473–476.

Estes, J. A., A. Burdin, and D. F. Doak. 2016. Sea otters, kelp forests, and the extinction of Steller's sea cow. *Proceedings of the National Academy of Sciences* 113(4):880–885.

Estes, James A., and John F. Palmisano. 1974. Sea otters: their role in structuring nearshore communities. *Science* 185(4156):1058–1060.

Estes, J. A., et al. 2011. Trophic downgrading of planet Earth. *Science* 333(6040):301–306.

Fancourt, B. A., P. Cremasco, C. Wilson, and M. N. Gentle. 2019. Do introduced apex predators suppress introduced mesopredators? A multiscale spatiotemporal study of dingoes and feral cats in Australia suggests not. *Journal of Applied Ecology* 56(12):2584–2595.

Fariña, J. M., S. Salazar, K. P. Wallem, J. D. Witman, and J. C. Ellis. 2003. Nutrient exchanges between marine and terrestrial ecosystems: the case of the Galapagos sea lion Zalophus Wollebaecki. *Journal of Animal Ecology* 72(5):873–887.

Feng, Z., et al. 2012. Plant toxins and trophic cascades alter fire regime and succession on a boreal forest landscape. *Ecological Modelling* 244(October):79–92.

Fernández, N., L. M. Navarro, and H. M. Pereira. 2017. Rewilding: a call for boosting ecological complexity in conservation. *Conservation Letters* 10(3):276–278.

Finke, D. L. and R. F. Denno. 2004. Predator diversity dampens trophic cascades. *Nature* 429(6990):407–410.

Flagel, D. G., G. E. Belovsky, and D. E. Beyer. 2016. Natural and experimental tests of trophic cascades: gray wolves and white-tailed deer in a Great Lakes forest. *Oecologia* 180(4):1183–1194.

Folke, C., et al. 2004. Regime shifts, resilience, and biodiversity in ecosystem management. *Annual Review of Ecology, Evolution, and Systematics* 35(1):557–581.

Ford, A. T., and J. R. Goheen. 2015. Trophic cascades by large carnivores: a case for strong inference and mechanism. *Trends in Ecology & Evolution* 30(12):725–735.

Ford, A. T., et al. 2015. Recovery of African wild dogs suppresses prey but does not trigger a trophic cascade. *Ecology* 96(10):2705–2714.

Ford, A. T., et al. 2014. Large carnivores make savanna tree communities less thorny. *Science* 346(6207):346–349.

Gable, T. D., S. M. Johnson-Bice, A. T. Homkes, S. K. Windels, and J. K. Bump. 2020. Outsized effect of predation: wolves alter wetland creation and recolonization by killing ecosystem engineers. *Science Advances* 6(46):eabc5439.

Gilbert, S., N. Carter, and R. Naidoo. 2021. Predation services: quantifying societal effects of predators and their prey. *Frontiers in Ecology and the Environment* 19(5): 292–299.

Gilbert, S. L., et al. 2017. Socioeconomic benefits of large carnivore recolonization through reduced wildlife-vehicle collisions. *Conservation Letters* 10(4):431–439.

Gordon, C. E., D. J. Eldridge, W. J. Ripple, M. S. Crowther, B. D. Moore, and M. Letnic. 2017. Shrub encroachment is linked to extirpation of an apex predator. *Journal of Animal Ecology* 86(1):147–157.

Gordon, C. E., A. Feit, J. Grüber, and M. Letnic. 2015. Mesopredator suppression by an apex predator alleviates the risk of predation perceived by small prey. *Proceedings of the Royal Society B: Biological Sciences* 282(1802):20142870.

Gregr, E. J., et al. 2020. Cascading social-ecological costs and benefits triggered by a recovering keystone predator. *Science* 368(6496):1243–1247.

Grover, H. D., and H. B. Musick. 1990. Shrubland encroachment in southern New Mexico, U.S.A.: an analysis of desertification processes in the American southwest. *Climatic Change* 17(2):305–330.

Halaj, J. and D. H. Wise. 2001. Terrestrial trophic cascades: how much do they trickle?. *The American Naturalist* 157(3):262–281.

Harrer, L. E. F. and T. Levi. 2018. The primacy of bears as seed dispersers in salmon-bearing ecosystems. *Ecosphere* 9(1):e02076.

Haswell, P. M., J. Kusak, and M. W. Hayward. 2017. Large carnivore impacts are context-dependent. *Food Webs* 12(September):3–13.

Hebblewhite, M., et al. 2005. Human activity mediates a trophic cascade caused by wolves. *Ecology* 86(August):2135–2144.

Henke, S. E. and F. C. Bryant. 1999. Effects of coyote removal on the faunal community in Western Texas. *The Journal of Wildlife Management* 63(4):1066–1081.

Hessing-Lewis, M., et al. 2018. Ecosystem features determine seagrass community response to sea otter foraging. *Marine Pollution Bulletin* 134(September): 134–144.

Hobbs, N. T., D. B. Johnston, K. N. Marshall, E. C. Wolf, and D. J. Cooper. 2024. Does restoring apex predators to food webs restore ecosystems? Large carnivores in Yellowstone as a model system. *Ecological Monographs* 94 (2):e1598.

Hoeks, S., M. A. J. Huijbregts, M. Busana, M. B. J. Harfoot, J.-C. Svenning, and L. Santini. 2020. Mechanistic insights into the role of large carnivores for ecosystem structure and functioning. *Ecography* 43(12):1752–1763.

Holling, C S. 1973. Resilience and stability of ecological systems. *Annual Review of Ecology and Systematics* 4(1):1–23.

Hollings, T., M. Jones, N. Mooney, and H. McCallum. 2016. Disease-induced decline of an apex predator drives invasive dominated states and threatens biodiversity. Ecology 97 (2): 394–405.

Hoy, S. R., J. A. Vucetich, and R. O. Peterson. 2022. The role of wolves in regulating a chronic non-communicable disease, osteoarthritis, in prey populations. *Frontiers in Ecology and Evolution* 10:819137.

Hughes, B. B., et al. 2013. Recovery of a top predator mediates negative eutrophic effects on seagrass. *Proceedings of the National Academy of Sciences* 110(38):15313–15318.

Hughes, B. B., K. K. Hammerstrom, N. E. Grant, U. Hoshijima, R. Eby, and K. Wasson. 2016. Trophic cascades on the edge: fostering seagrass resilience via a novel pathway. *Oecologia* 182(1):231–241.

Hunter, D. O., M. Lagisz, V. Leo, S. Nakagawa, and M. Letnic. 2018. Not all predators are equal: a continent-scale analysis of the effects of predator control on Australian mammals. *Mammal Review* 48(2):108–122.

Ingeman, K. E., et al. 2022. Glimmers of hope in large carnivore recoveries. *Scientific Reports* 12(1):10005.

Johnson, C. N. and J. L. Isaac. 2009. Body mass and extinction risk in Australian marsupials: the 'critical weight range' revisited. *Austral Ecology* 34(1):35–40.

Kauffman, M. J., J. F. Brodie, and E. S. Jules. 2010. Are wolves saving Yellowstone's Aspen? A landscape-level test of a behaviorally mediated trophic cascade. *Ecology* 91(9):2742–2755.

Kohl, M. T., et al. 2018. Diel predator activity drives a dynamic landscape of fear. *Ecological Monographs* 88(4):638–652.

Kojola, I., et al. 2009. European wild forest reindeer and wolves: endangered prey and predators. *Annales Zoologici Fennici* 46(6):416–422.

Kreplins, T. L., M. S. Kennedy, R. A. O'Leary, P. J. Adams, S. J. Dundas, and P. A. Fleming. 2021. Fighting like cats and dogs? Dingoes do not constrain spatial and temporal movements of feral cats. *Food Webs* 27(June): e00173.

Kuijper, D. P. J., et al. 2016. Paws without claws? Ecological effects of large carnivores in anthropogenic landscapes. *Proceedings of the Royal Society B: Biological Sciences* 283(1841):20161625.

Lamelas-López, L., X. Pardavila, P. A. V. Borges, M. Santos-Reis, I. R. Amorim, and M. J. Santos. 2020. Modelling the distribution of *Mustela Nivalis* and *M. Putorius* in the Azores Archipelago based on native and introduced ranges. *PLoS One* 15(8):e0237216.

Lamelas-López, L. and M. João Santos. 2021. Factors influencing the relative abundance of invasive predators and omnivores on islands. *Biological Invasions* 23(9): 2819–2830.

Laundré, J. W., L. Hernández, and K. B. Altendorf. 2001. Wolves, elk, and bison: reestablishing the 'landscape of fear' in Yellowstone National Park, U.S.A. *Canadian Journal of Zoology* 79(8):1401–1409.

Letnic, M. and M. S. Crowther. 2013. Patterns in the abundance of kangaroo populations in arid Australia are consistent with the exploitation ecosystems hypothesis. *Oikos* 122(5):761–769.

Letnic, M. and F. Koch. 2010. Are dingoes a trophic regulator in arid Australia? A comparison of mammal communities on either side of the dingo fence. *Austral Ecology* 35(2):167–175.

Levi, T., A. Marm Kilpatrick, M. Mangel, and C. C. Wilmers. 2012. Deer, predators, and the emergence of Lyme disease. *Proceedings of the National Academy of Sciences* 109(27):10942–10947.

Lewis, P.-M., G. L. Burns, and D. Jones. 2017. Response and responsibility: humans as apex predators and ethical actors in a changing societal environment. *Food Webs* 12(September):49–55.

Lorimer, J., C. Sandom, P. Jepson, C. Doughty, M. Barua, and K. J. Kirby. 2015. Rewilding: science, practice, and politics. *Annual Review of Environment and Resources* 40(1):39–62.

Lundgren, E. J., et al. 2022. A novel trophic cascade between cougars and feral donkeys shapes desert wetlands. *Journal of Animal Ecology* 91(12):2348–2357.

Malhi, Y., C. E. Doughty, M. Galetti, F. A. Smith, J.-C. Svenning, and J. W. Terborgh. 2016. Megafauna and ecosystem function from the Pleistocene to the Anthropocene. *Proceedings of the National Academy of Sciences* 113(4):838–846.

Maron, J. L. and D. E. Pearson. 2011. Vertebrate predators have minimal cascading effects on plant production or seed predation in an intact grassland ecosystem. *Ecology Letters* 14(7):661–669.

Marshall, K. N., A. C. Stier, J. F. Samhouri, R. P. Kelly, and E. J. Ward. 2016. Conservation challenges of predator recovery. *Conservation Letters* 9(1):70–78.

McInturff, A., C. E. B. Cannon, P. S. Alagona, and D. N. Pellow. 2021. Meeting at the crossroads: an environmental justice framework for large carnivore reintroductions and recoveries. *Elementa: Science of the Anthropocene* 9(1):00172.

Meserve, P. L., D. A. Kelt, W. B. Milstead, and J. R. Gutiérrez. 2003. Thirteen years of shifting top-down and bottom-up control. *BioScience* 53(7):633–646.

Middleton, A. D., et al. 2013. Linking anti-predator behaviour to prey demography reveals limited risk effects of an actively hunting large carnivore. *Ecology Letters* 16(8):1023–1030.

Monk, J. D., and O. J. Schmitz. 2022. Landscapes shaped from the top down: predicting cascading predator effects on spatial biogeochemistry. *Oikos* 2022(5):e08554.

Montgomery, R. A., K. Borona, H. Kasozi, T. Mudumba, and M. Ogada. 2020. Positioning human heritage at the center of conservation practice. *Conservation Biology* 34(5):1122–1130.

Montgomery, R. A., R. J. Moll, E. Say-Sallaz, M. Valeix, and L. R. Prugh. 2019. A tendency to simplify complex systems. *Biological Conservation* 233(May):1–11.

Morris, T. and M. Letnic. 2017. Removal of an apex predator initiates a trophic cascade that extends from herbivores to vegetation and the soil nutrient pool. *Proceedings of the Royal Society B: Biological Sciences* 284(1854):20170111.

Motta Junior, J. C. and K. Martins. 2002. The frugivorous diet of the maned wolf, chrysocyon brachyurus, in Brazil: ecology and conservation. Seed Dispersal and Frugivory: Ecology, Evolution and Conservation. Third International Symposium-Workshop on Frugivores and Seed Dispersal, SO Pedro, Brazil, 6–11 August 2000, CABI Books, January, 291–303.

Mueller, E. N., J. Wainwright, and A. J. Parsons. 2007. The stability of vegetation boundaries and the propagation of desertification in the American southwest: a modelling approach. *Ecological Modelling* 208(2–4):91–101.

Muhly, T. B., M. Hebblewhite, D. Paton, J. A. Pitt, M. S. Boyce, and M. Musiani. 2013. Humans strengthen bottom-up effects and weaken trophic cascades in a terrestrial food web. *PLoS One* 8(5):e64311.

Newsome, T. M. and W. J. Ripple. 2015. A continental scale trophic cascade from wolves through coyotes to foxes. *Journal of Animal Ecology* 84(1):49–59.

O'Bryan, C. J., A. R. Braczkowski, H. L. Beyer, N. H. Carter, J. E. M. Watson, and E. McDonald-Madden. 2018. The contribution of predators and scavengers to human well-being. *Nature Ecology & Evolution* 2(2):229–236.

O'Bryan, C. J., M. H. Holden, and J. E. M. Watson. 2019. The mesoscavenger release hypothesis and implications for ecosystem and human well-being. *Ecology Letters* 22(9):1340–1348.

Ordiz, A., R. Bischof, and J. E. Swenson. 2013. Saving large carnivores, but losing the apex predator? *Biological Conservation* 168(December):128–133.

Oriol-Cotterill, Alayne, Marion Valeix, Laurence G. Frank, Corinna Riginos, and David W. Macdonald. 2015. Landscapes of coexistence for terrestrial carnivores: the ecological consequences of being downgraded from ultimate to penultimate predator by humans. *Oikos* 124(10):1263–1273.

Oro, D., M. Genovart, G. Tavecchia, M. S. Fowler, and A. Martínez-Abraín. 2013. Ecological and evolutionary implications of food subsidies from humans. *Ecology Letters* 16(12):1501–1514.

Österblom, H., et al. 2007. Human-induced trophic cascades and ecological regime shifts in the Baltic Sea. *Ecosystems* 10(6):877–889.

Ostfeld, R. S. and R. D. Holt. 2004. Are predators good for your health? Evaluating evidence for top-down regulation of zoonotic disease reservoirs. *Frontiers in Ecology and the Environment* 2(1):13–20.

Pace, M. L., J. J. Cole, S. R. Carpenter, and J. F. Kitchell. 1999. Trophic cascades revealed in diverse ecosystems. *Trends in Ecology & Evolution* 14(12):483–488.

Packer, C., R. D. Holt, P. J. Hudson, K. D. Lafferty, and A. P. Dobson. 2003. Keeping the herds healthy and alert: implications of predator control for infectious disease. *Ecology Letters* 6(9):797–802.

Páez-Rosas, D., X. Moreno-Sánchez, A. Tripp-Valdez, F. R. Elorriaga-Verplancken, and S. Carranco-Narváez. 2020. Changes in the Galapagos sea lion diet as a response to

El Niño-Southern oscillation. *Regional Studies in Marine Science* 40(November):101485.

Piovia-Scott, J., L. H. Yang, and A. N. Wright. 2017. Temporal variation in trophic cascades. *Annual Review of Ecology Evolution and Systematics* 48(1):1–20.

Power, M. E. 1990. Effects of fish in river food webs. *Science* 250(4982):811–814.

Preisser, E. L., J. L. Orrock, and O. J. Schmitz. 2007. Predator hunting mode and habitat domain alter nonconsumptive effects in predator–prey interactions. *Ecology* 88(11):2744–2751.

Pringle, R. M., et al. 2023. Impacts of large herbivores on terrestrial ecosystems. *Current Biology* 33(11):R584–610.

Proffitt, K. M., J. L. Grigg, K. L. Hamlin, and R. A. Garrott. 2009. Contrasting effects of wolves and human hunters on elk behavioral responses to predation risk. *Journal of Wildlife Management* 73(3):345–356.

Prowse, T. A. A., C. N. Johnson, P. Cassey, C. J. A. Bradshaw, and B. W. Brook. 2015. Ecological and economic benefits to cattle rangelands of restoring an apex predator. *Journal of Applied Ecology* 52(2):455–466.

Prugh, L. R., et al. 2009. The rise of the mesopredator. *BioScience* 59(9):779–791.

Raymond, W. W., B. B. Hughes, T. A. Stephens, C. R. Mattson, A. T. Bolwerk, and G. L. Eckert. 2021. Testing the generality of sea otter-mediated trophic cascades in seagrass meadows. *Oikos* 130(5):725–738.

Rezende, F., P. A. P. Antiqueira, O. L. Petchey, L. F. M. Velho, L. C. Rodrigues, and G. Q. Romero. 2021. Trophic downgrading decreases species asynchrony and community stability regardless of climate warming. *Ecology Letters* 24(12):2660–2673.

Ripple, W. J. and R. L. Beschta. 2003. Wolf reintroduction, predation risk, and cottonwood recovery in Yellowstone National Park. *Forest Ecology and Management* 184(1):299–313.

Ripple, W. J. and R. L. Beschta. 2004. Wolves and the ecology of fear: can predation risk structure ecosystems?. *BioScience* 54(8):755–766.

Ripple, W. J. and R. L. Beschta. 2006. Linking a cougar decline, trophic cascade, and catastrophic regime shift in Zion National Park. *Biological Conservation* 133(4):397–408.

Ripple, W. J. and R. L. Beschta. 2007. Restoring Yellowstone's aspen with wolves. *Biological Conservation* 138(3):514–519.

Ripple, W. J., et al. 2014. Status and ecological effects of the world's largest carnivores. *Science* 343(6167): 1241484.

Ripple, W. J., et al. 2016. What is a trophic cascade?. *Trends in Ecology & Evolution* 31(11):842–849.

Ritchie, E. G., B. Elmhagen, A. S. Glen, M. Letnic, G. Ludwig, and R. A. McDonald. 2012. Ecosystem restoration with teeth: what role for predators?. *Trends in Ecology & Evolution* 27(5):265–271.

Ritchie, E. and C. Johnson. 2009. Predator interactions, mesopredator release and biodiversity conservation. *Ecology Letters* 12(August):982–998.

Roffler, G. H., C. E. Eriksson, J. M. Allen, and T. Levi. 2023. Recovery of a marine keystone predator transforms terrestrial predator–prey dynamics. *Proceedings of the National Academy of Sciences* 120(5):e2209037120.

Rohr, J. R., E. S. Bernhardt, M. W. Cadotte, and W. H. Clements. 2018. The ecology and economics of restoration: when, what, where, and how to restore ecosystems. *Ecology and Society* 23(2):5636603.

Rominger, E. M. 2018. The Gordian knot of mountain lion predation and bighorn sheep. *Journal of Wildlife Management* 82(1):19–31.

Rose, J. R. and D. J. Cooper. 2017. The influence of floods and herbivory on cottonwood establishment and growth in Yellowstone National Park. *Ecohydrology* 10(1):e1768.

Rouet-Leduc, J., et al. 2021. Effects of large herbivores on fire regimes and wildfire mitigation. *Journal of Applied Ecology* 58(12):2690–2702.

Roux, E. le, G. I. H. Kerley, and J. P. G. M. Cromsigt. 2018. Megaherbivores modify trophic cascades triggered by fear of predation in an African savanna ecosystem. *Current Biology* 28(15):2493–2499.e3.

Rubenstein, D. R. and D. I. Rubenstein. 2016. From Pleistocene to trophic rewilding: a wolf in sheep's clothing. *Proceedings of the National Academy of Sciences of the United States of America* 113(1):E1.

Russell, M. B., C. W. Woodall, K. M. Potter, B. F. Walters, G. M. Domke, and C. M. Oswalt. 2017. Interactions between white-tailed deer density and the composition of forest understories in the northern United States. *Forest Ecology and Management* 384(January): 26–33.

Sahlén, V., A. Ordiz, J. E. Swenson, and O. G. Støen. 2015. Behavioural differences between single Scandinavian brown bears (*Ursus Arctos*) and females with dependent young when experimentally approached by humans. *PLoS One* 10(4):e0121576.

Scheffer, M., S. Carpenter, J. A. Foley, C. Folke, and B. Walker. 2001. Catastrophic shifts in ecosystems. 1*Nature* 413(6856):591–596.

Schmitz, O. J. 2008. Effects of predator hunting mode on grassland ecosystem function. *Science* 319(5865): 952–954.

Schmitz, O. J., P. A. Hambäck, A. P. Beckerman, S. D. Gaines, and M. A. Leibold. 2000. Trophic cascades in terrestrial systems: a review of the effects of carnivore removals on plants. *The American Naturalist* 155(2): 141–153.

Schweiger, A. H., I. Boulangeat, T. Conradi, M. Davis, and J.-C. Svenning. 2019. The importance of ecological memory for trophic rewilding as an ecosystem restoration approach. *Biological Reviews* 94(1):1–15.

Shurin, J. B., et al. 2002. A cross-ecosystem comparison of the strength of trophic cascades. *Ecology Letters* 5(6): 785–791.

Shurin, J. B. and E. W. Seabloom. 2005. The strength of trophic cascades across ecosystems: predictions from allometry and energetics. *Journal of Animal Ecology* 74(6):1029–1038.

Silliman, B. R., et al. 2018. Are the ghosts of nature's past haunting ecology today?. *Current Biology* 28(9): R532–R537.

Smith, F. A., E. A. E. Smith, A. Villaseñor, C. P. Tomé, S. K. Lyons, and S. D. Newsome. 2022. Late Pleistocene megafauna extinction leads to missing pieces of ecological space in a North American mammal community. *Proceedings of the National Academy of Sciences* 119(39):e2115015119.

Smith, J. A., Y. Wang, and C. C. Wilmers. 2015. Top carnivores increase their kill rates on prey as a response to human-induced fear. *Proceedings of the Royal Society B: Biological Sciences* 282(1802):20142711.

Stalmans, M. E., T. J. Massad, M. J. S. Peel, C. E. Tarnita, and R M. Pringle. 2019. War-induced collapse and asymmetric recovery of large-mammal populations in Gorongosa National Park, Mozambique. *PLoS One* 14(3):e0212864.

Stier, A C., et al. 2016. Ecosystem context and historical contingency in apex predator recoveries. *Science Advances* 2(5):e1501769.

Strong, D. R. 1992. Are trophic cascades all wet? Differentiation and donor-control in speciose ecosystems. *Ecology* 73(3):747–754.

Suraci, J. P., M. Clinchy, L. M. Dill, D. Roberts, and L. Y. Zanette. 2016. Fear of large carnivores causes a trophic cascade. *Nature Communications* 7(1):10698.

Svenning, J.-C., et al. 2016. Science for a wilder anthropocene: synthesis and future directions for trophic rewilding research. *Proceedings of the National Academy of Sciences* 113(4):898–906.

Tanner, E., A. White, P. Acevedo, A. Balseiro, J. Marcos, and C. Gortázar. 2019. Wolves contribute to disease control in a multi-host system. *Scientific Reports* 9(1):7940.

Terborgh, J. and J. A. Estes. 2013. *Trophic Cascades: Predators, Prey, and the Changing Dynamics of Nature.* Island Press, Washington, D. C.

Terborgh, J., et al. 2001. Ecological meltdown in predator-free forest fragments. *Science* 294(5548):1923–1926.

Thinley, P., et al. 2018. The ecological benefit of tigers (*Panthera Tigris*) to farmers in reducing crop and livestock losses in the Eastern Himalayas: implications for conservation of large apex predators. *Biological Conservation* 219(March):119–125.

Thomas, S., et al. 2023. Evaluating the performance of conservation translocations in large carnivores across the world. *Biological Conservation* 279(March):109909.

Van Langevelde, F., et al. 2003. Effects of fire and herbivory on the stability of savanna ecosystems. *Ecology* 84(2):337–350.

Weise, F. J., K. J. Stratford, and R. J. van Vuuren. 2014. Financial costs of large carnivore translocations—accounting for conservation. *PLoS One* 9(8):e105042.

Wilmers, C. C, J. A Estes, M. Edwards, K. L Laidre, and B. Konar. 2012. Do trophic cascades affect the storage and flux of atmospheric carbon? An analysis of sea otters and kelp forests. *Frontiers in Ecology and the Environment* 10(8):409–415.

Wilmers, C. C. and O. J. Schmitz. 2016. Effects of gray wolf-induced trophic cascades on ecosystem carbon cycling. *Ecosphere* 7(10):e01501.

Worm, B. and R. T. Paine. 2016. Humans as a hyperkeystone species. *Trends in Ecology & Evolution* 31(8): 600–607.

Yeats, M. 2010. *The Restoration of Gorongosa National Park: A Project To Conserve Wildlife Whilst Assisting Impoverished Communities.* Available at: https://biblioteca.biofund.org.mz/wp-content/uploads/2019/01/1548760933-F275.The%20restoration%20of%20Gorongosa%20National%20Park.pdf

Yirga, G., et al. 2015. Spotted hyena (*Crocuta Crocuta*) concentrate around urban waste dumps across Tigray, Northern Ethiopia. *Wildlife Research* 42(7):563–569.

Young, H. S., et al. 2015. Context-dependent effects of large-wildlife declines on small-mammal communities in central Kenya. *Ecological Applications* 25(2): 348–360.

SECTION FIVE

The Effects of Conservation on Carnivore Populations

CHAPTER 13

Contemporary Land-Sparing Initiatives and Intergenerational Equity in Carnivore Conservation

Fikirte Gebresenbet, Siewe Siewe, and Fidelis A. Atuo

Introduction

Conservation of biological diversity stands as one of the most pressing global concerns, particularly when it comes to large carnivore populations. These top predators occupy a unique ecological niche and hold a pivotal role as keystone species within their ecosystems, with many species facing severe threats, including habitat loss, human–wildlife conflicts, and illegal hunting (Ripple et al. 2014). In comparison to many other species, large carnivores require vast home ranges and expansive geographical territories to sustain their populations. It is imperative that carnivore conservation efforts underscore this inherent requirement of extensive home ranges and territories by safeguarding large areas of suitable habitat protected from harmful anthropogenic disturbances. Moreover, human–carnivore conflicts are a common issue, often arising when carnivores pose threats on properties, livestock, or human lives. One obvious solution lies in the establishment of protected areas, where human activities are either restricted or carefully managed. Whereas these protected areas can serve as crucial havens for carnivores and other wildlife, they can become sources of distress causing reprisal attacks of the very species they were established to protect when the core principles of equity are breached.

The Convention on Biological Diversity (CBD) has set specific targets for increasing the coverage of protected areas as a crucial strategy in the collective endeavour to protect biodiversity and conserve ecosystems. These targets have evolved over time, with the latest goal being the protection of at least 30% of the planet's land and ocean by 2030 (CBD 2020). Moreover, the CBD places significant emphasis on equitable governance within these protected areas, ensuring that the distribution of benefits and costs is fair and balanced among stakeholders. This inclusive approach is recognized by CBD to be inextricably linked to the success of protected area management (CBD 2020). Thus, in addition to ecological challenges, it's important to consider the social, cultural, and other human perspectives surrounding carnivore conservation. This involves acknowledging the identities and values associated with large carnivores and allowing local communities to have true participation in conservation and decision-making processes. In addition, addressing concerns such as intergenerational equity would form a cornerstone in bestowing a legacy of biodiversity for future generations. Building on the CBD goal, carnivore biologists must also explore contemporary land-sharing and sparing initiatives for effective carnivore conservation. These strategies involve setting aside specific areas for carnivore conservation while diligently minimizing human impact on these designated areas. Such initiatives come with their own set of challenges and opportunities, some of which are discussed in this chapter.

Fikirte Gebresenbet, Siewe Siewe, and Fidelis A. Atuo, *Contemporary Land-Sparing Initiatives and Intergenerational Equity in Carnivore Conservation*.
In: *Apex Predators in the Anthropocene*. Edited by: Robert A. Montgomery and Remington J. Moll, Oxford University Press.
 DOI: 10.1093/9780191996665.003.0013

Carnivore conservation and its interface with culture, politics, and identity

People are a critical piece of conservation and hold the 'power' to determine the success or failure of specific interventions. This point is receiving increasing attention over time and various organizations are involving local communities, including Indigenous people, to maximize conservation success. This complex interplay of ecological conditions, sociopolitical relations, and cultural practices results in distinct environments and particular systems of environmental governance (Nygren and Rikoon 2008). Attempts at conservation and the creation of protected areas had been described as 'a way of seeing, understanding, and (re)producing the world' (West et al. 2006). This 'way of seeing, understanding, and (re)producing the world' is what political ecologists regard as a form of control of resources, loaded with symbols, meanings, and aesthetics that all influence the way humans interpret their environments (Neumann 1996, Robbins 2012).

Political ecologists examine the origins, consequences, and context of relationships between varying levels of human structures and how these impact the environment (Robbins 2012). Most human societies are avaricious, placing the pursuit of wealth ahead of other motivations, exerting significant pressure on the environment (Gentina et al. 2021). This sets off a domino effect of environmental deterioration, with each piece 'domino' intensifying its pressure on the environment, and certain groups of society experiencing an unequal share of the burden, leading to their marginalization. Marginalization is a cyclical process of pushing disempowered individuals to inhabit precarious or economically unproductive areas causing them to exert excessive pressure on the ecosystem as they strive to meet their basic needs (Robbins 2012). Understanding the contemporary issues in human dimensions of carnivore conservation is the foundation of any successful conservation effort. This must start with recognizing not only the origins of conservation but also, of its underlying power dynamics and intricate relationship with identities.

The concept of identity has been extensively examined by social scientists, and it is a complex construct rooted in sociology and social psychology that can be used to define an individual's or a group's 'image' (Tajfel and Turner 1979). Identity encompasses how individuals or groups see themselves or are seen by others, their characteristics, beliefs, values, and the affiliations that make them distinct and distinguishes them from others (Hall 1990). Social identity theory allows individuals or groups to self-categorize themselves while expressing their identity through normative behaviours (Hogg 2016). Identity encompasses cultural, social, ethnic, national, administrative, and personal traits. Place identity is the inclusion of a place as one of the identity-defining factors/traits of an individual. Individuals who have a strong attachment to a specific place or generally the environment are more likely to take a step toward conservation efforts or environmental protection (Gooden 2019).

Individual and group identity have played, and continue to play, a significant role in conservation discourse and the design of conservation plans. In terms of cultural identities, Dawson et al. (2021) discuss the role of Indigenous peoples and local communities in achieving effective and equitable conservation. They emphasize the importance of Traditional Ecological Knowledge (TEK) in achieving effective and equitable conservation. This point is further supported by Diaz et al. (2015), who discuss the integration of traditional knowledge into conservation policies, highlighting its potential for enhancing adaptive management strategies. Conservation policies that respect Indigenous knowledge are critical to successful carnivore conservation. Lute et al. (2020) and Torrents-Tico et al. (2021) highlight the ethics surrounding carnivore conservation and emphasize the importance of aligning culturally sensitive conservation practices with Indigenous values.

Social identities involve areas of gender, ethnicity, and socioeconomic status. It is a person's sense of who they are, based on the group they belong to. Social identities affect the design, structure, and success of conservation policies. For example, Agarwal (2000) investigated the role of gender in the management of community forests in India and Nepal. The study examined gendered roles in collective action, social norms and power dynamic, and women's environmental knowledge, concluding that understanding the gendered dimensions of environmental collective action is essential in designing inclusive and effective environmental policies and initiatives. Agrawal (2000) concluded that households and communities are 'gendered units' and that neglecting the gender dimension can significantly impact the understanding of collective action in environmental initiatives. Bele et al. (2014) also showed that the roles of men and women within households and communities influence the management of resources because women often possess unique ecological knowledge and make significant contributions to conservation practices.

The impact of differing livelihood choices in a single area can also lead subsets of a community to have different opinions or attitudes towards a specific conservation component or effort. For example, in Gujjara (northern India), the preferences of pastoralists who make up about 86% of the Gujjar population differed from the remaining 14%, who had multiple sources of income, in their choice of coexistence strategies to support the tiger population recovery programme at the landscape level (Harihar et al 2015). The point here is simple: households and communities are diverse in their makeup, structure, and identity. Understanding these differences contributes to the framework, longevity, and overall success of carnivore conservation projects.

Intergenerational equity in biodiversity conservation

A sustainable world is characterized by the equitable fulfilment of current human needs while safeguarding the ability of future generations to meet their own needs. This vision of sustainability, however, hinges upon the invaluable ecosystem services furnished by the Earth's biodiversity and wild ecosystems, encompassing functions such as air and water purification, climate regulation, carbon sequestration, cultural and aesthetic amenities, disease and pest control, energy resources, fire regulation, food provision, habitat support, pharmaceutical resources, land stability, natural disaster mitigation, pollination services, wind resistance, and seed dispersal (Costanza et al. 1997, Hayward et al. 2022). Ensuring that future generations enjoy these resources is not only a responsibility the current generation owes, but an important precept of equity and justice.

Within the realm of conservation biology, the foundational principle of intergenerational equity underscores the imperative of fairly distributing resources and benefits across successive generations. It emphasizes a moral obligation for the present generation to manage natural resources judiciously, preserving their capacity to meet the needs of posterity. This requires developing and managing resources, social structures, contexts, and socioecological systems so that future generations can access natural resources in a way that fulfil their needs. Regrettably, the unsustainable exploitation of natural resources poses a formidable barrier to the realization of intergenerational equity, potentially hindering the prospect of subsequent generations enjoying a superior quality of life in comparison to their predecessors (Hayward et al. 2022). What is more, the concept of intergenerational equity is intrinsically interwoven with the sustainability of decisions impacting human well-being, as it necessitates a judicious redistribution of well-being across temporal and spatial dimensions (Summers and Smith 2014). Temporally, it puts a moral responsibility on current decision-makers to enact policies that regulate human utilization of nature's resources that cut across a gradient of generational thoughts. Spatially, efforts to improve the diversity of thoughts must seek to eliminate barriers that have either intentionally or inadvertently minimized access to nature based on race, ethnicity, gender, or socioeconomic status (Davis 2019, Miriti et al. 2022, Schmidt and Garroway 2022). The ideals of intergenerational equity thus confront numerous challenges, including short-term thinking, rampant overexploitation, and pervasive environmental degradation which have plagued biodiversity management planning for so long (Shi et al. 2019, Elsawy and Youssef 2023).

Effectively operationalizing intergenerational equity in carnivore conservation necessitates the adoption of proactive strategies. These strategies should encompass all stakeholders across generational, ethnic, gender, economic, and sociocultural divides (Munasinghe 2001, Shi et al. 2019) when addressing critical carnivore conservation issues like habitat restoration, creation of protected areas, or climate action. In many parts of the world, carnivores co-occur with people in multifunctional landscapes where cultural and spiritual connections between humans and nature have been forged for millennia. As overlaps between industrial societies and carnivore habitats continue to grow, there are legitimate concerns from all stakeholders (e.g. Indigenous communities, conservation advocates, animal welfare advocates, etc.) that could affect carnivore populations and alter their conservation for millennia to come. In many countries, traditional and Indigenous communities bear the brunt of externally (government and NGOs) imposed conservation policies, such as displacement, restricted resource use, wildlife attacks and damage, fines, arrests and killing of poachers, and high levels of disempowerment (Atuọ et al. 2020, Mkonyi 2022) often with little or no compensation. Young people in these regions grow up with a disdain for conservation policies and those that enforce them, leading to non-compliance and increased illegal killings. Such an attitude, if persisted, represents a glimpse into a bleak future that could be the reality of carnivore conservation.

In other areas, such as the United States, a supermajority of young people and people of colour live in urban areas where opportunities for outdoor nature experiences are limited. An increasing number of studies are already recording a decline in emotional and cognitive connection to biodiversity conservation among young people (Soga et al 2015, Soga and Gaston 2016, Colléony et al 2020), although the magnitude and generality of the trend are not fully understood (Cazalis et al 2023). Without this connection, future generations are less likely to prioritize biodiversity conservation, and their behaviour may exacerbate further destruction of natural species. Indeed, by candidly acknowledging the challenges of intergenerational equity and capitalizing on its attendant opportunities, the current generation can bequeath a legacy of environmental robustness and well-being to future generations.

Equity and carnivore conservation

Carnivores, as apex predators, play a pivotal role in ecological equilibrium (Mattson et al. 2006, Treves et al. 2017), yet their conservation has suffered several human-related shortfalls that vary across cultures. In several societies, several carnivores once occupied a venerable status, intricately woven into spiritual beliefs and folklore. Consider the Jaguar (*Panthera onca*), emblematic of political and military power in ancient American civilizations and deeply enshrined in the Mayan cosmology, where it bore connotations of the underworld and immortality (Hayward et al. 2022). Such venerations most likely provided multiple layers of protection that benefited the species. Today, the species is restricted to approximately 45% of its historical range and experiences persistent threats from habitat loss and fragmentation, illegal killing in retaliation for livestock depredation, as well as illegal pet and animal parts trade (Harrington 2015, Ceballos et al. 2021).

The relentless triad of land-use change, human–carnivore conflict, and trade on carnivore body parts are partly to blame. However, the larger issues of social and environmental justice, exclusion of local knowledge and opinions from conservation policies, as well as a restricted demography that discriminate young people, racial minorities, and certain genders from policy decisions possess a greater threat (Koot et al. 2022). Indeed, carnivores' substantial foraging territories render them exceptionally susceptible to the perturbations arising from land-use alterations. These dynamics frequently culminate in the encroachment of human settlements into natural habitats, which can provoke confrontations with carnivores. For example, from 2015 to 2017, such conflicts resulted in the killing of 23 lions by a Maasai community in the Tarangire–Manyara ecosystem of East Africa (Eustace and Tarimo 2019), predominantly recorded as retaliatory killings for livestock depredation. In some cases, these killings are perpetuated by young people (Manqele et al. 2018) and could also reflect a change in worldview, values, and belief systems. Further, social and economic inequalities that promote youth unemployment or reduce access to gainful employment for minority communities could precipitate an unregulated harvest of natural resources for subsistence. This is evident in many African and Asian countries where carnivores are killed or captured to feed the persistent quandary of demand for carnivore body parts (Karanth and Chellam 2009, Atuo et al. 2015, Lam et al. 2023), the pet trade (Harrington 2015, Hellinx 2020), and bush meat (Doughty et al. 2015). Therefore, isolation of this demography and their needs from conservation planning not only helps to foster resentment and further carnivore persecution but also eliminates a sense of ownership and emotional affinity essential for carnivore conservation in the future. Moreover, there is ample evidence that when properly educated, young adults are more likely to support carnivore conservation even in developing countries (Megaze et al. 2017, Felix et al. 2022).

In contrast to Africa and Asia, carnivores like wolves, e.g. the grey wolf (*Canis lupus*) and the red wolf (*Canis rufus*), that bore the brunt of relentless persecution in the annals of North America and Europe during the nineteenth century are beginning to see fervent advocacy for their protection and reintroduction, reflecting an evolving societal attitude that should be propagated and encouraged elsewhere. Intriguingly, the social lens through which many people view carnivore conservation in these continents has metamorphosed positively in recent times. This transformation, catalysed by evolving human–animal interactions, is seeing a shift from the paradigm of dominance to one of mutualism. Traditional convictions of humans exerting control over wildlife are gradually giving way to an ethos that perceives wildlife as co-members of a shared environment whose existence is intricately tied to humans' sustenance. Furthermore, acknowledgement of the intrinsic worth of carnivores as apex predators whose absence from an ecosystem could trigger cascading ecological repercussions has been demonstrated in places like Yellowstone National Park, United States, with great media celebration (Ripple and Beschta 2012). Nonetheless, elements of intergenerational and systemic racial inequalities persist and limit the diversity of ideas needed for inclusive conservation impacts. For example, significant disparities still exist in access to urban nature parks between white and racial minority communities due to historical settlement policies that relegated black people to settlements around highways, garbage dumps, and factories, with limited access to green space (Finney 2014). In cities, urban green space is often where people of all ages interact the most with nature and forming cognitive attachments and appreciation of resources therein. Consequently, ethnic minorities and low-income households are disproportionately less likely than Whites and stable-income households to visit national parks, forest reserves, and zoos in the United States due to reasons like ease of access or concerns about racial discrimination (O'Brien et al. 2020). They are also less likely to support nature conservation due to a general lack of awareness and financial capacity.

Be it in the developed or developing countries, the exclusion of certain demographics or income groupings from conservation planning presents a unique problem and illuminates how much economic status could considerably tilt the scales of carnivore conservation. Economic development has been shown to overshadow conservation objectives by spurring habitat degradation and unsustainable resource exploitation. These could be exacerbated when the diversity of opinions excludes the voice/opinions of those most affected by conservation decisions. Flagship models of wildlife conservation have been unequivocal on the role that carnivores play in promoting conservation acceptance and eliciting substantial support for biodiversity conservation (Linnell et al. 2000, Rozylowicz et al. 2011, Crespo-Gascón and Guerrero-Casado 2019). Yet, their incorporation as an iconic model of intergenerational equity in biodiversity conservation is underexplored. Such embracement could augment the prospect of carnivore preservation, securing biodiversity conservation benefits for posterity.

Sharing the burden

The concept of intergenerational equity underscores the need for a fair distribution of both the costs and benefits associated with conservation efforts across diverse age groups and socioeconomic strata. In this context, it is imperative to acknowledge that young people are not passive bystanders, but rather, active agents of change as advocated by Hart (2013). Central to Hart's framework is the idea that youth should not be relegated to mere recipients of adult decisions; instead, they should be actively engaged in shaping their communities and environment. This concept lays the groundwork for what is known as meaningful youth participation, a principle that can yield more sustainable and equitable outcomes. Today's conservation efforts could be better served by recognizing the significance of affording youths opportunities to connect with the natural world, emphasizing the positive impacts of such experiences on their personal development and commitment to environmental conservation. Through these meaningful interactions, these young individuals not only acquire knowledge but also forge a profound emotional connection with the natural world.

Despite growing recognition of the importance of early childhood interactions with nature, these opportunities remain exiguous for young people, racial minorities, and low-income households around the world. The emergence of youth-led conservation organizations, such as Youth for Wildlife Conservation, stands as a testament to the pivotal energy that young people bring to the cause of biodiversity protection. Youth-led global protests and marches advocating for climate action and wildlife conservation have captured international attention, exerting substantial pressure on governments and conservation organizations to prioritize environmental health. In the same way, youth engagement could play a pivotal role in advocating for carnivore conservation and could initiate transformative changes through activism and innovation (Ballard et al. 2017, Chen et al. 2019). Their unparalleled level of passion and idealism, coupled with unwavering dedication to issues of importance, could serve as a potent driving force within the carnivore conservation movement.

The use of social media to connect with other youths and galvanize support for carnivore conservation across sociocultural and economic backgrounds can be particularly helpful here. Although evidence abounds on the negative impact of digital media on youths' connection with nature (Edwards and Larson 2020, Truong and Clayton 2020), growing up in the digital era offers several opportunities. Young carnivore advocates can leverage technology to amplify their voices, heighten awareness, and mobilize support for carnivore conservation. Case studies featuring youth-led carnivore conservation initiatives, such as the Youth for Lion Conservation Initiative in South Africa, clearly underscore the power of digital media on youths' mobilization. Run primarily by youth ambassadors, most of whom are less than 20 years old, the Youth for Lion Conservation Initiative is an arm of the Blood Lions NPC, which encourages media campaigns through social media posts to curb the commercialization of carnivore breeding, trade on lion body parts, and illegal hunting (Waal et al. 2022). Members are awarded Bronze, Silver, and Gold membership ranks depending on the number of social media posts. Together with their parent organization, Youth for Lion Initiative, efforts resulted in the production of the award-winning documentary *Blood Lions*, which premiered in 2015, drawing global attention and resources to lion conservation. On a global scale, youth are at the vanguard of developing and implementing innovative technologies, exemplified by the creation of applications and platforms that facilitate community engagement and foster a sense of shared responsibility.

Senior citizens, as a demographic group, offer a reservoir of experience and historical perspective that can profoundly enrich the discourse surrounding conservation efforts. Many seniors have dedicated their lifetimes to observing and connecting with the natural world, and their deep-seated appreciation for nature frequently serves as a wellspring of inspiration for younger generations to embrace environmental consciousness. A significant number of seniors have borne witness to the transformative changes that have unfolded in the environment over the course of their lives. Their firsthand accounts provide invaluable historical context for comprehending environmental shifts. For example, in several African cultures, oral history was the primary means of knowledge transfer and information preservation. Such intergenerational transfer of knowledge has shaped landscapes and livelihood practices for millennia. In fact, much of the early exploratory knowledge in natural science during colonial explorations in Africa relied on knowledge of Indigenous people passed down through the ancestral tree (Shava 2013). However, the institutionalization of acquired knowledge in languages foreign to the Indigenous people excluded them from the scientific concepts developed within the prevailing perspectives. Amidst the increasing complexity of global conservation challenges, striving to diversify conservation strategies could benefit from the intermesh of intergenerational knowledge of Indigenous peoples and the accumulating scientific information. The coupling of these sources of information could open up new pathways for re-imagining a future that is not only sustainable but more equitable in the management of natural resources (O'Donoghue et al. 2019). While other conservation fields are beginning to embrace this concept, the persistent decline in carnivore abundance in the face of large-scale land-use changes and persecution by local people is perhaps a reminder that carnivore biologists are behind the curve. Further, seniors frequently exhibit sustainable living practices, encompassing activities such as gardening, composting, and energy conservation, which function as exemplars for environmentally responsible lifestyles (Wright and Wadsworth 2014, Moonstone 2016, Morse 2018). The reservoir of knowledge and perspectives they offer concerning conservation forms a robust foundation for advancing the cause of intergenerational conservation.

Carnivore conservation, as expected, is a global imperative, propelled by the need to safeguard apex predators and uphold ecological equilibrium. However, the efficacy of conservation endeavours is often intricately bound to the distribution of wealth, wisdom, and influence within societies. The pursuit of economic growth often precipitates habitat destruction and resource extraction, with adverse repercussions for carnivore populations (Chanchani et al. 2014, Singh et al. 2017). Therefore, the principle of intergenerational equity requires the coming together of individuals across a spectrum of socioeconomic divides. Affluent individuals and corporations could assume a pivotal role in funding conservation initiatives and supporting critical research. Those in positions of influence wield the power to shape policies that reverberate throughout the realm of carnivore conservation, encompassing domains like land use, hunting regulations, and protected area management. Nonetheless, local communities can partake in conservation decisions, which redistributes authority and fosters a collective sense of responsibility.

Carnivore biologists also need some mechanisms to promote thinking and practice at global levels. Given the accumulation of wealth in some world regions and the presence of larger potentials for large carnivore conservation in others, it would be necessary for wealth transfers to be considered for effective conservation. Particularly, Africa and Asia would need to receive performance-based funds/grants to offset immediate economic gains put on hold for ecological interests. The burden sharing would, thus, need to happen between different age groups, socioeconomic classes, and world regions.

Present decisions, future impacts: ethical concerns and moral responsibilities

As we emphasize the pivotal ecological niche that carnivores occupy in diverse ecosystems, we acknowledge that the mandate to safeguard them transcends the current era. The decisions taken today concerning carnivore management, habitat preservation, and human–carnivore coexistence carry far-reaching consequences for how future generations will interact with these species. Choices pertaining to the regulation of carnivore hunting and poaching wield direct influence over carnivore populations and the dividends that future generations will reap from the ecological services these creatures provide. The promotion of coexistence strategies, such as predator-friendly livestock practices, holds the power to mould the attitudes and behaviours of future generations toward carnivores. Investing in environmental education programmes today plays a crucial role in fostering a deep sense of respect and understanding among future generations for carnivores and their vital role in maintaining ecological balance. Nurturing this appreciation can ensure that tomorrow's leaders are well-informed about the significance of these predators in preserving biodiversity and supporting healthy ecosystems. Moreover, initiatives aimed at involving local communities in carnivore conservation serve to instil a sense of responsibility and stewardship that transcends generations.

Though some meaningful achievements have been made by the broader conservation field to integrate young minds into conservation efforts, youth's involvement in carnivore conservation is still comparatively behind demanding a deep ethical contemplation and sense of moral duty. The decisions made in the present concerning carnivore management, habitat preservation, and harmonious human–wildlife coexistence could reverberate across time, shaping the dynamics of how future generations will engage with these emblematic species. Ethics, in this context, emerges from the human inclination to discern what is morally right or wrong, good or bad, just or unjust, encompassing the ends that must be pursued and the appropriate means to attain them. Within this evolving process, carnivore biologists have an obligation to construct ethical arguments grounded in reason and evidence. These can further evolve into widely accepted moral principles for the review and modification of conduct toward other taxa. Developing conservation ethics and morals is a gradual process that requires time and effort. To advance this, carnivore biologists need to collaborate across disciplines and form partnerships with local educators, and the general public to call for urgent revisions to curricula to incorporate ecological concerns as central topics in education.

Contemporary land-use initiatives for conservation: land-sparing, SDGs, and 30×30 initiatives

Habitat loss and fragmentation are among the top threats to carnivore conservation (Minin et al. 2016, Ripple et al. 2014). These dynamics are particularly concerning because large carnivores tend to occur at comparatively low densities over large geographic ranges. Further, the expanding human population needs to be supported by space for the built environment, arable land, and livestock production. Land, in addition to being a physical entity, is a fundamental natural asset that is both a social reference and a source (and proof) of wealth in most global communities. It also holds a non-economic attachment, as is often the case for many Indigenous autochthonous communities. The alarmingly growing human population is coupled with the need for more arable and livable land, resulting in detrimental impacts on biodiversity and habitats. Current trends of increasing consumption rates, population growth, and ecological degradation will continue to have catastrophic implications for future generations. In this chapter, two approaches proposed to lessen the impacts on current and future generations are land-sparing and land-sharing (LSS) which target meeting food production needs without jeopardizing biological diversity.

Land-sparing intends to address the dual objectives of satisfying food production needs while safeguarding biodiversity. Specifically, land-sparing is a strategy that allocates extensive, contiguous land parcels solely for conservation, allowing for more intensive, high-yield agriculture on already converted (and relatively smaller) land areas (Balmford et al. 2005). As a result, this approach mitigates agriculture's detrimental effects on biodiversity by safeguarding habitats from conversion. Land-sharing, on the other hand, combines agricultural production and conservation of natural habitats on the same parcel of land. Although the success of each approach is context-dependent (e.g. Edwards et al. 2015, Law et al. 2015, Loos et al. 2018, Hasan et al. 2020), some generalizations can be made. One such generalization, for example, is the scale at which they result in landscape heterogeneity. Land-sparing spatially separates agricultural production and conservation with an impact at a coarser scale, while land-sharing does not separate the two but integrates them on the same land operating at a finer scale (Fischer et al. 2013, Salles et al. 2017).

LSS initiatives are often assessed from the perspectives of environmental sustainability, particularly by reconciling outcomes of agricultural productivity with biological diversity conservation, but it is also crucial to consider the implications on peoples' identities and intergenerational equity. In the remainder of this chapter, we will examine two global frameworks that have relevance to contemporary land-sparing initiatives—the Sustainable Development Goals (SDGs) and the 30×30 initiatives—by specifically exploring the human dimensions of carnivore conservation.

UN Sustainable Development Goals

The Sustainable Development Goals (SDGs) are a set of 17 global goals established by the United Nations in 2015 to address pressing social, economic, and environmental sustainability challenges. Several SDGs directly or indirectly relate to 'land-sparing' initiatives. Goals 1 and 15, in particular, focus on ending global hunger and protecting and restoring terrestrial ecosystems. Land-sparing strategies align with these goals and the targets within them by striving to maintain or enhance the ecological integrity of ecosystems while accommodating human development and feeding the global population. The 30×30 initiative calls for conserving 30% of the Earth's land and sea by 2030 to address the ongoing biodiversity crisis (CBD 2020). This initiative was proposed in the CBD meeting (CBD 2020), and in January 2021, President Biden issued an executive order that pledged to conserve 30% of the United States's lands and waters by 2030 (The White House 2021). The 30×30 initiative reflects a more quantifiable approach to land-sparing, emphasizing the significance of safeguarding natural habitats and biodiversity. These objectives align with the broader principles of the SDGs, emphasizing the interdependence of environment and human well-being.

The implementation of both of these initiatives is often assessed through the lens of environmental sustainability. Although there has been an increasing effort to include local peoples' values and voices in decision-making processes in the past three or so decades, there is still a lot more to be done (Infield et al. 2017). Garnering Indigenous knowledge and support through locals' engagement in conservation initiatives results in more effective conservation outcomes (Jiren et al. 2017). As a result, it is crucial to consider the implications of human identities and intergenerational equity while setting up criteria to set aside land under any of the initiatives to conduct land-sparing or land-sharing.

Carnivore conservation and land-sparing

Carnivore conservation is a critical element within land-sparing initiatives, as these top predators need large geographic extents, in general, and play essential roles in maintaining ecological balance. The human dimensions of carnivore conservation involve addressing the complex interactions between carnivores and local communities. These interactions could be negative (pre-emptive or retaliatory killing of large carnivores due to perceived or actual attacks on livestock or people, e.g. Ogada et al. 2003, Young et al. 2015, Srivathsa et al. 2019, Chatterjee et al. 2022, Parchizadeh et al. 2023), neutral (when the carnivores modify their behaviour to human presence and do not attack people or livestock; e.g. Gao et al. 2023), or positive (when carnivores contribute to preserving biological diversity through their ecological role, or when the carnivores have cultural significance (e.g. Ordiz et al. 2013, Arbieu et al. 2020). Most human–carnivore interaction studies focused on the conflicts (the negative outcomes), but there is a shift to understanding the complex dynamics of human–carnivore interactions (Bhatia et al. 2019).

With carnivores, proximity to people tends to be positively correlated with high levels of human–carnivore conflicts (Woodroffe 2000, Dhanwatey et al. 2013, Chinchilla et al. 2022). Therefore, a land-sharing initiative, where both carnivore conservation and agricultural production are prospering on the same land, will only be successful in places where local cultural values allow carnivores to thrive. For example, lions around Kafa Biosphere Reserve (a form of land-sharing approach (Ma et al. 2009)) do not get killed by locals because of the cultural value that prohibits the killing of lions, thereby promoting coexistence (Gebresenbet et al. 2018). Even in such landscapes, as large carnivore populations increase, the amount of negative interactions could surpass the level of tolerance by local people. In places that lack such cultural values that promote the coexistence of people and carnivores, land-sparing is the only option. More importantly, with increasing affluence and human populations, large carnivore conservation may be successful when people and predators are kept spatially separate (Packer et al. 2013). However, the nature versus human dichotomy should not follow the fortress conservation approach, but needs to allow the participation of local people. Fortress conservation isolates protected areas from human use, prioritizing nature over people (local communities) and cutting-off communities from their nearest resources. Instead, an approach that integrates local involvement ensures that conservation efforts are more sustainable and equitable, balancing ecological goals with human needs.

A successful land-sparing approach must consider the livelihoods and identities of local communities. It involves engaging with them in a participatory manner, acknowledging their traditional knowledge, and ensuring that the initiatives do not disproportionately negatively impact their well-being. Furthermore, recognizing any cultural and/or socioenvironmental significance of carnivores in a given landscape can be pivotal in garnering local support for conservation efforts. In many global contexts, carnivores hold a deep cultural and spiritual significance, and their preservation aligns with the preservation of cultural identities (e.g. Kellert et al. 1996, Gebresenbet et al. 2018, Arbieu et al. 2020, Dheer et al. 2021). Culturally important carnivores are often woven into local folklore, and their presence in the ecosystem is tied to broader cultural practices. Therefore, effective carnivore conservation must acknowledge and respect these connections and strive to coexist harmoniously with local beliefs and practices. Achieving the SDGs and the 30×30 objectives while ensuring the coexistence of humans and carnivores necessitates a comprehensive approach that respects local identities, promotes sustainable livelihoods, and safeguards the rights of present and future generations. Such interventions should not be top-down, but promote a truly equitable partnership between local communities, national conservation authorities, and global actors.

The implementation of land-sparing schemes is complex, requiring a high level of sensitivity and careful consideration of numerous socioenvironmental factors. To assist with this difficult task, scientists have proposed strategies to implement the contemporary LSS strategies in ways that incorporate social dimensions. For example, the International Institute for Environment and Development had proposed three low-cost tools by which the social impacts, governance, and equity of stakeholders can be assessed toward a given conservation effort. These are: i) Social Assessment for Protected and Conserved Areas (SAPA); ii) Governance Assessment for Protected and Conserved Areas (GAPA); and the most recent iii) Site-level Assessment of Governance and Equity (SAGE) for evaluating social impacts (IIED 2020). Additionally, place differences must be considered when strategies are designed for the implementation of land-sparing schemes. For instance, South-Southeast Asia, East Asia, and sub-Saharan Africa harbour the most significant number of carnivore species, with South-Southeast Asia and East Asia having the highest percentage of endangered and declining carnivore species, closely followed by North Asia (Fernández-Sepúlveda and Martín 2022). This distribution pattern, coupled with the high population growth and agricultural expansions that happened in the last few decades (Nyberg et al. 2012, Richards and Friess 2015, Chen et al. 2020, Bullock et al. 2021, Prabhakar 2021) are factors that must be given serious consideration during planning and implementation stages. In general, carnivore conservationists should incorporate the unique socioenvironmental contexts of local people and allow multidisciplinary approaches that 'fix' the systemic flaws by targeting the true sources of ecological and social injury rather than a Band-Aid approach of focusing on the carnivore ecosystem alone. Such an approach may exclude local people and fail to engage experts from important disciplines. As a result, its long-term sustainability will be questionable without the much-needed buy-in from local communities, be it due to a lack of understanding or resentment due to their exclusion.

Land-sparing initiatives, if implemented correctly, bring opportunities including promoting intergenerational equity. However, the opposite may lead to challenges like intensifying conflicts, heightening economic challenges, and escalating cultural dissonance. Failure in balancing conservation objectives with the needs of local communities can lead to conflicts over land use. In areas where people depend on natural resources for their sustenance, restrictions imposed by conservation initiatives can be met with resistance. Well-meaning conservation activities, such as transitioning to more sustainable practices from harmful ones (e.g. excessive harvesting, mining, or unsustainable agriculture) may pose economic challenges to those dependent on the later in the short term. Moreover, the clash between Western conservation models and local cultures in the global south can lead to cultural dissonance. The imposition of foreign conservation ideals may not align with Indigenous worldviews and can result in a loss of cultural identity. Involving local communities in decision-making processes and empowering them to participate in conservation efforts can lead to more effective outcomes in carnivore conservation (Durant et al. 2022). This requires a governance approach that takes into account the place-based contexts and livelihoods of local communities (Durant et al. 2022). Burden sharing schemes will help challenges related to land-use conflicts and economic disparities, while truly equitable partnerships (i.e. power dynamics, where the local norms and values are valued and appreciated) will mitigate risks relate to cultural dissonance.

Conclusion

Balancing conservation objectives with the needs of human populations can be a delicate and challenging task. The evolution of conservation and protected area designation do not necessarily make it easier, as they are tainted with their own colonial past and had denied the human agency of local communities for so long. Without healthier political ecology arenas, contemporary land-sparing initiatives will face challenges from various angles. These may include potential conflicts over land use, particularly in areas where local communities depend on natural resources for their sustenance. However, if land-sparing initiatives informed by decisions that incorporated local's voices and nuanced considerations tailored to communities' and carnivores needs, they could offer an opportunity to promote co-existence of people and large carnivores (Figure 13.1). The initiatives can strengthen community resilience and inclusivity by allowing for more tact, effective, and meaningful involvement of locals in decision-making processes. Conservationists must design strategies that tap into the place identity of local communities to promote wildlife and environmentally responsible actions and behaviours. Important cultural, political, and socioeconomic topics such as land reform, land value trends, intergenerational conflicts in real property law, and institutional rigidity and/or bureaucracy in land rights, and their relationship

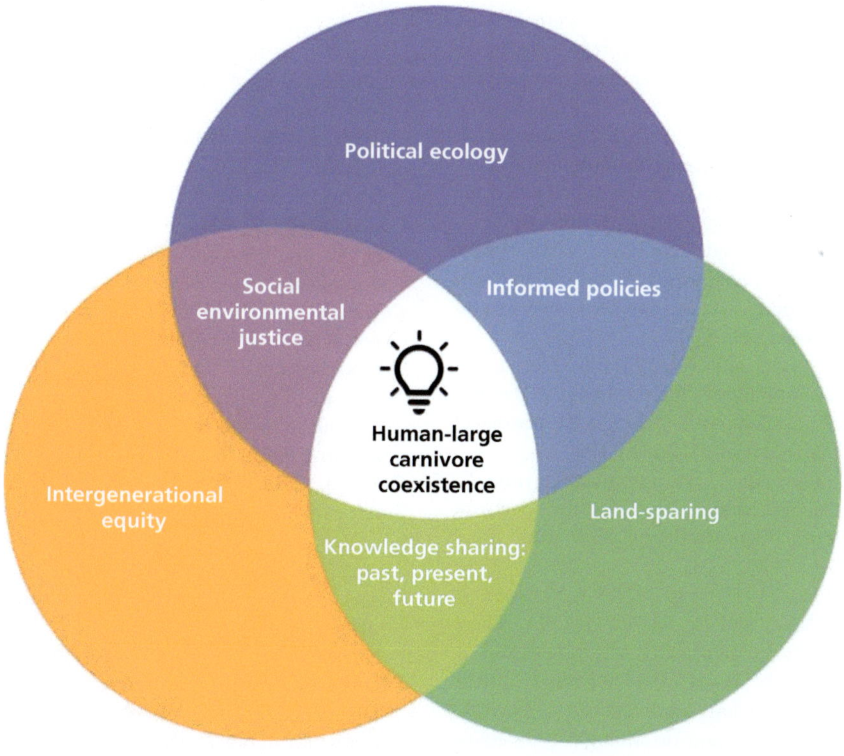

Figure 13.1 A diagram illustrating the interconnectedness of people, land, and large carnivores. A locally functional framework of political ecology is crucial for recognizing the human-agency of all stakeholders, and ensuring the true participation of local communities. A resilient political ecology leads to informed and sound decisions, and results in effective land sparing strategies that promote social and environmental justice. The prevalence of social and environmental justice fosters intergenerational equity, facilitating the sharing of ecological knowledge across generations.

with land-sparing strategies, presence/absence of TEK that promotes carnivore conservation must be addressed. Such focus on the human dimensions of large carnivore conservation is critical because, ultimately, it is the locals' tolerance and willingness to coexist that tilts the balance in favour of conservation success.

References

Agarwal, B. 2000. Conceptualizing environmental collective action: why gender matters. *Cambridge Journal of Economics* 24(3):283–310.

Arbieu, U., J. Albrecht, M. Mehring, N. Bunnefeld, I. Reinhardt, and T. Mueller. 2020. The positive experience of encountering wolves in the wild. *Conservation Science and Practice* 2:e184.

Atuo, F. A., P. A. Abanyam, and T. J. O'Connell. 2015. An assessment of socio-economic drivers of avian body parts trade in West African rainforests. *Biological Conservation* 191: 614–622.

Atuo, F. A., J. Fu., T. J. O'Connell., J. A. Agida, and J. A. Agaldo. 2020. Coupling law enforcement and community-based regulations in support of compliance with biodiversity conservation regulations. *Environmental Conservation* 47(2):104–112.

Balmford, A., R. E. Green, and J. P. W. Scharlemann. 2005. Sparing land for nature: exploring the potential impact of changes in agricultural yield on the area needed for crop production. *Global Change Biology* 11:1594–1605.

Ballard, H. L., C. G. Dixon, and E. M. Harris. 2017. Youth-focused citizen science: Examining the role of environmental science learning and agency for conservation. *Biological Conservation* 208:65–75.

Bele, M. Y, D. J. Sonwa, and A. M. Tiani. 2014. Local communities vulnerability to climate change and adaptation strategies in Bukavu in DR Congo. *The Journal of Environment & Development* 23(3):331–357.

Bhatia, S., S. M. Redpath, K. R. Suryawanshi, and C. Mishra. 2019. Beyond conflict: exploring the spectrum of human–wildlife interactions and their underlying mechanisms. *Oryx* 54:621–628.

Bullock, E. L., S. P. Healey, Z. Yang, P. Oduor, N. Gorelick, S. Omondi, E. Ouko, and W. B. Cohen. 2021. Three decades of land cover change in East Africa. *Land* 10:150.

CBD (Convention on Biological Diversity). 2020. Kunming-Montreal Global biodiversity framework. Secretariat of the Convention on Biological Diversity. Available at: https://www.cbd.int/doc/c/e6d3/cd1d/daf663719a03902a9b116c34/cop-15-l-25-en.pdf.

Cazalis, V., M. Loreau, and G. Barragan-Jason. 2023. A global synthesis of trends in human experience of nature. *Frontiers in Ecology and the Environment* 21(2): 85–93.

Ceballos, G., et al. 2021. Jaguar distribution, biological corridors and protected areas in Mexico: from science to public policies. *Landscape Ecology* 36:3287–3309.

Chanchani, P., et al. 2014. Tigers of the Transboundary Terai Arc landscape: status, distribution and movement in the Terai of India and Nepal. *National Tiger Conservation Authority, Government of India, and Department of National Park and Wildlife Conservation, Government of Nepal NTNC/DNPWC*, 3. Available at: https://globaltigerforum.org/wp-content/uploads/2017/05/208_Indo-Nepal-Joint-Monitoring-Report.pdf.

Chatterjee, M., N. Chatterjee, P. Chandel, T. Bhattacharya, and R. Kaul. 2022. Predicting negative human-tiger (*Panthera tigris*) interactions in mosaic landscapes around Dudhwa and Pilibhit tiger reserves in India. *Frontiers in Conservation Science* 3:999195.

Chen, D., A. Garmulewicz, C. Merner, C. Elphinstone. C. Leggott, and H. Dewar. 2019. Encouraging youth engagement in marine protected areas: a survey of best practices in Canada. *Aquatic Conservation: Marine and Freshwater Ecosystems* 29:223–232.

Chen, J., L. Yu, F. Yan, and S Zhang. 2020. Ecosystem service loss in response to agricultural expansion in the small Sanjiang Plain, Northeast China: process, driver and management. *Sustainability* 12(6):2430.

Chinchilla, S., E. V. D. Berghe, J. Polisar, C. Arévalo, and C. Bonacic. 2022. Livestock–carnivore coexistence: moving beyond preventive killing. *Animals* 12(4):479.

Colléony, A., R. Cohen-Seffer, and A. Shwartz. 2020. Unpacking the causes and consequences of the extinction of experience. *Biological Conservation* 251:108788.

Costanza, R., et al. 1997. The value of the world's ecosystem services and natural capital. *Nature* 387:253–260.

Crespo-Gascón, S. and J. Guerrero-Casado. 2019. The role of the spectacled bear (*Tremarctos ornatus*) as an umbrella species for Andean ecoregions. *Wildlife Research* 46(2):176–183.

Davis, J. 2019. Black faces, black spaces: Rethinking African American underrepresentation in wildland spaces and outdoor recreation. *Environment and Planning E: Nature and Space* 2(1):89–109.

Dawson, N. M., et al. 2021. The role of indigenous peoples and local communities in effective and equitable conservation. *Ecology and Society* 26(3):19.

Dhanwatey, H. S., J. C. Crawford, L. A. S. Abade, P. H. Dhanwatey, C. K. Nielsen, and C. Sillero-Zubiri. 2013. Large carnivore attacks on humans in central India: a case study from the Tadoba-Andhari Tiger Reserve. *Oryx* 47(2):221–227.

Dheer, A., E. Davidian, M. H. Jacobs, J. Ndorosa, T. M. Straka, and O. P. Höner. 2021. Emotions and cultural importance predict the acceptance of large carnivore management strategies by Maasai pastoralists. *Frontiers in Conservation Science* 2:691975.

Díaz, S., et al. 2015. The IPBES Conceptual Framework—connecting nature and people. *Current Opinion in Environmental Sustainability* 14:1–16.

Doughty, H. L., S. M. Karpanty, and H. M. Wilbur. 2015. Local hunting of carnivores in forested Africa: a meta-analysis. *Oryx* 49:88–95.

Durant, S. M., et al. 2022. Fostering coexistence between people and large carnivores in Africa: Using a theory of change to identify pathways to impact and their underlying assumptions. *Frontiers in Conservation Science* 2:698631.

Edwards, D. P., J. J. Gilroy, G. H. Thomas, C. A. M. Uribe, and T. Haugaasen. 2015. Land-sparing agriculture best protects avian phylogenetic diversity. *Current Biology* 25:2384–2391.

Edwards, R. C. and Larson, B.M. 2020. When screens replace backyards: strategies to connect digital-media-oriented young people to nature. *Environmental Education Research* 26(7):950–968.

Elsawy, M. and M. Youssef. 2023. Economic sustainability: meeting needs without compromising future generations. *International Journal of Economics and Finance* 15:1–23.

Eustace, A. and D. B. Tarimo. 2019. Maasai and lion killings in the Tarangire–Manyara ecosystem. *African Journal of Ecology* 57:140–143.

Felix, N., B. M. Kissui, L. Munishi, and A. C. Treydte. 2022. Retaliatory killing negatively affects African lion (*Panthera leo*) male coalitions in the Tarangire-Manyara Ecosystem, Tanzania. *PloS One* 17:e0272272.

Fernández-Sepúlveda, J. and C. A. Martín. 2022. Conservation status of the world's carnivorous mammals (order Carnivora). *Mammalian Biology* 102:1911–1925.

Finney, C. 2014. *Black Faces, White Spaces: Reimagining the Relationship of African Americans to the Great Outdoors*. The University of North Carolina Press, Chapel Hill, N. C.

Fischer, J., et al. 2013. Land sparing versus land sharing: moving forward. *Conservation Letters* 7(3):149–157.

Gao, Y., A. T. L. Lee, Y. Luo, J. S. Alexander, X. Shi, T. Sangpo, and S. G. Clark. 2023. Large carnivore encounters through the lens of mobile videos on social media. *Conservation Science and Practice* 5:e12907.

Gebresenbet F., B. Baraki, G. Yirga, C. Sillero-Zubiri, and H. Bauer. 2018. A culture of tolerance: coexisting with large carnivores in the Kafa Highlands, Ethiopia. *Oryx* 52:751–760.

Gentina, E., C. Daniel, and T. L.-P. Tang. 2021. Mindfulness reduces avaricious monetary attitudes and enhances ethical consumer beliefs: Mindfulness training, timing, and practicing matter. *Journal of Business Ethics* 173: 301–323.

Gooden, J. L. 2019. Cultivating identity through private land conservation. *People and Nature* 1:362–375.

Hall, S. 1990. Cultural Identity and diaspora. In *Identity: Community, Culture, Difference*, edited by J. Rutherford. Lawrence & Wishart Publishers, London, U. K.

Harihar, A., D. Veríssimo, and D. C. MacMillan. 2015. Beyond compensation: Integrating local communities' livelihood choices in large carnivore conservation. *Global Environmental Change* 33:122–130.

Harrington, L. A. 2015. International commercial trade in live carnivores and primates 2006–2012: response to Bush et al. 2014. *Conservation Biology* 29:293–296.

Hart, R. A. 2013. *Children's Participation: The Theory and Practice of Involving Young Citizens in Community Development and Environmental Care*. Routledge, London.

Hasan, A., M. M. Girona, G. Grosbois, N. Saha, M. A. Halim. 2020. Land sparing can maintain bird diversity in northeastern Bangladesh. *Sustainability* 12:6472.

Hayward, M. W., et al. 2022. Intergenerational inequity: stealing the joy and benefits of nature from our children. *Frontiers in Ecology and Evolution* 10:830830.

Hellinx, E. 2020. The CMS-CITES African carnivore initiative as an illustration of synergies between MEAs. *Frontiers in Ecology and Evolution* 8:10.

Hogg, M. A. 2016 Social identity theory. In: *Understanding Peace and Conflict Through Social Identity Theory*, edited by S. McKeown, R. Haji, and N. Ferguson. Peace Psychology Book Series, Springer, Cham.

IIED (International Institute for Environment and Development). 2020. Assessing the social impacts, governance and equity of conservation: SAPA, GAPA, or SAGE? Available at: https://www.iied.org/assessing-social-impacts-governance-equity-conservation-sapa-gapa-or-sage.

Infield, M., A. Entwistle, H. Anthem, A. Mugisha, and K. Phillips. 2017. Reflections on cultural values approaches to conservation: lessons from 20 years of implementation. *Oryx* 52:220–230.

Jiren, T. S., I. Dorresteijn, J. Schultner, and J. Fischer. 2017. The governance of land use strategies: institutional and social dimensions of land sparing and land sharing. *Conservation Letters* 11:e12429.

Karanth, K. U. and R. Chellam. 2009. Carnivore conservation at the crossroads. *Oryx* 43(1):1–2.

Kellert, S. R., M. Black, C. R. Rush, and A. J. Bath. 1996. Human culture and large carnivore conservation in North America. *Conservation Biology* 10(4):977–990.

Koot, S., B. Büscher, and L. Thakholi. 2022. The new green apartheid? Race, capital and logics of enclosure in South Africa's wildlife economy. *Environment and Planning E: Nature and Space* 7(1):123–140.

Lam, W. Y., et al. 2023. Using a crime prevention framework to evaluate tiger counter-poaching in a Southeast Asian rainforest. *Frontiers in Conservation Science* 4:1213552.

Law, E. A., E. Meijaard, B. A. Bryan, T. Mallawaarachchi, L. P. Koh, and K. A. Wilson. 2015. Better land-use allocation outperforms land sparing and land sharing approaches to conservation in Central Kalimantan, Indonesia. *Biological Conservation* 186:276–286.

Loos, J., and H. V. Wehrden. 2018. Beyond biodiversity conservation: land sharing constitutes sustainable agriculture in European cultural landscapes. *Sustainability* 10(5):1395.

Linnell, J. D., J. E. Swenson, and R. Andersen. 2000. Conservation of biodiversity in Scandinavian boreal forests: large carnivores as flagships, umbrellas, indicators, or keystones? *Biodiversity & Conservation* 9:857–868.

Lute, M. L., H. H. Carter, J. V. López-Bao, and J. D. Linnell. 2020. Conservation professionals' views on governing for coexistence with large carnivores. *Biological Conservation* 248:108668.

Ma, Z., B. Li, W. Li, N. Han, J. Chen, and A. R. Watkinson. 2009. Conflicts between biodiversity conservation and development in a biosphere reserve. *Journal of Applied Ecology* 46(3):527–535.

Manqele, N. S., J. A. Selier, T. R. Hill, and C. T. Downs. 2018. Drivers of the illegal hunting of serval (*Leptailurus serval*) and oribi (*Ourebia ourebi*) in the KwaZulu-Natal Midlands, South Africa. *African Journal of Wildlife Research* 48(2):1–18.

Mattson, D. J., K. L. Byrd, M. B. Rutherford, S. R. Brown, and T. W. Clark. 2006. Finding common ground in large carnivore conservation: mapping contending perspectives. *Environmental Science & Policy* 9(4):392–405.

Megaze, A., M. Balakrishnan, and G. Belay. 2017. Human–wildlife conflict and attitude of local people towards conservation of wildlife in Chebera Churchura National Park, Ethiopia. *African Zoology* 52:1–8.

Minin, E. D., et al. 2016. Global priorities for national carnivore conservation under land use change. *Scientific Reports* 6:23814.

Miriti, M. N., A. J. Rawson, and B. Mansfield. 2022. The history of natural history and race: Decolonizing human dimensions of ecology. *Ecological Applications 33*:e2748.

Mkonyi, F. J. 2022. An integrated approach for the management of human-carnivore conflict: A review of conflict management interventions in Tanzania. *Mammalian Biology* 102(5):2061–2081.

Moonstone, B. A. 2016. Nature *Experiences* and *Ecoliteracy*: The *Effect* of *Place Based Education Experiences, Adventure* and *Outdoor Education Experiences* and *Undirected Nature Experiences* on *Ecoliteracy*. MA thesis, Hamline University. Available at: https://digitalcommons.hamline.edu/hse_all/4147.

Morse, S. 2018. Mind the Gap: The Potential of Generational Wisdom Towards Sustainable Communities. MA thesis, Prescott College.

Munasinghe, M. 2001. Sustainable development and climate change: applying the sustainomics transdisciplinary meta-framework. *International Journal of Global Environmental Issues* 1:13–55.

Neumann, R. P. 1996. Dukes, earls, and ersatz edens: aristocratic nature preservationists in colonial Africa. *Environment and Planning D: Society and Space* 14(1): 79–98.

Nyberg, G., A. Bargués-Tobella, J. Kinyangi, and U. Ilstedt. 2012. Soil property changes over a 120-yr chronosequence from forest to agriculture in western Kenya. *Hydrology and Earth System Sciences* 16(7):2085–2094.

Nygren, A., and S. Rikoon. 2008. Political ecology revisited: Integration of politics and ecology does matter. *Society & Natural Resources* 21(9):767–782.

O'Brien, L. T., H. L. Bart, and D. M. Garcia. 2020. Why are there so few ethnic minorities in ecology and evolutionary biology? Challenges to inclusion and the role of sense of belonging. *Social Psychology of Education* 23:449–477.

O'Donoghue, R., J. C. A. Sandoval-Rivera, and U. Payyappallimana. 2019. Landscape, memory and learning to change in changing worlds: contemplating intergenerational learning and traditional knowledge practices within social-ecological landscapes of change. *Southern African Journal of Environmental Education* 35:28–31.

Ogada, M., R. Woodroffe, N. Oguge, and L. G. Frank. 2003. Limiting depredation by African carnivores: the role of livestock husbandry. *Conservation Biology* 17(6): 1521–1530.

Ordiz, A., R. Bischof, and J. E. Swenson. 2013. Saving large carnivores, but losing the apex predator? *Biological Conservation* 168:128–133.

Packer, C., et al. 2013. Conserving large carnivores: dollars and fence. *Ecology Letters* 16(5):635–641.

Parchizadeh, J., K. F. Kellner, J. E. Hurst, D. W. Kramer, and J. L Belant. 2023. Factors influencing frequency and severity of human-American black bear conflicts in New York, USA. *Plos One* 18(2):e0282322.

Prabhakar, S. V. 2021. A succinct review and analysis of drivers and impacts of agricultural land transformations in Asia. *Land Use Policy* 102:105238.

Richards, D. R. and D. A. Friess. 2015. Rates and drivers of mangrove deforestation in Southeast Asia, 2000–2012. *Proceedings of the National Academy of Sciences* 113(2): 344–349.

Ripple, W. J. and R. L. Beschta. 2012. Trophic cascades in Yellowstone: the first 15 years after wolf reintroduction. *Biological Conservation* 145(1):205–213.

Ripple, W. J, et al. 2014. Status and ecological effects of the world's largest carnivores. *Science* 343(6167):1241484.

Robbins, P. 2012. *Political Ecology: A Critical Introduction* (2nd edition). John Wiley & Sons Ltd, West Sussex, U. K.

Rozylowicz, L., V. D. Popescu, M. Pătroescu, and G. Chişamera. 2011. The potential of large carnivores as conservation surrogates in the Romanian Carpathians. *Biodiversity and Conservation* 20:561–579.

Salles, J.M., F. Teillard, M. Tichit, and M Zanella. 2017. Land sparing versus land sharing: an economist's perspective. *Regional Environmental Change* 17:1455–1465.

Schmidt, C. and C. J. Garroway. 2022. Systemic racism alters wildlife genetic diversity. *Proceedings of the National Academy of Sciences* 119(43):e2102860119.

Shava, S. 2013. The representations of indigenous knowledges. In *International Handbook of Research on Environmental Education*, edited by R. B. Stevenson, M. Brody, J. Dillon, and A. E. J. Wals. Routledge, New York.

Shi, L., L. Han, F. Yang, and L. Gao. 2019. The evolution of sustainable development theory: Types, goals, and research prospects. *Sustainability* 11(24):7158.

Singh, S. K., et al. 2017. Fine-scale population genetic structure of the Bengal tiger (*Panthera tigris tigris*) in a human-dominated western Terai Arc Landscape, India. *PloS One* 12(4):e0174371.

Soga, M. and K. J. Gaston. 2016. Extinction of experience: the loss of human–nature interactions. *Frontiers in Ecology and the Environment* 14(2):94–101.

Soga, M., Y. Yamaura, T. Aikoh, Y. Shoji, T. Kubo, and K. J. Gaston. 2015. Reducing the extinction of experience: association between urban form and recreational use of public greenspace. *Landscape and Urban Planning* 143: 69–75.

Srivathsa, A., M. Puri, K. K. Karanth, I. Patel, and N. S. Kumar. 2019. Examining human–carnivore interactions using a socio-ecological framework: sympatric wild canids in India as a case study. *Royal Society Open Science* 6(5):182008.

Summers, J. K. and L. M. Smith. 2014. The role of social and intergenerational equity in making changes in human well-being sustainable. *Ambio* 43(6):718–728.

Tajfel, H. and J. C. Turner. 1979. An integrative theory of intergroup conflict. In *The Social Psychology of Intergroup Relations*, edited by W. G. Austin and S. Worchel. Brooks/Cole Publishing Company, Pacific Grove, C. A.

The White House. 2021. Executive order on tackling the climate crisis at home and abroad. Available at: https://www.whitehouse.gov/briefing-room/presidential-actions/2021/01/27/executive-order-on-tackling-the-climate-crisis-at-home-and-abroad.

Treves, A., G. Chapron, J. V. López-Bao, C. Shoemaker, A. R. Goeckner, and J. T. Bruskotter. 2017. Predators and the public trust. *Biological Reviews* 92(1):248–270.

Torrents-Ticó, M., Á. Fernández-Llamazares, D. Burgas, and M. Cabeza. 2021. Convergences and divergences between scientific and indigenous and local knowledge contribute to inform carnivore conservation. *Ambio* 50:990–1002.

Truong, M.X.A. and S. Clayton, S., 2020. Technologically transformed experiences of nature: A challenge for environmental conservation? *Biological Conservation* 244:e108532.

Waal, L. D., C. Jakins, S. E. Klarmann, J. Green, and N. D'Cruze. 2022. The unregulated nature of the commercial captive predator industry in South Africa: Insights gained using the PAIA process. *Nature Conservation* 50:227–264.

West, P., J. Igoe, and D. Brockington. 2006. Parks and peoples: the social impact of protected areas. *Annual Review of Anthropology* 35:251–277.

Wright, S. D. and A. M. Wadsworth. 2014. Gray and green revisited: A multidisciplinary perspective of gardens, gardening, and the aging process. *Journal of Aging Research* 2014:283682.

Woodroffe, R. 2000. Predators and people: using human densities to interpret declines of large carnivores. *Animal Conservation* 3:165–173.

Young, J. K., Z. Ma, A. Laudati, and J. Berger. 2015. Human–carnivore interactions: lessons learned from communities in the American west. *Human Dimensions of Wildlife* 20(4):349–366.

CHAPTER 14

Measures to Promote Human–Carnivore Coexistence

Christine E. Wilkinson, Harshad Karandikar, and Mitchell W. Serota

Introduction

Large carnivores are critical for ecological functioning (Creel and Christianson 2009, Estes et al. 2011, Hoeks et al. 2020, Prugh et al. 2009) and provide myriad direct and indirect ecosystem services (see Chapters 10, 11, 12), yet face substantial global threats. Among these threats are habitat loss and degradation, overutilization, prey depletion, and persecution by people (Ripple et al. 2014). The latter of these is an integral component of human–carnivore conflict, which in the broader context of **human–wildlife conflict** is defined as 'struggles that emerge when the presence or behaviour of wildlife poses actual or perceived, direct, and recurring threats to human interests or needs, leading to disagreements between groups of people and negative impacts on people and/or wildlife' (IUCN SSC HWCTF 2020). Human conflicts with large carnivores can include carnivore predation of livestock and pets, occasional attacks on people, and lethal retaliation by people against carnivores for both real and perceived threats. Research on **human–wildlife interactions**, with a focus on conflict, has increased exponentially over the past three decades (Konig et al. 2020, Venumiere-Lefebvre et al. 2022), mirroring the increasing challenges to carnivore conservation posed by human–carnivore conflict. The importance of preventing and mitigating human–carnivore conflict is only more pressing with the exacerbation of climate change, which will continue to intensify human–wildlife conflict globally as people and wildlife must share progressively crowded spaces (Abrahms et al. 2023).

Research on human–wildlife interactions has now begun to shift towards a **coexistence** framing (Bergstrom 2017, Su et al. 2022). Researchers and managers recognize this intentional reframing towards coexistence as critical to being forward-thinking, nuanced, and empathetic in efforts to resolve human–wildlife conflict (Pooley et al. 2020). In fact, the recently released first edition of the human–wildlife conflict and coexistence guidelines created by the International Union for the Conservation of Nature (IUCN) heavily emphasizes coexistence and its social-ecological mechanisms (IUCN 2023). In this chapter we review global human–carnivore coexistence efforts focusing specifically on large carnivores, explain the rationale behind these efforts, and explore their mechanisms. Specifically, we

i. consider the foundational literature on human–carnivore coexistence and how it has evolved over time,
ii. outline the major categories of global efforts to promote coexistence with apex predators and review the current literature to provide examples and general takeaways from each human–carnivore coexistence approach, and

Christine E. Wilkinson, Harshad Karandikar, and Mitchell W. Serota, *Measures to Promote Human–Carnivore Coexistence*.
In: *Apex Predators in the Anthropocene*. Edited by: Robert A. Montgomery and Remington J. Moll, Oxford University Press. DOI: 10.1093/9780191996665.003.0014

iii. discuss the lessons learned, current challenges, and potential best practices for conserving and restoring carnivore populations through bolstering human–carnivore coexistence.

Human–carnivore coexistence as a dynamic social-ecological endeavour

Human–carnivore coexistence is a complex social-ecological endeavour (Figure 14.1). While the ecological factors and wildlife behaviours—such as a particular carnivore species' or individual's propensity for preying on livestock in different seasons or landscape types (e.g. Ugarte et al. 2019)—in a specific human–carnivore conflict situation dictate elements of those interactions, human–wildlife conflict is ultimately rooted in conflicts between human groups, i.e. human–human conflict (Peterson et al. 2010). Human societal values, perceptions, economics, livelihoods, and public health challenges are all integral components of whether and how human–carnivore interactions will be positive, neutral, or negative (Nyhus 2016). Thus, to be successful, human–carnivore coexistence efforts must take into account human values (Dickman 2010, Montgomery et al. 2020), human safety and perceptions thereof (Wilkinson et al. 2021a), economic impacts (Treves et al. 2006), historical and current interpersonal relationships and governance (Redpath et al. 2014), ecological needs (i.e. for sustainable carnivore population sizes and genetic stability, Lamb et al. 2020), and carnivore behaviours (i.e. employing tools that most effectively minimize human–carnivore conflict on the landscape, Wilkinson et al. 2020).

As human–carnivore interactions are dependent upon their social-ecological contexts, human–carnivore coexistence is usually a dynamic and adaptive process (Pooley et al. 2020) that is unlikely to remain static or stable (Carter and Linnell 2016). Human–carnivore coexistence may also be defined and experienced differently by the various human and non-human actors in these contexts (Pooley et al. 2016), and politicized through human–human conflict. For example, fears about, and threats towards, reintroduced carnivores can become more acute when local landowners are

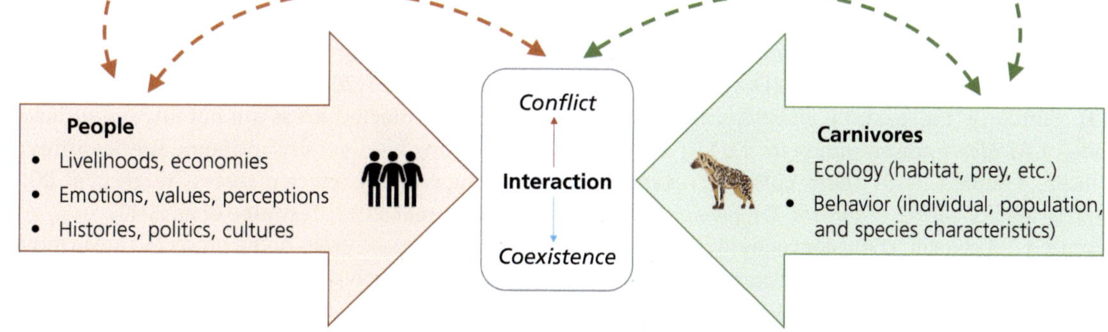

Figure 14.1 A framework depicting the social-ecological factors contributing to human-carnivore interactions (HCI). Dashed orange arrows (the two dashed arrows on the left side of the figure) represent social and policy mechanisms (the ways in which global environmental change and people impact one another, and the ways in which people and HCI impact one another). Dashed green arrows (the two dashed arrows on the right side of the figure) represent ecological mechanisms (the ways in which global environmental change impacts carnivores, and the ways in which carnivores and HCI impact one another).

already mistrustful of local and regional government entities who may be involved in the reintroductions (Dickman and Hazzah 2016). Current and historical socioeconomic disparities among human actors in the landscape can also drive disagreements in whether coexistence can occur and what may constitute coexistence, since these disparities contribute to inequities in who is affected most heavily by human–carnivore conflict (Braczkowski et al. 2023). To co-create effective and lasting coexistence measures, managers must recognize human–carnivore coexistence as ever-shifting, and meaningfully consider responsibility, equity, justice, and inclusion in the planning process (Harris et al. 2023).

While there is considerable global focus on alleviating conflict, some communities have found relative success in coexisting with their local large carnivores. For example, despite their role in livestock depredation, African lions (*Panthera leo*) are highly tolerated and not retaliated against in the Kafa region of Ethiopia due to people's cultural beliefs; indeed, in this region, lion predation on livestock can even signify good luck (Gebresenbet et al. 2018). Along with cultural beliefs, the length of time or generations that people have lived alongside large carnivores can also dictate the magnitude and strength of coexistence. For example, in Transylvania, Romania, brown bears (*Ursus arctos*) and people have successfully lived alongside one another in working landscapes for centuries. In this case, coexistence is influenced by the long land tenure of both bears and people, which is underpinned by people's tolerance for livestock loss, considerable bear habitat availability and connectivity throughout the working landscapes, and long-term, embedded use of traditional herding techniques (Dorresteijn et al. 2014). Similarly, the pastoral lifestyles and livelihoods of *Maldhari* communities in the Gir Forests of India, coupled with their cultural reverence towards Asiatic lions (*Panthera leo persica*), have supported long-term coexistence, with lion densities calculated to be higher within regions that have livestock (Banerjee et al. 2013). Meanwhile, some species, such as black bears (*Ursus americanus*), may seek anthropogenic subsidies (Newsome et al. 2014) for only small portions of the year (i.e. seasonally when resources are more scarce) while spatially and temporally avoiding people on fine scales (see Gaynor et al. 2019), otherwise preferring highly vegetated wild areas for much of the year (e.g. van Bommel et al. 2022). By learning from existing situations in which coexistence has been successful, and carefully considering how elements such as culture, land tenure, and embedded ways of knowing, ecology, and temporality influence human–carnivore coexistence, conservation and management activities can be better equipped to boost the effectiveness of global coexistence measures.

Current efforts, techniques, and tools to promote human–carnivore coexistence

Protected area establishment and management

Protected areas are a well-established tool for conserving biodiversity, and over the last few decades protected areas have expanded to cover over 15% of terrestrial and freshwater environments and 7.5% of marine environments (UNEP-WCMC and IUCN 2023). In general, protected areas set aside large areas of terrestrial or aquatic environments to safeguard biodiversity and can increase species abundances and richness (Coetzee et al. 2014, Gray et al. 2016, Watson et al. 2014). Protected areas can produce significant conservation benefits for large carnivores by dramatically reducing their susceptibility to human persecution (Ripple et al. 2014). Further, protected areas are expected to expand in the coming decades after the Convention on Biological Diversity (CBD) drafted a post-2020 global biodiversity conservation framework centred around conserving 30% of the Earth's land and sea by 2030 (Tsioumani 2020). Still, recent research indicates that protected areas are not always sufficient in protecting biodiversity and some species are even declining within protected areas (Craigie et al. 2010, Geldman et al. 2013, Terraube et al. 2020).

Despite the successes of the protected areas movement, large carnivore populations may be at risk both within parks and in the surrounding, often human-dominated, landscapes. Even within protected areas, for instance, carnivores may still be subject to poaching. For example, one study found that brown bears, Eurasian lynx (*Lynx lynx*), and wolverines (*Gulo gulo*) all suffered higher rates of poaching within protected areas than in adjacent unprotected landscapes, a trend the authors attributed to inadequate law enforcement within the park (Rauset et al. 2015). Efforts to increase patrolling or remove snares (e.g. the Snares to Wares Initiative in Uganda; see Montgomery et al. 2022a) can be effective mitigation strategies; however, in some cases, increased enforcement does very little to prevent illegal killings of wildlife (Jenks et al. 2012, Becker et al. 2013). Further, for wide-ranging species like many apex predators, protected areas are subject to edge effects and are often too small to sustain viable, connected populations (Lopez-Bao et al. 2015). At the edges of protected areas, carnivores suffer higher mortality rates and live at smaller densities (Balme et al. 2010). Therefore, while protected areas may be a refuge for carnivores, the human-dominated landscapes along the peripheries may act as a sink for some species (Lamb et al. 2020). One proposed solution to prevent edge effects and reduce conflict is the fencing of protected areas. Under certain conditions, fenced protected areas can actually support higher densities of carnivores (Packer et al. 2013). Fencing, however, may require extensive management to prevent negative impacts (e.g. loss of genetic diversity and overpopulation) (Smith et al. 2020). Fencing can also drastically alter animal movement patterns, fragment populations, prove costly to maintain, and ultimately, animals find a way to navigate them (Creel et al. 2013, McInturff et al. 2020, Wilkinson et al. 2021b). Since decreasing the number of people that live near protected areas or increasing the size of individual protected areas are unlikely to be viable solutions, conservation efforts should consider ways that improve the socioeconomic outcomes that drive human persecution of wildlife.

For both protected areas and human–carnivore coexistence, positive ecological outcomes are inextricably tied to social outcomes. A meta-analysis found that protected areas were more likely to have positive conservation outcomes when they also had positive socioeconomic outcomes, including empowerment of local people, reduced economic inequality, and the maintenance of cultural and livelihood benefits (Oldekop et al. 2015). Another review of 27 marine protected areas found that progress towards social equity in protected areas and the integration of local communities in their implementation and management were key factors in the protected area's success (Giakoumi and Pey 2017). Social factors can undermine the effectiveness of protected areas by driving conservation challenges (e.g. human–wildlife conflict) (Dickman 2010). Historically, the creation of protected areas has physically, economically, and culturally displaced people, and in some cases, led to disastrous and inequitable outcomes for local communities (Agrawal and Redford 2009). Even today, a recent survey of over 200 protected areas found inequitable management of protected areas (Zafra-Calvo et al. 2019). Therefore, prioritizing equity in the creation and management of protected areas could enhance their effectiveness.

Promoting carnivore habitat connectivity through establishing and maintaining corridors

Most apex predators typically require large geographic areas for their **home ranges**, often moving outside of the boundaries of protected areas (Carter and Linnell 2016). At the extreme end of the spectrum, the typical individual snow leopard (*Panthera uncia*) home range (Johansson et al. 2016, Rosenbaum et al. 2023) is substantially larger than many protected areas in their range. Over the last few decades, there has been an increasing understanding of the spatial limitations of protected areas in offering viable home ranges for sustainable large carnivore populations. Recent estimates suggest that almost 82% of the global extant large carnivore range falls outside protected areas (Braczkowski et al. 2023). In many cases, individual carnivore home ranges have large portions falling outside of protected areas, often overlapping with working lands. Most carnivore species also tend to disperse significant distances, especially as young adults, in search of suitable areas that could be used as a home range. These long-range dispersals are more prevalent from high-quality habitats at carrying capacity, influenced by the solitary nature of many carnivore species and a tendency to largely avoid contact with conspecifics, although this understanding continues to be challenged with recent studies that conclude that the nature and frequency of interactions between conspecific individuals may be considerably higher than previously understood (Elbroch et al. 2017).

Given the limited spatial extent of protected areas in many parts of the world (Athreya et al. 2013), wildlife conservation has seen a strong shift towards a landscape approach over the last few decades. Several other associated factors have also contributed to this trend. First, many carnivores move large distances, often entering and exiting protected areas in response to prey behaviour and seasonal migration (Hofer and East 1993, Ballard et al. 1997, Pierce et al. 1999, Walton et al. 2001). Second, connectivity between spatially segregated carnivore populations is critical to prevent **genetic drift** that can threaten the long-term survival of the species. For example, recent studies indicate significant genetic declines and increasing isolation between jaguar (*Panthera onca*) populations in Mesoamerica (Wultsch et al. 2016), the Atlantic Forest in Brazil (Haag et al. 2010), and in other parts of the species' range (Roques et al. 2015). Promoting carnivore habitat connectivity through establishing and maintaining **wildlife corridors** is thus critical for sustainable, long-term conservation gains for many of these iconic species (Wen et al. 2022). Well-designed wildlife corridors have the potential to reduce impacts on other adjacent land that is outside the protected area and identified corridors, given that many wildlife species prefer to minimize direct interactions with humans and will likely utilize areas that offer strong functional connectivity (Adams et al. 2015, Green et al. 2018). Corridor areas may also increase long-term resilience of species by offering a source of individuals that have the phenotypic flexibility to learn to adapt to human presence without conflict, a change that has the potential to transform the long-term relationship between the species and human communities (Lamb et al. 2020).

Corridor conservation, however, is an extremely difficult task. Most identified corridor areas comprise working landscapes or other areas that are economically or culturally important for human communities that already, in many cases, incur significant direct and indirect costs associated with living next to protected areas that host large wildlife (Barua et al. 2013). In most cases, the idea of formal recognition of additional land for conservation is likely to generate significant resistance across several different stakeholder groups. Also, when adjacent human-use areas offer important resources at high densities, corridors may result in a host of negative impacts that are likely to create additional longer-term challenges (Ogden 2015). Given the multi-use nature of land likely to be used to form corridors, identifying and operationalizing corridors comprises several challenges. Some of the difficulties associated with this stem from historic injustices, perpetrated by top-down decision-making, that have forced countless Indigenous peoples and local communities out of their lands and livelihoods in the name of wildlife conservation (West et al. 2006). An additional challenge is that, in most cases, wildlife corridors need to be legally treated separately from protected areas to factor in the significantly different nature of land-use, access rights, and other issues associated with these areas. This challenge is exacerbated by the lack of existing legal classifications or categories that support the formal creation of these entities in many parts of the world (Srivastava and Tyagi 2016).

Given the complexities associated with operationalizing wildlife corridors in different geographies, we offer a framework with prospective steps to consider (Figure 14.2). As a first step, we propose a high-level evaluation of whether active creation or management of wildlife corridors is justified in a given context, based on discussions with wildlife biologists, conservation practitioners, and community members in the region. Understanding existing animal movement through the region, human–wildlife interactions, and the likely isolation of key wildlife populations will help to answer this question. In essence, this is a prioritization exercise with an objective to determine if the social-ecological factors in a region under consideration can justify the investments and process. The next step would comprise a quantitative analysis to map the socially and ecologically ideal pathways for animal movement, using functional and structural connectivity, for example, the range-wide assessment of corridors undertaken for the jaguar in the Americas (Rabinowitz and Zeller 2010). The next step includes an array of activities: (i) identifying an entity that is willing to lead the process; (ii) working out the legal framework that can be used to support the corridor (for example, conservation easements in the United

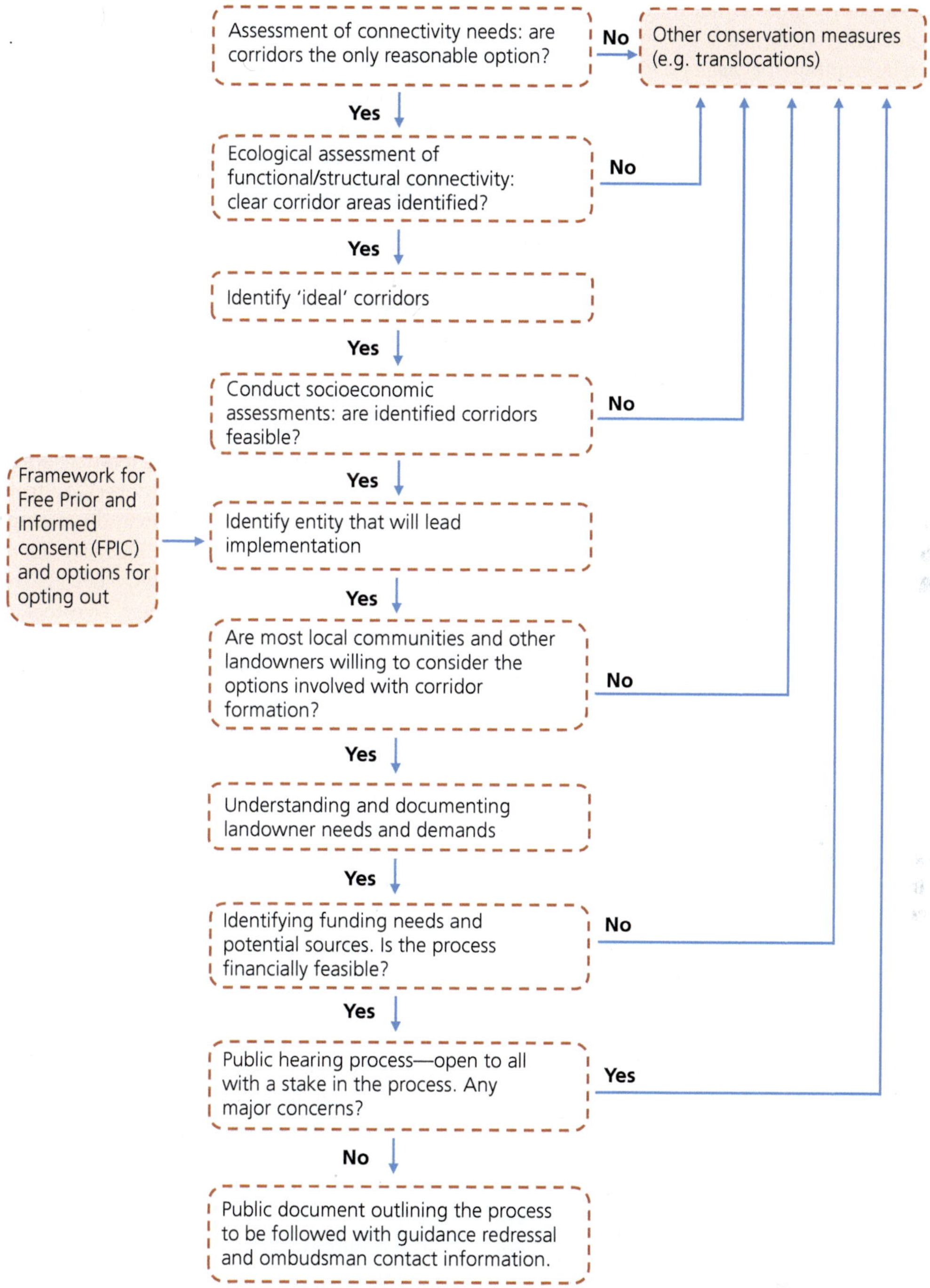

Figure 14.2 A framework for identifying and operationalizing wildlife corridors, comprising ecological and socioeconomic assessments and Free, Prior, and Informed Consent (FPIC) to evaluate the feasibility of establishing wildlife corridors.

States); (iii) identifying potential sources of funding for leasing or purchasing the land; (iv) understanding and documenting the needs and demands of landowners and other community members that have strong links to the land and natural resources in the region, with a focus on **Free, Prior, and Informed Consent (FPIC)** and an option to opt out; and (v) determining the role of government agencies in the process. A subsequent public hearing process that is open to all relevant actors can offer a public forum for all affected entities to voice their concerns and grievances. As a final step of the planning process, a comprehensive document that outlines the corridor design and land acquisition process and strategy should be publicly available in simple, accessible language. The use of **nonviolent communication** (Kansky and Maassarani 2022) is critical to the dialogue process, with an additional emphasis on the language used in all documents, dialogue, and other communication (Boronyak et al. 2021).

Prevention and mitigation of human–carnivore conflict through interventions and tools

The vast majority of practitioners focused on human–carnivore coexistence are concerned with interventions and physical tools, such as fences and other barriers, that can be implemented on the ground in particular landscapes or circumstances (Figure 14.3). While human social factors are key to understanding what works best for preventing and mitigating human–carnivore conflict (Konig et al. 2021), identifiable ecological mechanisms underlie many instances of realized human–carnivore conflict. Thus, an understanding of how ecological theory maps onto human–carnivore interactions is key to deciding which interventions will work best in any given context (Wilkinson et al. 2020). For example, contextualizing carnivore predation on domestic prey using **optimal foraging theory** (Brown et al. 1999) can allow practitioners to consider the real or perceived costs to carnivores of hunting livestock when choosing a predation prevention tool. Alternatively, managers can draw upon **predator-prey shell games theory** (see Mitchell 2009) and **behavioural response race** models (Laundre 2010, Sih 1984) to act as the prey-response role for livestock through altering livestock breed, demographics, and distributions across the landscape (Minnie et al. 2015).

The most prevalent form of human–carnivore conflict for which there are myriad interventions globally is carnivore depredation of livestock (Ripple et al. 2014), or **carnivore–livestock conflict.** Interventions to prevent or mitigate carnivore-livestock conflict tend to act on one or more of three main mechanisms: carnivore ecology, livestock ecology, and the ecological landscape. Of these, the most common are interventions that act directly on carnivore ecology, such as carnivore demographics, behaviour, physiology, and distribution, as well as wild prey availability and community interactions. For example, **carnivore deterrents** such as guardian dogs (Andelt 1992), human herders (Woodroffe et al. 2007), **fladry** (Davidson-Nelson and Gehring 2010), or flashing lights (Shivik et al. 2003) can act directly upon carnivore behaviour by increasing carnivores' perceived costs of preying upon livestock. Meanwhile, carnivore control, through hunting or targeted removal (Blejwas et al. 2002), translocation (Milligan et al. 2018, Thomas et al. 2023), or contraception (Boitani and Powell 2012), is intended to act upon demographics, populations, and distribution of carnivores. However, such measures (particularly hunting) may also influence carnivore behaviour through increasing carnivores' fear of people and human settlements (Oriol-Cotterill et al. 2015). While lethal removal has proven effective in some cases (e.g. removal of individual nuisance animals, Breck et al. 2017), such interventions have been shown to backfire in various instances (Bergstrom 2017), such as through mesopredator release (Ritchie and Johnson 2009) and resultant prey switching to other livestock species or increases in mortality of threatened wild ungulates (Berger et al. 2008), or even through exacerbating predation on livestock (Wielgus and Peebles 2014).

Interventions that act directly on **livestock ecology** take advantage of the fact that people are the primary directors of where livestock live and forage and how they defend themselves. Livestock management interventions usually act directly as human-curated livestock defences that in turn indirectly influence carnivore behaviours towards

Figure 14.3 Examples of interventions and tools designed to mitigate carnivore–livestock conflict. The San Martinero breed of criollo cattle is used to repel jaguars in various locations, including Colombia (pictured here), purportedly for their instinctive behaviour to protect their young and the herd (top left). An anti-predator electric fence constructed to exclude pumas (*Puma concolor*) from extensive grazing lands (top middle). A livestock guardian dog sits with goats beyond an electronet, with both interventions designed to deter coyotes (*Canis latrans*) and pumas as part of the Bay Area Carnivore-Livestock Interactions Project (top right). Turbo fladry is deployed to test its effectiveness against coyotes and pumas in the Santa Cruz Mountains (bottom left). A predator-proof enclosure with opaque mesh, located in Soysambu Conservancy, Kenya, protects cattle via physical and visual barriers (bottom middle). Nenguba Keitsumetse applies a stamp emulating eyes onto the back of a cow to experiment with deterring large carnivores (bottom right). Images courtesy of Matthew Hyde (criollo cattle and electric fence); Bobby-Jo Vial (eyes on cattle); Christine E. Wilkinson (predator-proof enclosure); Veronica Yovovich (livestock guardian dog and turbo fladry).

livestock. Such interventions include actively guarding livestock (Rigg et al. 2011), constructing and maintaining robust livestock enclosures (Kolowski and Holekamp 2006) and calving barns (Pimenta et al. 2017), keeping livestock within fenced ranges (Boitani and Powell 2012), and directing cattle to graze in low-predation risk areas (Johansson et al. 2015). Managers may also choose to keep cattle breeds that have retained reliable defence behaviours towards predators, such as the San Martinero criollo cattle used in parts of South America, which have a number of defence and calf-care attributes that foster their resilience towards predators such as jaguars (Quigley et al. 2015, Hoogesteijn et al. 2016). Meanwhile, interventions such as rotational grazing of livestock can increase forage quality and abundance, thus boosting wild prey populations which can, in some contexts, promote carnivore preferences for hunting wild prey (Mishra et al. 2003).

The ecological attributes of a particular landscape, such as vegetation, terrain, and water availability, are major contributing factors to the distribution and behaviour of carnivores and livestock. For example, vegetation type has been shown to influence livestock vulnerability to predation by leopards (*Panthera pardus*) (Minnie et al. 2015), while habitat composition (i.e. the presence and location of prime habitat patches for carnivores) can enhance the ability of leopards and tigers (*Panthera tigris*) to successfully stalk and kill livestock (Rostro-Garcia et al. 2016). Very few human–carnivore conflict interventions directly target the biophysical landscape, usually focusing instead on livestock ecology and carnivore ecology. This is likely because it can be much more difficult for people to predictably control or influence the biophysical landscape than to alter livestock husbandry patterns or implement a tangible physical carnivore-deterrent tool. The influence of landscape features on carnivore-livestock conflict can also be particularly context-specific, meaning scalable empirical studies on the anti-predation effects of biophysical landscape alteration may be challenging. Additionally, while vegetation removal is a common mitigation method for other pressing human–nature issues such as reducing wildfire danger (e.g. Nelson et al. 2003), and was also historically used as a preventative measure against wolf (*Canis lupus*) predation on livestock (Ruid et al. 2009), encouraging vegetation removal across a working landscape could be detrimental to wild prey populations, resulting in adverse impacts on human–carnivore coexistence efforts. However, some promising avenues for this work may lie in the ways in which carnivore coexistence efforts can be informed by ongoing evaluations of landscape-altering methods for preventing conflicts with other wildlife, such as large ungulates, i.e. digging trenches or planting less palatable crops like chilli (*Capsicum annuum*) or garlic (*Allium sativum*) to deter elephants (*Loxodonta spp.*) from crop raiding (Webber et al. 2011, Montgomery et al. 2022b).

Financial instruments for promoting human–carnivore coexistence

Human communities share a long history of coexisting and competing with large carnivores (Figure 14.4). Coexisting with large carnivores may offer other direct benefits to human populations, including limiting herbivore numbers that may otherwise result in negative impacts on crops, as documented in the case of olive baboons (*Papio anubis*) in Africa (Brashares et al. 2010), and disease regulation (Ripple et al. 2014), especially through the clearing of ungulate carcasses (Sonawane et al. 2021). This is especially important and valuable in the case of diseases that may be transmitted from wild ungulates to livestock. Additional benefits may include tourism revenues from wildlife watching and trophy hunting, reductions in wildlife–vehicle collisions (Gilbert et al. 2017), and other less tangible benefits. Despite these benefits, most of the conversation around interactions with large carnivores tends to be focused on the disadvantages and negative impacts arising from the presence of these animals in or adjacent to working lands (Rode et al. 2021). These range from carnivore attacks or the fear of attacks on humans to economic impacts due to attacks on livestock and other less predictable issues, such as a drop in school attendance and performance (Masago and Kweingoti 2018) and a drop in calories consumed by children (Braczkowski et al. 2023).

Addressing the financial aspect of these negative impacts, in the form of direct payments or other modes, remains a popular albeit extraordinarily complex and challenging tool that has been used across the world in several different forms. Programmes that aim to compensate livestock owners for losses caused by carnivore depredation are the most commonly used tool. Other methods that have been tried or continue to be used include insurance schemes and payments for predator presence. Dickman et al. (2011) describe these 'Payments to Encourage Coexistence (PEC)' as mechanisms to address what is described as a classic market failure problem in carnivore conservation. PECs are commonly seen as an essential tool to achieve coexistence, especially in working landscapes where affected landowners may have the legal right to use lethal control methods or where the magnitude of the issue is so large that legislation against lethal control is largely ignored.

Despite the widespread prevalence of compensation schemes, a few fundamental issues around them remain challenging (Nyhus et al. 2003, Nyhus 2005). Firstly, few compensation schemes are able to fully compensate for the loss of critical livelihood assets such as mature and productive livestock. Increasing the accuracy of estimating losses is complicated and often impractical. Secondly, compensation schemes often create perverse incentives or moral hazards (Bulte and Rondeau 2005, Dickman et al. 2011), disincentivizing communities from finding solutions to manage the issue. Thirdly, successes in large carnivore conservation measures that result in increased carnivore populations seriously threaten the long-term financial viability of these schemes. Lastly, once established, compensation schemes are extremely hard to stop due

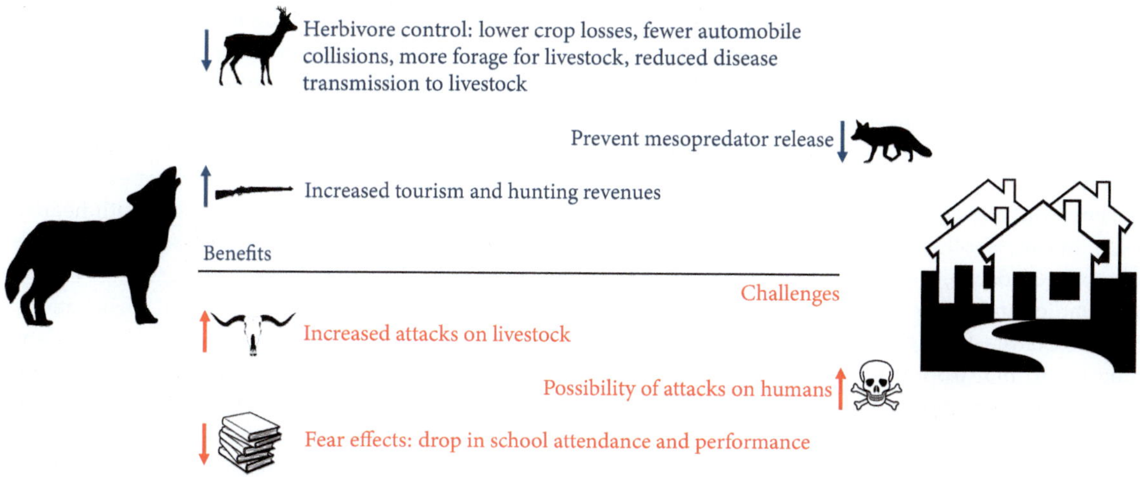

Figure 14.4 Some of the key benefits and challenges of living alongside apex predators. Arrows indicate whether an effect tends to increase or decrease with the presence of apex predators.

to the likely backlash and retaliatory killings of carnivores.

Insurance is another form of PEC that has been tested in a few contexts globally, including snow leopards in Pakistan (Loch-Temzelides 2021) and tigers in India (WWF 2016). Insurance programmes suffer from the same fundamental issues that plague compensation schemes, but have the potential to be more financially sustainable if properly designed. However, these programmes typically need substantial upfront investment in terms of transaction costs involved with assessing animal health and fair market value. Community-based insurance schemes that do not include a formal insurance provider may be a pragmatic option in some contexts. However, this entails the presence and participation of a strong community-based organization that is trusted and respected by the community. For example, the success of a livestock depredation insurance programme in Tost Bagh, Mongolia, can be attributed to the Snow Leopard Trust and the Snow Leopard Conservation Foundation building thoughtful and trustworthy relationships with community members and co-creating the insurance programme with livestock owners (Mishra et al. 2017, Alexander et al. 2021).

Recognizing the limitations of many of the current compensation programmes, other methods of ensuring that livestock owners and landowners receive fair compensation for the losses incurred by large carnivore activity and presence are currently being proposed or piloted. 'Payments for predator presence' is an idea under consideration in Siskiyou County in California, USA, that aims to compensate landowners for the mere presence of large wildlife on their property (Siskiyou County 2015). A similar model is employed through Lion Landscapes near Ruaha, Tanzania, in which villages that have the most detections of large carnivores or other vulnerable species captured on camera traps can receive periodic funding towards community initiatives or needs of their choice (Dickman et al. 2018). On the other hand, in parts of Wyoming, USA, livestock owners are compensated up to seven times the market value of each animal killed by wolves, based on the assumption that it may be impossible to accurately report and verify each incident of livestock depredation (Handwerk 2017). Whether the community-first approach of these programmes substantially improves tolerance for large carnivores on private property or in working landscapes remains to be seen, and the issue of moral hazards and perverse incentives should still be under consideration even if these may play out differently across scenarios.

Funding remains a major issue for most of these programmes intending to address carnivore depredation of livestock. The structure of wildlife and conservation funding and the use of revenues generated from tourism remain a bottleneck in many parts of the world, especially in North America. Existing systems prevent or limit the use of tourism revenues in national parks in the United States for addressing the negative impacts of large wildlife presence outside of these parks (Middleton et al. 2021). In several other parts of the world, visitation fees are mandated to be used locally for addressing conflict and other community needs (Uganda Wildlife Statute 1996, NTCA 2012). Recent bans or attempts to ban trophy hunting of large wildlife, while important from an animal rights perspective, may result in loss of local revenues and increased antagonism towards wildlife and wildlife conservation (Dickman et al. 2019, Hare et al. 2023). Keeping tourism revenues local would go a long way in addressing one of the biggest criticisms of existing wildlife conservation systems—that local communities tend to bear most of the costs while the benefits are reaped by the broader global community—and reduce the increasing urban–rural divide and polarization of opinion on large carnivore presence and reintroductions (Ditmer et al. 2022).

Translocations

Conservation translocations, or the intentional movement of organisms from one location to another, are an essential tool for carnivore conservation and management (Stepkovitch et al. 2022). Conservation translocations can be designed to reintroduce a species to an area where it had been previously extirpated, to reinforce a population by translocating individuals to an already existing population, or to introduce a species to a new location outside the species' historic range to mitigate the effects of climate change or habitat destruction. As a tool, conservation translocations are vital for addressing the decline or extinction of populations, as well as for promoting human–wildlife coexistence.

Reinforcements, or the addition of organisms to an existing population, are frequently used to bolster genetic diversity and resilience of carnivore populations. As habitats are becoming increasingly fragmented and degraded, the lack of genetic diversity in some wildlife populations has led to a decrease in fitness, and inbreeding depression has been documented for several apex predators including wolves, African lions, and Iberian lynx (*Lynx pardinus*) (Cronkrark and Roff 1999, Liberg et al. 2005, Trinkel et al. 2011, Ruiz-Lopez et al. 2012). Further, in human-dominated landscapes, mortality rates of carnivores tend to be high, with nearby protected areas supporting the source population (Lamb et al. 2020). Therefore, conservation efforts may need to increasingly rely on translocations from larger populations to smaller, more endangered populations to promote genetic introgression. For example, in South Florida, eight pumas (*Puma concolor*) were introduced in the mid-1990s to reduce inbreeding depression for the declining Florida panther (*Puma concolor couguar*). Following the Florida panther population's reinforcement, there has been an improvement in genetic diversity, a decrease in the occurrence of morphological and biomedical correlates associated with inbreeding depression, and an increase in population size (Van de Kerk et al. 2019).

In addition to bolstering endangered populations, translocations are often used to mitigate human–wildlife conflict. Conflicts with carnivores are expected to increase in the future due to loss of habitat, climate change, and even local recoveries of wildlife (Fonturbel and Simonetti 2011, Abrahms et al. 2023). In the past, when there was persistent human–carnivore conflict or nuisance and dangerous individuals, wildlife managers would resolve the conflict through culling or increased hunting. As public perception has shifted over the years, there is now an increased emphasis on the translocation of individuals to new environments where there is no conflict, as it is seen as a more humane and conservation-sensitive approach. The success of translocations to mitigate human–carnivore conflict has been mixed (Fonturbel and Simonetti 2011, Stuparyk et al. 2018). Previous studies of American black bears found that translocated bears have very low rates of recidivism and relatively high rates of survival (Bauder et al. 2020, Bauder et al. 2021). Conversely, in Botswana, only 18% of translocated cheetahs (*Acinonyx jubatus*) managed to survive for at least 1 year (Boast et al. 2016). In some cases, the translocation of problem individuals may just shift the problem to another location (Fernando et al. 2012). While there are certainly many sociopolitical and environmental-context-dependent factors affecting the success of a translocation, greater translocation distance has been found to increase success in some cases (Milligan et al. 2018, McManus et al. 2022).

In general, studies looking at the common driver of wildlife translocation success have largely focused on biological and ecological factors (e.g. Bubac et al. 2019, Fischer and Lindenmayer 2000). However, as conservation efforts are increasingly implemented in coupled human and natural systems, the success of translocations has become more reliant on coexistence with people. In fact, a recent study found that conservation translocation efforts that deliberately incorporated sociopolitical components (e.g. education programmes in local communities and economic benefits for landowners coexisting with wildlife) into their efforts were more likely to achieve positive outcomes (Serota et al. 2023). Conservation translocations can be critical for reversing declines in biodiversity and for mitigating conflict, but careful attention to human values, human safety, and local economic livelihoods will undoubtedly promote human–carnivore coexistence (Figure 14.5 and Box 14.1).

Common challenges across human–carnivore coexistence efforts

One of the main challenges for human–carnivore coexistence is a lack of empirical evidence supporting the effectiveness of many interventions (Eklund et al. 2017, Moreira-Arce et al. 2018, van Eeden et al. 2018). There are a myriad of reasons for this. First, human–carnivore conflict is highly context dependent, dictated by the unique integration of particular social, political, economic, and ecological factors within a landscape. Such context-dependency creates challenges for designing robust empirical studies on intervention effectiveness. For example, the unpredictability of livestock depredation events (Elbroch and Treves 2023) can require longitudinal (i.e. Strand et al. 2019), likely highly resource-intensive before-and-after control-impact studies. Meanwhile, developing robust 'gold standard' studies using randomized control-treatment design (see Treves et al. 2016) can be difficult due to limited time, finances, or ethical dilemmas (Lorand et al. 2022), such as when determining which sites are 'control' sites in locations where people's livelihoods are at risk from conflict with carnivores. Additionally, studying intervention effectiveness in the social-ecological systems within which conflict occurs requires multifaceted, interdisciplinary, and empathetic approaches. Yet wildlife scientists and conservation biologists are only in recent years beginning to be more broadly interested and trained in meaningful, ethical community engagement (e.g. Glikman et al. 2019, Chua et al. 2020, Pooley et al. 2020, Niemiec et al. 2021), which is critical to coexistence work and requires long-term relationship building and maintenance of trust (Brittain et al. 2020). Indeed, studies that have successfully employed randomized treatments for evaluating intervention effectiveness (e.g. assessing puma deterrents in Chile; Ohrens et al. 2019) placed participatory engagement and community buy-in at the forefront. Yet many funding structures, which are often short-term, highly disciplinary, and aimed at targeted questions (e.g. Keeley et al. 2019, Elsaway et al. 2020), compound the difficulties of conducting meaningful and iterative community-engaged, longitudinal, landscape-scale studies. Taken together, these elements fuel immense obstacles to both designing empirical studies on interventions and drawing broad, scalable conclusions about their results.

A second main challenge for achieving and maintaining human–carnivore coexistence globally is that researchers and practitioners are continuously 'reinventing the wheel' regarding interventions, rather than having streamlined avenues for collaboration and communication. For example, **range rider programmes,** which involve active cattle surveillance (often on horseback) to reduce predation, have been systematically implemented in the United States beginning as recently as 2003 (Parks and Messmer 2016) and are often touted as a novel solution to human–carnivore conflict despite the fact that the practice of actively herding livestock is both ancient (e.g. Bollig et al. 2013) and globally common. This seemingly obvious oversight in part stems from the historical decimation of large carnivore populations in favour of ranching in the United States. Now that carnivore populations are

Box 14.1 Trying to achieve coexistence—the translocation of brown bears in the Italian Alps

Due to increasing deforestation for farming, and human–wildlife conflict, brown bears (*Ursus arctos*) were driven to near extinction in the Italian Alps (Mustoni et al. 2003). By the late twentieth century, there were only three individuals left in the wild (Mustoni et al. 2003). Starting in 1999, 10 bears were released into the Trento (Trentino) Province of the Alps as part of the Life Ursus Project with the goal of reaching a population of 35–50 bears. Prior to the translocation, the Italian Institute for Environmental Protection and Research carried out a feasibility study to assess the environmental, organizational, and socioeconomic components of the translocation of bears (Soorae 2013). Importantly, the project sought the support of local stakeholders including the Hunting Association of the Trento Province, Trento World Wildlife Fund, as well as groups representing those significantly impacted by the presence of bears, including livestock farmers and beekeepers. Leaders of the reintroduction initiative established comprehensive guidelines for bear monitoring, managing nuisance bears, and personnel training, while ensuring financial backing for compensating losses from damage caused by reintroduced bears (Soorae 2013). After just two decades, the population of brown bears is estimated to be approximately 70 bears, which is a resounding success. However, the increase in the bear population has also coincided with an increase in human–wildlife conflict, mainly involving predation on livestock and the destruction of beehives. In 2015, economic damage was estimated at €41,374 for compensation and €23,527 for prevention (Tosi et al. 2015). Most troubling, however, a brown bear killed a jogger in Trento in early 2023, which has reignited debates around rewilding in Europe. While the reintroduction of brown bears is so far considered a coexistence success, the persistence of life-threatening danger and economic loss reinforces the notion that achieving coexistence is a dynamic (e.g. Harris et al. 2023) and ever-evolving endeavour.

Figure 14.5 A brown bear (*Ursus arctos*) in Katmai National Park, Alaska.
Photo by Vishal Subramanyan.

beginning to recover and some carnivore species have protected status in the United States, ranchers must now learn to coexist, often for the first time (LaRocque 2014), and range riding has emerged as a primary coexistence tool. There are numerous other examples of reinventing the wheel, especially in terms of physical tools to prevent and mitigate conflict. A lack of broad, global communication to and among the practitioners on the ground about effective (anecdotally and empirically) tools can slow innovation in promoting human–wildlife coexistence. Additionally, there remain gaps in communication among scientists working on human–wildlife interactions in urban locations and those working in rural regions, despite similar foundational principles in achieving coexistence. However, there are some efforts to improve cross-context communication, such as the newly released human–wildlife conflict and coexistence guidelines developed by the IUCN Human–Wildlife Conflict and Coexistence Task Force, and online platforms that compile effective interventions and management strategies, such as ENCOSH (http://encosh.org) and EviWild (https://eviwild.slu.se). Though these efforts provide an indication of progress, they are not yet capable of dynamically communicating real-time, streamlined information between diverse end users. On-the-ground practitioners, such as ranchers and herders, continue to be particularly underserved in this realm. Bridging the direct communication gap among practitioners across contexts will be critical to moving the needle on global human–carnivore coexistence.

While grassroots collaborations are key for achieving human–carnivore coexistence, such effects remain challenging to implement in practice. The issue here is a present inability to translate bottom-up progress to sustained, wider-scale policies. First, local and regional governance structures and resources can prevent coexistence efforts from proliferating across scales or being sustained, for instance, when local community interests conflict with government interests (see Nelson et al. 2020, Durant et al. 2022). Additionally, broader economic conditions can influence both conflict itself and conflict mitigation efforts. For instance, in regions of Namibia, conservation conflicts involving carnivores and poaching resulted from resource disparities caused by **apartheid** and also from global market forces impacting livestock farming profitability, such as the prices of livestock feed and fuel (Rust et al. 2016). However, Namibia has also implemented a suite of highly successful broad-scale governmental policies aimed at redirecting power over wildlife management to local communities (Nelson et al. 2020), enabling local coexistence efforts to have more sustained impacts. Government-run wildlife damage recompense efforts (i.e. compensation for livestock attacked by carnivores) can act as another economic lever. These efforts have been viewed by recipients as unsatisfactory (e.g. Mmopelwa and Mpolokeng 2008) or, more harmfully, can be temporarily halted or cancelled entirely due to governance and administration challenges (e.g. Mishra 1997). Importantly, the real or perceived inability to have a voice in local and regional carnivore coexistence policies can also create a negative feedback loop that fuels human–carnivore conflict. For example, in parts of Washington state, USA, a history of real and perceived top-down and outsourced decisions regarding wolf management intensified anti-wolf sentiment for some ranching communities (Young et al. 2015a).

One way that practitioners can begin to address the challenges outlined in this chapter will be to instil in common practice the development of a **Theory of Change** for any human–carnivore coexistence effort. The Theory of Change concept, which essentially describes how and why a particular initiative works, was initially proposed by Weiss (1995) to improve the planning and evaluation of community-based social and public health programmes. However, while developing a Theory of Change is common in international development, its use in conservation science is relatively recent (Rice et al. 2020). Since human–carnivore coexistence is inherently community-involved, practitioners are beginning to realize that this particular realm of conservation science could highly benefit from the use of this practice. Recent studies have noted that brainstorming and adhering to a Theory of Change can be critical for the success of both local-level human–carnivore coexistence projects, such as the Jaguars of Iguaçu Project (Marchini et al. 2021), and

Table 14.1 Mechanisms for improving upon global human–carnivore coexistence strategies

Current limitation in human–carnivore coexistence practices	Reasons why practice is limited	Potential mechanisms of improvement	Goal
A focus almost exclusively on context-specific, hyper-local solutions.	While hyper-local solutions are key components of human–carnivore coexistence, there are many occasions in which coexistence practice and policy can potentially extend beyond local situations but is limited by governance structures and economic incentives.	1. Change governance structures and incentives at local, regional, and country-level scales.	More practices that explicitly plan for broader scale engagement and impacts.
An emphasis on top-down approaches via government policies and/or NGO practices.	Top-down coexistence policies can exacerbate human–carnivore conflict (via human–human conflict). Top-down NGO practices can be both harmfully neocolonialist and limited in scope, making long-term coexistence unsustainable.	1. Prioritize diversity, equity, inclusion, and justice (DEIJ) in policy-making. 2. Make incorporation of Indigenous knowledge and ways of being common practice. 3. Require NGOs and other charity entities to co-create programmes with local actors.	Prioritization of bottom-up, decentralized, community-engaged coexistence practices that forefront Indigenous and local knowledge and provide pathways for communities to contribute to broader policies.
An overabundance of short-term, logistically and financially limited or fragile projects.	Even well-meaning, highly community-engaged coexistence programmes can backfire or fail due to limitations in funding amounts, timelines, and requirements, since funding is often from uncertain donor sources. Monitoring and evaluation of projects can be difficult and also limited due to limited available resources.	1. Change governance structures to connect coexistence projects with one another and with available resources. 2. Revise existing research and programme funding structures to prioritize long-term, community-engaged projects. 3. Consider global markets and create financial structures that encourage communities to fund coexistence measures rather than relying on charity. 4. Develop a Theory of Change for all new and existing coexistence programmes.	Structures and resources that encourage lasting, financially sustainable coexistence practices.
Many programmes and interventions are based on anecdotal evidence on effectiveness.	Intervention effectiveness can be difficult to measure due to the challenges (i.e. financial, logistical) of designing and implementing robust in-situ experiments. Local-scale intervention effectiveness is also often communicated solely within local contexts, due to logistical challenges in global or cross-cultural information dissemination.	1. Revise existing research and programme funding structures to prioritize long-term, community-engaged empirical research projects on human–carnivore conflict and coexistence. 2. Foster global communication resources and platforms regarding existing coexistence measures and programmes.	Coexistence programmes and interventions should be based on empirical evidence regarding effectiveness.

Current limitation in human–carnivore coexistence practices	Reasons why practice is limited	Potential mechanisms of improvement	Goal
Many coexistence programmes and practices continue to reinvent the wheel in regard to conflict interventions and mitigation–rather than learning from practitioners in similar contexts.	Local practitioners usually rely on either their neighbours or on government and NGO sources to receive information regarding intervention effectiveness. However, innovative interventions that are used in certain contexts may never reach practitioners who could also use them effectively in other contexts. There is a need for speed of and access to such communications.	1. Prioritize and improve global practitioner communication platforms so practitioners can learn from one another regarding which practices work and which do not, given different social-ecological contexts.	The effectiveness of coexistence practices in different contexts should be easily and broadly communicated among local practitioners who are in the most need of such information.

for fostering effective broader, continental-level, adaptive coexistence practices (Durant et al. 2022). Using Theory of Change approaches to brainstorm and describe the mechanisms for coexistence in a specific context or more broadly, such as pathways of community engagement and decentralization, governance barriers and opportunities, existing and potential incentives, and ecological factors, can have lasting impacts through increased transparency and through providing a clear roadmap for monitoring and evaluating coexistence projects (Table 14.1).

Conclusion

Human–carnivore coexistence is complex, context-specific, and an inherently social-ecological endeavour. To be successful over the long term, human–carnivore coexistence strategies will integrate a variety of ecologically informed, community-involved practices in combination with one another, while taking into account social-ecological contexts and dynamism. The most important broad challenges to overcome in creating lasting, environmentally and socially just coexistence policies will be meaningful co-creation of such policies with communities and scaling those policies across contexts while navigating global change. Climate change is bringing people and wildlife into closer proximity with one another (Abrahms et al. 2023), which will exacerbate existing inequities in conflict-related vulnerability (Braczkowski et al. 2023). Thus, to further bolster coexistence efforts, researchers and managers should consider prioritizing studies and outreach strategies that shift the narrative towards benefits that people receive from apex predators, such as disease removal, ecotourism gains, and population control for ungulates and other species (see Treves and Karanth 2003, Rode et al. 2021, Sonawane et al. 2021). By taking into account these benefits, along with human histories and lessons from existing context-specific coexistence successes, practitioners can better promote and plan for scalable coexistence between people and carnivores on a rapidly changing planet.

References

Abrahms, B., et al. 2023. Climate change as a global amplifier of human-wildlife conflict. *Nature Climate Change* 13:224–234.

Adams, R. V. and T. M. Burg. 2015. Gene flow of a forest-dependent bird across a fragmented landscape. *PLoS One* 10(11):0140938.

Agrawal, A. and K. Redford. 2009. Conservation and displacement: an overview. *Conservation and Society* 7(1): 1–10.

Alexander, J. S., et al. 2021. Assessing the effectiveness of a community-based livestock insurance program. *Environmental Management* 68:87–99.

Andelt, W. F. 1992. Effectiveness of livestock guarding dogs for reducing predation on domestic sheep. *Wildlife Society Bulletin* 20:55–62.

Athreya, V., M. Odden, J. D. Linnell, J. Krishnaswamy, and U. Karanth. 2013. Big cats in our backyards: persistence of large carnivores in a human dominated landscape in India. *PloS One* 8(3):e57872.

Balme, G. A., R. Slotow, and L. T. B. Hunter. 2010. Edge effects and the impact of non-protected areas in carnivore conservation: leopards in the Phinda-Mkhuze Complex, South Africa. *Animal Conservation* 13(3): 315–323.

Ballard, W. B., L. A. Ayres, P. R. Krausman, D. J. Reed, and S. G. Fancy. 1997. Ecology of wolves in relation to a migratory caribou herd in northwest Alaska. *Wildlife Monographs* 135:3–47.

Banerjee, K., Y. V. Jhala, K. S. Chauhan, C. V. Dave. 2013. Living with lions: the economics of coexistence in the Gir Forests, India. *PLoS One* 9(2):89708.

Barua, M., S. A. Bhagwat, and S. Jadhav. 2013. The hidden dimensions of human–wildlife conflict: health impacts, opportunity and transaction costs. *Biological Conservation* 157:309–316.

Bauder, J. M., D. Ruid, N. M. Roberts, B. Kohn, and A. L. Allen. 2021. Effects of translocation on survival of nuisance bears. *Animal Conservation* 24:820–831.

Bauder, J. M., N. M. Roberts, D. Ruid, B. Kohn, and M. L. Allen. 2020. Black bear translocations in response to nuisance behaviour indicate increased effectiveness by translocation distance and landscape context. *Wildlife Research* 47:426–435.

Becker, M., et al. 2013. Evaluating wire-snare poaching trends and the impacts of by-catch on elephants and large carnivores. *Biological Conservation* 158:26–36.

Berger, K. M., E. M. Gese, and J. Berger. 2008. Indirect effects and traditional trophic cascades: a test involving wolves, coyotes, and pronghorn. *Ecology* 89:818–828.

Bergstrom, B. J. 2017. Carnivore conservation: shifting the paradigm from control to coexistence. *Journal of Mammalogy* 98:1–6.

Blejwas, K. M., B. N. Sacks, M. M. Jaeger, and D. R. McCullough. 2002. The effectiveness of selective removal of breeding coyotes in reducing sheep predation. *Journal of Wildlife Management* 66:451–462.

Boast, L. K, K. Good, and R. Klein. 2016. Translocation of problem predators: is it an effective way to mitigate conflict between farmers and cheetahs acinonyx jubatus in Botswana? *Oryx* 50:537–544.

Boitani, L. and R. A. Powell, eds. 2012. *Carnivore Ecology and Conservation: A Handbook of Techniques*. Oxford University Press, New York.

Bollig, M., M. Schnegg, and H. P. Wotzka. 2013. *Pastoralism in Africa: Past, Present and Future*. Ebook Library, Berghahn, N. Y.

Boronyak, L., et al. 2021. Pathways toward coexistence with large carnivores in production systems. *Agriculture and Human Values* 39:47–64.

Braczkowski, A.R., et al. 2023. The unequal burden of human-wildlife conflict. *Commun Biology* 6:182.

Brashares, J. S., L. R. Prugh, C. J. Stoner, and C. W. Epps. 2010. Ecological and conservation implications of mesopredator release. In *Trophic Cascades: Predators, Prey, and the Changing Dynamics of Nature*, edited by J. Terborgh and J. A. Estes. Island Press, Washington, D. C.

Breck, S. W, S. A. Poessel, and M. Bonnel. 2017. Evaluating lethal and nonlethal management options for urban coyotes. *Human-Wildlife Interactions* 11(2): 133–145.

Brittain, S., et al. 2020. Ethical considerations when conservation research involves people. *Conservation Biology* 34(4):925–933.

Brown, J., J. Laundre, and M. Gurung. 1999. The ecology of fear: optimal foraging, game theory, and trophic interactions. *Journal of Mammalogy* 80:385–399.

Bubac, C. M., A. C. Johnson, J. A. Fox, and C. I. Cullingham. 2019. Conservation translocations and post-release monitoring: identifying trends in failures, biases, and challenges from around the world. *Biological Conservation* 238:108239.

Bulte, E. H. and D. Rondeau. 2005. Why compensating wildlife damages may be bad for conservation. *The Journal of Wildlife Management* 69(1):14–19.

Carter, N. H., J. D. C. Linnell. 2016. Co-adaptation is key to coexisting with large carnivores. *TREE* 31(8): 575–578. https://doi.org/10.1016/j.tree.2016.05.006

Chua, L., et al. 2020. Conservation and the social sciences: Beyond critique and co-optation. a case study from orangutan conservation. *People and Nature* 2(1): 42–60.

Coetzee, B. W. T., K. J. Gatson, and S. L. Chown. 2014. Local scale comparisons of biodiversity as a test for protected area ecological performance: a meta-analysis. *PLoS One* 9:e105824.

Craigie, I. D., et al. 2010. Large mammal population declines in Africa's protected areas. *Biological Conservation* 143:2221–2228.

Creel, S., et al. 2013. Conserving large populations of lions—the argument for fences has holes. *Ecology Letters* 16(11):1413, e3.

Creel, S. and D. Christianson. 2009. Relationships between direct predation and risk effects. *Trends in Ecology and Evolution* 23(4):194–201.

Crnokrak, P. and D. A. Roff. 1999. Inbreeding depression in the wild. *Heredity* 83:260–270.

Davidson-Nelson, S. J. and T. M. Gehring. 2010. Testing fladry as a nonlethal management tool for wolves and coyotes in Michigan. *Human-Wildlife Interactions* 4:87–94.

Dickman, A. J., et al. 2019. Trophy hunting bans imperil biodiversity. *Science* 365:6456.

Dickman, A. J., C. Begg, S. Bhalla, A. Cotterill, S. Dolrenry, L. Hazzah, and D. Macdonald. 2018. Incentives for lion conservation and financial tools for co-existence. *1st Meeting of Range State for the Joint CMS-CITES African Carnivore Initiative (ACI1)*, 6.9.

Dickman, A. J. and L. Hazzah. 2016. Money, myths and man-eaters: complexities of human-wildlife conflict. In *Problematic Wildlife: A Cross-Discplinary Approach*, edited by F. M. Angelici. Springer, Cham, Switzerland.

Dickman, A. J., E. A. Macdonald, and D. W. Macdonald. 2011. A review of financial instruments to pay for predator conservation and encourage human-carnivore coexistence. *Proceedings of the National Academy of Sciences U S A* 108(34):13937–13944

Dickman, A. J. 2010. Complexities of conflict: the importance of considering social factors for effectively resolving human-wildlife conflict. *Animal Conservation* 13(5):458–466.

Ditmer, M. A., R. M. Niemiec, G Wittemyer, and K. R. Crooks. 2022. Socio-ecological drivers of public conservation voting: restoring gray wolves to Colorado, USA. *Ecological Applications* 32(3):e2532.

Dorresteijn, I., et al. 2014. Human-carnivore coexistence in a traditional rural landscape. *Landscape Ecology* 29: 1145–1155.

Durant, S. M., et al. 2022. Fostering coexistence between people and large carnivores in Africa: using a theory of change to identify pathways to impact and their underlying assumptions. *Frontiers in Conservation Science* 2:698631.

Eklund, A., J. V. Lopez-Bao, M. Tourani, G. Chapron, and J. Frank. 2017. Limited evidence on the effectiveness of interventions to reduce livestock predation by large carnivores. *Nature Scientific Reports* 7(1): 2097.

Elbroch, L. M. and A. Treves. 2023. Perspective: why might removing carnivores maintain or increase risks for domestic animals? *Biological Conservation* 283: 110106.

Elbroch, L. M., M. Levy, M. Lubell, H. Quigley, and A. Caragiulo. 2017. Adaptive social strategies in a solitary carnivore. *Science Advances* 3(10):e1701218.

Elsaway, S., et al. 2020. Eight grand challenges in socio-environmental systems modeling. *Socio-Environmental Systems Modelling* 2:16226.

Estes, J. A., et al. 2011. Trophic downgrading of planet earth. *Science* 333:301–306.

Fernando, P., P. Leimgruber, T. Prasad, and J. Pastorini. 2012. Problem-elephant translocation: translocating the problem and the elephant? *PLoS One* 7(12):e50917.

Fischer, J. and D. B. Lindenmayer. 2000. An assessment of the published results of animal relocations. *Biological Conservation* 96:1–11.

Fonturbel, F. E. and J. A. Simonetti. 2011. Translocations and human-carnivore conflicts: problem solving or problem creating? *Wildlife Biology* 17(2): 217–224.

Gaynor, K. M., C. E. Hojnowski, N. H. Carter, and J. S. Brashares. 2019. The influence of human disturbance on wildlife nocturnality. *Science* 360(6394):1232–1235.

Gebresenbet, F., B. Baraki, G. Yirga, C. Sillero-Zubiri, and H. Bauer. 2018. A culture of tolerance: coexisting with large carnivores in the Kafa Highlands, Ethiopia. *Oryx* 52(4):751–760.

Geldmann, J., M. Barnes, L. Coad, I. D. Craigie, M. Hockings, and N. D. Burgess. 2013. Effectiveness of terrestrial protected areas in reducing habitat loss and population declines. *Biological Conservation* 161:230–238.

Giakoumi, S. and A. Pey. 2017. Assessing the effects of marine protected areas on biological invasions: a global review. *Frontiers in Marine Science* 4:1–6.

Gilbert, S. L., et al. 2017. Socioeconomic benefits of large carnivore recolonization through reduced wildlife-vehicle collisions. *Conservation Letters* 10:431–439.

Glikman, J. A., B. Frank, and S. Marchini. 2019. Multi-faceted approaches for turning conflict into coexistence. In *Human-Wildlife Interactions: Turning Conflict into Coexistence*, edited by B. Frank, J. A. Glikman, and S. Marchini. Cambridge University Press, Cambridge.

Gray, C. L., et al. 2016. Local biodiversity is higher inside than outside terrestrial protected areas worldwide. *Nature Communications* 7:12306.

Green, S. E., Z. Davidson, T. Kaaria, and C. P. Doncaster. 2018. Do wildlife corridors link or extend habitat? Insights from elephant use of a Kenyan wildlife corridor. *African Journal of Ecology* 56(4):860–871.

Handwerk, B. 2017. Predator insurance: when livestock becomes prey, conservationists pay. *National Geographic.* Available at: https://www.nationalgeographic.com/adventure/article/131210-wildlife-conservation-compensation-predators-science

Hare, D., et al. 2023. Trophy hunting is not one big thing. *Biodiversity and Conservation* 32:2149–2153.

Harris, N. C., C. E. Wilkinson, G. Fleury, and Z. N. Nhleko. 2023. Responsibility, equity, justice, and inclusion in dynamic human-wildlife interactions. *Frontiers in Ecology and the Environment* 21(8):380–387.

Haag, T., et al. 2010. Molecular tracking of jaguar melanism using faecal DNA. *Conservation Genetics* 11:1239–1242.

Hoeks, S., M. A. J. Huijbregts, M. Busana, M. B. J. Harfoot, J. Svenning, and L. Santini. 2020. Mechanistic insights into the role of large carnivores for ecosystem structure and functioning. *Ecography* 43(12):1752–1763.

Hofer, H. and M. L. East. 1993. The commuting system of Serengeti spotted hyaenas: how a predator copes with migratory prey. I. Social organization. *Animal Behaviour* 46(3):547–557.

Honnay, O. 2013. Genetic drift. In *Brenner's Encyclopedia of Genetics*, edited by S. Maloy and K. Hughes. Academic Press, San Diego, C. A.

Hoogesteijn, R., E. Payán, C. A. Valderrama-Vásquez, F. Tortato, and A. L. Hoogesteijn. 2016. Comportamiento del ganado criollo Sanmartinero y Pantaneiro: La experiencia brasileña y colombiana. In *Conflictos entre felinos y humanos en América Latina [Conflicts between felines and humans in Latin America]*, edited by C. Castaño-Uribe, C. A. Lasso, R. Hoogesteijn, A. D.-P. Diaz-Pulido, and E. Payán. Research Institute of Biological Resources Alexander von Humboldt, Bogotá, Colombia.

IUCN SSC HWCTF. 2023. *IUCN SSC Guidelines on Human-Wildlife Conflict and Coexistence*. First edition. IUCN, Gland, Switzerland.

IUCN SSC HWCTF. 2020. What is human-wildlife conflict? Briefing paper by the IUCN SSC Human-wildlife Conflict Task Force. Available at: https://www.hwctf.org/policies

Jenks, K. E., J. Howard, and P. Leimgruber. 2012. Do ranger stations deter poaching activity in national parks in Thailand? *Biotropica* 44(6): 826–833.

Johansson, O., T. McCarthy, G. Samelius, H. Andren, L. Tumursukh, and C. Mishra. 2015. Snow leopard predation in a livestock dominated landscape in Mongolia. *Biological Conservation* 184:251–258.

Johansson, Ö, G. R. Rauset, G. Samelius, T. McCarthy, H. Andrén, L. Tumursukh, and C. Mishra. 2016. Land sharing is essential for snow leopard conservation. *Biological Conservation* 203:1–7.

Kansky, R. and T. Maassarani. 2022. Teaching nonviolent communication to increase empathy between people and toward wildlife to promote human-wildlife coexistence. *Conservation Letters* 15:e12862.

Keeley, A. T. H., P. Beier, T. Creech, K. Jones, R. H. G. Jongman, C. Stonecipher, and G. M. Tabor. 2019. Thirty years of connectivity conservation planning: an assessment of factors influencing plan implementation. *Environmental Research Letters* 14:103001.

Kolowski, J. M. and K. E. Holekamp. 2006. Spatial, temporal, and physical characteristics of livestock depredations by large carnivores along a Kenyan reserve border. *Biological Conservation* 128:529–541.

Konig, H. J., N. Carter, S. Ceausu, C. Lamb, A. T. Ford, and C. Kiffner. 2021. Human-wildlife coexistence in science and practice. *Conservation Science and Practice* 3:e401.

Konig, H. J., C. Kiffner, S. Kramer-Schadt, C. First, O. Keuling, and A. T. Ford. 2020. Human-wildlife coexistence in a changing world. *Conservation Biology* 34(4):786–794.

Lamb, C. T., et al. 2020. The ecology of human-carnivore coexistence. *Proceedings of the National Academy of Sciences* 117(3):17876–17883.

LaRocque, O. 2014. Revisiting distinctions between ranching and pastoralism: a matter of interspecies relations between livestock, people, and predators. *Critique of Anthropology* 34(1):73–93.

Laundre, J. 2010. Behavioral response races, predator-prey shell games, ecology of fear, and path use of pumas and their ungulate prey. *Ecology* 91:2995–3007.

Leidy, R., C. R. Jorge, B. R. Elena, and B. F. Jordi. 2015. Comparative study of trophic behaviour and herd structure in wild and feral goats living in a Mediterranean island: management implications. *Applied Animal Behaviour Science* 165:81–87.

Liberg, O., et al. 2005. Severe inbreeding depression in a wild wolf (*Canis Lupus*) population. *Biology Letters* 1: 17–20.

Loch-Temzelides, T. 2021. Conservation, risk aversion, and livestock insurance: the case of the snow leopard. *Conservation Letters* 14(4):12793.

Lorand, C., A. Robert, A. Gastineau, J. Mihoub, and C. Bessa-Gomes. 2022. Effectiveness of interventions for managing human-large carnivore conflicts worldwide: scare them off, don't remove them. *Science of the Total Environment* 838:156195.

López-Bao, J., P. Kaczensky, J. D. C. Linnell, L. Boitani, and G. Chapron. 2015. Carnivore coexistence: wilderness not required. *Science* 348:871–872.

Marchini, S., et al. 2021. Planning for human-wildlife coexistence: conceptual framework, workshop process, and a model for transdisciplinary collaboration. *Frontiers in Conservation Science* 2:752953.

Masago, J. M. O. and R. G. Kweingoti. 2018. Education as a goldmine in Africa: a case study of Eastern Africa. *Academic Research Insight Journal* 1:30–45.

McInturff, A., W. Xu, C. E. Wilkinson, N. Dejid, and J. S. Brashares. 2020. Fence ecology: frameworks for understanding the ecological effects of fences. *BioScience* 70(11):971–985.

Mcmanus, J., et al. 2022. Assessment of leopard translocations in South Africa. *Frontiers in Conservation Science* 3:943078.

Middleton, A. D., T. Stoellinger, H. Karandikar, B. Leonard, H. Doremus, and C. Kremen. 2021. Harnessing visitors' enthusiasm for national parks to fund cooperative large-landscape conservation. *Conservation Science and Practice* 3(2):335.

Milligan, S., L. Brown, D. Hobson, P. Frame, and G. Stenhouse. 2018. Factors affecting the success of grizzly bear translocations. *Journal of Wildlife Management* 82:519–530

Minnie, L., A. Boshoff, and G. Kerley. 2015. Vegetation type influences livestock predation by leopards: implications for conservation in agro-ecosystems. *African Journal of Wildlife Research* 45:204–214.

Mishra, C., J. C. Young, M. Fiechter, B. Rutherford, and S. M. Redpath. 2017. Building partnerships with communities for biodiversity conservation: lessons from Asian mountains. *Journal of Applied Ecology* 54(6):1583–1591.

Mishra, C., P. Allen, T. McCarthy, M. D. Madhusudan, A. Bayarjargal, and H. H. T. Prins. 2003. The role of incentive programs in conserving the snow leopard. *Conservation Biology* 17:1512–1520.

Mishra, C. 1997. Livestock depredation by large carnivores in the Indian Trans-Himalaya: conflict perceptions and conservation prospects. *Environmental Conservation* 24(4):338–343.

Mitchell, W. A. 2009. Multi-behavioral strategies in a predator-prey game: an evolutionary algorithm analysis. *Oikos* 118:1073–1083.

Mmopelwa, G. and T. Mpolokeng. 2008. Attitudes and perceptions of livestock farmers on the adequacy of government compensation scheme: Human-carnivore conflict in Ngamiland. *Botswana Notes and Records* 4(40):147–158.

Montgomery, R. A., A. M. Pointer, S. Jingo, H. Kasozi, M. Ogada, and T. Mudumba. 2022a. Integrating social justice into higher education conservation science. *BioScience* 72(6):549–559.

Montgomery, R. A., J. Raupp, M. Mukhwana, A. Greenleaf, T. Mudumba, and T. Muruthi. 2022b. The efficacy of interventions to protect crops from raiding elephants. *Ambio* 51:716–727.

Montgomery, R. A., K. Borona, H. Kasozi, T. Mudumba, and M. Ogada. 2020. Positioning human heritage at the center of conservation practice. *Conservation Biology* 34(5):1122–1130.

Moreira-Arce, D., C. S. Ugarte, F. Zorondo-Rodriguez, and J. A. Simonetti. 2018. Management tools to reduce carnivore-livestock conflicts: current gap and future challenges. *Rangeland Ecology & Management* 71(3): 389–394.

Mustoni, A., et al. 2003. Planning the Brown bear Ursus arctos re-introduction in the Adamello Brenta Natural Park. A tool to establish a metapopulation in the Central-Eastern Alps. *Hystrix* 14:3–27.

National Tiger Conservation Authority (NTCA). 2012. F. No. 15–31/2012-NTCA, Govt. of India, 15 October 2012. Available at: https://ntca.gov.in/assets/uploads/guidelines/Gazette-08-11-12.pdf

Nelson, K. C., M. C. Monroe, and J. F. Johnson. 2003. The look of the land: homeowner landscape management and wildfire preparedness in Minnesota and Florida. *Society & Natural Resources* 18(4):321–336.

Nelson, F., P. Muyamwa-Mupeta, S. Muyengwa, E. Sulle, and D. Kaelo. 2020. Progress or regression? Institutional evolutions of community-based conservation in eastern and southern Africa. *Conservation Science and Practice* 3(1):e302.

Newsome, T. M., et al. 2014. The ecological effects of providing resource subsidies to predators. *Global Ecology and Biogeography* 24(1):1–11.

Niemiec, R. M., et al. 2021. Integrating social science into conservation planning. *Biological Conservation* 262:109298.

Nyhus, P., H. Fischer, F. Madden, and S. Osofsky. 2003. Taking the bite out of wildlife damage the challenges of Wildlife Compensation Schemes. *Conservation in Practice* 4:37–43.

Nyhus, P. J., S. A. Osofsky, P. Ferraro, F. Madden and H. Fischer. 2005. Bearing the costs of human-wildlife conflict: the challenges of compensation schemes. In *People and Wildlife, Conflict or Co-existence?*, edited by R. Woodroffe, S. Thirgood, and A. Rabinowitz. Cambridge University Press, Cambridge, U. K.

Nyhus, P. J. 2016. Human-wildlife conflict and coexistence. *Annual Review of Environment and Resources* 41:143–171.

Ohrens, O., C. Bonacic, and A. Treves. 2019. Non-lethal defense of livestock against predators: flashing lights deter puma attacks in Chile. *Frontiers in Ecology and the Environment* 17(1):32–38.

Ogden, L. E. 2015. Do wildlife corridors have a downside? *Bioscience* 65(4):452–452.

Oldekop, J. A., G. Holmes, W. E. Harris, and K. L. Evans. 2015. Global assessment of the social and conservation outcomes of protected areas. *Conservation Biology* 30:133–141.

Oriol-Cotterill, A., M. Valeix, L. G. Frank, C. Riginos, and D. W. Macdonald. 2015. Landscapes of coexistence for terrestrial carnivores: the ecological consequences of being downgraded from ultimate to penultimate predator by humans. *Oikos* 124(10):1263–1273.

Packer, C., et al. 2013. Conserving large carnivores: dollars and fence. *Ecology Letters* 16(5):635–641.

Parks, M. and T. Messmer. 2016. Participant perceptions of Range Rider Programs operating to mitigate wolf-livestock conflicts in the western United States. *Wildlife Society Bulletin* 40(3):514–524.

Peterson, M. N., J. L. Birckhead, K. Leong, M. J. Peterson, and T. R. Peterson. 2010. Rearticulating the myth of human-wildlife conflict. *Conservation Letters* 3(2):74–82.

Pierce, B. M., V. C. Bleich, J. D. Wehausen, and R. T. Bowyer. 1999. Migratory patterns of mountain lions: implications for social regulation and conservation. *Journal of Mammalogy 80*(3):986–992.

Pimenta, V., I. Barroso, L. Boitani, and P. Beja. 2017. Wolf predation on cattle in Portugal: assessing the effects of husbandry systems. *Biol Cons* 207:17–26.

Pooley, S., S. Bhatia, and A. Vasava. 2020. Rethinking the study of human-wildlife coexistence. *Conservation Biology* 35(3):784–793.

Pooley, S., et al. 2016. An interdisciplinary review of current and future approaches to improving human-predator relations. *Conservation Biology* 31(3): 513–523.

Prugh, L. R., C. J. Stoner, C. W. Epps, W. T. Bean, W. J. Ripple, A. Laliberte, and J. S. Brashares. 2009. The rise of the mesopredator. *BioScience* 59(9):779–791.

Quigley, H., et al. 2015. Observations and preliminary testing of jaguar depredation reduction techniques in and between core jaguar populations. *Parks* 21:63–72.

Rabinowitz, A. and K. A. Zeller. 2010. A range-wide model of landscape connectivity and conservation for the jaguar, *Panthera onca*. *Biological Conservation* 143(4): 939–945.

Rauset, G. R., et al. 2015. National parks in northern Sweden as refuges for illegal killing of large carnivores. *Conservation Letters* 9(5):334–341.

Rosenbaum, B., et al. 2023. Seasonal space use and habitat selection of GPS collared snow leopards (*Panthera uncia*) in the Mongolian Altai range. *PloS One* 18(1):e0280011.

Redpath, S. M., S. Bhatia, and J. Young. 2014. Tilting at wildlife: reconsidering human-wildlife conflict. *Oryx* 49(2):222–225.

Rice, W. S., M. R. Sowman, and M. Bavinck. 2020. Using theory of change to improve post-2020 conservation: a proposed framework and recommendations for use. *Conservation Science and Practice* 2(12):e301.

Rigg, R., S. Findo, M. Wechselberger, M. L. Gorman, C. Sillero-Zubiri, and D. W. Macdonald. 2011. Mitigating carnivore-livestock conflict in Europe: lessons from Slovakia. *Oryx* 45:272–280.

Ripple W. J., et al. 2014. Status and ecological effects of the world's largest carnivores. *Science* 343(6167):1241484.

Ritchie, E. G. and C. N. Johnson. 2009. Predator interactions, mesopredator release and biodiversity conservation. *Ecology Letters* 12(9):982–998.

Rode, J., L. Flinzberger, R. Karutz, A. Berghofer, and C. Schroter-Schlaack. 2021. Why so negative? Exploring the socio-economic impacts of large carnivores from a European perspective. *Biological Conservation* 255:108918.

Roques, S., et al. 2015. Effects of habitat deterioration on the population genetics and conservation of the jaguar. *Conservation Genetics* 17:125–139.

Rostro-Garcia, S., L. Tharchen, L. Abade, C. Astaras, S. A. Cushman, and D. W. Macdonald. 2016. Scale dependence of felid predation risk: identifying predictors of livestock kills by tiger and leopard in Bhutan. *Landscape Ecology* 31:1277–1298.

Ruid, D. B., et al. 2009. Wolf-human conflicts and management in Minnesota, Wisconsin, and Michigan. In *Recovery of Gray Wolves in the Great Lakes Region of the United States*, edited by A. P. Wydeven, T. R. Van Deelen, and E. J. Heske. Springer, New York.

Ruiz-López, M., et al. 2012. Heterozygosity-fitness correlations and inbreeding. *Conservation Biology* 26(6): 1121–1129.

Rust, N. A., J. Tzanopoulos, T. Humle, and D. C. MacMillan. 2016. Why has human-carnivore conflict not been resolved in Namibia? *Society & Natural Resources* 29(9):1079–1094.

Serota, M. W., et al. 2023. Incorporating human dimensions is associated with better wildlife translocation outcomes. *Nature Communications* 14:2119.

Shivik, J. A., A. Treves, and P. Callahan. 2003. Nonlethal techniques for managing predation: primary and secondary repellents. *Conservation Biology* 17:1531–1537.

Sih, A. 1984. The behavioral response race between predators and prey. *American Naturalist* 123:143–150.

Siskiyou County. 2015. Grazing policy. Available at: https://www.co.siskiyou.ca.us/sites/default/files/fileattachments/natural_resources/page/7831/nr-2015_final_grazing_policy.pdf

Smith, D., R. King, and B. L. Allen. 2020. Impacts of exclusion fencing on target and non-target fauna: a global review. *Biological Reviews* 95(6):1590–1606.

Sonawane, C., G. Yirga, and N. H. Carter. 2021. Public health and economic benefits of spotted hyenas *Crocuta crocuta* in a peri-urban system. *Journal of Applied Ecology* 58(12):2892–2902.

Soorae, P. S. 2013. *Global Re-Introduction Perspectives, 2013: Further Case Studies from Around the Globe.* IUCN Species Survival Commission, Abu Dhabi, U. A. E.

Srivastava, R. and R. Tyagi. 2016. Wildlife corridors in India: viable legal tools for species conservation? *Environmental Law Review* 18(3):205–223.

Stepkovitch, B., R. T. Kingsford, and K. E. Moseby. 2022. A comprehensive review of mammalian carnivore translocations. *Mammal Review* 52(4):554–572.

Strand, G., I. Hansen, A. de Boon, and C. Sandstrom. 2019. Carnivore management zones and their impact on sheep farming in Norway. *Environmental Management* 64:537–552.

Stuparyk, B., C. J. Horn, S. Karabatsos, and J. Arteaga Torres. 2018. A meta-analysis of animal survival following translocation: comparisons between conflict and conservation efforts. *Canadian Wildlife Biology & Management* 7(1):3–17.

Su, K., H. Zhang, L. Lin, Y. Hou, and Y. Wen. 2022. Bibliometric analysis of human-wildlife conflict: From conflict to coexistence. *Ecological Informatics* 68:101531.

Terraube, J., J. Van Doninck, P. Helle, and M. Cabeza. 2020. Assessing the effectiveness of a national protected area network for carnivore conservation. *Nature Communications* 11:2957.

Thomas, S., et al. 2023. Evaluating the performance of conservation translocations in large carnivores across the world. *Biological Conservation* 279:109909.

Tosi, G., et al. 2015. Brown bear reintroduction in the Southern Alps: To what extent are expectations being met? *Journal for Nature Conservation* 26:9–19.

Treves, A., M. Krofel, and J. McManus. 2016. Predator control should not be a shot in the dark. *Frontiers in Ecology and the Environment* 14(7):380–388.

Treves, A., R. B. Wallace, L. Naughton-Treves, and A. Morales. 2006. Co-managing human-wildlife conflicts: a review. *Human Dimensions of Wildlife* 11(6): 383–396.

Treves, A., and K. U. Karanth. 2003. Human-carnivore conflict and perspectives on carnivore management worldwide. *Conservation Biology* 17(6):1491–1499.

Trinkel, M., D. Cooper, C. Packer, R. Slotow. 2011. Inbreeding depression increases susceptibility to bovine tuberculosis in lions: an experimental test using an inbred-outbred contrast through translocation. *Journal of Wildlife Diseases* 47(3):494–500.

Tsioumani, E. 2020. Convention on biological diversity: a review of the post-2020 global biodiversity framework working group negotiations. *Environmental Policy and Law* 50:55–59.

Uganda Wildlife Statute. No. 14. 1996. Available at: http://extwprlegs1.fao.org/docs/pdf/uga9000.pdf

Ugarte, C. S., D. Moreira-Arce, and J. A. Simonetti. 2019. Ecological attributes of carnivore-livestock conflict. *Frontiers in Ecology and Evolution* 7:433.

UNEP-WCMC and IUCN. 2023. *Protected Planet: The World Database of Protected Areas (WDPA)*. UNEP-WCMC and IUCN, Cambridge U. K.

van Bommel, J. K., C. Sun, A. T. Ford, M. Todd, and A. C. Burton. 2022. Coexistence or conflict: black bear habitat use along an urban-wildland gradient. *PLoS One* 17(11):e0276448.

van Eeden, L. M., et al. 2018. Carnivore conservation needs evidence-based livestock protection. *PLoS Biology* 16(9):e2005577.

van de Kerk, M., D. P. Onorato, J. A. Hostetler, B. M. Bolker, and M. K. Oli. 2019. Dynamics, persistence, and genetic management of the endangered florida panther population. *Wildlife Monographs* 203: 3–35.

Venumiere-Lefebvre, C. C., S. W. Breck, and K. R. Crooks. 2022. A systematic map of human-carnivore coexistence. *Biological Conservation* 268:109515.

Walton, L. R., H. D. Cluff, P. C. Paquet, and M. A. Ramsay. 2001. Movement patterns of barren-ground wolves in the central Canadian Arctic. *Journal of Mammalogy* 82(3):867–876.

Watson, J. E. M., N. Dudley, D. B. Segan, and M. Hockings. 2014. The performance and potential of protected areas. *Nature* 515:67–73.

Webber, C. E., T. Sereivathana, M. P. Maltby, and P. C. Lee. 2011. Elephant crop-raiding and human-elephant conflict in Cambodia: crop selection and seasonal timings of raids. *Oryx* 45(2):243–251.

Weiss, C. H. 1995. Nothing as practical as good theory: Exploring theory-based evaluation for comprehensive community initiatives for children and families. In *New Approaches to Evaluating Community Initiatives: Concepts, Methods, and Contexts*, edited by J. Connell, et al. Aspen Institute, Washington, D. C.

Wen, D., et al. 2022. Conservation potentials and limitations of large carnivores in protected areas: a case study in Northeast China. *Conservation Science and Practice* 4(6):e12693.

West, P., J. Igoe, and D. Brockington. 2006. Parks and peoples: the social impact of protected areas. *Annual Review of Anthropology* 35:251–277.

Wielgus, R. B. and K. A. Peebles. 2014. Effects of wolf mortality on livestock depredations. *PLoS One* 9: e113505.

Wilkinson, C. E., J. S. Brashares, A. C. Bett, and M. Kelly. 2021a. Examining drivers of divergence in recorded and perceived human-carnivore conflict hotspots by integrating participatory and ecological data. *Frontiers in Conservation Science* 2:681769.

Wilkinson, C. E., A. McInturff, M. Kelly, and J. S. Brashares. 2021b. Quantifying wildlife responses to conservation fencing in East Africa. *Biological Conservation* 256:109071.

Wilkinson, C. E., et al. 2020. An ecological framework for contextualizing carnivore-livestock conflict. *Conservation Biology* 34(4):854–867.

Woodroffe, R., L. G. Frank, P. A. Lindsey, S. M. K. Ole Ranah, and S. Romanach. 2007. Livestock husbandry as a tool for carnivore conservation in Africa's community rangelands: a case-control study. *Biodiversity and Conservation* 16:1245–1260.

Wultsch, C., A. Caragiulo, I. Dias-Freedman, H. Quigley, S. Rabinowitz, and G. Amato. 2016. Genetic diversity and population structure of Mesoamerican jaguars (*Panthera onca*): implications for conservation and management. *PloS One* 11(10):e0162377.

WWF-India. 2016. 'TigerSafe', a pro-conservation livestock insurance scheme launched in Uttarakhand. Available at: https://www.wwfindia.org/?15721/Insurance-Scheme-launched-in-Uttarakhand

Young. J. K., Z. Ma, A. Laudati, and J. Berger. 2015a. Human-carnivore interactions: lessons learned from communities in the American West. *Human Dimensions of Wildlife* 20(4):349–366.

Zafra-Calvo, N., et al. 2019. Progress toward equitably managed protected areas in aichi target 11: a global survey. *BioScience* 69:191–197.

CHAPTER 15

The Future of Human–Carnivore Interactions in the Anthropocene

Nyeema C. Harris, Evie Jones, Pallavi Ghaskadbi, and Mariela G. Gantchoff

Introduction

Large carnivores are regulators across ecosystems, evoking complex interactions with human societies due to their feeding habits, movement patterns, and life history characteristics (Figure 15.1). Tangible benefits for people from carnivores are undeniable, from cultural value and resultant contributions to national heritage and spirituality to economic opportunities derived from ecotourism activities (Buijs and Jacobs 2021). Via the top-down control of ungulate populations, large carnivores play a role in; *i*) minimizing the socioeconomic impact of ungulate-related damage to agriculture and forestry (Davoli et al. 2022, Chapter 14), *ii*) mitigating ungulate-vehicle collisions (Gilbert et al. 2017), and *iii*) lowering disease prevalence among ungulate populations, thereby reducing costs associated with spillover effects on livestock (Richards et al. 2022). Concurrently, communities living with large carnivores can experience burdens that compromise their livelihoods and well-being (Bombieri et al. 2023). These interactions between humans and large carnivores do not simply exist along a continuum from conflict to coexistence based on categorizations of outcomes being positive, negative, or neutral. Instead, Harris and colleagues argued for considering human-wildlife interactions within a lifecycle to capture the dynamisms of these interactions, highlighting how governing principles centred on responsibility, equity, justice and inclusion could influence the outcome and duration of interaction within a positive or negative state (Harris et al. 2023b).

Throughout this book, we've been unpacking the lethal (Chapter 4) and non-lethal (Chapter 9), the conspicuous and inconspicuous, the direct and indirect effects stemming from human sympatry with large carnivores. While humans exist within a network of interactions, they have a unique dominance that transforms biomes (Chapter 1). And ultimately, how ecosystems function and communities are structured will be governed by the effects humans exert on large carnivores. We saw in Chapters 10–12 how changes in the presence, behaviour, and distribution of carnivores cascade to prey, plants, and even phosphorus, such that their removal has massive consequences for healthy natural systems. In exploring the historic (Chapter 2) and present-day (Chapter 3) interactions between humans and carnivores, we recognize that these patterns may provide some insight but not the complete picture in anticipating future human–carnivore interactions.

In this final chapter, we elucidate the vulnerability and dynamism of human–carnivore interactions amid an era of intense global anthropogenic change. Human populations are expected to expand by two billion in the next 30 years, increasing pressure and modifying the configuration and quality of lands and oceans that ultimately alter the accessibility and selection of resources for carnivores. Demand for meat, and therefore livestock production, is increasing globally, intensifying the

Nyeema C. Harris et al., *The Future of Human–Carnivore Interactions in the Anthropocene*. In: *Apex Predators in the Anthropocene*. Edited by: Robert A. Montgomery and Remington J. Moll, Oxford University Press. © Oxford University Press (2026). DOI: 10.1093/9780191996665.003.0015

Figure 15.1 Diverse human–carnivore interactions. Conflict outcomes (a–c): road mortality of golden jackal (Credit: Eliane Küpfer, available under Creative Commons Attribution-Share Alike International 4.0 (CC BY-SA 4.0) license https://creativecommons.org/licenses/by-sa/4.0/); Tigress (*Panthera tigris*) kills domestic cow (Credit: Pallavi Ghaskadbi); Himalayan brown bear (*Ursus arctos isabellinus*) feeding at dump site (Credit: Niaz Khan/WLPD Kargil). Positive benefits derived from sympatry with large carnivores (d–f): Hindu shrine of a deity Waghoba (Credit: AJT Johnsingh, available under Creative Commons Attribution-Share Alike International 4.0 (CC BY-SA 4.0) license https://creativecommons.org/licenses/by-sa/4.0/); Ecotourism spurred by African lions (*Panthera leo*, Credit: Nyeema C. Harris); Local entrepreneurship for snow leopard conservation (*Panthera uncia*, Credit: Nyeema C. Harris).

likelihood of conflict with carnivores. Climatic environmental changes are accelerating and intensifying, synergistically influencing humans, carnivores, prey, and their hypervolume of interactions, amplifying the urgency of mitigation and adaptation strategies to promote coexistence (Chapter 14). Here, our chapter envisions the future of human–carnivore interactions in a changing world. The first section examines the outlook for carnivores that are using increasingly human-modified landscapes. Within this context, we explore how humans may identify ways to coexist with carnivores in expanding modified landscapes of agricultural and urbanization. The second section outlines the impacts of climate change and other threats on carnivores and how this will change human–carnivore interactions, including altered predator and prey distributions, effects on carnivore behaviour, and effects on people that are likely to influence tolerance for carnivores. We conclude with a discussion of future directions including carnivore species trajectories and emerging research needs that can provide more effective conservation outcomes.

Landscape transformations

Historical habitat transformation (i.e. habitat fragmentation, degradation, and loss) has compromised the ability of many landscapes to accommodate contiguous, large populations of carnivores. Globally, 15 out of the 25 largest terrestrial carnivore species have lost 50% or more of their historical range, with six species having lost over 90% (Wolf and Ripple 2017). Projections of land-use change suggest that agriculture will continue to intensify along with increased urbanization in some of the world's hotspots of biodiversity and carbon stocks (e.g. India, Papua New Guinea, Indonesia, Southern Africa, and Central and South America (Wolff et al. 2018). Throughout Africa, which supports approximately one-third of all terrestrial carnivore species (Harris et al. 2023a), an anticipated twofold and more human population increase coupled with rapid demands for economic growth between 2054–2100 will expectedly result in increased deforestation rates (Vollset et al. 2020). Where deforestation has been decelerated, a reduction in the extinction risks for carnivores was observed, as previously demonstrated in southeast Asia in the past 40 years (Visconti et al. 2016).

The ultimate source of habitat transformation originates from patterns of human mobility that subsequently requires infrastructural support for recreation and transportation needs (Ellis-Soto et al. 2023). Road networks are expanding in growing economies that bisect habitats, compromising animal movement and gene flow, with an additional 25 million km of roads predicted by 2050 (Laurance 2018). These roads undoubtedly will yield some societal benefits while fragmenting habitats, increasing edge effects, and causing direct mortality for carnivores (see Chapter 6, this volume). A large global analysis found that carnivore populations most vulnerable to population declines due to roadkill include leopards (*Panthera pardus*) in northern India and southern Africa, brown hyenas (*Hyaena brunnea*) and lions (*Panthera leo*) in southern Africa, tigers (*Panthera tigris*) in eastern Asia, and maned wolf (*Chrysocyon brachyurus*) and oncilla (*Leopardus tigrinus*) in south-eastern South America (Grilo et al. 2021).

Coupled with the negative direct consequences of habitat transformation on carnivores are indirect effects caused by prey depletion (Wolf and Ripple 2016). Prey depletion represents a significant threat to carnivore population viability with ramifications to human interactions (Wolf and Ripple 2016). The decline of wild prey resulting from agricultural expansion, hunting pressures, and other factors increases the likelihood of livestock depredation, and social intolerance coupled with ineffective compensatory schemes results in retaliatory killings (Jordan et al. 2020). Such patterns have been reported in carnivore–human interactions around the globe, from snow leopards (*Panthera uncia*) in the Himalayas (Suryawanshi et al. 2017) to Madagascar's apex predator, the fosa (*Cryptoprocta ferox*) (Merson et al. 2019), to mountain lions in the Western United States (Benson et al. 2023). One of the primary predictors of livestock depredation by some large carnivores appears to be the biomass of available wild prey (e.g. Khorozyan et al. 2015). For many carnivore species, retaliatory killings following livestock depredations are a primary factor limiting their population recovery. This is the case for Persian leopard (*Panthera pardus saxicolor*) in the Caucasus (Bleyhl et al. 2021) and for jaguars in northern Argentina (Lodeiro Ocampo and Nigro 2020). Some large carnivore populations, like pumas in Patagonia, are showing dietary flexibility by relying on smaller prey, including non-native prey, when there is lower availability of larger native prey, which may play a crucial role in mitigating the risk of local extinction (Osorio et al. 2020).

While many studies emphasize the negative consequences emergent from land-use changes, some species are experiencing benefits from converted habitats. The urbanization surge, expectedly resulting in ~20 million hectares of natural habitat loss by 2100, and substantial fragmentation under different climate and development scenarios will certainly be exploited by some generalist species (Li et al. 2022). In North America, the quintessential exemplar of this phenomenon is the coyote (*Canis latrans*), an ascended apex predator and exceptional case study in community ecology and ecological adaptation in the Anthropocene. Before European colonization, coyotes primarily inhabited non-forested habitats. The rapid expansion of coyotes in North and Central America is attributed mostly to the conversion of forested landscapes to agriculture and the extirpation of top carnivores like wolves and cougars (Hody and Kays 2018). The recent arrival of coyotes to Panama (Méndez-Carvajal and Moreno 2014) has positioned them to colonize South America in future decades, with unknown and uncertain implications for tropical ecosystems.

Increasingly, evidence demonstrates that large and medium-sized carnivores can fare well in urban and suburban environments due to exhibited behavioural and dietary adaptations. Some smaller carnivores may benefit from the human-driven decline of larger carnivores through mesopredator release (Ritchie and Johnson 2009). Habitat generalists and omnivorous species tend to cope better with anthropogenic changes than habitat specialists and hypercarnivores (Ferreira et al. 2018, Jones et al. 2023). American black bears (*Ursus americanus*) in urban areas become more nocturnal to avoid human encounters and often shift their diet to include human foods (Zeller et al. 2019). Versatile leopards inhabit areas with very high human densities in the outskirts of urban centres like Nairobi and Mumbai (Surve et al. 2022); similarly, spotted hyenas (*Crocuta crocuta*) have been found thriving close to humans in Ethiopia (Yirga et al. 2013). Carnivores occurring in cities can leverage roads for communication (e.g. urine spraying, scat depositing), foraging (finding prey and carrion), movement (e.g. easier movement in thick vegetation or deep snow), refugia (e.g. denning), and thermoregulation (e.g. culverts) (Hill et al. 2021). However, urban areas do pose a myriad of risks for carnivores such as road mortality, exposure to pesticides and other chemicals, and direct encounters with humans and their pets (Chapter 3, 9). In Brazil, human population density, the built environment, and road development are key drivers of carnivore occurrence in urban landscapes, though the directions of these effects are complex (Pereira et al. 2023). A higher tolerance to human presence (i.e. habituation) with increasing densities will also inevitably lead to increased human–carnivore conflicts of various risk levels (i.e. from garbage raiding to attacks on humans), which could in some cases create source-sink dynamics for carnivore populations.

Political regimes and legal infrastructure greatly influence carnivore persistence. In the United States, administrative oscillations of regulation and deregulation alter the expansion and effectiveness of protected areas and legal status of carnivores, in turn impacting hunting opportunities (e.g. Hejny 2018). The eastern population of Steller sea lions (*Eumetopias jubatus*) in North America showed a ~40% increase between 2000 and 2015 (IUCN 2023) that was in part attributed to the protection of critical areas. Large terrestrial carnivores that experienced decades of systematic extermination are now recovering due to habitat restoration and increased legal protection. Classic examples of this include American black bears (*Ursus americanus*) in northern Mexico (Medellín et al. 2005) and the Midwestern USA (Gantchoff et al. 2022), grey wolves (*Canis lupus*) in the Great Lakes and Rocky Mountains region (Ausband and Mech 2023, van den Bosch et al. 2023), and mountain lions (*Puma concolor*) in the eastern USA (Larue et al. 2012, Gantchoff et al. 2021). To mitigate the unprecedented rate of human-induced habitat transformations, there is a pressing need for large scale restoration initiatives that include protective schemes and connectivity enhancements. For example, the Bonn Challenge (bonnchallenge.org) initiated by the Global Partnership on Forest Landscape Restoration aims to restore 350 million hectares of deforested and degraded land through agroforestry and forest landscape restoration by 2030.

Area abandonment from agricultural practices are yielding new opportunities for habitat restoration and natural recolonization by carnivores (Mantero et al. 2020). Throughout Europe, lands abandoned from agricultural practices are reverting to rich forests to support the rewilding of carnivores with active and passive restoration efforts (Pereira and Navarro 2015, Hart et al. 2023). In response to this availability of habitat, wild ungulates return and large carnivores soon follow as observed in Bulgaria (Fletcher and Toncheva 2021). Europe has experienced recolonization of grey wolves, brown bears *(U. arctos*), Eurasian lynx (*Lynx lynx*), and wolverines (*Gulo gulo*) across several countries (Chapron et al. 2014, Milanesi et al. 2017). A less recognized example is the golden jackal (*Canis aureus)* that is both recolonizing historical ranges following population reductions and extirpation and expanding into new areas (Box 15.1, Arnold et al. 2012, Kowalczyk et al. 2020). Natural (i.e. passive) recolonization and recovery has so far been less common outside of North America and Europe, but some examples include mountain lions in Brazil (Mazzolli 2012), African wild dogs in Malawi (Sievert et al. 2023), and Eurasian otters (*Lutra lutra*) in southern China (Lau et al. 2010). Increasingly, restoration acknowledges the return of ecosystem function and not simply individual species occurrence. Successful restoration extends beyond vegetation-only improvements and must incorporate trophic interactions with an emphasis on the restoration of ecosystem function and complexity (Loch et al. 2020).

Climate change threats

The well-being, survival, and geographical distribution of any species depend upon its ability to respond to environmental change (Fox et al. 2019). Animals respond to climate change through shifts in behaviour, physiology, vital rates, and survival rates (Bastille-Rousseau et al. 2018, Wells et al. 2022)

Box 15.1 Carnivore natural range expansion

The golden jackal (*Canis aureus*) has undergone rapid geographic expansion in Europe. Yet, the species has not attracted much attention compared to other large carnivore species (Trouwborst et al. 2015). The factors that facilitate their expansion are unclear, but their generalist nature, tolerance to human disturbance, land-use changes, climate change, and lack of competition from other canids have been suggested (Arnold et al. 2012; Cunze and Klimpel 2022). The golden jackal plays a prominent role in the ecosystem as a canid mesopredator. Thus, its range expansion may lead to considerable changes in ecosystems, such as predator–prey dynamics and competition with other mesopredators.

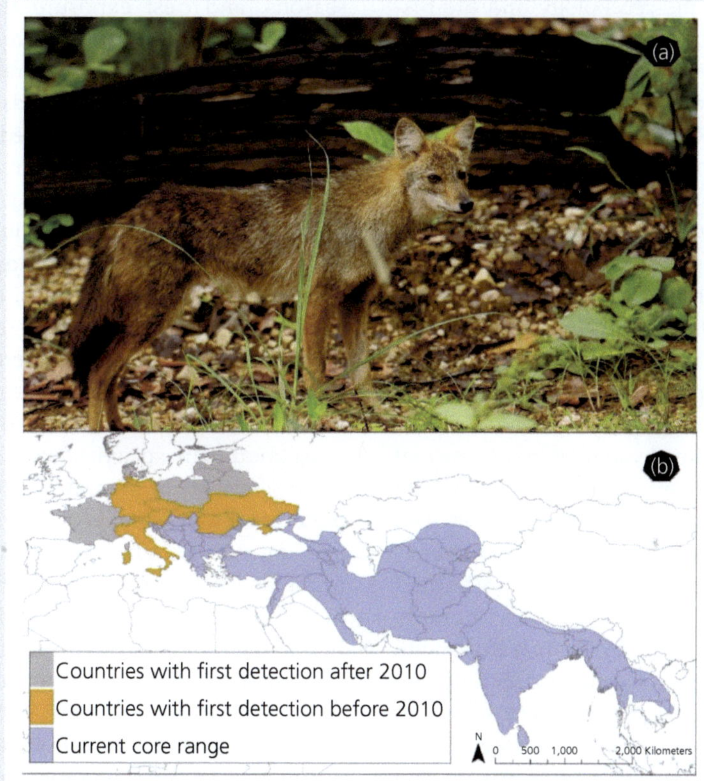

(a) A golden jackal standing in a forest, (b) A map of golden jackal range expansion across Europe before and after 2010

Source: (a) Tontan Travel, available under a Creative Commons CC BY-SA 2.0 license https://creativecommons.org/licenses/by-sa/4.0/; (b) Adapted from Lanszki et al. 2022, available under a Creative Commons CC BY-NC 4.0 license https://creativecommons.org/licenses/by-nc/4.0/.

and often the most behaviourally plastic are the most successful (Beever et al. 2017). However, rapid anthropogenic climate change has become a major threat to biodiversity as well as the well-being of humans (Chapter 10). Coupled with habitat transformations, climate change profoundly shapes environments and creates vulnerabilities for carnivores. Carnivore distributions are shifting, shuffling communities and altering subsequent patterns of predation, herbivory, and competition (Gilman et al. 2010, Pecl et al. 2017, Abrahms et al. 2022). As temperatures rise, previously distinct habitats are undergoing transformations, forcing carnivores to adapt or relocate. For example, Arctic regions, once dominated by polar bears (*Ursus maritimus*), are witnessing the encroachment of brown and black bears as they venture further north in search of suitable habitats (Lan et al. 2022). Consequently, this increased range overlap may result in hybridization risks and disrupted predator–prey dynamics that typically serve as a balance for ecologically sensitive ecosystems.

Increasingly, the warming of the climate has been proven to affect the physiology and life history traits of carnivores. Rising air temperature has caused reductions in meerkat (*Suricata suricatta*) pup growth and survival (Van de Ven et al. 2020) and also has been linked to pup survivorship and population decline in African wild dogs (Lycaon pictus) (Woodroffe et al. 2017). Long-term warming has also led to changes in birthing and denning dates in African wild dogs, potentially leading to lower reproductive success (Abrahms et al. 2022). Eurasian otters (*Lutra lutra*) were reported to exhibit an increase in body size across the final decades of the twentieth century, potentially due to energy savings associated with global warming (Yom-Tov et al. 2006). Conversely, spotted hyaenas (*Crocuta crocuta*) appear to be adapting well to climate change, with maternal den presence unaffected by climate-driven changes in prey availability within their ranges (Gicquel et al. 2022). Aquatic ecosystems are vulnerable to climate change as metabolic rates of organisms are directly regulated by temperature with a limited ability to adapt to changes. Studies on aquatic mammalian carnivores like otters, seals, and sea lions suggest the impacts are increasing and are expected to be severe, especially for ice-associated or mountain-dwelling species (Hamilton et al. 2016, Jamwal et al. 2022). Notably, climate change will not induce universally negative consequences for carnivores, as evident from the recent northward range expansion of golden jackals in Europe (Trouwborst and Blackmore 2020). Despite widespread negative and popularized effects of climate change on polar bears, empirical evidence suggests transient positive effects with increased range sizes, improved body condition across age classes, and stable reproductive metrics in a Kane Basin population (Laidre et al. 2020). The changing physical environment, driven by climate-induced glacial retreat, has also facilitated the colonization of suitable habitats by sea otters in Alaska (Williams et al. 2019). As perennial Arctic sea ice declines, marine species may encroach into northerly latitudes, potentially leading to competition over resources with implications that can re-reshape ecosystems.

Humans within these changing climates are also experiencing shifts in availability and access to resources, including water availability, growing seasons, and pest exposure (IPCC 2018). These changes are particularly concerning for communities heavily reliant on natural resources to sustain their traditional livelihoods and cultural practices. The IPCC has pinpointed the world's most climate-vulnerable regions, forecasting a displacement of 150 million people by 2050 (IPCC 2018). Although humans are reconciling with the changing climate and its effects, adaptation capacities vary across countries, regions, and socioeconomic groups, necessitating flexible and people-centric governance. Further, climate change may become an indirect driver of social instability as observed in the Middle East and North Africa, where indigenous mountain societies were forced into involuntary settlement migration due to extreme climate conditions and geological factors (Sofuoğlu and Ay 2020). India has established a National Mission on Biodiversity and Human Well-Being, a holistic framework for biodiversity conservation which integrates a variety of factors such as climate change, ecosystem services and human dimensions (Bawa et al. 2021). Mitigation strategies include migration, irrigation, water conservation techniques, land reclamation, and adapting the cultivation of plants in terms of timing, location, and elevation to cope with biodiversity loss and adapt to climate change (Macchi et al. 2015). Ultimately, the mitigation and adaptation strategies enacted by human populations to handle changing climates in the future have implications for carnivore persistence and will alter human–carnivore interactions.

The fluctuations induced by climate change in the abundance and distribution of prey species may trigger carnivores to modify their hunting strategies and alter consumptive patterns (Hallam and

Harris 2023). Unpredictable climate change, such as rainfall or extreme conditions like drought, can lead to changes in herbivore abundance, subsequently influencing the carnivore populations dependent on prey abundance (Morin et al. 2022). Growing evidence suggests climate variability impacts the foraging behaviour of predators. Increased snowfall prompted grey wolves to hunt in larger packs, resulting in a decrease in moose (*Alces alces*) abundance; subsequently leading to an increase in *Abies balsamea*, the primary forage species of moose (Post et al. 1999). African wild dogs showed lower daytime activity and higher nocturnal activity with increasing temperatures, potentially constraining their hunting activity time under climate change (Rabaiotti and Woodroffe 2019). In some cases, future climate will exacerbate conflicts between carnivores and people, as competition over a scarcity of shared resources mounts in stressed environments (Abrahms et al. 2023, Newsom et al. 2023). Scientists caution that reduced hibernation time in Asiatic black bears (*Ursus thibetanus*) has implications for conflict as bears extend their foraging season in Japan (Honda and Kozakai 2020). Reliance on domestic ungulates by carnivores remains a primary driver of retaliatory killing worldwide (van Eeden et al. 2018a, Lorand et al. 2022, Chapter 4). In the Maasai Mara, a lower abundance of natural prey due to climate change-induced migration led to predators such as leopards, lions and spotted hyaenas increasingly shifting their primary prey source to livestock (Mukeka et al. 2020). Similarly, dholes (*Cuon alpinus*) predate more on livestock when the abundance of their primary wild prey is diminished across their range (Srivathsa et al. 2020). In the Himalayan region, vegetation changes at high elevations due to gradual warming drives herbivores to lower elevations, and as the snow leopard follows, new opportunities to exploit domestic prey emerge (Aryal et al. 2014, Karki and Panthi 2021). Simultaneously, glacial retreat under climate change disrupts water availability, causing seasonal herders and their livestock to roam wider, increasing encounters with wolves and advancing into previously inaccessible higher mountain regions, including snow leopard core activity zones (Aryal et al. 2014, Lozano et al. 2019). Climatic changes that destabilize ecosystem functioning in sensitive regions, such as the Himalayas, can pose serious challenges for balancing the interests of wildlife biodiversity and resource use by local human populations (Mishra et al. 2004).

The effects of climate and land-use change on human–carnivore interactions are interlinked. Climate change tends to compound ecological threats such as food and water insecurity, causing mass displacement of people (Peace 2023). Most displaced people will move to cities, contributing to rapid urbanization. Climate change will also have a strong influence on suitable growing conditions, driving the distribution and intensity of agricultural land use (Gregory et al. 2005, Gomez-Zavaglia et al. 2020). Conversely, changes in land use will drive altered fossil fuel emissions and sequestration, affecting the rate of climate change (Pielke 2005). Increased wildfires and temperatures due to climate change, along with landscape fragmentation from human activities, have together decreased available habitat for the critically endangered Amur leopard across their range (*Panthera pardus orientalis*) (Li et al. 2023). Subsequent ecosystem degradation has led to lower prey availability and increased competition with Amur tigers (*Panthera tigris altaica*), driving leopards to depredate livestock, increasing the likelihood of human-caused mortality of leopards. Ultimately, mitigating the effects of a changing climate requires a comprehensive understanding of the intricacies involved and a global commitment to fostering resilient ecosystems that can withstand the inherent challenges.

Synergistic threats and interactions

Anthropogenic influences on carnivores, including land modification, climate change, and other threats, do not act in isolation. Threatening processes co-occur and interact in complex and dynamic ways to influence species and biodiversity (Geary et al. 2019). To understand how these 'threat webs' affect human–carnivore interactions (Figure 15.2), we must consider the synergistic effects of these pressures alongside human dimensions of Indigenous knowledge and culture, policy, tolerance, and socioeconomic factors. Landscape

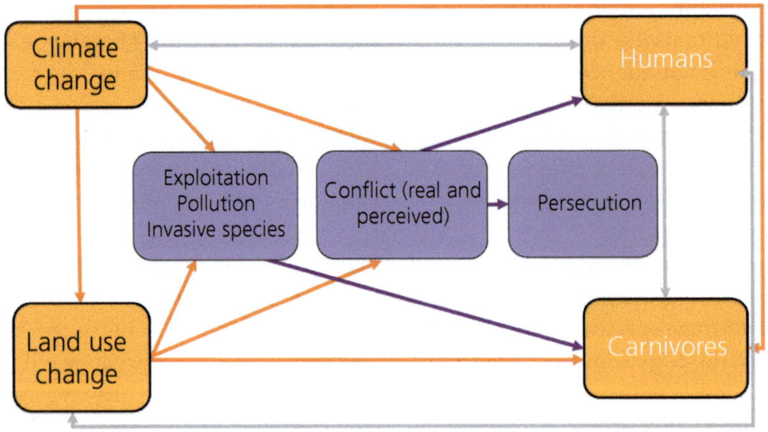

Figure 15.2 A conceptual diagram or 'threat web' of interactions among anthropogenic pressures affecting human–carnivore interactions.

transformations and climate change, while key drivers of human–carnivore interactions in the future, are not the only anthropogenic pressures affecting carnivores. Persecution, discussed in Chapter 4, is a major driver of global carnivore declines (Ripple et al. 2014). Additionally, as highlighted in Chapter 3 and 8, exploitation, disease, pollution, and invasive species cause ecological and evolutionary changes in carnivores.

Not surprisingly, persecution and exploitation will continue to be major threats to carnivores in future. Exploitation of carnivores for bushmeat and the illegal wildlife trade persist, particularly in Southeast Asia (Willcox 2020). However, as carnivore populations decline and carnivore communities are fragmented, the targets of exploitation—both species and regions—are likely to change. Demand for tiger bone, for example, is one of the drivers of tiger declines, and with declining tiger populations in some areas—along with stronger tiger protection efforts—the focus is shifting to using African lion bones as a substitute (Williams et al. 2017). Similar to the importance of anticipating hotspots of future depredation events, predicting locations where overutilization may occur would be fruitful for guiding future conservation efforts. The socioeconomic pressures of climate change on people, such as food insecurity, droughts, and disasters, will likely drive many to find alternative sources of income, exacerbating the exploitation of carnivores and other wildlife. The same process will increase the economic cost of depredated livestock (Braczkowski et al. 2023), which is predicted to intensify persecution of carnivores (Chapter 3).

Human development and land-use changes, particularly for livestock production, amplify disease transmission risks across human–domestic–wildlife interfaces, including viruses (Wells et al. 2020) and parasites (Clark et al. 2018). The high trophic level of large carnivores increases disease exposure risk due to their consumption of other endothermic vertebrates (Brearley et al. 2013). Urban expansion is increasing disease risk to species such as raccoons (*Procyon lotor*), coyotes and San Joaquin kit foxes (*Vulpes macrotis*) (Shultz et al. 2023). Climate change is also driving changes in disease patterns in carnivores, such as increased pathogen seroprevalence in polar bears (Pilfold et al. 2021). In the case of human-leopard conflict in Iran, aridization expected in the twenty-first century may decrease disease and conflict risks, but further research is needed to understand these relationships and the role of wild prey (Khorozyan et al. 2015).

Pollution from trace metals (lead and mercury), pesticides (insecticides and rodenticides), and industrial sources likely contributed to carnivore population declines in the past (Rodriguez-Jorquera et al. 2017, Rodríguez-Estival and Mateo 2019), and will continue to do so without changes in human behaviour and regulation. With expanding landscape modification, new areas will be exposed to anthropogenic pollutants, intensifying this threat to carnivores and other wildlife (Chapter 3, Box 15.2). Contamination from lead ammunition has recently been found to be a threat to many mammalian carnivores that scavenge on lead-shot carcasses (e.g. American black bears, Brown et al. 2023)). This, along with the established threat of lead ammunition to humans and wildlife, has driven global calls from scientists for a transition to lead-free ammunition (Bellinger et al. 2013, Arnemo et al. 2019). Some regions, particularly in Europe, have banned the use of lead ammunition (Arnemo et al. 2019) and others may follow suit, which will reduce lead exposure to carnivores in these areas. Ongoing declines in hunting culture in some regions such as North America may also contribute to reduced exposure to lead, as well as reduced carnivore deaths from shooting (Winkler and Warnke 2013). Anticoagulant rodenticides will also continue to be a significant threat to carnivores into the future, as ingestion through predation leads to secondary poisoning (Elliott et al. 2016). Some regions, such as parts of Canada and the USA, have begun restricting the use of these rodenticides, but effective alternatives must now be posed as new models of integrated pest management (van den Brink et al. 2018, Dara 2019).

Anthropogenic pressures will not only drive changes in carnivore communities into the future, but also species traits such as demographics, social structure, behaviour, and genetics, as discussed across several previous chapters. One process by which anthropogenic changes can influence a species' genetics and conservation status is through hybridization. Landscape modification (Howard-McCombe et al. 2023) and lethal control (Bohling and Waits 2015) can increase hybridization in carnivores. Climate change may exacerbate hybridization between species due to range changes, as discussed earlier with Arctic ursids. Hybridization may become a growing threat to carnivore species. The Scottish wildcat (*Felis sylvestris*), for instance, was recently declared 'genomically extinct' in the wild due to introgression from domestic cats. This 'genetic swamping' was driven by human encroachment on their habitat, coinciding with disease-induced prey declines (Howard-McCombe et al. 2023). Despite negative impacts on some individual species, a recent review found that hybridization is generally not a threat to the genetic integrity of carnivores (Tensen and Fischer 2024). Benefits of hybridization include novel genetic diversity, adaptation to extreme environments, and increased reproductive fitness, allowing species to cope with genetic drift and adapt to the Anthropocene. In the long term, this increased hybridization may even lead to the creation of new species, as the Asiatic black bear, African golden wolf (*Canis lupaster*), and red wolf (*Canis rufus*) likely originated from hybrids (Tensen and Fischer 2024).

Carnivore species trajectories

The fate of carnivores continuing to deliver their myriad of services remains uncertain under future environmental scenarios, despite their ecological and socioecological importance and clear vulnerabilities that guide ongoing conservation efforts. With the global trajectory of human–carnivore interactions, we are likely to see continued declines in carnivores without major interventions. Over a quarter of carnivore species are threatened and nearly half have declining populations (Fernández-Sepúlveda and Martín 2022). East and South Asia, which support some of the world's most diverse guilds of carnivores, also have the highest proportion of threatened and declining species (Fernández-Sepúlveda and Martín 2022). All four critically endangered Carnivora species (European mink *Mustela lutreola*, pygmy raccoon *Procyon pygmaeus*, red wolf *Canis rufus* and Malabar civet *Viverra civettina*) are in decline, along with all but one endangered terrestrial carnivore (Iberian lynx *Lynx pardinus*) (IUCN 2023). On the current path, several of these species may become extinct within the next few decades. Projections of agricultural expansion by 2050 suggest that for large carnivores, the largest range contractions will occur within genus Canis, Leopardus, and Panthera (Williams et al. 2021). In a continental analysis of African carnivores, authors found that the range for Ethiopian wolves (*Canis simensis*) was at risk of contraction, as was on average 15% of ranges for other species (Harris et al. 2023a, Figure 15.3).

Variations in conservation and research effort towards different carnivore species will influence future population trajectories. This effort is

Box 15.2 Polar bears and pollution

Compounds such as persistent organic pollutants (POPs) and mercury are a major threat to polar bears. High concentrations of these pollutants have been found in bear tissues, with myriad impacts on health including reproduction, bone density, and immune, endocrine and neurological function, as well as carcinogenic effects (Dietz et al. 2013, 2015, Sonne 2010, Sonne et al. 2012). This likely affects polar bear fitness, population growth and survival (Desforges et al. 2016, Sonne et al. 2012). These pollutants have similar impacts on other marine and freshwater carnivores (Desforges et al. 2016, Dietz et al. 2013, Martin et al. 2017). With expanding human activities, exposure to these pollutants is likely to increase in the future.

Three polar bears sleeping in the snow
Source: E. Jones

not evenly distributed among carnivore species, however, nor is it allocated according to where conservation is needed most. Carnivores in countries with lower wealth and less modernization receive worse conservation outcomes (Bruskotter et al. 2017, Dickman et al. 2015). Relatively little research on biodiversity conservation is undertaken in the world's most biodiverse countries (Wilson et al. 2016). Conservation and research attention is skewed towards large, charismatic carnivores rather than threatened species (Brodie 2009, Tensen 2018). Carnivore species that are valuable to people, be it for social or economic reasons, receive more research funding than those that are reviled or uncharismatic (Restani and Marzluff 2002, Strampelli et al. 2022). Endangered giant pandas (*Ailuropoda melanoleuca*), for example, are subject to extensive research and conservation efforts (Wei et al. 2015) and their cultural and political importance ensures they are unlikely to become extinct (Collard 2013). Large carnivores are often touted as umbrella and flagship species because of public appeal, aiding in garnering financial support. Choosing the right species, however, is crucial for achieving conservation goals. While the emblematic giant panda has shown marked population increases, other carnivore species, specifically leopards, snow leopards, grey wolves, and dholes,

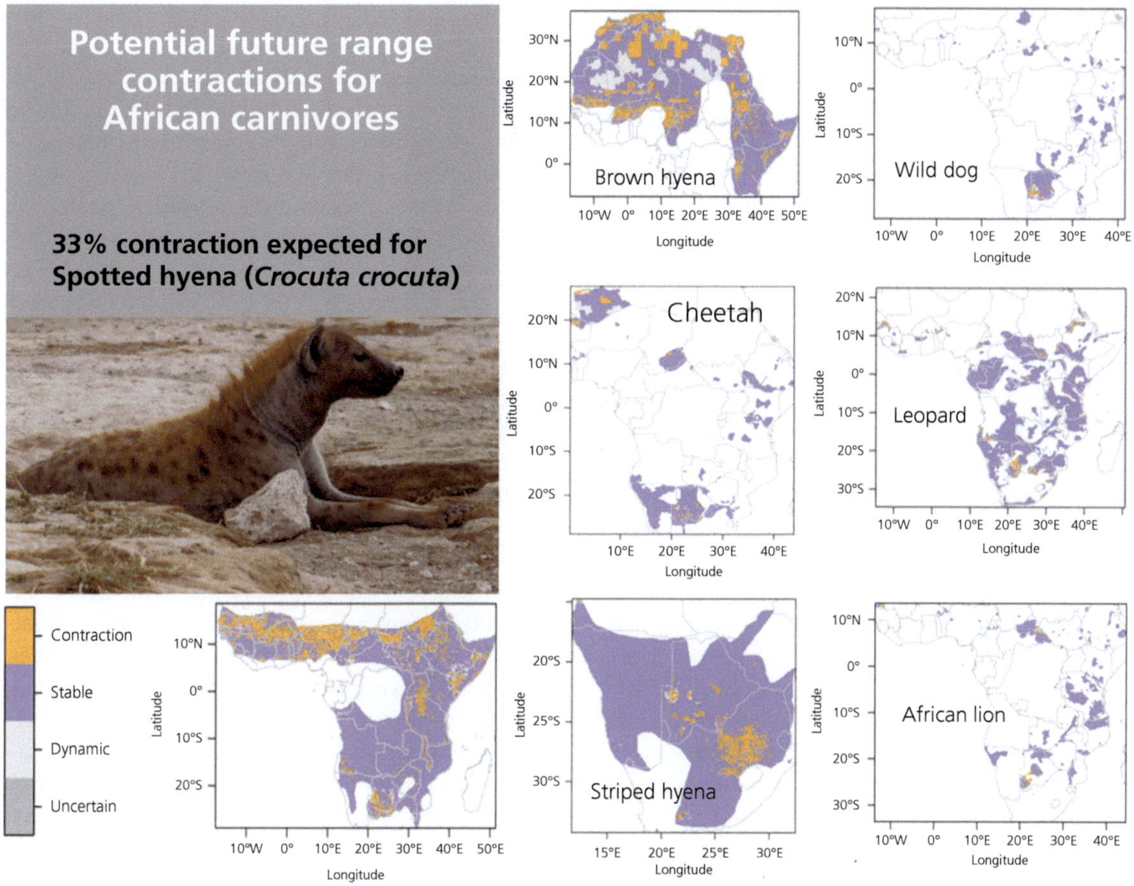

Figure 15.3 Vulnerability of range contractions for African carnivores modified from Harris et al. (2023a). Range contractions are designated from a geospatial assessment where the distribution of threats (e.g. drought, hunting) exceeded the distribution of resources (e.g. cultural diversity, protected areas) throughout a species range. Photo credit: Amboseli National Park, Kenya; NC Harris.

have lost >75% of their range within panda reserves since the 1960s (Li et al. 2020). This challenges the notion that species-specific interventions can have positive sympatric community benefits.

One group of carnivores seemingly thriving in the Anthropocene is invasive species, including red foxes (*Vulpes vulpes*) and stoats (*Mustela erminea*), as well as domestic cats (*Felis catus*) and dogs (*Canis familiaris*). Invasive species have devastating effects on native carnivores and biodiversity more broadly (Doherty et al. 2015), while causing a slew of problems for people, including livestock loss (Fleming et al. 2021) and spread of disease such as canine distemper, West Nile virus, and roundworm to humans and native carnivores (Beltrán-Beck et al. 2012, Sepúlveda et al. 2014). The distribution and impacts of invasive carnivores will likely intensify in the future as they benefit from anthropogenic changes such as land clearing, climate change and native predator control (Doherty et al. 2015). Eradication of invasive carnivores has been successful in some areas, particularly small islands, and will continue (Jones et al. 2016). New Zealand, for example, has a goal of complete invasive predator eradication (Russell et al. 2015). However, where eradication is unfeasible, populations of invasive carnivores will persist and even expand in some cases. This will necessitate ongoing management to mitigate impacts on people and wildlife, including native carnivores.

Levers of change and scenario pathways

As threats mount that exacerbate carnivore vulnerabilities to extirpation and extinction, interventions require increased scrutiny to ensure species persistence. The current global protected area network will be unable to prevent most local vertebrate extinctions, particularly for large carnivores (Williams et al. 2022). In addition, protected areas are largely failing to deliver on conservation outcomes to sustain viable populations of wide-ranging carnivore species (Carter and Linnell 2016), a challenge likely to persist under projected global change scenarios (Arthur 2014). Nonetheless, more than 100 countries publicly committed to support a goal to protect at least 30% of the global land and ocean by 2030 ('30 by 30'; 76th United Nations General Assembly in New York). Many national parks are poorly enforced ('paper parks'), which exhibit low levels of compliance with park restrictions, reducing their effectiveness in achieving conservation aims. Therefore, conservation efforts along a business-as-usual path will unlikely yield outcomes that are effective over time. A combination of successful predator deterrents (e.g. electrified fences, lights, non-lethal deterrents, and other emergent technologies), effective economic compensations (not just 'paper compensations'), and restoration of native prey populations is often argued as necessary to facilitate long-term carnivore persistence in these shared landscapes (van Eeden et al. 2018b). Even apparent successes in species recovery may be short-lived in these emergent and dynamic environments with a myriad of novel socioecological conditions. Individuals may succumb to these new environments or be forced to adapt—by shifting behaviours, distributions, and life history strategies, altering diets, and fostering new interactions with varying consequences as discussed earlier. Inevitably, some survival attempts will be unsuccessful, and more species will be lost. Amid these natural processes, where wildlife species are clawing for their survival, the fate of biodiversity remains ultimately burdened by the decisions, values, and activities of humans. The spatial and temporal heterogeneity of human actions and inactions can trigger cascading effects across scales.

Human–carnivore coexistence depends upon social factors, including the collective psyche of tolerance of people and assignments of responsibility (Expósito-Granados et al. 2019, Harris et al. 2023b). Tolerance tends to be related to perceived risk of conflict, which is not always correlated with actual risk (Thorn et al. 2015, Inskip et al. 2016) and is driven by cultural norms (Chapters 3 and 13). With increasing periods of scarcity and stress in the future, we anticipate decreasing tolerance of carnivores by humans as an adaptation strategy. We can further hypothesize that these dynamics will serve to intensify human–carnivore conflict. Systems where tangible benefits from sympatry with carnivores are common, particularly due to monetary returns emergent from ecotourism or hunting revenue will likely be buffered by fewer fluctuations in tolerance (Rode et al. 2021). In Germany, just positive experiences from wild encounters with grey wolves (*Canis lupus*) were deemed beneficial (Arbieu et al. 2020). Modernization (such as increased wealth, education, agricultural mechanization, technology, and urbanization) appears to be driving more positive attitudes towards wildlife (Manfredo et al. 2009, Dietsch et al. 2016). People in cities tend to have more favourable attitudes to carnivores than those in rural areas (Franchini et al. 2021), so the ongoing urbanization of human populations globally may increase carnivore tolerance and support policies favourable to carnivores, even when tolerance in rural areas is low (Bruskotter et al. 2017). This, however, is worrisome as the people living with carnivores are not the ones driving the management decisions, intensifying the rural–urban divide and leading to local resistance to policies (Tauli-Corpuz et al. 2020). Perpetuating adversarial 'trade-offs' between economic growth and conservation remain common (Johnson et al. 2023), yet problematic for securing the health of carnivore populations and their ecosystem services in the future. Inequalities in power enable privileged populations to promote carnivore conservation implicitly over improving the living standards of people, particularly vulnerable populations (Dietsch et al. 2021). Moving forward, wild and wilderness must be redefined as strictly protected areas representing a small

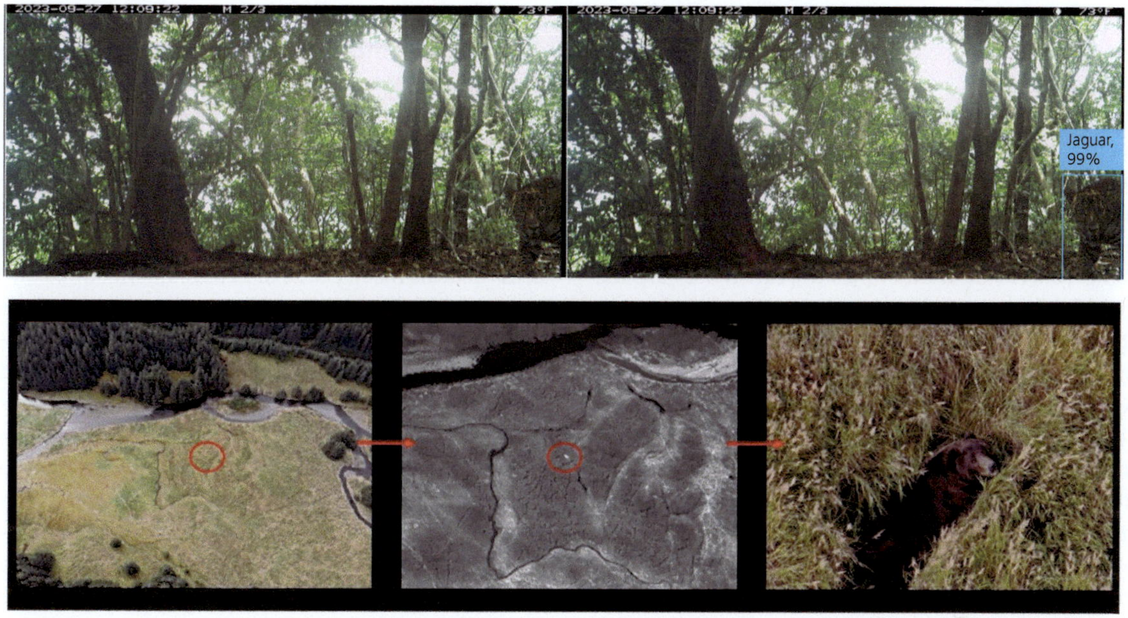

Figure 15.4 Examples of technological advancements aiding in the monitoring and management of conservation. Top row: Jaguar detection from camera trap survey using machine learning for species assignment (Photo credit: El Triunfo Biosphere Reserve, Mexico, Applied Wildlife Ecology Lab, NC Harris). Bottom row: Unmanned aerial vehicles coupled with thermal imagery used for population estimation of brown bears. Photo credit: Koniag Alaskan Native Cooperation, S. Finnegan.

fraction of the conservation estate, and as such, all areas should be both managed and studied as coupled human-natural ecosystems (Ghoddousi et al. 2022).

History, and not just throughout the duration of formalized conservation biology as a discipline, provides evidence that interventions are compulsory to protect carnivores. Therefore, in tackling the complexities of carnivore coexistence in the Anthropocene Era, looking forward, it is increasingly apparent that the practice of conservation must be implemented differently. While quibbles continue on the effectiveness of protected areas and the need for more recovery plans for species, both of these conservation strategies remain flawed and require reconstruction (Schwartz 2008, Malcom and Li 2015, Williams et al. 2022). As such, a more fruitful future endeavour could involve devising 'persistent plans' for carnivores across conservation threat levels and based on projected rates of land use, climate, and human demographic changes. Science continues to innovate in ways that have broad implications for society and species conservation. Advancements in technology afford improved methods for both data collection in monitoring and data processing. For example, new tracking devices (e.g. subdermal implants), unmanned aerial vehicles (i.e. 'drones'), hormone analysis, environmental DNA (eDNA) approaches, and applications of machine learning algorithms are providing new scientific insights (Chapter 8, Figure 15.4). However, as conservation tools and technologies continue to develop, we urge caution as these advancements could be used in a manner that harms a conservation agenda. For example, the use of artificial intelligence and machine learning will become more pervasive in both the conservation and exploitation of natural resources. As a management tool, park authorities and scientists can collaboratively use artificial intelligence (AI) to detect poacher habits and allocate patrol units in a systematic manner (Fang et al. 2019). Far too often, criminal enterprises (i.e. organized poaching syndicates) are better resourced than government rangers, as evidenced by stronger weaponry (Kahler and Gore 2017). Application of AI could also help to better inform target locations of vulnerable species for exploitation or guide plans to maximize trapping/hunting efforts for their counter benefit.

Given the global distribution of carnivores interfacing with diverse human societies, these human dimensions of how carnivores are studied and managed will affect the persistence of species and interactions with humans in the future. Implicit in realizing conservation outcomes for carnivores are the confounding considerations of scale, transferability, and ethics (Pienkowski et al. 2024). Once a successful practice is implemented in a specific location and confirmed, that strategy can then be tested in other locations for which similar results are anticipated. In this context, considerations and adjustments based on idiosyncrasies induced by climatic considerations, cultural norms, harmful societal legacies, instructional infrastructure, and governance are required or desired outcomes may be unrealized. For example, strategies implemented in the Global North should never be blanketly applied or viewed as superior to those originating from the Global South (Browne-Nuñez and Jonker 2008, Tan 2021). Adoption through transfer is often a luxury for the privileged subset with power and contingent upon adequate benefit and knowledge sharing (Hill et al. 2020). Such an approach can exacerbate inequities in access to resources and the erasure of Indigenous knowledge. The call to decolonize science, more broadly, requires not just acknowledgement of harmful practices (e.g. racist terminology, exclusion), but an explicit change in professional standards and disciplinary epistemologies (Trisos et al. 2021, Reed and Rudman 2023).

Conclusion

Ultimately, carnivores and their services will be contingent upon not only how the biophysical environment changes and their corresponding adaptive strategies, but also how socioeconomic systems change. Human attitudes, the extent of their pro-environmental behaviours, and how inclusive and responsive governance infrastructure is for conflict mitigation will determine conservation outcomes (Chapter 13). Carnivore abundances and ranges will expectedly continue to decline and contract without major intervention. In future, strategies for the conservation and recovery of large carnivores in the Anthropocene must not only target landscapes with minimum human influence but also address ecosystems that concurrently support human livelihoods. It is in these shared environments that the most substantial and crucial impacts of large carnivore recovery on human communities become evident. With the United Nations General Assembly designating 2021–2030 as the 'Decade of Ecosystem Restoration', a compelling opportunity arises for policymakers and decision-makers to give prominence to discussions on large carnivore recovery within the context of biodiversity goals. As we embark on this dedicated decade, it can act as a catalyst for shaping new policies and strategies that actively contribute to the restoration and preservation of ecosystems globally. The challenge ahead is to promote land management practices that meet human needs while remaining accessible and accommodating for large carnivores and other wildlife.

Technology, inclusivity via diverse partnerships, knowledge transfer, and land management will be crucial for maintaining and supplementing carnivore populations. Coupling human behaviour with animal adaptation, the future of human–carnivore interactions has a myriad of complex potential outcomes. The business-as-usual scenarios will result in continued carnivore extirpations, minute conservation successes, and degraded ecosystems that yield fewer services. However, as discussed throughout this chapter and the book, many intervention pathways are known to help ensure the preservation of carnivores and their socioecological benefits across ecosystems.

References

Abrahms, B., et al. 2023. Climate change as a global amplifier of human-wildlife conflict. *Nature Climate Change* 13:224–234.

Abrahms, B., K. Rafiq, N. R. Jordan, and J. W. McNutt. 2022. Long-term, climate-driven phenological shift in a tropical large carnivore. *Proceedings of the National Academy of Sciences of the United States of America* 119:e2121667119.

Arbieu, U., J. Albrecht, M. Mehring, N. Bunnefeld, I. Reinhardt, and T. Mueller. 2020. The positive experience of encountering wolves in the wild. *Conservation Science and Practice* 2:e184.

Arnemo, J. M., R. Cromie, A. D. Fox, N. Kanstrup, R. Mateo, D. J. Pain, and V. G. Thomas. 2019. Transition to lead-free ammunition benefits all. *Ambio* 48: 1097–1098.

Arnold, J., A. Humer, M. Heltai, D. Murariu, N. Spassov, and K. Hackländer. 2012. Current status and distribution of golden jackals *Canis aureus* in Europe. *Mammal Review* 42:1–11.

Arthur, J. 2014. *The Utility of Protected Areas for Large Carnivore Conservation*. Imperial College London, London.

Aryal, S., G. Cockfield, and T. N. Maraseni. 2014. Vulnerability of Himalayan transhumant communities to climate change. *Climatic Change* 125:193–208.

Ausband, D. E., and L. D. Mech. 2023. The challenges of success: future wolf conservation and management in the United States. *BioScience* 73:587–591.

Bastille-Rousseau, G., et al. 2018. Climate change can alter predator-prey dynamics and population viability of prey. *Oecologia* 186:141–150.

Bawa, K. S., et al. 2021. Securing biodiversity, securing our future: a national mission on biodiversity and human well-being for India. *Biological Conservation* 253: 108867.

Beever, E. A., et al. 2017. Behavioral flexibility as a mechanism for coping with climate change. *Frontiers in Ecology and the Environment* 15:299–308.

Bellinger, D. C., et al. 2013. Health risks from lead-based ammunition in the environment. *Environmental Health Perspectives* 121(6):a178–a179.

Beltrán-Beck, B., F. J. García, and C. Gortázar. 2012. Raccoons in Europe: disease hazards due to the establishment of an invasive species. *European Journal of Wildlife Research* 58(1):5–15.

Benson, J. F., et al. 2023. The ecology of human-caused mortality for a protected large carnivore. *Proceedings of the National Academy of Sciences* 120:e2220030120.

Bleyhl, B., et al. 2021. Reducing persecution is more effective for restoring large carnivores than restoring their prey. *Ecological Applications* 31:e02338.

Bohling, J. H. and L. P. Waits. 2015. Factors influencing red wolf–coyote hybridization in eastern North Carolina, USA. *Biological Conservation* 184:108–116.

Bombieri, G., et al. 2023. A worldwide perspective on large carnivore attacks on humans. *PLoS Biology* 21:e3001946.

Braczkowski, A. R., et al. 2023. The unequal burden of human-wildlife conflict. *Communications Biology* 6(1):182.

Brearley, G., et al. 2013. Wildlife disease prevalence in human-modified landscapes. *Biological Reviews* 88:427–442.

Brodie, J. F. 2009. Is research effort allocated efficiently for conservation? Felidae as a global case study. *Biodiversity and Conservation* 18:2927–2939.

Brown, L., et al. 2023. Lead exposure in brown bears is linked to environmental levels and the distribution of moose kills. *Science of the Total Environment* 873:162099.

Browne-Nuñez, C. and S. A. Jonker. 2008. Attitudes toward wildlife and conservation across Africa: a review of survey research. *Human Dimensions of Wildlife* 13:47–70.

Bruskotter, J. T., et al. 2017. Modernization, risk, and conservation of the world's largest carnivores. *BioScience* 67:646–655.

Buijs, A. and M. Jacobs. 2021. Avoiding negativity bias: towards a positive psychology of human-wildlife relationships. *Ambio* 50:281–288.

Carter, N. H. and J. D. C. Linnell. 2016. Co-adaptation is key to coexisting with large carnivores. *Trends in Ecology & Evolution* 31:575–578.

Chapron, G., et al. 2014. Recovery of large carnivores in Europe's modern human-dominated landscapes. *Science* 346:1517–1519.

Clark, N. J., J. M. Seddon, J. Slapeta, and K. Wells. 2018. Parasite spread at the domestic animal–wildlife interface: anthropogenic habitat use, phylogeny and body mass drive risk of cat and dog flea (*Ctenocephalides* spp.) infestation in wild mammals. *Parasites & Vectors* 11:8.

Collard, R.-C. 2013. Panda politics. *Canadian Geographies/Géographies canadiennes* 57:226–232.

Cunze, S. and S. Klimpel. 2022. From the Balkan towards Western Europe: Range expansion of the golden jackal (*Canis aureus*)—a climatic niche modeling approach. *Ecology & Evolution* 12(7):e9141.

Dara, S. K. 2019. The new integrated pest management paradigm for the modern age. *Journal of Integrated Pest Management* 10(1):12.

Davoli, M., A. Ghoddousi, F. M. Sabatini, E. Fabbri, R. Caniglia, and T. Kuemmerle. 2022. Changing patterns of conflict between humans, carnivores and crop-raiding prey as large carnivores recolonize human-dominated landscapes. *Biological Conservation* 269:109553.

Desforges, J. P. W., C. Sonne, M. Levin, U. Siebert, S. De Guise, and R. Dietz. 2016. Immunotoxic effects of environmental pollutants in marine mammals. *Environment International* 86:126–139.

Dietz, R., et al. 2013. What are the toxicological effects of mercury in Arctic biota?. *Science of the Total Environment* 443:775–790.

Dietz, R., et al. 2015. Physiologically-based pharmacokinetic modelling of immune, reproductive and carcinogenic effects from contaminant exposure in polar bears (*Ursus maritimus*) across the Arctic. *Environmental Research* 140:45–55.

Dietsch, A. M., T. L. Teel, and M. J. Manfredo. 2016. Social values and biodiversity conservation in a dynamic world. *Conservation Biology* 30:1212–1221.

Dietsch, A. M., D. M. Wald, M. J. Stern, and B. Tully. 2021. An understanding of trust, identity, and power can enhance equitable and resilient conservation partnerships and processes. *Conservation Science and Practice* 3:e421.

Doherty, T. S., C. R. Dickman, D. G. Nimmo, and E. G. Ritchie. 2015. Multiple threats, or multiplying the threats? Interactions between invasive predators and other ecological disturbances. *Biological Conservation* 190:60–68.

Elliott, J. E., B. A. Rattner, R. F. Shore, and N. W. Van Den Brink. 2016. Paying the pipers: mitigating the impact of anticoagulant rodenticides on predators and scavengers. *BioScience* 66:401–407.

Ellis-Soto, D., et al. 2023. A vision for incorporating human mobility in the study of human-wildlife interactions. *Nature Ecology & Evolution* 7:1362–1372.

Expósito-Granados, M., et al. 2019. Human-carnivore relations: conflicts, tolerance and coexistence in the American West. *Environmental Research Letters* 14:123005.

Fang, F., M. Tambe, B. Dilkina, and A. J. Plumptre. 2019. *Artificial Intelligence and Conservation*. Cambridge University Press, Cambridge, U. K.

Fernández-Sepúlveda, J. and C. A. Martín. 2022. Conservation status of the world's carnivorous mammals (order Carnivora). *Mammalian Biology* 102:1911–1925.

Ferreira, A. S., C. A. Peres, J. A. Bogoni, and C. R. Cassano. 2018. Use of agroecosystem matrix habitats by mammalian carnivores (Carnivora): a global-scale analysis. *Mammal Review* 48:312–327.

Fleming, P. A., et al. 2021. Diet of the introduced red fox Vulpes vulpes in Australia: analysis of temporal and spatial patterns. *Mammal Review* 51:508–527.

Fletcher, R. and S. Toncheva. 2021. The political economy of human-wildlife conflict and coexistence. *Biological Conservation* 260:109216.

Fox, R. J., J. M. Donelson, C. Schunter, T. Ravasi, and J. D. Gaitán-Espitia. 2019. Beyond buying time: the role of plasticity in phenotypic adaptation to rapid environmental change. *Philosophical Transactions of the Royal Society B: Biological Sciences* 374:20180174.

Franchini, M., M. Corazzin, S. Bovolenta, and S. Filacorda. 2021. The return of large carnivores and extensive farming systems: a review of stakeholders' perception at an EU level. *Animals* 11:1735.

Gantchoff, M. G., L. Conlee, M. R. Boudreau, R. B. Iglay, C. Anderson, and J. L. Belant. 2022. Spatially-explicit population modeling to predict large carnivore recovery and expansion. *Ecological Modelling* 470:110033.

Gantchoff, M. G., et al. 2021. Potential distribution and connectivity for recolonizing cougars in the Great Lakes region, USA. *Biological Conservation* 257:109144.

Geary, W. L., D. G. Nimmo, T. S. Doherty, E. G. Ritchie, and A. I. Tulloch. 2019. Threat webs: reframing the co-occurrence and interactions of threats to biodiversity. *Journal of Applied Ecology* 56(8):1992–1997.

Ghoddousi, A., J. Loos, and T. Kuemmerle. 2022. An outcome-oriented, social-ecological framework for assessing protected area effectiveness. *BioScience* 72:201–212.

Gicquel, M., M. L. East, H. Hofer, S. Cubaynes, and S. Benhaiem. 2022. Climate change does not decouple interactions between a central-place-foraging predator and its migratory prey. *Ecosphere* 13:e4012.

Gilbert, S. L., et al. 2017. Socioeconomic benefits of large carnivore recolonization through reduced wildlife-vehicle collisions. *Conservation Letters* 10:431–439.

Gilman, S. E., M. C. Urban, J. Tewksbury, G. W. Gilchrist, and R. D. Holt. 2010. A framework for community interactions under climate change. *Trends in Ecology & Evolution* 25:325–331.

Gomez-Zavaglia, A., J. C. Mejuto, and J. Simal-Gandara. 2020. Mitigation of emerging implications of climate change on food production systems. *Food Research International* 134:109256.

Gregory, P. J., J. S. Ingram, and M. Brklacich. 2005. Climate change and food security. *Philosophical Transactions of the Royal Society of London: Series B, Biological Sciences* 360:2139–2148.

Grilo, C., L., et al. 2021. Conservation threats from roadkill in the global road network. *Global Ecology and Biogeography* 30:2200–2210.

Hallam, J. and N. C. Harris. 2023. What's going to be on the menu with global environmental changes? *Global Change Biology* 29:5744–5759.

Hamilton, L. C., K. Saito, P. A. Loring, R. B. Lammers, and H. P. Huntington. 2016. Climigration? Population and climate change in Arctic Alaska. *Population and Environment* 38:115–133.

Harris, N. C., A. Murphy, A. R. Green, S. Gámez, D. M. Mwamidi, and G. C. Nunez-Mir. 2023a. Socio-ecological gap analysis to forecast species range contractions for conservation. *Proceedings of the National Academy of Sciences* 120:e2201942119.

Harris, N. C., C. E. Wilkinson, G. Fleury, and Z. N. Nhleko. 2023b. Responsibility, equity, justice, and inclusion in dynamic human–wildlife interactions. *Frontiers in Ecology and the Environment* 21:380–387.

Hart, E. E., A. Haigh, and S. Ciuti. 2023. A scoping review of the scientific evidence base for rewilding in Europe. *Biological Conservation* 285:110243.

Hejny, J. 2018. The Trump Administration and environmental policy: Reagan redux? *Journal of Environmental Studies and Sciences* 8:197–211.

Hill, J. E., T. L. De Vault, and J. L. Belant. 2021. A review of ecological factors promoting road use by mammals. *Mammal Review* 51:214–227.

Hill, R., et al. 2020. Working with Indigenous, local and scientific knowledge in assessments of nature and nature's linkages with people. *Current Opinion in Environmental Sustainability* 43:8–20.

Hody, J. W. and R. Kays. 2018. Mapping the expansion of coyotes (Canis latrans) across North and Central America. ZooKeys 759:81.

Honda, T. and C. Kozakai. 2020. Mechanisms of human-black bear conflicts in Japan: In preparation for climate change. *Science of the Total Environment* 739:140028.

Howard-McCombe, J., et al. 2023. Genetic swamping of the critically endangered Scottish wildcat was recent and accelerated by disease. *Current Biology* 33:4761–+.

Inskip, C., N. Carter, S. Riley, T. Roberts, and D. MacMillan. 2016. Toward human-carnivore coexistence: understanding tolerance for tigers in Bangladesh. *PLoS One* 11:e0145913.

IPCC. 2018. Special Report on Global Warming of 1.5°C (SR1.5). Intergovernmental Panel on Climate Change.

IUCN. 2023. IUCN Red List of Threatened Species.

Jamwal, P. S., M. Di Febbraro, M. L. Carranza, M. Savage, and A. Loy. 2022. Global change on the roof of the world: Vulnerability of Himalayan otter species to land use and climate alterations. *Diversity and Distributions* 28:1635–1649.

Johnson, T. F., N. J. B. Isaac, A. Paviolo, and M. González-Suárez. 2023. Socioeconomic factors predict population changes of large carnivores better than climate change or habitat loss. *Nature Communications* 14:74.

Jones, E. M., A. J. Koch, R. K. Hamede, and M. E. Jones. 2023. A systematic global review of mammalian carnivore responses to production forests. *Mammal Review* 54(2):133–149.

Jones, H. P., et al. 2016. Invasive mammal eradication on islands results in substantial conservation gains. *Proceedings of the National Academy of Sciences* 113:4033–4038.

Jordan, N. R., B. P. Smith, R. G. Appleby, L. M. van Eeden, and H. S. Webster. 2020. Addressing inequality and intolerance in human-wildlife coexistence. *Conservation Biology* 34:803–810.

Kahler, J. S., and M. L. Gore. 2017. Conservation crime science. In *Conservation Criminology*, edited by M. L. Gore. Wiley, Hoboken, N. J.

Karki, A. and S. Panthi. 2021. Factors affecting livestock depredation by snow leopards (*Panthera uncia*) in the Himalayan region of Nepal. *PeerJ* 9:e11575.

Khorozyan, I., M. Soofi, A. Ghoddousi, A. Khaleghi Hamidi, and M. Waltert. 2015. The relationship between climate, diseases of domestic animals and human-carnivore conflicts. *Basic and Applied Ecology* 16:703–713.

Kowalczyk, R., M. Wudarczyk, J. M. Wójcik, and H. Okarma. 2020. Northernmost record of reproduction of the expanding golden jackal population. *Mammalian Biology* 100:107–111.

Laidre, K. L., et al. 2020. Transient benefits of climate change for a high-Arctic polar bear (*Ursus maritimus*) subpopulation. *Global Change Biology* 26:6251–6265.

Lan, T. Y., et al. 2022. Insights into bear evolution from a Pleistocene polar bear genome. *Proceedings of the National Academy of Sciences of the United States of America* 119(24):e2200016119.

Lanszki, J., M. W. Hayward, N. Ranc, and A. Zalewski. 2022. Dietary flexibility promotes range expansion: the case of golden jackals in Eurasia. *Journal of Biogeography* 49(6):993–1005.

Larue, M. A., C. K. Nielsen, M. Dowling, K. Miller, B. Wilson, H. Shaw, and C. R. Anderson. 2012. Cougars are recolonizing the midwest: Analysis of cougar confirmations during 1990–2008. *Journal of Wildlife Management* 76:1364–1369.

Lau, M. W.-N., J. R. Fellowes, and B. P. L. Chan. 2010. Carnivores (Mammalia: Carnivora) in South China: a status review with notes on the commercial trade. *Mammal Review* 40:247–292.

Laurance, W. 2018. If you can't build well, then build nothing at all. *Nature* 563(7731):295.

Li, G., et al. 2022. Global impacts of future urban expansion on terrestrial vertebrate diversity. *Nature Communications* 13:1628.

Li, S., et al. 2020. Retreat of large carnivores across the giant panda distribution range. *Nature Ecology & Evolution* 4:1327–1331.

Li, X., S. S. Lam, C. Xia, H. Zhong, and C. Sonne. 2023. Climate change puts Amur leopard at risk. *Science* 382:1007.

Loch, J. M. H., L. J. Walters, and G. S. Cook. 2020. Recovering trophic structure through habitat restoration: a review. *Food Webs* 25:00162.

Lodeiro Ocampo, N., and N. Nigro. 2020. Características de depredación de yaguareté (Panthera onca) a vacunos en el centro de la provincia de Misiones, República Argentina: propuestas para su identificación y documentación. Notas sobre Mamíferos Sudamericanos.

Lorand, C., A. Robert, A. Gastineau, J. B. Mihoub, and C. Bessa-Gomes. 2022. Effectiveness of interventions for managing human-large carnivore conflicts worldwide: Scare them off, don't remove them. *Science of the Total Environment* 838:156195.

Lozano, J., et al. 2019. Human-carnivore relations: a systematic review. *Biological Conservation* 237:480–492.

Macchi, M., A. M. Gurung, and B. Hoermann. 2015. Community perceptions and responses to climate variability and change in the Himalayas. *Climate and Development* 7:414–425.

Malcom, J. W., and Y.-W. Li. 2015. Data contradict common perceptions about a controversial provision of the US Endangered Species Act. *Proceedings of the National Academy of Sciences* 112:15844–15849.

Manfredo, M. J., T. L. Teel, and K. L. Henry. 2009. Linking society and environment: a multilevel model of shifting wildlife value orientations in the western United States*. *Social Science Quarterly* 90:407–427.

Mantero, G., D. Morresi, R. Marzano, R. Motta, D. J. Mladenoff, and M. Garbarino. 2020. The influence of land abandonment on forest disturbance regimes: a global review. *Landscape Ecology* 35:2723–2744.

Mazzolli, M. 2012. Natural recolonization and suburban presence of pumas (*Puma concolor*) in Brazil. *Journal of Ecology and the Natural Environment* 4 (14): 344–362.

Medellín, R. A., C. Manterola, M. Valdéz, D. G. Hewitt, D. Doan-Crider, and T. E. Fulbright. 2005. History, ecology, and conservation of the pronghorn antelope, bighorn sheep, and black bear in Mexico. In *Biodiversity, Ecosystems, and Conservation in Northern Mexico*, edited by J.-L. E. Cartron, G. Ceballos, and R. S. Felger. Oxford University Press, Oxford.

Méndez-Carvajal, P. and R. Moreno. 2014. Mammalia, Carnivora, Canidae, Canis latrans (Say, 1823): actual distribution in Panama. *Check List* 10:376–379.

Merson, S. D., L. J. Dollar, P. J. Johnson, and D. W. Macdonald. 2019. Retaliatory killing and human perceptions of Madagascar's largest carnivore and livestock predator, the fosa (*Cryptoprocta ferox*). *PLoS One* 14:e0213341.

Milanesi, P., F. T. Breiner, F. Puopolo, and R. Holderegger. 2017. European human-dominated landscapes provide ample space for the recolonization of large carnivore populations under future land change scenarios. *Ecography* 40:1359–1368.

Mishra, C., S. E. Van Wieren, P. Ketner, I. M. A. Heitkönig, and H. H. T. Prins. 2004. Competition between domestic livestock and wild bharal *Pseudois nayaur* in the Indian Trans-Himalaya. *Journal of Applied Ecology* 41: 344–354.

Morin, D. J., D. B. Lesmeister, C. K. Nielsen, and E. M. Schauber. 2022. Asymmetrical intraguild interactions with coyotes, red foxes, and domestic dogs may contribute to competitive exclusion of declining gray foxes. *Ecology & Evolution* 12:e9074.

Mukeka, J. M., J. O. Ogutu, E. Kanga, and E. Roskaft. 2020. Spatial and temporal dynamics of human-wildlife conflicts in the Kenya Greater Tsavo Ecosystem. *Human-Wildlife Interactions* 14:255–272.

Newsom, A., Z. Sebesvari, and I. Dorresteijn. 2023. Climate change influences the risk of physically harmful human-wildlife interactions. *Biological Conservation* 286: 110255.

Osorio, C., A. Muñoz, N. Guarda, C. Bonacic, and M. Kelly. 2020. Exotic prey facilitate coexistence between pumas and culpeo foxes in the Andes of Central Chile. Diversity-Basel 12(9):317.

Peace, I. f. E. 2023. *Ecological Threat Report 2023: Analysing Ecological Threats, Resilience & Peace*. Institute for Economics & Peace, Sydney.

Pecl, G. T., et al. 2017. Biodiversity redistribution under climate change: Impacts on ecosystems and human well-being. *Science* 355(6332):eaai9214.

Pereira, A. D., N. C. Iliuk, K. L. Kuryluk, and J. A. Bogoni. 2023. Human density, development, and roads are the main drivers of carnivore presence in urban areas. *Perspectives in Ecology and Conservation* 22(1): 55–62.

Pereira, H. M., and L. M. Navarro. 2015. *Rewilding European Landscapes*. Springer Nature, Paris, France.

Pielke, R. A., Sr. 2005. Atmospheric science. Land use and climate change. *Science* 310:1625–1626.

Pienkowski, T., et al. 2024. Five lessons for avoiding failure when scaling in conservation. *Nature Ecology & Evolution* 8:1804–1814.

Pilfold, N. W., et al. 2021. Long-term increases in pathogen seroprevalence in polar bears (*Ursus maritimus*) influenced by climate change. *Global Change Biology* 27: 4481–4497.

Post, E., R. O. Peterson, N. C. Stenseth, and B. E. McLaren. 1999. Ecosystem consequences of wolf behavioural response to climate. *Nature* 401:905–907.

Rabaiotti, D. and Woodroffe, R. 2019. Coping with climate change: limited behavioral responses to hot weather in a tropical carnivore. *Oecologia* 189(3):587–599.

Reed, M. S. and H. Rudman. 2023. Re-thinking research impact: voice, context and power at the interface of science, policy and practice. *Sustainability Science* 18: 967–981.

Restani, M., and J. M. Marzluff. 2002. Funding extinction? Biological needs and political realities in the allocation of resources to endangered species recovery: an existing priority system, which should guide the Fish and Wildlife Service in endangered species recovery, is ineffective, and current spending patterns decrease long-term viability of island species. *BioScience* 52: 169–177.

Richards, R. L., J. M. Drake, and V. O. Ezenwa. 2022. Do predators keep prey healthy or make them sicker? A meta-analysis. *Ecology Letters* 25:278–294.

Ripple, W. J., et al. 2014. Status and ecological effects of the world's largest carnivores. *Science* 343:1241484.

Ritchie, E. G. and C. N. Johnson. 2009. Predator interactions, mesopredator release and biodiversity conservation. *Ecology Letters* 12:982–998.

Rode, J., L. Flinzberger, R. Karutz, A. Berghöfer, and C. Schröter-Schlaack. 2021. Why so negative? Exploring the socio-economic impacts of large carnivores from a European perspective. *Biological Conservation* 255:108918.

Rodríguez-Estival, J. and R. Mateo. 2019. Exposure to anthropogenic chemicals in wild carnivores: a silent conservation threat demanding long-term surveillance. *Current Opinion in Environmental Science & Health* 11: 21–25.

Rodriguez-Jorquera, I., et al. 2017. Contamination of the upper class: occurrence and effects of chemical pollutants in terrestrial top predators. *Current Pollution Reports* 3:206–219.

Russell, J. C., J. G. Innes, P. H. Brown, and A. E. Byrom. 2015. Predator-free New Zealand: conservation country. *BioScience* 65:520–525.

Schwartz, M. W. 2008. The performance of the Endangered Species Act. *Annual Review of Ecology, Evolution, and Systematics* 39:279–299.

Sepulveda, M. A., R. S. Singer, E. A. Silva-Rodríguez, A. Eguren, P. Stowhas, and K. Pelican. 2014. Invasive American mink: linking pathogen risk between domestic and endangered carnivores. *EcoHealth* 11(3): 409–419.

Shultz, L., et al. 2023. Vector-borne disease in wild mammals impacted by urban expansion and climate change. *Ecohealth* 20:286–299.

Sievert, O., M. Hammer, E. Comley, B. Hintz, W. O. Mgoola, and R. S. Davis. 2023. A novel record of African wild dogs (*Lycaon pictus*) in Vwaza Marsh Wildlife Reserve, Malawi. *Ecology & Evolution* 13(11):e10671.

Sofuoğlu, E. and A. Ay. 2020. The relationship between climate change and political instability: the case of MENA countries (1985:01–2016:12). *Environmental Science and Pollution Research* 27:14033–14043.

Sonne, C. 2010. Health effects from long-range transported contaminants in Arctic top predators: an integrated review based on studies of polar bears and relevant model species. *Environment International* 36(5):461–491.

Sonne, C., et al. 2012. Temporal monitoring of liver and kidney lesions in contaminated East Greenland polar bears (Ursus maritimus) during 1999–2010. *Environment International* 48:143–149.

Srivathsa, A., et al. 2020. Opportunities for prioritizing and expanding conservation enterprise in India using a guild of carnivores as flagships. *Environmental Research Letters* 15(6):064009.

Strampelli, P., L. A. D. Campbell, P. Henschel, S. K. Nicholson, D. W. Macdonald, and A. J. Dickman. 2022. Trends and biases in African large carnivore population assessments: identifying priorities and opportunities from a systematic review of two decades of research. *PeerJ* 10:e14354.

Surve, N. S., S. Sathyakumar, K. Sankar, D. Jathanna, V. Gupta, and V. Athreya. 2022. Leopards in the city: the tale of Sanjay Gandhi National Park and Tungareshwar Wildlife Sanctuary, two protected areas in and adjacent to Mumbai, India. *Frontiers in Conservation Science* 3:787031.

Suryawanshi, K. R., et al. 2017. Impact of wild prey availability on livestock predation by snow leopards. *Royal Society Open Science* 4:170026.

Tan, K. C. 2021. Just conservation: the question of justice in global wildlife conservation. *Philosophy Compass* 16:e12720.

Tauli-Corpuz, V., J. Alcorn, A. Molnar, C. Healy, and E. Barrow. 2020. Cornered by PAs: adopting rights-based approaches to enable cost-effective conservation and climate action. *World Development* 130:104923.

Tensen, L. 2018. Biases in wildlife and conservation research, using felids and canids as a case study. *Global Ecology and Conservation* 15:e00423.

Tensen, L. and K. Fischer. 2024. Evaluating hybrid speciation and swamping in wild carnivores with a decision-tree approach. *Conservation Biology* 38(1):e14197.

Thorn, M., M. Green, K. Marnewick, and D. M. Scott. 2015. Determinants of attitudes to carnivores: implications for mitigating human–carnivore conflict on South African farmland. *Oryx* 49:270–277.

Trisos, C. H., J. Auerbach, and M. Katti. 2021. Decoloniality and anti-oppressive practices for a more ethical ecology. *Nature Ecology & Evolution* 5:1205–1212.

Trouwborst, A., M. Krofel, and J. D. Linnell. 2015. Legal implications of range expansions in a terrestrial carnivore: the case of the golden jackal (*Canis aureus*) in Europe. *Biodiversity and Conservation* 24(10):2593–2610.

Trouwborst, A. and A. Blackmore. 2020. Hot dogs, hungry bears, and wolves running out of mountain—international wildlife law and the effects of climate change on large carnivores. *Journal of International Wildlife Law & Policy* 23:212–238.

Van de Ven, T., A. Fuller, and T. H. Clutton-Brock. 2020. Effects of climate change on pup growth and survival in a cooperative mammal, the meerkat. *Functional Ecology* 34:194–202.

van den Bosch, M., et al. 2023. Gray wolf range in the western Great Lakes region under forecasted land use and climate change. *Ecosphere* 14:e4630.

van den Brink, N. W., J. E. Elliott, R. F. Shore, and B. A. Rattner. 2018. Anticoagulant rodenticides and wildlife: concluding remarks. In *Anticoagulant Rodenticides and Wildlife*, edited by N. W. van den Brink, J. E. Elliott,

R. F. Shore, and B. A. Rattner. Springer International Publishing, Cham, Switzerland.

van Eeden, L. M., et al. 2018a. Managing conflict between large carnivores and livestock. *Conservation Biology* 32:26–34.

van Eeden, L. M., et al. 2018b. Carnivore conservation needs evidence-based livestock protection. *PLoS Biology* 16:e2005577.

Visconti, P., et al. 2016. Projecting global biodiversity indicators under future development scenarios. *Conservation Letters* 9:5–13.

Vollset, S. E., et al. 2020. Fertility, mortality, migration, and population scenarios for 195 countries and territories from 2017 to 2100: a forecasting analysis for the Global Burden of Disease Study. *The Lancet* 396:1285–1306.

Wei, F., et al. 2015. Progress in the ecology and conservation of giant pandas. *Conservation Biology* 29:1497–1507.

Wells, C. P., R. Barbier, S. Nelson, R. Kanaziz, and L. M. Aubry. 2022. Life history consequences of climate change in hibernating mammals: a review. *Ecography* 2022(6):e06056.

Wells, K., S. Morand, M. Wardeh, and M. Baylis. 2020. Distinct spread of DNA and RNA viruses among mammals amid prominent role of domestic species. *Global Ecology and Biogeography* 29:470–481.

Willcox, D. 2020. Conservation status, ex situ priorities and emerging threats to small carnivores. *International Zoo Yearbook* 54:19–34.

Williams, D. R., M. Clark, G. M. Buchanan, G. F. Ficetola, C. Rondinini, and D. Tilman. 2021. Proactive conservation to prevent habitat losses to agricultural expansion. *Nature Sustainability* 4:314–322.

Williams, D. R., C. Rondinini, and D. Tilman. 2022. Global protected areas seem insufficient to safeguard half of the world's mammals from human-induced extinction. *Proceedings of the National Academy of Sciences of the United States of America* 119(24):e2200118119.

Williams, P. J., M. B. Hooten, G. G. Esslinger, J. N. Womble, J. L. Bodkin, and M. R. Bower. 2019. The rise of an apex predator following deglaciation. *Diversity and Distributions* 25:895–908.

Williams, V. L., A. J. Loveridge, D. J. Newton, and D. W. Macdonald. 2017. A roaring trade? The legal trade in *Panthera leo* bones from Africa to East-Southeast Asia. *PLoS One* 12:e0185996.

Wilson, K. A., et al. 2016. Conservation research is not happening where it is most needed. *PLoS Biology* 14:e1002413.

Winkler, R. and K. Warnke. 2013. The future of hunting: an age-period-cohort analysis of deer hunter decline. *Population and Environment* 34:460–480.

Wolf, C. and W. J. Ripple. 2016. Prey depletion as a threat to the world's large carnivores. *Royal Society Open Science* 3(8):160252.

Wolf, C. and W. J. Ripple. 2017. Range contractions of the world's large carnivores. Royal Society Open Science 4(7):170052.

Wolff, S., E. A. Schrammeijer, C. J. E. Schulp, and P. H. Verburg. 2018. Meeting global land restoration and protection targets: what would the world look like in 2050? *Global Environmental Change* 52:259–272.

Woodroffe, R., R. Groom, and J. W. McNutt. 2017. Hot dogs: high ambient temperatures impact reproductive success in a tropical carnivore. *Journal of Animal Ecology* 86:1329–1338.

Yirga, G., W. Ersino, H. H. De Iongh, H. Leirs, K. Gebrehiwot, J. Deckers, and H. Bauer. 2013. Spotted hyena (*Crocuta crocuta*) coexisting at high density with people in Wukro district, northern Ethiopia. *Mammalian Biology* 78:193–197.

Yom-Tov, Y., T. M. Heggberget, O. Wiig, and S. Yom-Tov. 2006. Body size changes among otters, *Lutra lutra*, in Norway: the possible effects of food availability and global warming. *Oecologia* 150:155–160.

Zeller, K. A., D. W. Wattles, L. Conlee, and S. DeStefano. 2019. Black bears alter movements in response to anthropogenic features with time of day and season. *Movement Ecology* 7:19.

Glossary of Terms

Apartheid—a system of institutionalized social stratification based on racial segregation, which took place from 1948 into the 1990s within South Africa and Namibia.

Allee effects—a density-dependent ecological phenomenon that describes the relationship between population size and individual fitness.

Anthropocene—our current, proposed name for the geological epoch directly following the Holocene, marked by a period of intensive human activity which serves as the dominant influence on climate and the environment.

Behavioural response race—a theory that predicts that predators and prey respond to varying predation risk levels across habitat patches on the landscape.

Bioaccumulation—the process by which chemicals, substances, or contaminants build up in a living organism over time via direct ingestion and/or uptake from the surrounding medium.

Biomagnification—the increasing concentration of toxins in organismal tissues as such compounds travel up successively higher levels in the food chain.

Carnivore–livestock conflict—predation (or attempts thereof) of livestock by wild carnivores, which comprises a large portion of the focus for global human–carnivore conflict.

Carnivore deterrents—interventions that are designed to exclude carnivores from certain areas or to prevent predation on livestock or other assets. Carnivore deterrents can be physical, visual, chemical, and/or auditory, and include tools such as fences, fladry, noisemakers, flashing lights, etc.

Conservation translocations—the intentional movement of organisms from one location to another for conservation.

Density-mediated effects—effects that occur when the impacts of a species in an ecosystem are modified by changes in the abundance of an intermediary species with which it directly interacts.

Disturbance ecology—the scientific study of how extreme events (e.g. wildfires, storms, outbreaks, etc.) impact biological communities by disrupting the structure and composition of an ecosystem. Such disruptions can alter species interactions and responses to disturbances, as well as affect overall landscape dynamics.

Ecology—the study of the relationships of organisms to one another, and to their physical environment.

Ecological trap—scenarios in which animals mistakenly prefer habitats of lower or poor quality due to environmental cues that suggest habitat suitability yet result in decreased fitness outcomes.

Fladry—a visual stimulant designed to deter wolves and coyotes from crossing barriers, consisting of a strand of flags (often red) hanging on a nylon rope.

Free, Prior, and Informed Consent (FPIC)—a specific right granted to Indigenous Peoples recognized in the UN Declaration on the Rights of Indigenous Peoples (UNDRIP), which aligns with their universal right to self-determination. FPIC allows Indigenous peoples to provide or withhold/withdraw consent, at any point, regarding projects impacting their territories. FPIC allows Indigenous Peoples to engage in negotiations to shape the design, implementation, monitoring, and evaluation of projects (FAO).

Fundamental niche—represents the full range of environmental conditions where a species can live.

Genetic drift—the random variation in allele frequencies between generations in finite populations, due to sampling error.

Home range—that area traversed by the individual in its normal activities of food gathering, mating and caring for young.

Human–carnivore coexistence—when people and their activities spatially and/or temporally co-occur with sustainable carnivore populations and have minimal conflicts between people and carnivores or among people regarding carnivores.

Human–wildlife conflict—struggles that emerge when the presence or behaviour of wildlife poses actual or perceived, direct, and recurring threats to human interests or needs, leading to disagreements between groups of people and negative impacts on people and/or wildlife.

Human–wildlife interactions—the dynamic spatial and temporal juxtaposition of humans and wildlife and their activities, where wildlife, humans, or both are affected. Human–wildlife interactions can be positive, negative, or neutral, and their interactions and perceptions are dependent on and can vary among the humans and non-humans involved.

Human shield hypothesis—the notion that prey use areas with humans and human infrastructure as a buffer against predation risk.

Kleptoparasitism—the theft of already procured food by one individual from another.

Landscape of fear—a concept that describes how animals perceive the risk of predation, and how this risk changes across space and time.

Life history strategies—the way an organism allocates its energy and resources throughout its lifespan, including how it balances growth, survival, and reproduction, which incurs variable fitness trade-offs.

Livestock ecology—a combination of human management actions and the traits intrinsic to livestock that determine livestock use and navigation of the landscape, reproductive status, and behaviour, body condition, and demographic status.

Mesopredator release hypothesis—when a mesopredator species undergoes population increases following the removal of larger, more dominant predators.

Nonviolent communication—a method of communication that was proposed in the 1960s and prioritizes empathy for oneself and others. Nonviolent communication recognizes that feelings and thoughts are different from one another, encourages participants to distil their feelings and thoughts to the universal human need from which they are emerging, and supports making requests—rather than demands—of others.

Optimal foraging theory—prey and predators balance foraging opportunities and costs to select food resources that provide the greatest net benefit for survival and persistence.

Predation—the biological interaction where one species kills and consumes another organism (i.e. its prey).

Predator–prey shell games theory—a theory that predicts that prey purposely make their locations unpredictable by moving around the landscape to avoid detection by predators, and predators make subsequent decisions based on previous prey movements.

Prey depletion—the significant reduction of available prey species.

Prey switching—also known as apostatic selection, this phenomenon describes when a predator preferentially consumes the most common prey species when others are more scarce.

Range rider programmes—a nonlethal management strategy that involves using human surveillance—often on horseback—of livestock herds, to deter wolves from preying upon cattle.

Realized niche—the space that a species occupies in the environment, accounting for species interactions.

Reinforcements—the addition of organisms to an already existing population.

Sublethal effects—biological, physiological, demographic, or behavioural effects on an individual or population that survives exposure to a substance at a lethal or sublethal concentration.

Theory of change—an approach that promotes understanding of how and why a particular intervention or programme works by describing the processes that produce change.

Top-down control—the process by which species at higher trophic levels regulate species at lower trophic levels.

Trait-mediated effects—effects that occur when the impacts of one species in an ecosystem are modified by changes in the phenotype (behaviour, morphology, physiology) of a second, intermediary species with which it interacts.

Trait-mediated indirect interaction—when the secondary species with an altered trait or phenotype in response to a perception of predation risk affects a third species.

Trophic cascades—the ecological phenomenon that occurs when the abundance or behaviour of a predator population affects the abundance of a prey population, which in turn affects the abundance of the next lower trophic level in a food web.

Trophic downgrading—the process of removing apex predators from a large geographic area, with significant consequences for ecosystem functioning.

Wildlife corridors—area used by animals to pass from one 'habitat patch' to another; or an area that connects two patches of suitable habitat by passing through a matrix of unsuitable habitat.

Index

Tables, figures, and boxes are indicated by an italic *t*, *f*, or *b*.

30×30 238, 239

accelerometer data 140, 141, 145, 146, 151
activity budget 43*t*, 138, 142*f*, 165, 202
additive mortality 69, 89–90, 90*f*
Africa 18, 19, 21, 25–8, 41, 43*f*, 47*b*, 68, 82, 84, 87, 94, 103, 107*t*, 108, 113*f*, 113–114, 114*f*, 130, 170, 234–7, 240, 254, 270, 271, 274
African buffalo (Syncerus caffer) 50, 221*b*
African wild dog (Lycaon pictus) 22, 46, 47*b*, 87, 103, 106*t*, 112, 127, 212, 274
age structure 65, 69
aggression 49, 51, 112, 163
agricultural production 39, 238, 239
agriculture 3, 10, 39, 44, 124, 131, 172, 187, 220, 238, 269, 270, 271
Alaska, USA 108, 184, 188, 190, 191, 210*b*, 213, 258*f*, 274, 281*f*
Allee effect 47*b*
Amur tiger (Panthera tigris altaica) 67, 275
animal biologgers 139–41, 142*f*, 146
animal-borne camera 140, 142*f*
animal-borne sound recorder 141
anthropogenic noise. *See* noise
anthropogenic resources 44, 201*t*
anthropogenic subsidies 44, 82, 143, 149, 151, 167, 181, 188, 220, 248
antipredator behaviour 125, 126, 138, 140, 162–65
antipredator behaviour of livestock 85
artificial light at night (ALAN) 49
artificial selection 65, 69–71, 73
Asia 18, 26, 67, 103, 107*t*, 108, 113*f*, 113–4, 114*f*, 235, 237, 240, 271, 276, 277
aspen (Populus tremuloides) 5*f*, 183, 216*b*
attacks of carnivores on humans 12, 46, 85, 124, 239, 246, 254, 255*f*, 272
attacks of carnivores on livestock 85–87, 202, 239, 254, 255*f*
attitudes of humans towards carnivores 51, 85–87, 220–221, 237, 239, 280, 282
attractive sinks 91–93
auditory playback experiment. *See* experiments
Australia 18, 19, 24–31, 66, 93, 187–188, 191, 211, 214, 218*b*, 222
autocorrelation 144
avoidance behavior, of carnivores 39–41, 42*b*, 43*f*, 44, 75, 92, 93, 127, 129–33, 143–45, 147–49, 150*f*, 163–167, 181, 201*t*

baboon (Papio anubis) 215, 254
badger (Meles meles) 49, 126
bears 20, 21, 41, 42*b*, 43*f*, 44, 46, 48, 66, 67, 69–75, 83, 88, 92, 107–11, 110*t*, 114, 126, 129–33, 146, 148, 149, 151, 166, 167, 181, 183, 184, 189, 198, 200, 201*t*, 202, 203, 212, 248, 256, 258*b*, 258*f*, 271, 272, 274–7, 278*b*, 281*f*
  American black bear (Ursus americanus) 41, 66, 106*t*, 107–9, 110*t*, 129, 167, 181, 248, 272
  Asiatic black bear (Ursus thibetanus) 46, 106*t*, 110*t*, 275
  brown bear. *See* grizzly bear
  cave bears (Ursus spelaeus) 21
  grizzly bear (Ursus arctos) 21, 41, 66, 83, 106*t*, 107–09, 110*t*, 126, 146, 166, 181, 212, 217, 248, 258*b*, 258*f*, 272
  Himalayan brown bear (Ursus arctos isabellinus) 270*f*
  human hunting of 67, 170
  polar bear (Ursus maritimus) 46, 69, 106*t*, 110, 146, 202, 274, 278*b*
  short-faced bear (Arctodus simus) 18
  sloth bear (Melursus ursinus) 106*t*, 114
  spectacled bear (Tremarctos ornatus) 106*t*
  sun bear (Helarctos malayanus) 106*t*
beaver (Castor canadensis) 214
behavioural change point analysis 145
behavioural response race 252
behavioural state 145–46, 149
beliefs about carnivores 86, 87, 232, 234, 240, 248
bharal (Pseudois nayaur) 45, 190
Białowieża Forest 188
bimodal activity pattern 128
bioaccumulation 48
biodiversity 6, 7, 12, 13, 18, 25, 31, 38, 39, 40*f*, 44, 50, 75, 83, 94, 108, 112, 143, 160, 190, 218*b*, 221, 231, 233–35, 237–39, 248, 257, 270, 273–5, 278–80, 282
  loss 38–39, 40*f*, 75, 274
black-tailed deer (Odocoileus hemionus) 49, 128, 210*b*
blue sheep (Pseduois nayaur) 45, 190
bobcat (Lynx rufus) 48, 126, 185
body condition 43*t*, 45, 48, 92, 138, 162, 165, 274
body mass. *See* body size of carnivores

body size of carnivores 18, 20, 31, 40*f*, 43*f*, 44, 65, 111, 140, 162, 163, 165, 200
Botswana 43*f*, 46, 92, 130, 202, 256
bottleneck, genetic 40, 71
bottom-up human impacts 183, 185
bottom-up *vs.* top-down system 185, 186*f*, 216, 217, 220
bottom-up control 5*f*, 6, 30, 188, 190, 211*f*, 213, 217, 218*b*, 222, 259, 260*t*
Brazil 108, 114, 250, 272
breeding site 73, 124
Britain 126
British Columbia 67, 109, 166
brown world 183, 184*f*
bushbuck (Tragelaphus sylvaticus) 221*b*
bushmeat 44, 130, 276

California sea lions (Zalophus californianus) 166
camera trap 104*f*, 130, 133, 141, 142*f*, 146, 147, 149, 150*f*, 255, 281*f*
capybara (Hydrochoerus hydrochaeris) 128
caracals (Caracal caracal) 41
carcass 21, 23, 26, 82, 103, 105–13, 106*t*, 107*t*, 113*f*, 128–29, 188–190, 189*f*, 201*t*, 202, 212, 254, 277
caribou (Rangifer tarandus) 45, 105*b*, 187, 198
carnivore deterrents 252
carnivore–livestock conflict 252, 253*f*
carrion pulse of kangaroos 188
carrion subsidies provided by carnivores 188, 189*f*
carrying capacity 3, 30, 31, 91, 110, 145, 151, 191, 249
cheetah (Acinonyx jubatus) 19, 86, 106*t*, 130, 163, 256, 272
chimpanzee (Pan troglodytes) 4, 24
circadian rhythm 49
civet
  civet (Viverra tangalunga) 140
  Malabar civet (Viverra civettina) 277
climate change 19, 26–30, 32, 39, 40*f*, 43*f*, 45, 46, 140, 151, 190, 191, 199*f*, 202, 213, 214, 246, 256, 261, 270, 272, 273*b*, 274–7, 279
  effects on carnivores 19, 39, 40*f*, 43*f*, 45, 46, 140, 151, 190, 202, 213, 246, 261, 270, 272, 273*b*, 274–7
  effects on trophic systems 46, 190, 191, 199*f*, 202, 213, 214, 274–6
  as hypothesis for Pleistocene extinction 19, 27–32
Colorado, USA 9, 42*b*, 46, 127, 167
commensalism 6, 38
community science 142*f*
compensatory mortality 90*f*, 91
compensatory natality 90*f*, 91
competition 4, 6, 12, 22, 27, 30, 32, 38–41, 40*f*, 44–46, 47*b*, 89, 91, 94, 161, 163, 183, 184*f*, 187, 188, 210*b*, 211, 213, 217, 218*b*, 221*b*, 273*b*, 273–5
  apparent 6, 45, 48, 183–84, 184*f*, 187, 210*b*
  exploitative 6
  interspecific 30, 163, 164, 203
competitive exclusion (under species interactions) 40
continuous time-velocity models 142*f*, 145
connectivity, habitat 217, 248–50, 251*f*, 272
conspecific interaction 89, 140, 142*f*, 145, 249
Convention on Biological Diversity (CBD) 231, 239, 248
cooperative hunting 22, 23
corridor, wildlife 250, 251*f*, 252
cougar (Puma concolor) 41, 70, 86, 106*t*, 107–09, 110*t*, 123, 146, 150*f*, 164, 166, 181, 200, 214, 253*f*, 256, 272
coupled human and natural systems 4–6, 10, 13
COVID-19 31, 129, 143
cowbell 131
coyote (Canis latrans) 50, 82, 107, 130, 151, 162, 184, 200, 212, 253*f*, 271
cues of risk 8–9, 91, 92, 150–51, 163, 165, 168, 170
culling of carnivores 7, 46, 88, 108, 256
cultural impacts of conflict 84–86

Dall sheep (Ovis dalli dalli) 184, 191
deer mice (Peromyscus spp.) 215
demographic rates 46, 72
denning 43*f*, 46, 47*b*, 48, 110, 129, 140, 151, 190, 272, 274
density dependence 72, 89
density-independent effects 191
density-mediated effects 197, 200, 220
dhole (Cuon alpinus) 106*t*, 114, 215, 275
diel activity 147, 163, 168, 171
diet analysis 139, 142*f*
dietary preference 128, 189
dingo (dingoe) (Canis dingo, C. lupus dingo) 93, 106*t*, 211, 218*b*
disease, in animals 11, 12, 40*f*, 47*b*, 66, 68, 85, 89, 93, 212, 215, 216*b*, 254, 276, 279
disease, transmission by humans or domestic animals 31, 47*b*, 51, 145, 276, 277
dispersal 10, 32, 41, 42*b*, 46, 74, 112, 143, 144, 217, 221*b*, 249
DNA metabarcoding 142*f*, 143
domestic cat (Felis catus) 93, 187, 279
domestic dog (Canis lupus familiaris) 18, 162
domestication or domesticated species, include cats 3, 24, 31, 50, 51, 67, 75, 85
drought 29, 40, 42*b*, 45, 46, 151, 188, 190, 191, 275, 276, 279*f*

ear tags 140
ecological memory 219
ecological traps 41, 70, 92–93, 185, 187
ecosystem consequences of human interventions to reduce conflict 87, 89, 93–94
ecosystem engineer. *See* niche, constructor
ecosystem, regulation of 123, 209
ecosystem stability 39
ecotourism. *See* tourism
elephant (Loxodonta africana) 9, 50, 127, 221*b*, 254
elk (Cervus canadensis or C. elaphus) 5*f*, 11, 128, 162, 183, 200, 214
energetic cost 127, 138, 146, 161*f*, 167
error management theory (EMT) 161, 170, 171
Ethiopia 84, 86, 106*t*, 202, 215, 248, 272, 277
ethogram 139
Eurasian cave lion (Panthera spelaea) 4, 21
Eurasian lynx (Lynx lynx) 21, 83, 88, 106*t*, 108, 109, 110*t*, 111, 131, 151, 163, 185, 213, 248, 272
Eurasian otters (Lutra lutra) 272, 274
European mink (Mustela lutreola) 277
evolution, rapid 30, 38, 72, 83, 151
experiments 9, 123, 125, 126, 133, 152, 165, 204, 210, 260*t*

experiments, auditory playback 125, 126, 133, 150
experiments, field 9, 123
experiments, human vocalization playback 126, 127, 129, 133
exploitation ecosystem hypothesis 217
extinction in the Pleistocene 18, 25
extirpation of carnivores 49, 65, 138, 160, 183, 187, 209, 211–17, 218*b*, 219*f*, 272, 280, 282
extreme weather events 138, 191

fear. *See* landscape of fear
fecal glucocorticoid 142*f*
fecundity 40*f*, 43*f*, 69, 73, 90*f*, 92, 160, 162, 164, 166, 167
feeding station 170
feeding time 82, 128–29, 150, 167
fencing 40, 94, 105*b*, 200, 249
feral cat. *See* domestic cat
feral donkey (Equus asinus) 214
ferrets (Mustela putorius furo) 213
financial impacts of conflict 84–85
financial instruments 254
finfish fisheries 221
fitness consequences of risk effects on carnivores 7–9, 71, 142*f*, 150*f*, 151, 160–65, 161*f*, 167, 171, 189
fitness consequences of risk effects on humans 8, 12, 161, 166
fladry 88, 252, 253*f*
fleeing behaviour 133, 138, 141, 150*f*
Florida panther (Puma concolor coryi) 51
fosa (Cryptoprocta ferox) 83, 271
fossil fuel 48, 275
fox
  gray fox (Urocyon cinereoargenteus) 103, 112
  Joaquin kit foxes (Vulpes macrotis) 276
  red fox (Vulpes vulpes) 50, 88, 133, 187, 200, 212, 279
Free, Prior, and Informed Consent (FPIC) 251*f*, 252

Galápagos sea lions (Zalophus wollebaecki) 213
game reserves 131
genetic diversity responses to reinforcement 256, 277
geopolitical conflict 52
glacial cycle 18, 28
Global Biodiversity Information Facility [GBIF] 143, 148, 168
global South 3, 85, 94, 111, 113, 221, 240, 282
glucocorticoids 142*f*, 143, 162
golden age of biologging, the 139
golden jackal (Canis aureus) 270*f*, 272, 273*b*
Gorongosa National Park 221*b*, 222
Governance Assessment for Protected and Conserved Areas (GAPA) 240
GPS collars 107, 140, 142*f*, 150*f*
green world hypothesis 183, 184*f*
greenhouse gas 45
gunshots as stimuli 127, 141

habitat 3, 4, 10, 12, 25, 26, 30, 32, 39–41, 42*b*, 44–6, 49–51, 66, 68, 70, 72, 73, 75, 94, 104, 105*b*, 108, 110, 111, 115, 123, 124, 126*f*, 127, 130, 131, 133, 138–41, 142*f*, 143–9, 150*f*, 151, 161, 163–6, 168, 172, 183, 185–8, 190, 199, 200, 201*t*, 203, 209, 214, 216, 217, 220, 221, 231, 233–5, 237–9, 246, 247*f*, 248–50, 253, 256, 270–5, 277
  connectivity 249, 250
  domain 163, 169, 172
  fragmentation 10, 32, 39, 66, 103, 110, 124, 183, 185, 234, 238, 270, 271, 275
  grassland 130, 212, 219*f*, 221*b*
  loss 39, 66, 68, 209, 231, 234, 238, 246, 271
  selection, change in response to risk 70, 73, 75, 108, 127–33, 138, 139, 141, 142*f*, 143–5, 149, 150*f*, 151, 166
  utilization 138, 142*f*, 143–5, 150*f*
habituation to non-lethal interventions 87–9, 91–3
hair snares 142*f*, 143, 146, 148
harbour seals (Phoca vitulina) 45
harp seals (Pagophilus groenlandicus) 217
harvest, accidental take 107
harvest of carnivores, legal 68
harvest-induced selection. *See* artificial selection
heart rate monitor 141, 142*f*
herbivory 50, 183, 210, 211*f*, 211–14, 216, 217, 222, 273
hiking, human hiker 8, 9, 129, 165
hiking trail 127, 199*f*
hippopotamus (Hippopotamus amphibius) 127, 221*b*
Holocene 22
home range 149, 150*f*, 151, 209, 220, 231, 249
  local convex hull (LoCoH) 143, 144
  minimum convex polygon (MCP) 143
hominin 3, 4, 7, 25
Homo erectus 21
Homo sapiens 3
housing density 128, 147, 166–67
human 3–13, 5*f*, 18–32, 38–41, 42*b*, 43*f*, 44–46, 47*b*, 48–52, 65–75, 82–95, 90*f*, 103, 104, 105*b*, 107–112, 109*f*, 115, 123–34, 126*f*, 138–141, 142*f*, 143–52, 150*f*, 160, 161, 161*f*, 163–72, 169*f*, 171*f*, 181–5, 182*f*, 184*f*, 186*f*, 187–91, 189*f*, 197–204, 199*f*, 201*t*, 209, 211*f*, 214, 215, 218, 219*f*, 220, 221, 231–5, 237–42, 241*f*, 246–50, 247*f*, 252–254, 255*f*, 256, 257, 258*b*, 259, 260*t*, 261*t*, 261, 269–77, 270*f*, 273*b*, 276*f*, 278*b*, 279–282
  ecological roles of 5*f*, 6
  Human Footprint Index 168, 169*f*
  overkill hypothesis of megafaunal extinction 19, 25, 27, 29, 30, 32
  shield hypothesis 197, 200, 202
human-carnivore coexistence 246, 247
human-carnivore conflict 6, 45, 202
human-carnivore interactions 5, 6, 51, 52, 247*f*
human-human conflict 247, 260*t*
human-induced selection. *See* artificial selection
human shield hypothesis 197, 200, 202
human wildlife conflict 44, 94, 134, 190, 247, 258*b*
human-wildlife interactions 269
hunting dogs 24
hunting mode 8, 9, 163, 169, 172
hybridization 107, 274, 277
hyena 31, 47*b*, 49, 51, 83, 85–7, 93, 106*t*, 109, 110, 127, 130–2, 163, 164, 166, 189, 199*f*, 201*t*, 202, 212, 215, 221*b*, 271, 272, 279*f*
  brown hyena (Parahyaena brunnea) 106*t*, 271, 279*f*
  long-legged hyaena (Chasmaporthetes) 18
  spotted hyena (Crocuta crocuta) 21, 49, 82, 106*t*, 127, 163, 164, 202, 212, 272, 274, 279*f*

hyena (*Continued*)
  striped hyena (Hyaena hyaena) 106*t*, 109
hypercarnivores 29, 271
hyperphagia 130

Iberian lynx (Lynx pardinus) 43*f*, 51, 212, 256, 277
impala (Aepyceros melampus) 221*b*
iNaturalist 143, 148
inbreeding 40, 71, 256
independent and identically distributed location data (IID) 143, 144
indigenous people 5, 232, 236, 250
individual and group identity 232
inducible defences 7, 8
infanticide, sexually selected infanticide (SSI) 70, 74, 91
intelligence 21, 22, 26, 197, 281
interdisciplinarity 5, 95
intergenerational equity 231–41, 241*f*
Intergovernmental Science-Policy Platform on Biodiversity and Ecosystem Services (IPBES) 39
International Union for the Conservation of Nature (IUCN) 19, 106*t*, 107, 110, 114*f*, 168, 169*f*, 246, 259
interventions to reduce conflict 6, 75, 84, 87–95
intra-guild killing 188
Iran 45, 86, 108, 110, 111, 276
Isle Royale 71, 214
Italy 43*b*, 45, 70, 258*b*
IUCN Human–Wildlife Conflict and Coexistence Task Force 259
IUCN Red List Classification 19, 106*t*, 114*f*, 169*f*
invasive species 18, 27, 40*f*, 43*f*, 50, 276*f*, 276, 279

jaguar (Panthera onca) 32, 82, 106*t*, 234, 250

Kenya 9, 43*f*, 50, 68, 86–87, 92, 131, 166, 187, 215, 253*f*, 279*f*
keystone species 6, 10–11, 13, 45, 127, 204, 209, 210*f*, 231
  hyperkeystone, humans as 6, 10–11, 13, 127
kill rate. *See* predation rate
kill site fidelity 128
killing of carnivores (preventive or retaliatory) 46, 50, 68, 87–93, 146, 234, 255, 271, 275
kleptoparasitism 4, 23, 164
Kruger National Park 170

land-sparing 231–42
landscape of fear 41, 75, 92, 126*f*, 129, 160, 197–200, 216*b*, 221*b*
landscape of refugia 200
least weasel (Mustela nivalis) 213
leopard
  Amur leopard (Panthera pardus orientalis) 275
  clouded leopard (Neofelis nebulosa) 106*t*
  Diardi's clouded leopard (Neofelis diardi) 106*t*
  Panthera pardus 19, 41, 84, 106*t*, 109, 127, 146, 215, 253, 271
  snow leopard (Panthera uncia) 45, 106*t*, 190, 249, 270, 271
lethal control 65, 67, 88, 90, 91, 93, 191, 254, 277
lethal exploitation. *See* harvest of carnivores
life history trait 30, 86, 110, 115, 217, 269, 274
lion
  African lion (Panthera leo) 19, 30, 41, 66, 84, 106*t*, 127, 140, 163, 202, 215, 248, 270*f*, 271
  Asiatic lion (Panthera leo persica) 66, 68, 248
  Australian marsupial lion (Thyalcoleo carnifex) 18
  cave lion (Panthera leo spelaea) 21
litter size 40*f*, 43*f*, 44, 48, 69, 91, 162, 164, 166
livestock, depredation of 43*f*, 45, 83–95, 190, 202, 234, 248, 252, 255, 271
livestock husbandry 10, 254
local knowledge 143, 234, 260*t*
Lyme disease 215
lynx (Lynx canadensis) 7, 146, 162

Maasai people 9, 131, 234
machine learning 142*f*, 145, 281*f*, 281
male coalition 70
marine megafauna 210*b*
marten (Martes americana) 130
Masai Mara National Reserve 131
Masailand 68
maternal care 72
meat consumption by humans 44
meerkat (Suricata suricatta) 274
megacarnivore 218
megaherbivore 26–29, 217, 221*b*
mesocarnivore behavior 41, 126, 27, 133
meso-carnivore release hypothesis 200
mesopredator release 50, 93, 115, 187, 188, 212, 252, 255*f*
meso-predator release hypothesis 200
mesoscavenger release hypothesis 190
meta-analyses 107, 110–12, 124, 125, 133
Mexico 66, 86, 87, 272, 281*f*
microplastics 49
migration 124, 200, 250, 274, 275
Minnesota 71, 214
moonlight 132
moose (Alces alces) 45, 75, 104*f*, 165, 187, 197, 213, 275
mountain lion. *See* cougar
movement cost 127, 145, 190
  India 41, 68, 84, 88, 106*t*, 181, 215, 232, 233, 248, 255, 270, 271, 274
mule deer (Odocoileus hemionus) 49, 128, 210*b*
mutualism 6, 12, 38, 50, 235

Namibia 68, 259
natural system 4, 5, 5*f*, 6, 10, 13, 82, 185, 203, 269
Neanderthal (Homo neanderthalensis) 3, 21
neonate implant transmitter 142*f*
Nepal 43*f*, 45, 114, 190, 232
net-squared displacement (NSD) 144
New Zealand 25–28, 279
niche
  constructor 6, 10
  fundamental and realized 40
  partitioning 38, 40*f*, 41, 201*t*
nocturnality 40*f*, 41, 42*b*, 43*f*, 46, 47*b*, 125–133, 150*f*, 166, 168
noise 43*f*, 48–50, 88, 165, 204
noise pollution 48, 49
non-consumptive effect 7, 138, 142*f*, 198
non-invasive sampling 141, 151
non-lethal 6, 9, 87–9, 91–4, 111, 115, 151, 161, 165–8, 269, 280
nonviolent communication 252
North America 5, 11, 18, 19, 25, 27–29, 41, 46, 50, 66–68, 71, 72, 88, 103,

107–8, 107*t*, 111, 113–5, 113*f*, 114*f*, 186, 190, 211, 214, 222, 235, 256, 271, 272, 277
numerical effect 5*f*, 65, 74*f*

object-based image analysis (OBIA) 143
omnivory 27, 183
oncilla (Leopardus tigrinus) 271
optimal foraging theory 252
orca (Orcinus orca) 11, 45, 210*b*
Oregon, USA 164, 188
overbrowsing 181
overexploitation 44, 45, 233

pack size 47*b*, 70
Panama 18, 271
participatory science 112, 143
parturition 40*f*, 43*f*, 46, 47*b*, 48
Payments for predator presence 254, 255
Payments to Encourage Coexistence (PEC) 254, 255
perceived risk of depredation 86
persecution of carnivores 66, 124, 181, 186*f*, 235, 276*f*, 276
Persian leopards (Panthera pardus tulliana or P. p. saxicolor) 45, 51, 271
Peruvian National Park 132
phenology 40*f*, 46, 190
physiological responses of carnivores to humans 138, 142*f*, 161–67
pine martens (Martes martes) 88, 212
plains zebra (Equus quagga) 50, 187, 221*b*
playback. *See* experiments
Pleistocene megafaunal extinctions 19
poaching 67–70, 107–12, 160, 181, 191, 237, 248, 249, 259, 281
poisoning 48, 66, 67, 82, 87, 93, 110, 111, 277
Poland 43*f*, 45, 188
political affiliation 51
pollution 39, 40*f*, 43*f*, 48, 49, 204, 214, 276*f*, 276, 278*b*
population growth rate 44, 68–72, 94, 160, 166
predation rate 45, 128, 166, 167, 184, 185, 187, 191, 220
predation services 215
predator-prey dynamics 8, 32, 38, 39, 40*f*, 43*f*, 49, 65, 220
predator-prey shell games theory 252
prey availability 21, 104, 112, 187, 190, 202, 252, 274, 275
prey depletion 23, 45, 66, 68, 83, 94, 151, 181, 246, 271
prey-trap hypothesis 105*b*
primary productivity 185, 187, 190, 217
problem individuals 88, 256
proximity sensor 140, 142*f*
pseudomutualism 12

raccoon
  pygmy raccoon (Procyon pygmaeus) 277
  raccoon (Procyon lotor) 50, 83, 276
radio telemetry 107, 108, 109*f*, 110*f*, 111, 113*f*. *See also* VHF telemetry
range rider programmes 257
recolonization of carnivores 70–72, 172, 217, 272
recreation, motorized 49, 103, 132–33, 149, 166, 271
recruitment, animal 48, 69, 70, 92, 161–63, 166
red kangaroo (Macropus rufus) 214
remote technology 139
renewable energy 48
reproductive success 46, 49, 70, 110, 138, 141, 142*f*, 164, 190, 274
residency status (of carnivores) 127
residential areas 129, 130, 133, 202
resource subsidies. *See* anthropogenic subsidies
resting site 129, 141
retaliation 6, 87, 234, 246
rewilding 6, 217, 258*b*, 272
rhinoceros (Ceratotherium simum) 127
ringed seal (Pusa hispida) 166
risk of human encounter 125, 128, 130, 133, 166, 271
risk-avoidance behaviour 141, 163, 164
risk-disturbance hypothesis 8, 9, 161
risk-induced trait responses of carnivores to humans 138, 160
road density 73, 104–11, 114
roadkill 107–8, 110–15, 188, 189, 271
road mortality 103–115
rodenticide 43*f*, 48, 276
roe deer (Capreolus capreolus) 163, 185
Romania 185, 248

safety 84, 85, 214, 247, 257
Santa Cruz mountains 126, 129, 150*f*, 200, 202, 253*f*
Santa Monica 42*b*
sapling herbivory 211
scale (spatial and temporal) 6, 148
Scandinavia 44, 67, 70–73, 75, 83, 191
scat collection 139
scavenging 21, 23, 27, 87, 164, 183, 188–91
scent-marking 129, 140
scimitar-toothed cats (Smilodon and Homotherium) 18, 19, 30
sea otter (Enhydra lutris) 11, 45, 210*b*
sea urchins (Strongylocentrotus spp.) 210*b*
secondary sexual trait 72
seed dispersal 32, 211*f*, 212, 233
selection functions, resource and step 144
selective harvesting 73
sex ratio. *See* sex structure
sex structure 69–70
Site-level Assessment of Governance and Equity (SAGE) 240
size-selective extinction 18
Social Assessment for Protected and Conserved Areas (SAPA) 240
social media 236
snares 30, 68, 142*f*, 143, 146, 148, 249
snow tracking 139
social impacts of conflict 85
sociality 168
sociocultural 51, 233, 236
solitary 65, 69, 70, 128, 132, 249
source-sink dynamics 91, 181, 272
South Africa 41, 47*b*, 87, 132, 170, 236
South America 18, 27, 103, 107*t*, 108, 113, 113*f*, 114*f*, 253, 270, 271
southern elephant seal (Mirounga leonina) 166
Spain 24, 43*f*, 70–72, 93, 212, 217
squirrel
  grey squirrel (Sciurus carolinensis) 212
  red squirrel (Sciurus vulgaris) 188, 212
Stellar's sea cows (Hydrodamalis gigas) 32, 210*b*
Steller sea lion (Eumetopias jubatus) 45, 272
sterilisation 87, 89, 92
stoats (Mustela erminea) 279
stone martens (Martes foina) 212
stress response 138, 161

striped skunk (Mephitis mephitis), 126
strongholds of lion populations 68
sublethal effects of pollution 48
superadditive mortality 90*f*, 91
survival, of cubs 70, 73, 167
survival, of pups 70
sustainable development goals (SDGs) 238–40
sustainable yield 65
Sweden 43*f*, 44, 69, 70, 88, 187
synchronous breeding 70

Tarangire–Manyara ecosystem 234
Tasmanian devil (Sarcophilus harrisii) 187, 213
temperature logger 140
temporal partitioning 166–68
Thailand 132
threshold of risk hypothesis 172
thylacine (Thylacinus cynocephalus) 66
tiger (Panthera tigris) 51, 85, 106*t*, 114, 215, 253, 270*f*, 271
timber harvest 186
time budget 139
tool and weaponry use for harvest 19–21
top-down effect 183, 184, 198, 220
top-down forcing hypothesis 27
top-down regulation 191, 212
tourism revenue 254–55
Toxoplasma gondii 51
trait-mediated effects 197–99, 202–4
trait-mediated indirect interaction 197
translocation 11–13, 87, 89, 91–94, 251*f*, 252, 256, 257, 258*b*
tree recruitment 211, 214
triangular trophic interactions 184
trophic cascades 5*f*, 6, 11, 31, 40*f*, 183, 187, 200, 209–19
trophic downgrading 40*f*, 44, 115, 209–23
trophic triangle. *See* triangular trophic interactions
trophic upgrading 209–23
trophy hunting 68, 70, 71, 256

ultra-high frequency (UHF) collar 140
unmanned aerial vehicles (UAVs) 142*f*, 143, 281*f*, 281
urbanization 39, 48, 49, 124, 128, 133, 148, 168, 188, 270, 271, 275, 280
  development gradient 126–28
  land use and rate of change 39
utilization distribution 145

vehicle collisions 11, 103–15, 131, 146, 181, 182, 182*f*, 185, 215, 269
VHF telemetry 108, 139
vigilance 12, 40*f*, 43*f*, 49, 85, 92, 123, 124, 126, 130, 131, 133, 138, 140, 141, 142*f*, 161
Virginia opossum (Didelphis virginiana), 126
visual 8, 9, 125, 141, 149, 150, 151, 182*f*, 185, 253*f*

warrah. *See* wolf, Falklands wolf
Washington, USA 185, 259
Wasteful consumption 201
waterbuck (Kobus ellipsiprymnus) 221*b*
white-tailed deer (Odocoileus virginianus) 7, 162, 185, 211
Wisconsin, USA 108, 113, 170
wild prey 45, 67, 75, 84, 85, 95, 252–4, 271, 275, 276
wildlife reserve 84, 128, 131
wolf
  African golden wolf (Canis lupaster) 277
  dire wolf (Canis dirus) 18
  domestication of 23–24
  Ethiopian wolf (Canis simensis) 106*t*, 277
  Falklands wolf (Dusicyon australis) 26
  Finnish wolf population 68
  grey wolf (Canis lupus) 5*f*, 21, 39, 66, 82, 103, 106*t*, 110*t*, 112, 126, 139, 162, 181, 198, 235, 254, 272, 280
  Iberian wolf (Canis lupus signatus) 72
  maned wolf (Chrysocyon brachyurus) 106*t*, 212
  Red wolf (Canis rufus) 106*t*, 213, 235, 277
wolverine (Gulo gulo) 21, 66, 130, 149, 166, 217, 248, 272

Yellowstone National Park, USA 67, 130, 183, 216

Zimbabwe 43*f*, 68